Thomas Kirke Rose

The Metallurgy of Gold

Thomas Kirke Rose

The Metallurgy of Gold

ISBN/EAN: 9783743322172

Manufactured in Europe, USA, Canada, Australia, Japa

Cover: Foto ©ninafisch / pixelio.de

Manufactured and distributed by brebook publishing software (www.brebook.com)

Thomas Kirke Rose

The Metallurgy of Gold

THE METALLURGY OF GOLD.

NEW METALLURGICAL SERIES

EDITED BY

W. C. ROBERTS-AUSTEN, C.B., F.R.S.,

Chemist and Assayer of the Royal Mint; Professor of Metallurgy in the Royal College of Science.

In Large 8vo, Handsome Cloth. With Illustrations.

---

1. **INTRODUCTION to the STUDY of METALLURGY.** By the EDITOR. THIRD EDITION.

    "No English text-book at all approaches this in the COMPLETENESS with which the most modern views on the subject are dealt with. Professor Austen's volume will be INVALUABLE."—*Chemical News.*

2. **GOLD (The Metallurgy of).** By T. KIRKE ROSE, D.Sc., Assoc. R.S.M., F.C.S., Assist.-Assayer of the Royal Mint. SECOND EDITION.

    "The Four chapters on *Chlorination*, written from the point of view alike of the practical man and the chemist, TEEM WITH CONSIDERATIONS HITHERTO UNRECOGNISED, and constitute an addition to the literature of Metallurgy, which will prove to be of classical value."—*Nature.*

3. **COPPER (The Metallurgy of).** By THOMAS GIBB, Assoc. R.S.M.

4. **IRON and STEEL (The Metallurgy of).** By THOMAS TURNER, Assoc. R.S.M., F.I.C. Vol. I., IRON.

    "A MOST VALUABLE SUMMARY of useful knowledge relating to every method and stage in the manufacture of cast and wrought iron down to the present moment . . . particularly rich in chemical details. . . . An EXHAUSTIVE and REALLY NEEDED compilation by a MOST CAPABLE and THOROUGHLY UP-TO-DATE metallurgical authority."—*Bulletin of the American Iron and Steel Association.*

5. **METALLURGICAL MACHINERY**: the application of Engineering to Metallurgical Problems. By HENRY CHARLES JENKINS, Wh.Sc., Assoc. R.S.M., Assoc. M.Inst.C.E., of the Royal Mint.

6. **ALLOYS.** By the Editor.

*** Other Volumes in Preparation.

The 40-Stamp Battery of the New Heriot Company, showing the Stamps, Water-Supply, and Copper Tables.

THE
METALLURGY OF GOLD.

BY

T. KIRKE ROSE, D.Sc.,

ASSOCIATE OF THE ROYAL SCHOOL OF MINES; FELLOW OF THE CHEMICAL SOCIETY;
ASSISTANT ASSAYER OF THE ROYAL MINT.

*BEING ONE OF A SERIES OF TREATISES ON METALLURGY,
WRITTEN BY ASSOCIATES*

OF THE

ROYAL SCHOOL OF MINES.

EDITED BY

Prof. W. C. Roberts-Austen, C.B., F.R.S.

WITH NUMEROUS ILLUSTRATIONS.

SECOND EDITION.

LONDON:
CHARLES GRIFFIN AND COMPANY, LIMITED.
PHILADELPHIA: J. B. LIPPINCOTT COMPANY.
1896.

[*All Rights Reserved.*]

PREFACE.

In preparing this edition, the whole book has been carefully revised, due attention being paid to some useful suggestions contained in the various reviews of the first edition. Eight new illustrations have been added, including the frontispiece, which is from a photograph kindly lent by Messrs. Fraser & Chalmers. The rapid progress made in improving the cyanide process has rendered it necessary to make considerable alterations in, and additions to, the chapters devoted to that subject, and a new chapter on Economic Considerations has also been added. The author desires to express his thanks to the old students of the Royal School of Mines and others, who have kindly sent information on work connected with the metallurgy of gold, now in progress in various parts of the world.

Royal Mint,
August, 1896.

INTRODUCTION TO THE FIRST EDITION.

As this is the first of a series of treatises devoted to individual metals, which is being prepared under my guidance it may be well to offer a few introductory remarks.

Associates of the Royal School of Mines have taken their full share in conducting mining and metallurgical operations in all parts of the world, but, notwithstanding the wide experience they have gained, no treatise claiming to give a general account of the Metallurgy of Gold and adequately dealing with modern processes could hitherto have been attributed to a Student of the School. It may be claimed that in this country Dr. Percy founded the literature of Metallurgy, but his volume on "*Silver and Gold*," although unrivalled in accuracy of detail, is only a splendid fragment, and gold is alone dealt with in the sections devoted to the refining of bullion and to assaying. A large amount of valuable information concerning new methods and machinery used in the treatment of gold ores has appeared in various official publications during the last twenty years, and much is either scattered over the pages of the scientific press or incorporated in the proceedings of various learned societies. Any attempt, however, to study the subject as a whole from these sources of information would be hopeless for those whose time is limited, or who have not good libraries at their disposal.

Mr. Rose gained his practical experience of gold and silver extraction in the Western States of America, and has written a volume which should prove to be very useful, as its careful and conscientious preparation entitles it to confidence.

W. C. ROBERTS-AUSTEN.

ROYAL MINT, *March*, 1894.

PREFACE TO THE FIRST EDITION.

In the present volume an effort has been made to supply a succinct summary of the existing condition of the metallurgy of gold, for the use of students and others who are interested in the industries connected with the precious metals. It has been said that mill-managers as a class have little to learn from books; but, even if this be so, they will still need to keep themselves acquainted with the progress of their art in far distant countries. To them the bibliography appended will be useful.

Although, in this book, brief accounts are given of some typical machines, attention has been directed rather to methods of procedure than to details of machinery, which will be dealt with in a separate volume in the series. The practice in particular extraction works has been described at some length, as such information, wherever it is available, is among the most valuable that can be given. Some recently devised methods of great importance in the metallurgy of gold, such as the MacArthur-Forrest cyanide process, the new barrel chlorination process, and the improved Gutzkow parting process, are given for the first time in a manual. Particular attention has been paid to the assay of gold bullion, the system in use at the Royal Mint and the precautions necessary to ensure the highest attainable accuracy being described. Little space is devoted to the geographical and geological distribution of gold ores, as this ground has been amply covered by other works. Moreover the processes of smelting, leaching, and pan-amalgamation used in the treatment of silver ores, whether they contain gold or not, have either been omitted or merely sketched in outline, as belonging to the metallurgy of silver.

I am indebted to the President and Council of the Institute of Civil Engineers for leave to reproduce Figs. Nos. 14, 16, 17, 32, 33, 35, and 36, and to Mr. John Murray, for Figs. 58, 60, 61, 62, 63, 64, 66, and 67, which are copied from Dr. Percy's *Metallurgy*. I also take this opportunity of expressing my thanks to my colleague, Mr. F. W. Bayly, and to other friends for their kind assistance while certain sections were going through the press.

In conclusion, I desire gratefully to acknowledge the kindness of Prof. Roberts-Austen in giving valuable advice throughout the progress of the work. It was at his suggestion that the task was undertaken, and it is hoped that it will contribute to the realisation of his wish that the experience of School of Mines men should be collected in a number of works which will together form a comprehensive metallurgical series.

T. K. ROSE.

ROYAL MINT, *March*, 1894.

CONTENTS.

CHAPTER I.—THE PROPERTIES OF GOLD AND ITS ALLOYS.

	PAGE		PAGE
Physical and chemical properties of gold,	1	Magnetism,	3
Introduction,	1	Conductivity and expansion,	4
Colour,	1	Atomic weight and volume,	4
Malleability and ductibility,	2	Volatility,	4
Hardness,	2	Crystallisation,	8
Tenacity,	2	Solubility,	9
Specific gravity,	3	Preparation of pure gold,	10
Cohesion,	3	Allotropic forms of gold,	11
Specific heat,	3	Alloys of gold,	12
Fusibility,	3	Amalgams,	14
Spectrum,	3	Gold and silver,	15
Latent heat,	3	,, copper,	16
		Liquation,	17

CHAPTER II.—CHEMISTRY OF THE COMPOUNDS OF GOLD.

	PAGE		PAGE
Compounds of gold,	19	Aurous cyanide,	27
Haloid compounds,	20	Auro-cyanide of potassium,	27
Chlorides of gold,	20	Auri-cyanide of potassium,	28
Proto-chloride,	20	Oxides of gold,	28
Auro-aurichloride,	20	Aurous oxide,	28
Trichloride,	20	Auric oxide,	28
Chlor-auric acid,	26	Aurates,	29
Bromides of gold,	26	Fulminating gold,	29
Proto-bromide,	26	Sulphites of gold,	29
Auro-auric bromide,	27	Hyposulphites of gold,	30
Tribromide,	27	Silicates of gold,	32
Iodides of gold,	27	Sulphides of gold,	32
Cyanides of gold,	27	Purple of Cassius,	33

CHAPTER III.—MODE OF OCCURRENCE AND DISTRIBUTION OF GOLD.

	PAGE		PAGE
Forms in which gold occurs in nature,	34	Petzite,	37
Vein gold,	34	Nagyagite,	37
Placer gold and nuggets,	36	Composition of native gold,	38
Calaverite,	37	Geographical distribution of gold,	38
Sylvanite,	37	Origin of gold ores,	41

CHAPTER IV.—PLACER MINING—SHALLOW DEPOSITS.

Placer deposits,	42	Tail-race,	53
Methods of obtaining gravel from shallow placers,	43	Ground-sluice,	53
		Booming,	53
Appliances used in washing the gravel,	44	Dry diggings,	54
		Cement gravels,	54
Pan,	44	Tail sluices,	54
Batea,	44	Fly catchers,	55
Prospecting trough,	45	River mining,	55
Horn-spoons,	45	1. River mining proper,	55
Cradle,	45	2. Dredging,	57
Long-tom,	46	3. Deep bar mining,	57
Puddling-tub,	46	Methods of working Siberian placers,	58
Siberian trough,	47		
Sluice,	49	1. Siberian sluice,	59
Sluicing,	50	2. Trommel,	60
Use of drops,	51	3. Pan,	60
„ undercurrents,	51	Beach mining,	61
„ mercury,	52	New method of working shallow placer deposits,	61
Cleaning-up,	52		

CHAPTER V.—DEEP PLACER DEPOSITS.

Nature and mode of origin of deposits,	62	Supply of water,	71
		Breaking down the bank,	74
Distribution of gold in the gravels,	66	Sluicing gravel,	76
		Cleaning-up,	79
Origin of gold in the gravels,	67	Disposal of tailings,	80
Minerals occurring in the gravels,	69	Drift mining,	81
		Shaft „	81
Methods of working,	70	Hydraulic elevator,	83
"Hydraulicking,"	70	Economic conditions,	85
Commencement of operations,	71	Shallow placer deposits,	85
		Deep „ „	87

CHAPTER VI.—QUARTZ CRUSHING IN THE STAMP BATTERY.

	PAGE
Primitive methods of crushing and amalgamating,	88
Mortar,	89
Maray,	89
Chilian mill,	90
Arrastra,	90
Iron prospecting arrastra,	93
Stamp battery,	93
Rock-breakers,	96
Blake,	96
Dodge,	97
Gates,	98
Schranz,	99
Material employed for crushing surfaces,	100
Position and use of rock-breakers,	101
Battery proper,	102
Foundations,	103
Framework,	103
Mortar,	103
Splash-box,	105
Dies,	105
Cam-shaft,	106
Cams,	106
Tappets,	107
Stamp,	108
Guides,	109
Lever,	109
Screens,	110
Order of fall of stamps,	113
Automatic feeding,	113
Stanford's feeder,	114
Challenge ,,	115
Water supply,	116
Amalgamated plates,	117
Mill site,	119
General arrangement of the stamp mill,	120

CHAPTER VII.—AMALGAMATION IN THE STAMP BATTERY.

	PAGE
Treatment of the amalgamated plates,	122
Discoloration of the copper plates,	123
Chemicals used to promote amalgamation,	124
Use of mercury,	125
Grade of plates,	125
Muntz metal plates,	126
Shaking copper plates,	127
Corrugated plates,	128
Mercury wells,	128
Galvanic action in amalgamation,	128
Designolle process,	129
Cleaning-up,	130
The clean-up barrel,	131
,, ,, pan,	131
Retorting,	133
Loss of mercury,	135
,, gold,	138
Non-amalgamable gold,	144
Gold in pyrites,	145

CHAPTER VIII.—OTHER FORMS OF CRUSHING AND AMALGAMATING MACHINERY.

	PAGE
Special forms of stamps,	148
Husband,	148
Steam,	149
Elephant,	151
Huntington mill,	152
Crawford ball mill,	156
Other ball mills,	158
Pan-amalgamation,	158
The old system,	159
The Boss continuous system,	159
Treatment of concentrates,	163
Molloy's hydrogen amalgamator,	164
Jordan's amalgamator,	165

Chapter IX.—Concentration in Stamp Mills.

	PAGE		PAGE
Concentration,	166	Duncan's,	174
Settling boxes,	168	Percussion tables,	175
Classification according to size,	169	Gilpin County concentrator,	175
Screens,	169	Frue vanner,	176
Pointed boxes,	170	Method of working,	180
Early concentrating machinery,	171	Riffle belted,	183
Blanket strakes,	172	Embrey concentrator,	183
Riffled sluices,	173	Lührig vanner,	184
Raising-gate concentrator,	173	Hartz jigs,	187
Round buddle,	173	Pneumatic jig,	188
Centrifugal concentrators,	174	Clarkson and Standfield's concentrator,	188
Hendy's,	174		

Chapter X.—Stamp Battery Practice in Particular Localities.

	PAGE		PAGE
In California,	190	In the Thames Valley, New Zealand,	203
In Colorado,	195	In Dakota,	206
On free-milling ores in Australia and New Zealand,	199	In the Transvaal,	207

Chapter XI.—Chlorination: The Preparation of Ore for Treatment.

	PAGE		PAGE
The Plattner process,	215	Elimination of arsenic and antimony,	232
Origin,	215	Use of salt,	233
Method of working at Reichenstein,	216	Losses of gold,	235
Modern practice in chlorination,	217	Mechanical furnaces,	237
Crushing,	218	1. With mechanical stirrers,	238
Krom's rolls,	219	(a) O'Hara,	238
Rolls at Rapid City, Dakota,	222	(b) Spence,	239
Comparison between rolls and stamps,	223	(c) Pearce Turret,	240
		2. With rotating bed,	240
Drying the ore,	224	3. Revolving cylinders,	243
Roasting,	225	(a) Brückner,	243
Reverberatory furnace,	226	(b) Hofmann,	244
Chemistry of oxidising roasting,	229	(c) White,	246
Decomposition of various minerals,	231	(d) White-Howell,	247
		Use of producer gas in roasting,	248

Chapter XII.—Chlorination: The Vat Process.

	PAGE		PAGE
Construction of vats,	249	Amount of chlorine required,	257
Charging-in,	251	Leaching,	258
Generation of chlorine,	252	Precipitation of gold,	259
Impregnation,	254	Cost of working,	261
Reactions in vat,	255		

CHAPTER XIII.—CHLORINATION: THE BARREL PROCESS.

	PAGE		PAGE
History,	262	Organic substances,	275
Mears process,	263	Sulphuretted hydrogen,	275
Thies process,	265	Sulphurous acid,	276
Construction of barrel,	268	2. Solid precipitants,	277
Charging-in,	268	Charcoal,	277
Amounts of chemicals required,	270	Insoluble sulphides,	278
Method of leaching,	271	Metals,	279
Mechanical difficulties in leaching,	272	Modern patent processes of chlorination,	280
Leaching by pressure,	272	Newbery-Vautin,	280
Riotte's method,	273	Pollok,	280
Precipitation of gold,	273	Swedish,	281
1. Soluble precipitants,	274	Cassel,	283
Ferrous sulphate,	274	Julian,	283
		Greenwood,	283

CHAPTER XIV.—CHLORINATION: PRACTICE IN PARTICULAR MILLS.

Vat process,	284	Cost of working,	291
1. Butters' Mill, Kennel, California,	284	3. Modern barrel process at the Golden Reward Works, Deadwood, Dakota,	291
2. Plymouth Mill, Amador Co., California,	285	Cost of working,	298
3. Treadwell Mine, Alaska,	286	4. Bromination Mill, Rapid City, Dakota,	299
Barrel process—		Cost of working,	302
1. Mears process at Deloro, Canada,	287	Future of chlorination,	304
2. Thies process in Carolina,	290		

CHAPTER XV.—THE CYANIDE PROCESS.

History,	306	Molloy's method of precipitation,	321
MacArthur-Forrest process,	308	Plant required,	322
1. Preparation of ore,	308	Treatment of ore slimes,	323
2. Dissolving the gold,	309	Testing of ores,	323
Vats,	309	Direct treatment of Rand ore,	324
Charging-in,	310	Use of cyanide in stamp battery,	325
Separation of slimes,	311	Method of treatment—	
Leaching,	311	At the Robinson mine,	326
Agitation,	312	,, Sylvia mine,	326
Strength of cyanide solution,	315	,, New Primrose mine,	328
Disposal of tailings,	316	Siemens-Halske process,	329
3. Precipitation of gold,	317	Results of process—	
Cleaning-up,	318	In South Africa,	331
4. Production of bullion from precipitate,	319	In other parts of the world,	332

Chapter XVI.—Chemistry of the Cyanide Process.

	PAGE		PAGE
Action of potassium cyanide on gold and other metals,	333	Re-precipitation of gold and silver in leaching vats,	345
Decomposition of potassium cyanide,	338	Testing strength of solution,	345
		Strength of solution required,	347
Decomposition in the zinc boxes,	339	Consumption of cyanide,	347
Action of potassium cyanide—		Methods of increasing the speed of action of potassium cyanide,	348
On metallic salts and minerals,	340		
On oxidised pyrites,	343	The Hood process,	349
The soda solution,	344	The Sulman-Teed process,	351
		Ores suitable to process,	352

Chapter XVII.

Pyritic smelting, ... 353

Chapter XVIII.—The Refining and Parting of Gold Bullion.

General considerations,	356	5. Reduction of silver chloride,	374
Refining,	357	Parting by sulphuric acid,	375
Composition of bullion,	357	1. Mixing and granulating alloys,	375
Melting furnace,	359		
Crucibles,	360	2. Dissolving silver,	376
Melting,	360	3. Melting gold residue,	378
Refining,	361	4. Precipitation of silver,	379
Toughening,	363	5. Crystallisation of sulphate of copper,	379
Casting,	364		
Losses of bullion,	366	Combined process,	380
Refining by sulphur,	368	Gutzkow process,	381
Osmiridium in gold bars,	368	New Gutzkow process,	383
Parting processes,	369	Miller's chlorine process,	386
Cementation,	370	Original method,	387
Parting by sulphide of antimony,	370	Later improvements,	389
		Modern practice at Melbourne,	392
Parting by sulphur,	370		
Nitric acid process,	371	Modern practice at Sydney,	401
1. Granulation of alloys,	371	Refining brittle gold by chlorine,	403
2. Dissolving granulations,	372		
3. Treatment of gold residues,	373	Parting by electrolysis,	404
4. Treatment of silver solution,	374		

Chapter XIX.—The Assay of Gold Ores.

General considerations,	407	Fluxes,	412
Blowpipe assay,	407	Methods of operation,	414
Sampling and crushing the ore,	408	Roasting before fusion,	416
Metallics,	409	Cleaning slag,	417
Crucible method of assay,	410	Treatment of base ores,	417
Fusion,	410	Cupellation,	417
Assay-ton weights,	411	Influence of base metals,	420
General charges,	411		

Chapter XIX.—*Continued.*

	PAGE		PAGE
Inquartation and parting,	421	2. Amalgamation,	428
Examination of assay materials,	423	3. Chlorination,	428
		4. Whitehead's method,	430
Examination of cupel,	423	5. Assay of pyrites,	430
Assay by scorification,	424	(a) Swartz's method,	430
Detection of gold in minerals,	426	(b) Stapff's ,,	430
Estimation of gold in dilute solution,	427	6. Assay of purple of Cassius,	431
		7. ,, a Mint sweep,	431
Special methods of assay,	427		
1. Mixed wet and dry method,	427		

Chapter XX.—The Assay of Gold Bullion.

	PAGE		PAGE
Parting assay,	431	Assay of alloys of gold, silver, and copper,	452
1. Selection of sample,	432	Effects of other metals,	453
2. Preparation of assay-piece for cupellation,	433	Assay of various gold alloys,	453
3. Cupellation,	436	A. Alloys requiring scorification,	453
Assay furnace,	436	Arsenic and antimony,	453
Cupels,	438	Iron and manganese,	454
Method of operation,	438	Cobalt and nickel,	454
Flashing,	439	Zinc,	454
Temperature of muffle,	440	Tin,	454
4. Preparation of cupelled buttons for parting,	443	Aluminium,	454
		B. Amalgams,	455
5. Parting,	444	C. Platinum group,	455
(a) In parting flasks,	444	(1) Platinum,	455
(b) In platinum trays,	445	(2) Palladium,	456
Relative advantages of these,	446	(3) Rhodium and iridium,	456
6. Weighing the cornets,	447	D. Tellurium compounds,	457
Losses of gold in bullion assaying,	447	Wet methods of assay,	457
		Other methods,	458
Silver retained in cornets,	449	1. Touchstone,	458
Occluded gases,	450	2. Colour and hardness,	459
Checks or proofs,	450	3. Density,	459
Limits of accuracy in assay,	451	4. Spectroscope,	459
Parting by sulphuric acid,	451	5. Electrolysis,	460
Preliminary assay,	452	6. Induction balance,	460
Assay by cadmium,	452		

Chapter XXI.—Economic Considerations.

	PAGE		PAGE
Management of gold mills,	462	Table of production in different countries,	467
Cost of production of gold,	462		
Annual production of gold, past and present,	464	Amount produced by different processes,	469
		Consumption of gold,	469

BIBLIOGRAPHY.

	PAGE		PAGE
Periodicals,	471	Roasting,	478
General,	472	Chlorination,	479
Properties of gold and its alloys,	473	Cyanide process,	480
Distribution of gold,	475	Pyritic smelting,	480
Placer mining,	476	Refining and parting of bullion,	480
Crushing and amalgamation,	477	Assaying,	480
Concentration,	478		

INDEX, ... 483

THE METALLURGY OF GOLD.

CHAPTER I.

THE PROPERTIES OF GOLD AND ITS ALLOYS.

PHYSICAL AND CHEMICAL PROPERTIES OF GOLD.

Introduction.—Some of the more obvious physical properties of gold were already well known and taken advantage of at a very early period in the history of man, long before any exact science could be said to exist, and it is of interest to the metallurgist to remember that the earliest dawn of the science of chemistry was heralded by the study of the properties of gold, and by the efforts which were made to invest other matter with these properties. From the fourth to the fifteenth century, chemistry, which was first called "chemia," and then "alchemy," was defined as the art of transmuting base metals into gold and silver, almost all the labours of philosophers being intended to aid directly or indirectly in solving this problem. At the end of this period, while Paracelsus was giving to chemistry a new aim—that of investigating the composition of drugs, and their effect on the human body—Agricola was reducing to order the numerous empirical facts which together made up the art of metallurgy, and although alchemy died hard, its era of usefulness may be said to have ended here. Gold has been accused, with some justice, of having been the cause of many of the wars and marauding expeditions from which the world has suffered, but it may be asserted with equal truth that it has been instrumental, in a far greater degree than most other commodities, in promoting the growth of civilisation, the efforts of the alchemists having laid the foundations of the science of chemistry, and those of the gold-seekers having resulted in the discovery of new countries, and in the spread of knowledge of all kinds.

Colour.—The lustre and fine colour of gold have given rise to most of the words which are used to denote it in different languages. The word "*gold*" is probably connected with the Sanscrit word "*jvalita*," which is derived from the verb "*jval*," to shine. It is the only metal which has a yellow colour when in mass and in a state of purity. Impurities greatly modify this colour, small quantities of silver lowering the tint, while copper raises it. In a finely divided state, when prepared by volatilisation or precipitation, gold assumes various colours, such as deep violet, ruby and reddish-purple, the tint varying to brownish-purple and thence to dark brown and black. This purple colour has been supposed by some experimenters (viz., Guyton de Morveau, Büchner, Desmarest, Creuzbourg and Berzelius) to be due to the formation of a coloured oxide of gold of unknown composition, but Buisson, Proust, Figuier and, more recently, Krüss have shown that no oxygen can be obtained from this coloured material, and that it probably consists of metallic gold. Similar colours are seen in purple of Cassius, and in Roberts-Austen's purple alloy of aluminium and gold, the colour in each case being probably due to a particular form of finely divided gold. The ruby coloured gold, suspended in a liquid in which it has been precipitated by ether and phosphorus, is present in the most finely divided state obtainable, and perhaps it then approaches the condition of separate atoms. Such gold shows no tendency to settle by gravity even after being kept undisturbed for several years. Somewhat less finely divided gold gives a faint blue tinge to light transmitted through the liquid in which it is suspended. Gold precipitated from its solution as bromide is in a different molecular condition from that formed from chloride, giving out 3·2 calories in passing into the latter state.* The surface colour of small particles of native gold is often apparently reddened by being coated with translucent films of oxides of iron. Very thin plates of gold are translucent, and appear green by transmitted light, while remaining yellow by reflected light. On heating, the green colour changes to ruby red, but is restored by the pressure of a hard substance by which the state of aggregation is again altered (Faraday). Molten gold is green, and its vapour is also probably greenish.

Malleability and Ductility.—Malleability and ductility are possessed by gold at all temperatures to a far higher degree than by any other metal. A single grain of gold can be drawn out into a wire over 500 feet long, and leaves of not more than $\frac{1}{300000}$ of an inch in thickness can be obtained by beating. Faraday has shown that the thickness of these leaves may be still further reduced by floating them in a dilute solution of potassic cyanide by which they are partly dissolved.

Hardness.—The hardness of gold lies between that of aluminium and that of silver, corresponding to the number 979 in Bottone's scale, in which the diamond is 3,010.

Tenacity.—The purest gold obtainable has a tenacity of 7 tons

---

* Thomsen, *Thermochemische Untersuchungen*, vol. iii., p. 412.

per square inch and an elongation of 30·8 per cent., but the presence of minute traces—*i.e.*, $\frac{1}{2000}$ of other elements, especially those having high atomic volumes—*e.g.*, bismuth, tellurium, lead, &c., greatly lowers these constants, as well as the malleability and ductility of the metal, while its hardness is increased.* Gold containing $\frac{1}{2000}$ of bismuth can almost be crumbled in the fingers.

Specific Gravity.—The specific gravity of gold when precipitated from solution by oxalic acid is 19·49 (G. Rose);† when cast it varies from 19·29 to 19·37, but this can be raised by compression to over 19·48.‡ Henry Louis has shown§ that the specific gravity of unannealed "parted" gold (*i.e.*, the residue left after boiling silver-gold alloys in nitric acid) is 20·3, its density being lowered by the process of annealing, a fact which Louis considers is indicative of the existence of an allotropic modification of the metal.

Cohesion.—On heating, gold can be welded like iron below the point of fusion, and finely divided gold agglomerates on heating without being subjected to pressure. Pressure alone is also sufficient to make gold dust cohere, while a true flow of the particles of gold can be induced in the case of the pure metal and some of its alloys.

Specific Heat.—The specific heat of gold is ·0324 (Regnault) or ·0316 (Violle).

Fusibility.—Gold fuses, after passing through a pasty stage, at a clear cherry-red heat, just below the fusing point of copper and much above that of silver. The metal expands considerably on fusing and contracts again on solidifying. Carnelley gives the temperature of fusion as 1,037°, Violle as 1,045°, whilst recent determinations have given it as 1,061·7° (Heycock and Neville), 1,050° to 1,060° (Le Chatelier), and 1,072° (Holborn and Wien).

Spectrum.—In the gold spectrum Huggins saw 23 lines, the wave lengths of the most important ones being 523·1, 583·5, and 627·6 respectively.∥

Latent Heat.—The latent heat of fusion of gold is 16·3, and the normal lowering of the freezing point for 1 atom of impurity in 100 atoms of gold is 10°·6, but the presence of from 0·1 to 4·0 per cent. silver does not cause any alteration in the freezing point.

Magnetism.—Gold is diamagnetic, its specific magnetism being 3·47 (Becquerel), if that of iron is taken as 100.

---

* Roberts-Austen, *Phil. Trans. Royal Soc.*, vol. clxxix. (1888), p. 339.
† *Pogg. Ann.*, vol. lxxiii. (1848), p. 1, and vol. lxxv. (1848), p. 403.
‡ *Eighth Report of the Royal Mint*, 1877, p. 43.
§ *Trans. Am. Inst. of Mng. Eng.*, Chicago Meeting, 1893.
∥ For full information on the spectroscopic characteristics of gold, see Lockyer and Roberts, *Phil. Trans. Royal Soc.*, vol. clxiv. (1874), part ii., p. 495; and Frémy, *Ency. Chim.*, vol. iii. (1888), L'or, p. 40.

Conductivity and Expansion.—Its electrical conductivity is 76·7, and its thermal conductivity 53·2 (or, according to Gray, 77), that of silver being 100 in each case. Its coefficient of linear expansion is 0·0000144 between 0° and 100°.

Atomic Weight and Volume.—Its atomic weight, formerly believed to be about 196·2, has been more recently given by Krüss as 196·64, by Thorpe and Laurie as 196·85, and by Mallet as 196·79. It is probable that 196·8 is not far from the truth. The atomic volume of gold is 10·2.*

Volatilisation of Gold.—The boiling point of pure gold has not been determined; calculated according to Wiebe's formula it would be about 2,240°, or nearly 500° above the melting point of platinum.† However, contrary to the belief of the older experimenters (Gasto Claveus, and others), it is sensibly volatile in air at far lower temperatures. Robert Boyle was unaware of this fact, but Homburg gilded a silver plate in 1709 by holding it over gold strongly heated in the focus of a burning mirror (*Encyclopædia Britannica*, 1778, and *Gmelin's Handbuch*), and St. Claire Deville volatilised and again condensed gold when melting it with platinum. The rapid volatilisation of gold, when heated by an ordinary blowpipe, was first proved in 1802 by Dr. Robert Hare, of Philadelphia (*Tilloch's Magazine*), a purple stain being thus produced on bone-ash in a few seconds. Lastly, a discharge of high-tension electricity from gold points causes its volatilisation, and if the discharge is sent through a fine gold wire stretched on paper, it converts it into a purple streak of finely divided condensed particles of the metal. The rapid distillation of gold caused by heating it in a current of air of considerable velocity, such as that furnished by a blow-pipe, by which the liquid is thrown into waves, may be shown at any time by heating a fragment of the precious metal of the size of a pin's head on a bone-ash cupel in the oxidising flame of a good mouth blowpipe. Almost immediately after the fusion is complete, a purple stain of condensed gold begins to form on the outer margin of the cupel. The author has found that a piece of gold weighing 0·5 gramme loses half its weight in an hour, if heated on a cupel by a foot-blowpipe (the temperature attained being probably less than 1,300°), and only a few minute beads are observable, detached from the main button. Alloys of copper and gold disappear much more rapidly. No doubt most of the gold passes off as spray, but perhaps part of the loss may be due to rapid volatilisation, and could not be correctly described as mechanical loss.

The volatility of gold, both when pure and when alloyed with silver and copper, has been investigated by Napier,‡ who found

---

* *Introd. to the Study of Metallurgy*, by Prof. Roberts-Austen, p. 58.
† L. Meyer, *Mod. Theories of Chemistry*, p. 134.
‡ *Chem. Soc. Journ.*, vol. x. (1859), p. 229; and vol. xi., p. 168.

that an alloy of 100 parts gold to 12 parts of copper, if kept for six hours at a temperature just high enough to keep it melted, lost 0·234 per cent. of its gold contents, and at the highest temperature attainable in an assay muffle, it lost 0·8 per cent. in six hours. An increase in the amount of copper present caused an increase in the loss of gold. In the simple operation of pouring about 30 lbs. of a gold-copper alloy from a graphite crucible into moulds, fumes were given off, of which the part condensed in a wet glass beaker held above the crucible contained 4·5 grains of gold. Napier also found that gold does not appear to volatilise so readily when alloyed with silver only, as when copper is also present.

Makins found that gold volatilises sensibly along with silver and lead, when melted with these metals in a muffle in an ordinary bullion assay. The loss of gold by volatilisation on melting its copper alloy is the common experience of mints. At the Sydney Mint it was estimated to be 0·017 per cent., or £170 per million sterling melted, and is probably seldom less than 0·01 per cent., or one part in ten thousand. At the same Mint, Leibius found that the sweepings from the top coping-stone of a chimney 70 feet high contained 1·46 per cent. of gold and 6·06 per cent. of silver. Similar results have been obtained at some other mints.

A number of experiments were recently made at the Royal Mint by the author,* with the view of determining the effect of variations in the temperature and other conditions on the volatility of gold and some of its alloys. The test pieces were heated in a muffle furnace, the temperature of which was determined by the optical pyrometer, and by the Le Chatelier thermo-couple, which consists of platinum and rhodio-platinum wires. The results of some of the experiments on fine gold, and certain alloys of gold and copper, are given in the table on the next page.

The following conclusions may be drawn from the table:—

1. The loss of gold on heating the pure metal rises with the temperature, being four times as great at 1,250° as at 1,100°, whilst it is insignificant at 1,075° and probably *nil* at 1,045°, the melting point (Violle).

2. An atmosphere consisting largely of carbonic oxide is apparently favourable to the volatilisation of gold, the rate being six times greater than in an atmosphere of coal gas at the same temperature. The increased loss when the graphite crucible was substituted for a cupel may be noticed in this connection; clay crucibles of similar shape to the graphite one have an opposite effect, the loss being less than on cupels.

3. The increase of loss of gold alloyed with copper, when the percentage of the latter metal is increased, which is observable in one of Napier's experiments, is confirmed.

* *Chem. Soc. Journ.*, vol. lxiii. (1893), p. 714.

No. of Exp.	Nature of the Atmosphere	Time of Fusion	Temperature, in Degrees above 1045° (m. p. of gold).		Fine Gold.			Copper-gold Alloys.			
			Mean.	Max.	Weight.	Loss per 1000.	Loss per 1000 per Hour.	Composition.	Weight.	Losses per 1000 of Alloys.	
										Gold.	Copper.
1	Current of air	1 hr. 20 min.	+30°*	+45°*	Grammes. 0·5	0·12	0·09	gold, 952 copper, 48	Grammes. 0·5	0·55	?
2	CO and N (stationary)‡	40 min.	+45°†	+75°†	,,	0·35	0·52	gold, 900 copper,100	2·0	1·35	?4·5
3	CO and N (stationary)	57 min.	+47°*	+61°*	,,	0·60	0·63	gold, 916·6 copper, 83·3	,,	3·6	?1·5
4	CO and N (stationary)	21 min.	+90°*	+128°*	2·0	0·35	1·00	gold, 900 copper,100	,,	3·6	...
5	Current of air	45 min.	+125°†	+170°†	0·5	0·80	1·07	...	...	...	...
6	,,	1 hr. 55 min.	1st hr. +100° 2nd hr. +200°	+240°*†	2·0	2·57	1·32	...	...	...	...
7	CO and N (stationary)	1 hr. 15 min.	1 hr. at +200°†	,,	0·5	3·25	2·60	...	...	...	...
8	Current of coal gas	1 hr.	¾ hr. at +200°	...	,,	0·42	0·42	gold, 950 copper, 50 gold, 820 copper,180	0·5 ,,	0·54 2·62	?3·0 ?8·0

* By platinum and rhodio-platinum couple. † By optical pyrometer.

‡ That is, free from currents.

The volatilisation of gold-copper alloys merits further investigation, as it is of great industrial importance. The high percentage of the losses shown in the table is, no doubt, due to the smallness of the masses treated, and to the comparatively large surfaces exposed in consequence of this.

A number of experiments were also made at the same time on various gold alloys to determine the relative effect of the presence of other metals on the volatilisation of gold. The temperatures and conditions used were similar to those described above. It was found that the volatility of gold was increased by the presence of any metallic impurity, even by the non-volatile metals, such as platinum. The tellurium alloys suffered the heaviest losses, gold containing five per cent. of tellurium losing from 16·5 to 39·5 parts per 1,000 by volatilisation per hour at a temperature of 1,245° (*i.e.*, 200° above the melting point of pure gold), a point of much importance in connection with the metallurgy of telluric gold. Lead and platinum had a very slight effect in increasing the volatility of gold; copper and zinc a more marked effect, while five per cent. of antimony or mercury caused losses amounting to about two parts per 1,000 of gold per hour at 1,245°. Further conclusions which were drawn were as follows :—

1. Although the temperatures employed were higher than those at which pure zinc, cadmium and tellurium, and probably antimony and bismuth, are distilled, yet these elements were never completely volatilised. The "temperature of dissociation of the alloys," as in the case of the bismuth-arsenic alloy investigated by Edward Matthey in a paper read before the Royal Society, January 26, 1893, is much higher than the ordinary distillation point of the more volatile constituent. Thus zinc boils at 950°, but its gold alloy loses little or no zinc at 1,120°, and still retains part at 1,250°, whilst antimony was not driven off to any extent by the highest temperature attained. Copper appears to pass off more easily than some of the so-called volatile metals, perhaps because the dissociation of its alloy with gold may not form an initial stage of the operation.

2. The amount of gold lost depends partly on the volatility of the alloying metal, but perhaps too much stress has been laid on this factor in the past, the results of heating alloys of mercury, zinc, antimony and copper pointing to that conclusion. A metal with a strong attraction for gold, such as copper, may carry it off, perhaps as a vaporised alloy, more easily than one which mixes with it less intimately.

3. It is observable that those impurities which reduce the surface tension of a button of liquid gold appear to increase the vapour pressure of the metal as indicated by the loss on heating. This was to be expected.

4. A current of air or coal gas insufficient to disturb the

surface of the liquid metal does not appear to increase the volatilisation.

It may be added that Hellot stated that if an alloy of one part of gold and seven parts of zinc is heated in air, the whole of the gold comes off in the fumes.

Crystallisation of Gold.—Gold crystallises in the cubic system, occurring frequently in nature in the form of cubes, octahedra and rhombic dodecahedra. Cleavage is never exhibited. Single detached crystals are comparatively rare, and the crystals are usually attached end to end, forming strings, and branching, arborescent, or moss-like masses, which are composed of micro-

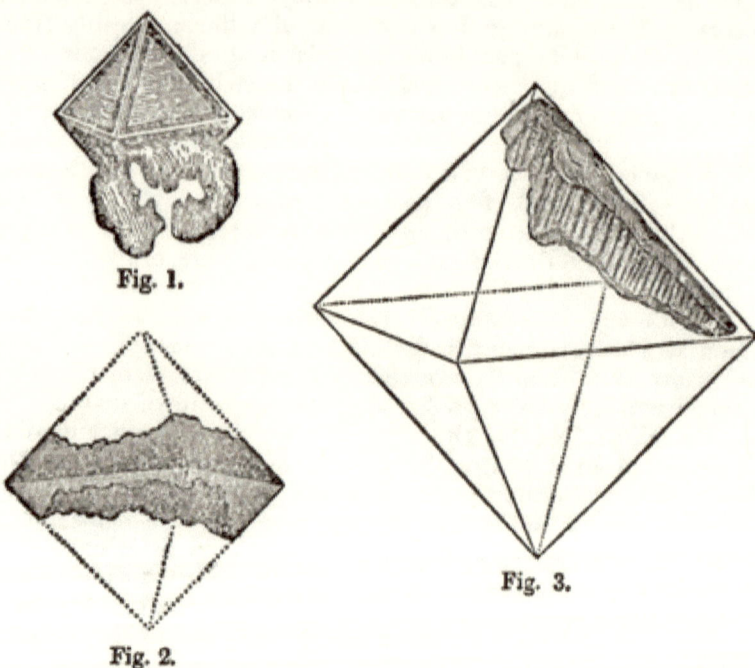

Fig. 1.

Fig. 2.

Fig. 3.

scopic crystals, usually octahedra. These forms occur frequently in quartz veins, but the single crystals, which are usually of larger size, viz., from $\frac{1}{4}$ inch to $1\frac{1}{2}$ inch in diameter, are mainly found in drift deposits. They are rarely perfect or of brilliant lustre, although such crystals were found at the Princeton Gold Mine, Mariposa County, California, but occur more frequently with rounded angles, raised edges, and cavernous faces, which are often marked with parallel striations, and possess little or no lustre (Fig. 1). The octahedra found in California are usually flattened parallel to two opposite faces, or elongated, or otherwise distorted. Still more frequently they are only partially developed, as in Figs. 2 and 3. In all these cases "the incomplete crystals have the appearance of a failure for lack of material" (W. P. Blake). Crystals of greater complexity, containing many modifying faces, occur chiefly in Siberia, Transylvania and Brazil. The most common forms occurring naturally in Australia are the octahedron and the rhombic dodecahedron.*

Artificial crystals can be obtained in several ways, but with great difficulty. The slow cooling of an ingot of gold from fusion usually gives faces and sometimes angles of octahedra on the surface of the metal.† The presence of small quantities of copper prevents this crystallisation. Feathery crystalline plates are precipitated in the electrolysis of a solution of chloride of gold and ammonium. By keeping an amalgam containing 5 per cent. of gold at a temperature of 80° for eight days, and then digesting it at 80° with nitric acid of specific gravity 1·35, and subsequently subjecting the residue to a red heat, bright crystals of gold which are, however, usually microscopic can be obtained.‡ In the Percy collection are some gold crystals found in the mercury troughs at the foot of the "blanket strakes," in an amalgamation mill. The troughs are placed so as to catch any stray particles of gold that may pass the blankets. As the amount of gold recovered in this way is very small, it is not worth while to clean out the troughs frequently, and in this case they had remained undisturbed for nine months, at the end of which time all the amalgam was found to be crystallised. The mercury has been dissolved off by nitric acid, and the gold crystals remain. The smaller crystals are rather indefinite in shape, but amongst the larger ones (which are about half the size of a pea) are well-defined combinations of the octahedron, rhombic dodecahedron, and cube.

Solubility of Gold.—Gold is readily soluble in aqua regia, or in any other mixture producing nascent chlorine, among such mixtures being solutions of (1) nitrates, chlorides, and sulphates —*e.g.*, bisulphate of soda, nitrate of soda, and common salt; (2) chlorides and some sulphates—*e.g.*, ferric sulphate; (3) hydrochloric acid and potassium chlorate; (4) bleaching powder and acids, or salts such as bicarbonate of soda. The action is much more rapid if heat is applied or if the gold is alloyed with one of the base metals than if it is pure. The presence of silver in the gold retards the process, a scale of insoluble chloride of silver being formed over the metal, and the action may eventually be completely stopped if the percentage of silver present is large. Gold is also dissolved by chlorine and bromine, but the action of both of these is much slower than that of aqua regia, and subject to the same difficulties if silver is present;

---

* For a full account of the crystalline forms of native gold, see a Paper by W. P. Blake, in *Precious Metals of the U.S.A.*, 1884, p. 573.
† Chester in *Am. Journ. of Science and Arts*, vol. xvi., July, 1878, p. 29.
‡ Krafft, *Encyclopædia Brit.*, article "Crystallisation."

heat assists the dissolution. Iodine only dissolves gold if it is nascent, or if heated with gold and water in a sealed tube to 50°. Metallic gold dissolves in hot strong sulphuric acid, especially if a little nitric acid is added (the precipitated metal dissolving most readily), forming a yellow liquid, which, when diluted with water, deposits the metal as a violet or brown powder. The solution also becomes covered with a shining film of reduced metal on exposure to moist air. On addition of hydrochloric acid or a metallic chloride, auric chloride is formed, no longer precipitable by water.* Gold is also attacked when used as the positive pole of the battery in the electrolysis of strong sulphuric acid, but is immediately reduced again by the evolved hydrogen.†

According to Nicklès,‡ the easily decomposable metallic perchlorides, perbromides and periodides are capable of dissolving gold, lower chlorides, &c., of the base metals being formed, and gold chloride, &c., produced. The higher chlorides and bromides of manganese and cobalt (Co_2Cl_6, &c.) act well, and a hot strong aqueous solution of ferric bromide or of ferric chloride will also readily dissolve gold. Ferric iodide is decomposed by gold under ordinary conditions, aurous iodide being produced. Some other haloid compounds only attack gold in the presence of ether, in which case even hydriodic acid itself is decomposed and aurous iodide formed. Selenic and iodic acids have also been mentioned as solvents for gold, and the effect of a mixture of nitric and nitrous acids is described in Chap. xx. Alkaline sulphides attack gold slowly in the cold, and more rapidly if heated, producing sulphide of gold which is subsequently dissolved. Ditte's observations, however, tend to show that alkaline polysulphides have no action on metallic gold.

Spring has shown § that gold is soluble in hydrochloric acid if heated with it to 150° in a closed tube, and is subsequently reduced by the liberated hydrogen and deposited as microscopic crystals on the side of the tube. C. Lossen pointed out in 1895 that if a solution of potassium bromide is electrolysed, the resulting alkaline solution, containing hypobromite and bromate of potassium, is capable of dissolving gold. Gold is dissolved by aqueous solutions of simple cyanides and by certain double cyanides, such as sulphocyanides and ferrocyanides, which act very slowly except in presence of oxidising agents and with the aid of heat.

Preparation of Pure Gold.—The purest gold obtainable is required for use as standards or check pieces in the assay of gold bullion. The following method of preparing it was adopted by Roberts-Austen in the manufacture of the Trial-Plate, by

---

* Watt's *Dict. of Chem.*, Supplement, p. 652.
† Spiller, *Chem. News*, vol. x., p. 178.
‡ *Ann. Ch. Phys.* [4], vol. x., p. 318.
§ *Zeitschr. anorgan. Chem.*, vol. i. (1893), p. 240.

which the imperial gold coinage is tested.* Gold assay cornets, from the purest gold which can be obtained, are dissolved in nitrohydrochloric acid, the excess of acid expelled, and alcohol and chloride of potassium added to precipitate traces of platinum. The chloride of gold is then dissolved in distilled water in the proportion of about half an ounce of the metal to one gallon, and the solution allowed to stand for three weeks. At the end of this time the whole of the precipitated silver chloride will have subsided to the bottom, and the supernatant liquid is removed by a glass siphon. Crystals of oxalic acid are then added from time to time, and the liquid gently warmed until it becomes colourless, when precipitation is complete, a point reached in three or four days if ten-gallon vessels are used. The spongy and scaly gold so obtained is washed repeatedly with hydrochloric acid, distilled water, ammonia, and distilled water again, until no reaction for silver or chlorine can be obtained, after which it is melted in a clay crucible with bisulphate of potash and borax, and poured into a stone mould. Lack of care in any one of the operations will result in gold containing one or two parts of impurity in ten thousand.

With regard to the above method, it may be observed that carefully purified sulphurous acid gas is a more convenient precipitant than oxalic acid, and may be substituted for it without any ill effects, as any foreign metals that may be present are in such small quantities that their sulphites, even if formed, would remain dissolved. It should be added that, according to the recent researches of Kohlrausch, Rose, and Holleman, silver chloride is soluble in 600,000 to 700,000 parts of pure water at the ordinary temperature. It follows that, under the conditions given by Roberts-Austen, the solution contains at least 0·3 to 0·4 part of silver per 1,000 parts of gold, and this proportion is doubtless higher in practice, owing to the greater solubility of silver chloride in solutions of other chlorides than in pure water. There can be no doubt that part of this silver is precipitated with the gold, and that, by redissolving and reprecipitating, a purer product is obtained. The amount of silver remaining in solution can, moreover, be reduced to about one-fifth of the amount noted above by adding a small quantity of hydrobromic acid to the solution, silver bromide being far less soluble than silver chloride. Another additional precaution is to remove the gases taken up by the gold during the process of melting by heating it to redness *in vacuo*.

Allotropic Forms of Gold.—Little is known of these. The marked influence of traces of other metals on the properties of gold has already been touched on; from this and from the variations in colour and other properties the existence of several allotropic modifications of gold might be inferred. In

* *Fourth Annual Report of the Mint*, 1873, p. 46.

alloys containing appreciable quantities of other metals, evidences of allotropy are not met with so frequently. The potassium alloy, however, containing 10 per cent. of gold, on being attacked by water, leaves a black finely-divided gold powder, and there is reason to believe that this combines with water to form a hydrate.*

Wilm † states that if gold is dissolved in dilute sodium amalgam under water, the aqueous liquid becomes dark violet, and when this is acidulated with hydrochloric acid, a black precipitate of pure gold is obtained. The black gold differs from the ordinary modifications in its extreme lightness; moreover, it is soluble in alkaline solutions, and does not amalgamate with mercury or with sodium amalgam. When heated, it yields the ordinary modification as a violet red powder. This form of gold appears, from Wilm's account, to resemble the black precipitate obtained on digesting certain aluminium-gold alloys with hydrochloric acid.

ALLOYS OF GOLD.

Gold can be made to alloy with almost all other metals, but most of the bodies thus formed are of little or no practical importance. Tin, zinc, arsenic and antimony unite with gold with contraction, and form pale yellow or grey coloured, hard, brittle and easily fusible alloys, of which all, except those containing zinc, are soluble with difficulty in aqua regia. The arsenic and antimony alloys are slowly decomposed by mercury, the base metal being separated as a black powder, which consists in part of arsenide or antimonide of mercury. Lead and iron alloy with gold with expansion, while in the case of copper no change of volume takes place.

Gold alloyed with a small percentage of lead is a hard, brittle, pale-yellow substance, which can be crumbled with the fingers. If more than about 4 per cent. of lead is present, there is marked segregation on solidification, and this also takes place in the case of the zinc alloy and of some others.

Heycock and Neville have shown ‡ that the freezing point of lead is lowered by the addition of gold to it in accordance with the general law. Thus, the freezing point of pure lead being 327°, an addition of 3·8 per cent. of gold reduces it to 301°, and Roberts-Austen has recently found that the eutectic alloy of gold and lead, which contains about 13 per cent. of the former metal, melts somewhere between 190° and 198°. Similarly, by adding 6·9 per cent. of gold to thallium, the freezing point of the latter is lowered from 301° to 261°.

* *Introd. to Study of Met.*, p. 91.
† *Zeitschr. anorgan. Chem.*, vol. iv. (1893), p. 325.
‡ *Chem. Soc. Journ.*, vol. lxv. (1894), p. 72.

Several alloys of gold with other metals in molecular proportions have been isolated, or their existence proved in various ways. Thus, for example, the compound AuSn was recognised by Mathiessen, from the curve of electric conductivity of the gold-tin series of alloys, and this substance was more recently detected by A. P. Laurie by observing the E. M. F. developed by alloys of different compositions when dipped into a solution of $SnCl_2$. Heycock and Neville succeeded in isolating the compound AuCd in 1892, after having suspected its existence for some time. In the same year, Mylius and Fromm prepared a number of gold alloys by precipitation from solution.* Thus a gold-zinc alloy, containing equal weights of the two metals, and approximating in composition to $AuZn_3$, was obtained in the form of black, spongy flocks, by adding a solution of gold sulphate to water in which a zinc plate was placed. The gold-zinc slimes obtained in the cyanide process (*q.v.*) may be compared with this. Gold-cadmium, similarly obtained, is a lead-grey crystalline precipitate, having the composition $AuCd_3$. On heating, it is converted into gold-mono-cadmium, AuCd (see above). If the gold-zinc alloy is shaken with a solution containing a cadmium salt, the gold-cadmium alloy and a zinc salt are obtained. Similarly, a copper plate, acting on solutions containing gold, yields a black, spongy compound of gold and copper; and gold-lead and gold-tin alloys in the form of black slimes are also readily prepared.

The aluminium alloys have been investigated by Roberts-Austen.† These alloys are remarkable for their intense colour, varying from yellowish-green to purple, and some of them appear to present the characteristics of true chemical compounds. A white alloy containing 10 per cent. of aluminium is very hard, and has a melting point no less than 417° lower than that of pure gold; but a deep purple alloy, containing 22 per cent. of aluminium (corresponding to the formula $AuAl_2$), appears to melt at a temperature between 1,065° and 1,070°, or over 20° higher than the melting point of pure gold. It presents, therefore, the extremely rare case of an alloy, the fusing point of which is higher than that of the least fusible of its constituents, and this fact affords strong evidence that it is a true compound of gold and aluminium. It is hardly necessary to point out that the melting points of ordinary chemical compounds are often much higher than the melting point of the least fusible constituent. There is a strong tendency for this purple alloy to be formed when gold is melted with an excess of aluminium, and the result is that the alloys rich in aluminium usually show a marked lack of homogeneity. Evidence of the existence of the compound $AuAl_2$ was also obtained by Heycock and Neville ‡

---

* *Chem. Soc. Journ.*, vol. lxvi., part 2 (1894), p. 236.   *Nature*, vol. xliv. (1891), p. 111.
† *Proc. Roy. Soc.*, vol. l. (1892), p. 367.   ‡ *Loc. cit.*

by noting the effect on the freezing point of tin caused by additions of gold and aluminium in various proportions. A crystalline alloy of gold and bismuth, containing gold 68·22 per cent., bismuth 31·78 per cent., has been prepared by Pearce by means of liquation. By melting this with silver, a crystalline alloy corresponding to the formula AuAg was produced, and crystalline alloys of gold, silver and copper were also obtained.* Specimens of these products are in the Percy collection.

The diffusion of gold into other metals, both liquid and solid, has lately been investigated by Roberts-Austen.† The rates of diffusion in liquid metals are of the same order as those of soluble salts in water, but the diffusion in solids is very slow. If gold is placed at the base of a cylinder of solid lead 70 mm. high, some is found to have reached the top in thirty days, the temperature being kept at 251°. The rate of diffusion is still measurable at 100°, but is almost inappreciable at ordinary temperatures.

The alloys which are most important to the metallurgist are those which gold forms with mercury, with copper, and with silver.

Amalgams.—Mercury rapidly dissolves gold at ordinary temperatures, forming liquid, pasty, or solid amalgams, according to the proportions of the metals present, and their purity. A piece of gold rubbed with mercury is immediately penetrated by it and becomes exceedingly brittle. The ductility is not always restored when the mercury is removed by distillation, a crystalline structure being often induced. Some forms of precipitated gold are not readily taken up by mercury, the particles tending to float on the surface of the latter. An amalgam containing 90 per cent. of mercury is liquid, that containing 87·5 per cent. is pasty, and that containing 85 per cent. crystallises in yellowish-white easily fusible prisms. On dissolving precipitated gold in mercury heated to 120°, and then cooling the mass, white crystalline plates having a composition corresponding to the formula $AuHg_4$ separate out.‡ Amalgams with smaller proportions of mercury can be obtained in various ways by heating gold and mercury to different temperatures up to a low red heat, and acting on the products with nitric acid. Gold amalgam dissolves readily in excess of mercury, forming a liquid mass from which it may be partially separated by straining through chamois leather, when a white pasty amalgam containing about 33 per cent. gold remains behind, while the mercury which filters through contains some gold, the amount varying with the temperature, but not with the pressure applied. Kasentseff has shown § that this liquid amalgam contains 0·11 per cent. of gold

---

* *Trans. Am. Inst. Mng. Eng.*, 1885.
† *Proc. Roy. Soc.*, vol. lix., p. 281. Bakerian Lecture, *Phil. Trans.*, 1896.
‡ Frémy, *Ency. Chim.*, vol. iii., L'or, p. 99.
§ *Bull. Soc. Chim.* [2], vol. xxx., p. 20.

if it is filtered at 0°, 0·126 per cent. at 20°, and 0·650 per cent. at 100° C.; these amalgams, therefore, behave like aqueous solutions.

When amalgams are gradually heated, the mercury is distilled off by degrees, the action soon ceasing if the temperature is allowed to become stationary, and distillation recommencing if it is again raised. At 440° (somewhat below a red heat), an amalgam containing about three parts of gold to one of mercury is obtained, and at a bright red heat almost all the mercury is expelled, and if the heating has not been pushed too rapidly the vapours contain but little gold. The gold obstinately retains about 0·1 per cent. of mercury, which is not driven off below the melting point of gold.

Gold and Silver.—Gold and silver unite in all proportions, yielding alloys which are harder, more fusible, and more elastic than either metal. The hardest is that containing two parts of gold to one of silver. The colour of gold is sensibly lowered by the addition of very small quantities of silver, and on increasing the proportion of the latter, the colour changes by tints of a greenish-yellow (when from 20 to 40 per cent. of silver is present) to white, with a scarcely perceptible yellow tinge (when 50 per cent. of silver is present), and silver-white (when more than 60 per cent. of silver is present). Pearce has obtained in regular octahedra the alloys corresponding to the formulæ Au_8Ag, Au_6Ag, and Au_2Ag by liquation, and Levol has obtained $AuAg$, $AuAg_2$, and $AuAg_5$ in perfectly definite homogeneous forms. The alloys containing small quantities (less than 20 per cent.) of gold liquate readily if kept for some time in a state of quiet fusion, an alloy containing one part of gold to five parts of silver ($AuAg_9$) sinking to the bottom, and slightly auriferous silver floating at the top. The silver-gold alloys most used in jewellery are *green gold* (silver 25, gold 75), *dead-leaf gold* (silver 30, gold 70), and the alloy containing 40 per cent. of silver. Triple alloys of gold, silver, and copper are employed far more frequently by English jewellers than those last mentioned; of these the alloys, consisting of 22-, 18-, 15-, 12-, and 9-carat gold respectively, can be hall marked.*
Alloys of gold and silver were much used for coinage before the methods of parting became well known and inexpensive.

Electrum, which includes pale yellow alloys with from 15 to 35 per cent. of silver, and which occurs native, was much used for ornaments and coins by the Greeks and Romans, and by the nations which acquired their arts. The use of silver in the gold-copper coinage alloys was not discontinued until quite recently, all English guineas and the Australian sovereigns manufactured at Sydney up to the year 1871 containing some

* 22-carat gold contains $\frac{22}{24}$ by weight of fine gold, and so on. For details of these alloys see Gee's *Goldsmith's Handbook*, pp. 41-52.

of it. Both nitric and sulphuric acids attack silver-gold alloys, almost completely dissolving out the silver if it is present in amounts variously stated at from 60 to 75 per cent. or over, while, if the proportion falls below 60 per cent., some of the silver is left undissolved with the gold. Hydrochloric acid scarcely attacks these alloys, and the action of aqua regia is soon arrested if the proportion of silver is considerable.

Gold and Copper.—These metals dissolve in one another in all proportions, forming a complete series of homogeneous alloys, which are less malleable, harder, more elastic, and more fusible than gold, and possess a reddish tint. Those with less than 12 per cent. of copper are fairly malleable; when more than this is present they are more difficult to work owing to their hardness. Since no change of volume occurs when these alloys are formed, their densities may be calculated from those of gold and copper. The density of standard gold is 17·48, and that of the alloy containing gold 900, copper 100 is 17·16. For the densities of other gold-copper alloys, see Rigg in the *Report of the Mint*, 1876, p. 46. Many of the alloys have been used for coinage at various times. The Greeks and Romans, after electrum had fallen into disuse, employed the purest gold they could procure, viz., that from 990 to 997 fine. Under the Roman Emperors, however, copper was intentionally added, and in the two centuries preceding the fall of Rome very base alloys were used, some containing only 2 per cent. of gold, or even less.* In the middle ages these base alloys were discarded, and the "byzant" of Constantinople and the "florin" of Florence were both nearly pure gold, while the first gold coins struck by the nations of Western Europe were also intended to be absolutely fine. The standard 916·6 or $\frac{11}{12}$ (*i.e.*, 916·6 parts gold in 1,000) was adopted by England in the year 1526, the standard of 994·8, which had been introduced in 1343, being finally abandoned in 1637; the 900 standard was introduced in France in 1794, and subsequently adopted in other countries. These two standards are now those most commonly used, the English standard being employed by Russia, Portugal, India, and Turkey, and the French standard by most other civilised countries; the Austrian ducat, however, has a fineness of 986 and that of Holland a fineness of 983, while the Egyptian standard is only 875. Of all these alloys the 900 and 916·6 standards are those best adapted for coinage, keeping their colour fairly well, and resisting wear better than richer alloys. The 900-alloy is harder and wears better than the 916·6-alloy, but the difference is not great, the rate of wear depending less on such small differences of composition than on the mechanical and thermal treatment of the alloys during the operation of coining.† The alloys used in coinage generally

---
* *La Monnaie dans l'Antiquité.* Paris, 1878.
† *Fourteenth Report of the Royal Mint*, 1884, p. 45 and seq.

contain from eight to twelve parts of silver per 1000 in addition to the gold.

Gold-copper alloys tarnish on exposure to air owing to oxidation of the copper, and blacken on heating in air from the same cause. This oxidised coating may be removed and the colour of fine gold (not that of the original alloy) produced by plunging the metal into dilute acids or alkaline solutions, the operation being technically known as "blanching." The colour of alloys may be improved without previous oxidation by dissolving out some copper by acids, a film of pure gold being thus left on the outside which can be burnished. French jewellers use a hot solution of two parts of nitrate of potash, one part of alum, and one of common salt for this purpose.

Nitric or sulphuric acid dissolves out the copper from gold-copper alloys under conditions similar to those under which it removes silver from silver-gold alloys. If the copper falls below 6·5 per cent. the alloy is not attacked by these acids (Pearce). Aqua regia dissolves all the alloys completely.

Liquation of Gold Alloys.—The subject of liquation generally, including that observed in gold alloys, has been discussed in the volume introductory to this series.* Experiments of Roberts-Austen and of Peligot are there described which tend to prove that liquation does not occur in the gold-silver and gold-copper alloys rich in gold and free from all impurities. Levol had previously come to the same conclusion as Peligot with regard to gold-copper alloys, but gave no details of his experiments, remarking, however, that the oxidation of the copper made the research difficult. Roberts-Austen has also cited† the evidence afforded in the preparation of the standard gold trial plate made in the Royal Mint in 1873 as conclusive in proving that considerable masses of standard gold can be obtained of uniform composition. A mass of standard gold weighing 72 ozs. was cast in a suitable mould, and rolled into a plate 37 inches long and 6·5 inches wide; portions of metal were cut from different parts of it, and when these were assayed it was found that the greatest variation between any two assays was $\frac{2}{10000}$, there being no evidence of concentration of the precious metal anywhere. In this case the gold and copper used to make the alloy were of exceptional purity. In 1889, moreover, this chemist ‡ examined the composition of two large ingots, weighing 400 ozs. each, which had been sent to the Mint for coinage by the Bank of England. The fineness of these ingots was 896·2 and 978·5 respectively, and the results showed that no definite liquation had taken place in them. In the case of many of the ordinary trade ingots, however, the discrepancies between assays on pieces taken from opposite ends of the bars

---

* Roberts-Austen, *Introd. to the Study of Met.*, p. 72.
† *Nineteenth Report of the Royal Mint*, 1888, p. 54.   ‡ *Loc. cit.*

prove that the composition is not uniform, but this lack of uniformity is doubtless due to the presence of some other element in addition to gold, silver and copper.

In support of this view it may be mentioned that Louis Janin Jr. instances the case of three ingots from an Idaho mine, which were melted with borax in plumbago pots, and on cooling showed evidences of liquation.* Dip samples were taken after vigorous stirring, and granulated, and assay cuts were also taken from the diagonally opposite corners of the top and bottom faces of the bar after pouring. On assaying these samples the following results were obtained :—

	Gold.	Silver.	Total.
Bar No. 1—			
Cuts,	808	181	989
Dips,	784	191	975
Bar No. 2—			
Cuts,	809	180	989
Dips,	786	184	970
Bar No. 3—			
Cuts,	805	190	995
Dips,	778	201	979

No further experiments were made on the composition of these bars, so that the nature of the base metals present was not determined, although it is probable that the cause of liquation was to be looked for here. It will be observed that silver moved towards the centre in these cases, and gold in a far greater degree towards the outside of the ingots.

James C. Booth, of the U.S. Mint, Philadelphia, has stated † that liquation of gold-copper alloys is induced by the presence of small quantities of base metals, the greatest effect being produced by antimony, bismuth and arsenic, while lead, tin and zinc act in the same way to a less degree, but no direct experiments have been adduced in support of this view. Alloys containing much lead, bismuth and zinc cannot be obtained perfectly homogeneous.

It has been pointed out by the author ‡ that segregation takes place in standard gold rendered crystalline by small quantities of lead or bismuth, the presence of 0·2 per cent. of either of these metals causing the centre of a sphere 3 inches in diameter to be enriched in gold to the extent of about one part per thousand. This is due to the fact, proved by Roberts-Austen, that an eutectic alloy of gold and lead remains molten after the remainder of the mass has solidified, and is consequently driven towards

---

* *Eng. and Mng. Journ.*, vol. liv., 1892, p. 317.
† *Condemnation of Ingot Melts.* Washington, 1875.
‡ *Chem. Soc. Journ.*, vol. lxvii., 1895, p. 552.

the centre of the sphere, and that this fusible alloy contains much gold but very little copper. Arnold's micrographic results* tend to confirm the view which follows from this that the brittleness of crystalline gold is due to the presence of films composed of such eutectic alloys separating the crystals of gold from each other.

It has been shown by Edward Matthey† that when gold ingots containing members of the platinum group are cooled from a state of fusion an alloy rich in the more fusible element (gold) falls out first, driving the less fusible constituent to the centre. Thus the assay of an outside cut of such an ingot gives a result too high in gold, sometimes by several per cent. It has long been known, moreover, that iridium and osmium become concentrated towards the bottom of the mass. The reason for this is that, at the temperature of fusion of gold, these refractory elements, either free or alloyed with gold, sink in the molten metal and are left in the state of small crystalline particles.

CHAPTER II.

CHEMISTRY OF THE COMPOUNDS OF GOLD.

Compounds of Gold.—Gold is characterised chemically by an extreme indifference to the action of all bodies usually met with in nature. Its compounds are formed with difficulty, and decompose readily, their heat of combination being in general small, while some are formed by endothermic actions. The result of this condition of things is that gold is found in nature chiefly in the metallic form, and the mineralogist has, therefore, few compounds to consider. Nevertheless, the laws governing the formation of artificially-prepared compounds of gold are of great importance to the metallurgist, as the processes of extraction are all based on these laws, and in particular, some knowledge of the reactions and general behaviour of gold in its compounds is essential to those who are engaged in extracting gold by wet processes. A short account of some of the compounds of gold is, therefore, appended, special attention being paid to those bodies which are most likely to be of interest to the metallurgist.

Gold forms two series of compounds, having the general formulæ AuR and AuR_3, while doubtful compounds corresponding to AuR_2, AuR_4, and AuR_5, have been declared to exist by Thomsen, Prat, Figuier, and others. These two series are denominated *aurous* and *auric* respectively.

---

* *Engineering*, vol. lxi. (1896), p. 176.
† *Proc. Roy. Soc.*, vol. li. (1892), p. 447, and *Phil. Trans. Roy. Soc.*, vol. clxxxiii. (1892), A. pp. 629-652. See also *Proc. Roy. Soc.*, 1896.

Compounds of Gold with the Halogens.—Gold forms two series of compounds with the halogens, the general formulæ being AuR and AuR$_3$ respectively. Supposed compounds having the formula AuR$_2$ have been described, but are probably mixtures of the series denoted by AuR and AuR$_3$. All these bodies are very unstable, existing throughout only low ranges of temperature, whether in the dry state or in aqueous solution. The chlorides are the most easily formed and the least unstable, the bromides coming next, as might be predicted from the heats of combination which are given in the subjoined table in calories:—

	Chloride.	Bromide.	Iodide.
AuR, solid,	+ 5·8	− 0·1	− 5·3
AuR$_3$, solid,	+ 22·8	?	?
AuR$_3$, in solution,	+ 27·3	+ 5·1	?

CHLORIDES OF GOLD.

Gold Monochloride or Aurous Chloride, AuCl.—This salt is prepared by heating the trichloride to 185° in air for twelve hours. It is non-volatile and unaltered at ordinary temperatures and pressures by dry air, even when exposed to light, but begins to decompose at temperatures above 160°, and the decomposition is complete if it is heated at 175° to 180° for six days, or at 250° for one hour. Its density is 7·4. Water converts aurous chloride into a mixture of gold and gold trichloride. It is a citron-yellow amorphous powder.

Auro-Aurichloride, Au$_2$Cl$_4$.—A dark red compound having this composition is said, by Thomsen, to be obtained by heating finely-divided gold in a current of chlorine to 140° to 170°. According to Krüss and Lindet it is merely a mixture of AuCl and AuCl$_3$. It yields gold and AuCl$_3$ if brought in contact with water, ether, or alcohol.

Auric Chloride or Gold Trichloride, AuCl$_3$.—*Preparation.*—Trichloride of gold can be prepared, according to Debray,* by heating finely-divided gold in a current of pure dry chlorine in a glass tube at a temperature of 300°. The chloride formed sublimes at this temperature and is deposited in the cooler part of the tube in fine red prisms and needles crystallising in the triclinic system. Krüss tried to repeat Debray's experiments without confirming his results.† He found that at first brownish vapours of AuCl$_3$ were given off freely at 140° to 150°, and were condensed in the cold part of the tube, but that after some time the sublimation ceased, and the whole mass was transformed to brownish-red trichloride. On continuing to raise the

* *Compt. Rend.*, vol. xix., p. 985. † *Berichte*, vol. xx., p. 212.

temperature $AuCl_3$ began to decompose at about 180° in spite of the presence of an excess of chlorine, and protochloride of gold, $AuCl$, of a greenish-yellow colour was formed. At 220° the $AuCl$ was completely decomposed, leaving metallic gold, while up to this temperature a little $AuCl_3$ continued to be sublimed, but towards 300° all action ceased, and the gold remained unattacked by the chlorine at all higher temperatures. On lowering the temperature, Krüss observed in succession the formation of $AuCl$ at 220°, and of $AuCl_3$ at 180°.

The exact point at which gold ceases to be attacked by chlorine and the rate of volatilisation of the chloride are of great importance in connection with the loss of gold on roasting auriferous materials with salt.* The matter has, therefore, been lately investigated by the author† with the following results. Gold unites with chlorine if placed in the gas at atmospheric pressure at all temperatures up to a white heat, but the subsequent decomposition of the chloride is rapid above 300°. The absorption of chlorine by gold with the formation of chlorides at first increases in rapidity as the temperature rises, and reaches its maximum at about 225°. The fact that gold is attacked by chlorine, and that the chloride is subsequently volatilised at all temperatures between 180° and 1,100°, was proved by means of Deville's hot and cold tubes, which enable part of the sublimed chloride to be collected. The rate of volatilisation at various temperatures is as follows:—

Number.	Temperature.	Gold Volatilised.		
		Lost from Porcelain Vessel containing the Gold.	Recovered from Water Tube.	Percentage Volatilised in 30 minutes.
		Grammes.	Grammes.	
1	180°	...	0·00105	0·007
2	230°	...	0·0522	0·35
3	300°	...	0·3473	2·32
4	390°	0·2731	0·1748	1·82
5	480°	0·1325	0·0884	0·88
6	580°	0·0907	0·0624	0·60
7	590°	0·0864	0·0590	0·58
8	805°	0·0753	0·0518	0·50
9	965°	0·2441	0·1672	1·63
10	1100°	0·2895	0·2037	1·93

For purposes of comparison, it may be added that when gold is heated in air or coal gas, no gold is volatilised below 1,050°,

---

* See Prof. Christy's experiments, Section on "Roasting," Chap. XI.
† *Chem. Soc. Journ.*, vol. lxvii. (1895), p. 881.

and only about 0·02 per cent. in 30 minutes at 1,100°,* or about one-hundredth part of that volatilised in chlorine.

In the table the amounts "lost" by volatilisation include the amounts recovered from the water tube and the gold condensed on the inside of the outer tube; the latter sublimate was not recovered separately after each experiment.

In every case the gold recovered from the water tube was associated with a little less chlorine than was required to form gold trichloride, and therefore presumably contained either some metallic gold or AuCl, or both; these doubtless resulted from the dissociation of some of the trichloride when in the form of vapour.

The amounts volatilised vary according to two different factors —(1) The vapour pressure of gold trichloride, $AuCl_3$, which of

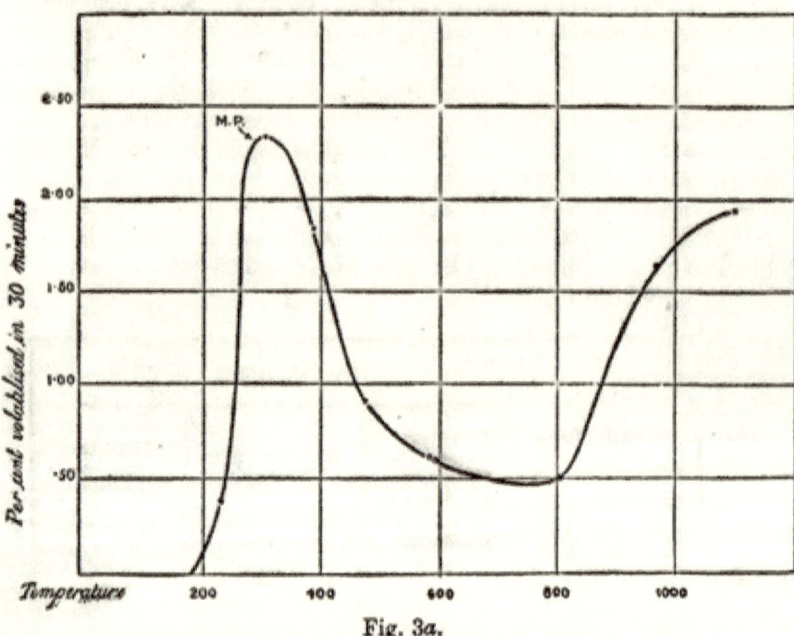

Fig. 3a.

course increases continuously as the temperature rises; and (2) the pressure of dissociation of the trichloride, which also rises continuously with the temperature, but not at the same rate as the vapour pressure. The rise of vapour pressure tends to raise, and that of the pressure of dissociation to reduce, the amount of gold volatilised as chloride. The vapour pressure increases more rapidly than the pressure of dissociation at temperatures below 300°, and also above 900°, but less rapidly at intermediate temperatures. Hence the curve (Fig. 3a) showing the variation of volatilisation with temperature is irregular, passing through a

* *Chem. Soc. Journ.*, vol. lxiii. (1893), p. 717.

maximum near 300°, and a minimum at a point somewhere below the melting point of gold. The first-named change in the direction of the curve possibly occurs at the melting point of the chloride, namely 288°. The second change is perhaps caused by the change of sign of the heat of formation of the trichloride $AuCl_3$; when this becomes negative, the pressure of dissociation of the compound would decrease, in accordance with the law of van't Hoff and Le Chatelier. However this may be, it is certain that when gold is heated in chlorine at atmospheric pressure, trichloride of gold is formed and volatilised at all temperatures above 180°, up to, and probably far beyond, 1,100°.

The usual method adopted for the preparation of auric chloride is to dissolve gold in aqua regia, and then to evaporate the liquid to dryness, keeping the temperature above 100° to prevent the formation of the hydrochloride $AuCl_3HCl$. A brownish-red mass is thus formed, consisting of $AuCl_3$ mixed with more or less of the protochloride and of hydrochloric acid. On taking up with water the protochloride is decomposed into gold and trichloride, but the hydrochloric acid can only be eliminated by shaking with ether, which withdraws the trichloride from its solution in water. If an attempt is made to drive off the hydrochloric acid by heat, a partial decomposition of the trichloride results.

Auric chloride exists both in the anhydrous state and in combination with two equivalents of water, $AuCl_3 . 2H_2O$, when it occurs in orange-red crystals. The anhydrous salt is of a brilliant-red colour, crystallising in needles belonging to the triclinic system and melting at 288° under a pressure of two atmospheres of chlorine. It can be prepared by drying the hydrated salt at 150°. The anhydrous and hydrated salts are both hygroscopic, and dissolve readily in water with elevation of temperature; they are also soluble in alcohol and ether, and in some acid chlorides, such as $AsCl_3$, $SbCl_5$, $SnCl_2$, $SiCl_4$, &c.

Auric chloride is readily decomposed by heat. Lowe states[*] that 4 grammes of the trichloride, when heated in a porcelain basin on a boiling water bath, can be completely transformed into the monochloride, although not until after the lapse of several days. On the other hand, as has already been mentioned, Krüss states that the decomposition of auric chloride, in an atmosphere of chlorine, begins at 180°. According to the experiments of the author,[†] auric chloride is observed to suffer slow decomposition at as low a temperature as 165° in an atmosphere consisting of chlorine, about 1·6 per cent. being converted into monochloride in four hours at this temperature; the decomposition is about five times more rapid at 190°. The decomposition in air can be readily observed at 100°, although it does not seem to be so rapid as was indicated by Lowe. In

[*] *Dingler's polyt. Journ.*, 1891, vol. cclxxix., p. 167.
[†] *Chem. Soc. Journ.*, vol. lxvii. (1895), p. 902.

seven days only 6·6 per cent. of the trichloride was decomposed, the initial rate of decomposition being 0·041 per cent. per hour. At 165°, however, the initial rate of decomposition appeared to be 3·2 per cent. per hour, and the conversion into monochloride was complete in four or five days at 160° and in ten hours at 190°. The rate of decomposition of the trichloride in air at various temperatures can be calculated from the above data by the help of Harcourt and Esson's formula $\alpha_1/\alpha_2 = \left(\tau_1/\tau_2\right)^m$, where α_1, α_2 are the rates of decomposition at the absolute temperatures τ_1, τ_2 respectively. The value of the constant m for this chemical action is found to be about 27, and by substituting this number for m in the equation, the rate of decomposition of gold trichloride is calculated to be 0·365 per cent. in a year at 15°. The decomposition begins to be observable at 70° in air, when the monochloride is formed, about twenty-five years, however, being required for the complete change at this temperature. The observed rate of decomposition shows that a similar change would require about 1,000 days at 100°, while it results from calculation, using Harcourt and Esson's formula, that at 200°, thirty-six hours, and at the melting point—namely, 288°—less than one minute suffices for the complete decomposition of $AuCl_3$ in air.

Whether dry or in aqueous solution, the trichloride is also decomposed by light, gold being deposited in scales in the latter case, but the presence of free hydrochloric acid prevents this decomposition. Weak voltaic currents precipitate metallic gold from the solution of the trichloride upon the negative pole. The solution of trichloride of gold is also decomposed by many reducing agents, such as most organic substances, metals and protosalts; heating the solution in every case hastens the decomposition. The reduction by organic matter is assisted by the action of light, which is especially efficacious in the presence of starchy and saccharine compounds, or of charcoal or ether. In direct sunlight the last reagent deposits a bright mirror of metallic gold, but under ordinary conditions excessively finely divided gold (Faraday's gold) is precipitated. Alkalies also quicken the action of organic matter, and it may be said that all organic compounds reduce gold chloride on boiling in the presence of potash or soda,* while Müller states that a mixture of glycerine and soda lye is one of the best precipitants for gold chloride, separating the metal completely in highly dilute solutions. According to Krüss, if potash and soda are quite free from organic matter, they have no action on solutions of auric chloride, whether cold or hot. If a small quantity of organic matter is present, sub-oxide of gold is precipitated; if larger quantities are present, both metallic gold and sub-oxide

* Frémy, *Ency. Chim.*, vol. iii., 16e Cahier, p. 74.

are precipitated in the cold, but gold alone at boiling point. Alkaline carbonates are without action on cold solutions, but if they are hot, then half the gold is precipitated as hydrate, while the other half remains in solution in the form of a double chloride of gold and the alkali. The precipitation by means of charcoal is of especial importance in view of its adoption in practice.

It has been stated that a current of hydrogen gas will precipitate gold completely, especially on boiling the solution, but Krüss has proved that, if the hydrogen is quite pure, it has no effect either on cold or hot solutions. Sulphur, selenium, phosphorus and arsenic all precipitate gold on boiling the solution of the trichloride. Many metals reduce chloride of gold, the action being, of course, most rapid in the case of the most highly electro-positive metals, such as zinc and iron. It seems probable from the author's experiments* that turnings of these metals would make good precipitants for use on a commercial scale in the chlorination process. Lead sometimes gives fine dendritic plates of gold. Sulphuretted hydrogen precipitates sulphide of gold from both neutral and acid solutions, all traces of gold being readily removed from a solution by this reagent, whilst phosphoretted, arseniuretted, and antimoniuretted hydrogen all precipitate metallic gold. The lower oxides of nitrogen, nitrous acid and many other "*ous*" acids and oxides effect the same decomposition. Sulphur dioxide is a convenient reagent, and is often used in the laboratory, being almost equally efficacious in cold and hot solutions. The reaction is

$$2AuCl_3 + 3SO_2 + 6H_2O = 2Au + 6HCl + 3H_2SO_4.$$

Various protosalts also reduce trichloride of gold. Ferrous sulphate is often used to detect the presence of gold in solution as chloride; this reagent gives the solution a pale blue colour by transmitted light, and brown by reflected light, owing to the formation of finely divided precipitated gold. The reaction is represented by the following equation—

$$2AuCl_3 + 6FeSO_4 = 2Au + 2Fe_2(SO_4)_3 + Fe_2Cl_6.$$

To test a dilute solution for gold, a test tube filled with the liquid is held in the hand side by side with a test tube filled with distilled water and a few drops of a clear solution of ferrous sulphate are added to each. On looking down through the length of the test tubes from above, with a white surface as background, any slight changes of colour may be detected by comparison and the liquids may also be compared with the original solution in a test tube. In this way, by a little practice, the presence of gold in the proportion of only $\frac{1}{720000}$ (1 dwt. per ton of water), or even less can be detected. The

* See Section on "Precipitation of Gold," Chap. xiii.

method is often used in the chlorination process, but it is better to use protochloride of tin, $SnCl_2$. This substance gives a brown precipitate of variable composition in concentrated solutions, but if mixed with the tetrachloride, $SnCl_4$, it gives a precipitate of purple of Cassius. The reaction is very sensitive, and by its means a violet coloration by transmitted light can be obtained in a solution containing 1 part of gold in 500,000 parts of water, while by special means the presence of 1 part of gold in 100,000,000 parts of water can be detected, as described below:*—The liquid supposed to contain gold is raised to boiling, and poured suddenly into a large beaker containing 5 to 10 c.c. of saturated solution of stannous chloride, and the liquids agitated so as to effect complete mixture. A yellowish-white precipitate of tin hydrate forms, which settles rapidly, and can be readily separated from the bulk of the liquid by decantation. If the solution originally contained at least 1 part of gold in 5,000,000 of water ($3\frac{1}{2}$ grs. per ton), the precipitate is coloured purplish-red or blackish-purple, according to the nature of the solution, and the condition of the precipitant. The colour can be seen without comparing it with other precipitates. If less gold than this is present it is better to compare the precipitate with one obtained by the use of boiling distilled water, and to increase the quantity of liquid used while adhering to the same amount of stannous chloride. In this way the presence of 1 part of gold in 100,000,000 parts of water (1 grain of gold in 6 tons of water) can be detected, the amount of liquid required in this case being about 3 litres. The gold is concentrated in the precipitate in which a distinct colour is caused by less than 0·05 per cent. of the precious metal.

Chlor-auric Acid, $HAuCl_4$.—Gold trichloride in the presence of free hydrochloric acid is supposed to form this compound, which crystallises out on evaporation in vacuo in long, yellow needles, having the composition $HAuCl_4 + 4H_2O$, and since gold chloride unites with many other soluble chlorides to form double chlorides, this hydrochloric acid compound is regarded as an acid. It is more stable than gold trichloride. The chloraurates, having the general formula $M'AuCl_4$, or $AuCl_3 . M'Cl$, are readily soluble bodies which can be crystallised, and which decompose with about the same readiness as chlor-auric acid.

BROMIDES OF GOLD.

Gold Protobromide, $AuBr$, is a yellowish-green powder obtained by heating the tribromide to about 140°. It is insoluble in water, but is decomposed by it, metallic gold and the tribro-

---

* T. K. Rose, in *Chem. News*, 1892, vol. lxvi., p. 271.

mide being formed; the change is especially rapid on boiling, and is hastened by the presence of hydrobromic acid.

Auro-auric Bromide, Au_2Br_4, is produced by the action of bromine on finely-divided gold in the cold, some tribromide being simultaneously formed. Water breaks up this bromide into AuBr and $AuBr_3$, and, according to some authorities, it is only a mixture of these bodies.

Gold Tribromide, $AuBr_3$, is produced by the action of a mixture of bromine and water on gold, particularly on the application of heat. Auric tribromide resembles the trichloride in most of its properties. It crystallises in blackish needles or scarlet plates. It is deliquescent, and very soluble in water, and suffers decompositions similar to those noted in describing $AuCl_3$, its solutions being still less stable than those of the chloride. A solution of gold tribromide is gradually decolorised by sulphur dioxide, being completely reduced to the state of monobromide before any precipitate of metallic gold is formed. It is prepared in a pure state by heating finely-divided gold in sealed tubes with bromine and arsenic bromide, $AsBr_3$, to 126°. Gold tribromide forms intensely coloured brownish-red aqueous solutions, the presence of a mere trace of the salt in a solution being observable in this way. Double bromides exist analogous to the chlor-aurates. The bromides are not volatile.

The **Iodides of Gold** are of little interest to the metallurgist. They are prepared with difficulty, and decompose more readily than the chlorides and bromides and are not likely to become of practical importance. The tri-iodide is formed if gold is heated with water and iodine to 50°, particularly in direct sunshine.

CYANIDES OF GOLD.[*]

Cyanogen and gold unite in two proportions forming aurous and auric cyanides, but the latter is only known with certainty in combination.

Aurous Cyanide, AuCy, is obtained by heating aurocyanide of potassium, $KAuCy_2$, with hydrochloric or nitric acid. It is a lemon-yellow crystalline powder, insoluble in water, and unaltered by exposure to air. It is decomposed by heat, yielding metallic gold and cyanogen, and is soluble in ammonia, in alkaline cyanides, and in hyposulphite of soda. It is unattacked by the mineral acids, except by aqua regia, but is decomposed when boiled with potash, metallic gold being thrown down.

Aurocyanide of Potassium, $KAuCy_2$, is obtained by crystallisation from its solution, which is prepared by dissolving metallic gold, auric oxide or aurous cyanide in a solution of potassium cyanide. It is slightly soluble in water and the

[*] See also chapter xvi.

aqueous solution, especially if hot, gilds copper or silver without the agency of a battery, the gold being replaced in solution by the other metal. Precipitates are also formed on the addition of salts of zinc, tin, iron, or silver, no precipitates being formed if potassium cyanide is present in excess.

Auricyanide of Potassium, $AuCy_3 . KCy$, is formed by adding potassium cyanide to a solution of trichloride of gold, the precipitate first formed being redissolved. The solution is completely decolorised, and on cooling deposits colourless crystals of $AuCy_3 . KCy + 3H_2O$. These effloresce in air, giving up two molecules of water; and, on heating, the third molecule of water and some cyanogen are given off, aurocyanide of potassium being formed, and this in its turn is decomposed at a slightly higher temperature.

OXIDES OF GOLD.

Aurous Oxide, Au_2O.—This oxide is prepared by decomposing aurous chloride, $AuCl$, or the corresponding bromide by potash in the cold (Berzelius), when a violet precipitate forms which is blackish when moist, but greyish when dry. When freshly precipitated it is soluble both in alkalies and in cold water, forming an indigo blue solution, with brownish fluorescence, and on warming the solution slightly the corresponding hydrate is precipitated. It is also prepared by the action of nitrate of mercury on the trichloride, and by boiling aurate of potash with organic compounds, such as citrates or tartrates, or by boiling a solution of the trichloride with the potassium salts of these acids. When prepared according to these methods, aurous oxide always contains a certain proportion of metallic gold. Krüss obtained the oxide pure by reducing brom-aurate of potassium at 0° by SO_2, passing in the gas only until the solution became colourless, after which an excess of gas would have precipitated metallic gold. Aurous hydrate is then precipitated by potash, and, after being agglomerated by boiling, it is filtered, washed with cold water, dried, and heated to 200° to expel the water of hydration. At 250° it is resolved into gold and oxygen. Hydrochloric acid decomposes aurous oxide into metallic gold and auric salts, slowly in the cold, quickly at a boiling temperature; aqua regia dissolves the oxide, but sulphuric and nitric acids are without action on it, while weak bases at once decompose it.

An intermediate oxide, AuO, is prepared as a black powder by dissolving metallic gold in aqua regia containing an excess of hydrochloric acid, then adding an excess of carbonate of potash, and afterwards filtering and drying the precipitate. It has been little studied, but the temperature at which it decomposes has been fixed at 205° and its hydrate has been prepared.

Auric Oxide, Au_2O_3.—This, the best known oxide, is a black powder when anhydrous, and is precipitated from solutions of auric chloride in the form of a hydrate by the caustic alkalies, the carbonates of the alkalies, and hydrates of the alkaline earths or zinc. The readiest method of preparation of this compound is to add caustic potash, little by little, to a hot solution of gold chloride, until the yellow precipitate first formed is dissolved to a brown liquid. Then a slight excess of sulphuric acid or some Glauber's salt is added, the precipitate filtered off, washed and purified from potash by being redissolved in concentrated nitric acid, and reprecipitated by dilution with water. On drying this precipitate in vacuo, the hydrate $Au_2O_3.H_2O$, an ochreous powder, results. If it is heated to 110°, oxygen begins to be given off. At 160°, AuO remains, and on heating for some time at 250°, metallic gold remains. Trioxide of gold dissolves in concentrated sulphuric and nitric acids, from which it is partly reprecipitated on boiling or on dilution, and these solutions are supposed to contain sulphates and nitrates of gold respectively. Double nitrates of gold and the alkalies have been obtained as crystals. Hydrochloric and hydrobromic acids dissolve the trioxide forming the haloid salts, but hydriodic acid decomposes it on boiling, giving iodine and metallic gold. Gold trioxide dissolves in boiling solutions of alkaline chlorides, giving aurates and chlor-aurates, while it also combines with metallic oxides to form aurates.

It is easily reduced by hydrogen, carbon and carbonic oxide, with the aid of very gentle heat. Boiling alcohol reduces it, yielding minute spangles of gold which were formerly used in miniature painting.

Aurates.—The aurates of potash and soda have the general formula $Au_2O_3.R'_2O$ or $R'_2Au_2O_4$ assigned to them. They are readily soluble, crystallisable compounds, and are formed when alkalies are added in excess to solutions of gold chloride. The aurates of calcium, magnesium and zinc are insoluble in water, but soluble in hydrochloric acid.

Fulminating Gold is a compound of auric oxide with ammonia, $Au_2O_3(NH_3)_4$, which is formed by precipitating gold chloride with ammonia or its carbonate, or by the action of ammonia on gold trioxide. When prepared by the former method its composition is variable, but the fulminate is always a fearful explosive, decomposing with violence at 145°, or on being struck, and sometimes even spontaneously. It is decomposed without explosion by sulphuretted hydrogen, and by protochloride of tin. It is a grey pulverulent powder, insoluble in water, but soluble in potassium cyanide, auricyanide of potassium being formed.

Sulphites of Gold.—Alkaline sulphites, or sulphur dioxide, which reduce gold trichloride easily, do not produce the same effect on a solution of an alkaline aurate. If sodic bisulphite is added to a boiling solution of sodic aurate ($NaAuO_2$) a yellowish precipitate is formed, soluble in excess of sodic bisulphite, and consisting of a double sulphite of gold and sodium, or *sodic auro-sulphite*, having the composition $3Na_2SO_3 . Au_2SO_3 + 3H_2O$. It is obtained pure by precipitating the corresponding baric salt with $BaCl_2$, and decomposing the precipitate with the minimum quantity of sodic carbonate. Double sulphites of potassium and ammonium with gold also exist. These salts are decomposed by acids, sulphite of gold being deposited, and also on boiling their aqueous solutions, but the addition of sulphuretted hydrogen or alkaline sulphides has no effect on them.

Hyposulphites of Gold.—These compounds are now called thiosulphates by chemists, but the old name is retained as being universally employed by metallurgists. They are especially interesting to the metallurgist, as on their formation depends the extraction of gold from auriferous silver ores, when these are treated by the ordinary hyposulphite, or by the Russell process. The soluble double hyposulphites of gold with the alkalies and alkaline earths have the general formula $3R''S_2O_3 . Au_2S_2O_3 + 4H_2O$. The double compounds of gold with sodium, potassium, calcium, magnesium and barium, are all known. The sodic salt is prepared by adding a dilute solution of gold trichloride little by little to a concentrated solution of sodic hyposulphite, when the following reaction occurs:—

$$8Na_2S_2O_3 + 2AuCl_3 = Au_2S_2O_3 . 3Na_2S_2O_3 + 2Na_2S_4O_6 + 6NaCl.$$

The double hyposulphite may be separated by precipitation with strong alcohol, with which it is also washed, or it may be purified by repeated solution in water and precipitation with alcohol. Thus prepared, it consists of colourless crystalline needles, highly soluble in water, but almost insoluble in alcohol. The solution, which possesses a sweetish taste, decomposes under the influence of heat, the action being much more rapid when nitric acid is present; metallic gold and sulphate of soda are formed. Gold, however, is not reduced from its solution as double hyposulphite by either stannous chloride, ferrous sulphate or oxalic acid, although sulphuretted hydrogen and alkaline sulphides give a black precipitate of Au_2S_3. The addition of hydrochloric acid or of dilute sulphuric acid does not immediately cause an evolution of sulphur dioxide and a deposit of sulphur, as in the case of ordinary hyposulphites. Since, therefore, the double sulphite of soda and gold does not present the characteristics of either aurous salts or of hyposulphurous acid, it has been suggested that it contains a compound radical, and has a composition expressed by either $Na_3S_4O_6Au$ or $Na_3S_4O_6Au + 2H_2O$. The addition of any dilute acid soon effects the decomposition of this body in solution, gold sulphide being precipitated; the reaction is accelerated by heat.

The double hyposulphites of potassium, calcium, barium and magnesium present similar characteristics. If the barium salt is treated with the amount of sulphuric acid required by theory, a solution of the acid auro-hyposulphite, $3H_2S_2O_3 . Au_2S_2O_3$, is obtained, but it cannot be crystallised. It has been supposed that the calcium salt is more easily formed than the sodic salt, and, therefore, that calcium hyposulphite is more suitable than sodium hyposulphite for use in the leaching process, whenever gold is present in perceptible quantities. According to a series of experiments conducted by Russell,[*] this is not the case, very little difference existing in the case of formation of the calcic, sodic, magnesic and potassic salts.

Russell has demonstrated [†] that finely divided gold is soluble to a limited extent (*i.e.*, 0·002 gramme in 1,000 c.c. in 48 hours), in solutions of sodic hyposulphite of all degrees of concentration. The action depends on the oxidation of the gold by the air present in the solution, the soluble double hyposulphite, $Au_2S_2O_3 . 3Na_2S_2O_3 + 4H_2O$, and caustic soda being formed.

The formation of this hyposulphite by the action of the sodic salt on gold sulphide is far more complete and rapid. In 24 hours in the cold, 0·066 gramme of gold, and in 2 hours at 65°, 0·117 gramme of gold were dissolved in dilute solutions. Since an alkaline sulphide will again precipitate Au_2S_3 from the solution, these facts seem to be at variance, and Stetefeldt suggests that the gold sulphide originally contained a small quantity of metallic gold in an excessively fine state of division, and that on heating more gold was set free. He instances Levol's statement that sulphuretted hydrogen precipitates metallic gold, and not gold sulphide, from a boiling solution of the trichloride. He believes [‡] that it is only the free gold which is attacked, and that the results obtained on attacking the sulphide were better than those in which metallic gold was used, owing to the much finer state of division in which the gold existed in the former case. Another reason for the inverse reactions may of course exist in the influence of mass, the small quantity of sodic sulphide which is formed by the reaction being insufficient to precipitate any gold, which comes down on further additions of the same reagent.

In some further experiments which Russell conducted, it was proved that gold sulphide is far more soluble in a solution containing the molecular quantities, $4Na_2S_2O_3 . 3Cu_2S_2O_3$ (hyposulphite of copper and sodium), than in sodic hyposulphite alone, especially if the solutions are cold. If the solution is heated the reduction of the gold sulphide to metallic gold renders it less soluble, and less rapid dissolution occurs. The composition of the double salts obtained by the use of this "extra" solution has not been worked out.

[*] Stetefeldt's *Lixiviation of Silver Ores*, New York, 1888, p. 90.
[†] *Ibid.*, p. 19.   [‡] *Ibid.*, p. 22.

SILICATES OF GOLD.

The existence of auro-silicates is now admitted without dispute, and gold has for centuries been used to impart colour to glasses, the method used being as follows:—A solution of chloride of gold is added to a mixture of sand with alkalies and alkaline earths or lead, and the whole is then fused, and colourless or yellow transparent silicates of gold thus formed. These are decomposed by being reheated gently to low redness, oxides of gold, or more probably metallic gold, being set free, and red or purple colorations thus obtained. The occurrence of silicates of gold in nature seems to be doubtful.

Experiments conducted by E. Cumenge* tend to show that the alkaline auro-silicates, obtained in the wet way, may have played an important part in the formation of auriferous quartz. The following conclusions have been established by these investigations:—

1. If an alkaline aurate, obtained by dissolving auric sesquioxide in caustic soda, is mixed with an alkaline solution of silicate of soda (soluble glass), the mixture may be concentrated by evaporation until it has attained a syrupy consistency without being decomposed. Auro-silicate of soda is, therefore, fairly stable, so long as there is an excess of alkali present.

2. The decomposition of this auro-silicate is effected by the addition of hydrochloric acid to it, by which gelatinous silica is precipitated. This carries down a certain proportion of gold which gives a rose colour to the white magma.

3. This decomposition may also be completely effected by the action of an aqueous solution of carbonic acid under pressure. Thus, if the syrupy, alkaline auro-silicate is introduced into a bottle of seltzer water, which is then hermetically closed, the decomposition can be seen to be gradually going on without the semi-fluid mass being dissolved, and the latter is replaced at the end of some days by coherent silica, which, on exposure to the air, assumes a white opaline appearance tinged with rose colour.

4. When gelatinous silica, obtained by the decomposition of an alkaline auro-silicate, is heated to redness in a current of steam, it assumes either a beautiful, unalterable rose colour, or a reddish tint with visible grains of gold, according to the proportion of precious metal present, and the conditions under which the precipitation has been effected.

SULPHIDES OF GOLD.

These compounds are prepared as brown or black precipitates by passing sulphuretted hydrogen through a solution of gold

* Frémy, *Ency. Chim.*, vol. iii., L'or, p. 62.

chloride. The exact composition of the precipitate varies with the temperature and degree of concentration of the solution, and the amount of free acid present. Levol and Krüss state that Au_2S is precipitated in the cold, but that only metallic gold and free sulphur are thrown down from boiling solutions. According to others, Au_2S_3 is precipitated from cold and Au_2S from boiling solutions. It seems probable, however, that free sulphur is always formed in considerable quantities, and whether the solutions are hot or cold, dilute or concentrated, definite compounds are not precipitated, variable mixtures of the two sulphides with free sulphur and metallic gold being formed. Similar precipitates are formed by alkaline sulphides and by sulphides of most of the heavy metals. The sulphides are soluble to some extent in a saturated solution of sulphuretted hydrogen, and are easily soluble in hot solutions of alkaline sulphides or alkalies, forming double salts so that precipitation from alkaline solutions is never complete. The sulphides are readily decomposed into gold and sulphur by the action of heat, the decomposition being complete below 270°. Sulphide of gold is also dissolved at ordinary temperatures by potassic cyanide, and is slowly attacked by mercury with formation of mercury sulphide.

Purple of Cassius.—This body was discovered by Cassius of Leyden in 1683. It contains gold and oxide of tin, and is used to colour glass and glazes, various shades of violet, red and purple being thus obtainable. Several methods of preparation are used, of which the following is that employed at the factory at Sèvres*:—Half a gramme of gold is dissolved in aqua regia composed of 16·8 grammes of hydrochloric and 10·2 grammes of nitric acid, and the solution is then diluted with 14 litres of water. To this solution is added, drop by drop, a solution of a mixture of protochloride and tetrachloride of tin, prepared as follows:—3 grammes of finely divided tin is dropped, little by little, into 18 grammes of aqua regia (constituted as above, with the addition of 5 c.c. water), the reaction is checked by cooling if it is too violent, and the solution of chloride of tin formed is allowed to cool. The precipitate of purple oxide thus obtained is finely coloured when it has been washed with boiling water. The purple precipitate obtained by Müller, by reducing chloride of gold with glucose in an alkaline solution containing tin oxide in suspension, differs from that prepared by the foregoing method in losing its colour at a red heat, while the true purple of Cassius becomes brick-red under such conditions. The true colour is seen when metallic tin acts on trichloride of gold, or when alloys of gold and tin are attacked with nitric acid.

The composition of purple of Cassius has been the subject of much discussion. Some chemists have considered it to be a compound of aurous oxide with the oxides of tin. Debray

* Frémy's *Ency. Chim.*, vol. v., L'or, p. 63.

regards it as a lake of stannic acid coloured by finely divided gold. If this latter view is correct then the gold may be present in an allotropic modification, since (1) the purple of Cassius is completely dissolved by ammonia, a purple solution being formed, and the solution may be kept unaltered for weeks although it is decomposed by light, or on heating, becoming bluish, and finally depositing metallic gold; the ammonia may be removed by dialysis, leaving a purple aqueous solution which contains both gold and stannic oxide;* (2) the purple of Cassius does not yield any gold on treatment with mercury. Müller confirms Debray's views, showing that fine purple compounds can be made with gold and magnesia, lime, baryta, sulphate of barium, &c., the colour depending on the presence of finely divided gold and not on the other constituent. The purple colour possessed by (possibly allotropic) gold, when in a finely divided condition, is further attested by the purple stain given to the fingers by a solution of gold trichloride, and by the colour of Roberts-Austen's aluminium alloy, $AuAl_2$.

CHAPTER III.

MODE OF OCCURRENCE AND DISTRIBUTION OF GOLD.

Forms in which Gold occurs in Nature.—Gold is obtained from two very different sources, viz., (1) from veins in rock formations, and (2) from placers, or the alluvial deposits of ancient and modern streams.

Vein Gold.—In this case the metal, whenever it is present in visible grains or masses, has sharp angular edges, and although usually not distinctly crystalline, it frequently penetrates the rock irregularly in various directions, and is completely interwoven with, and attached to the matrix, usually quartz, so that the metal cannot be separated from the rock without crushing the latter.

The gold in lodes is sometimes in the form of crystallisations, which are, however, exceedingly rare, and crystals of gold are still probably unknown to most miners, although they occur more frequently in placer deposits. Arborescent branching and dendritic masses of crystalline gold are more common than single crystals in both quartz lodes and placer deposits. The crystalline forms met with have already been described, p. 8. In the Transylvanian lodes, gold occurs chiefly in thin sheets or plates, often as

* *Chem. Soc. Journ.*, vol. lxiv. (1893) p. 575.

much as from half an inch to two or more inches in breadth. Such plates are rarely thicker than a visiting card, and are generally covered with crystalline lines and markings, revealing a distinct geometrical structure. Gold also occurs in wire-like forms, sometimes penetrating crystals of other minerals, such as calcite and dolomite.

The matrix in which the gold is contained is usually quartz, intersecting as veins or interlaminated with sub-crystalline, slaty, or schistose rocks, especially hydromica and chloritic slates. Gold also occurs sparingly in similar veins in granite and gneiss, and has been detected in the trachytes of Colorado, and in silurian and carboniferous trachytes, as well as in some limestones.

With regard to the distribution of gold, W. P. Blake observes*—" There is a much greater dissemination of gold in a ragged granular condition, *in situ*, in fine particles in the midst of rock formations, and without any obvious connection with veins, than is generally supposed. Prominent examples are found in the belts or zones of layers of soft slate in Georgia, and in North Carolina. . . . The Boly-Fields gold vein, Lumpkin County, Georgia, is an example of the occurrence of coarse ragged gold in the midst of a mass of slate, without any defined quartz vein. The gold is closely associated with bornite, pyrites, and dolomite." The dissemination of gold in the schistose rocks of North Carolina has also been noted by Professor Kerr,† and by Dr. Emmons, and similar occurrences have been observed in Texas, Nova Scotia, and in other parts of the world. At the Contention Mine, Tombstone, Arizona, free metallic gold is found in the thin cracks and cleavage surfaces of partially decayed porphyry, and appears to have been deposited there from solution, and not mechanically. It occurs in thin subcrystalline flakes and scales, and may have been derived from the decomposition of the iron pyrites with which the adjoining sedimentary formations are charged. Gold also occurs in small quantities (1 part in 1,124,000) in the bed of clay on which the city of Philadelphia is built. Its occurrence in solution in sea water has been proved by Sonstadt.‡

The wide distribution of gold in minute quantities throughout the world was pointed out by W. E. Dubois, an Assayer in the United States Mint, in 1861,§ and is further attested by a large number of specimens now in the Percy collection. These consist of small specks of gold of different sizes which have been obtained from the most varied sources. Thus, samples of Pattinson's crystallised and uncrystallised lead, pig lead from all

---

* *Prod. Gold and Silver in United States*, 1884, p. 581.
† *Trans. Am. Inst. Mng. Eng.*, vol. x., p. 475.
‡ *Chem. News*, vol. xxvi., p. 159.
§ *Journ. Am. Phil. Soc.*, June 1861.

countries, lead fume, red lead, litharge, white lead, precipitated carbonate of lead and acetate of lead were all found to contain gold, which seems to be invariably present in galena. Moveover, it appears to be impossible to procure samples of copper in which gold cannot be detected, although the Lake Superior copper contains less than 1 part in 1,000,000; the bronze and copper coins of all nations are usually found to contain much greater quantities of gold than this. Similar evidence has been adduced which tends to show that all ores of silver, antimony, and bismuth contain gold.*

Placer Gold and Nuggets.—Placer gold is usually in the form of small scales, but pellets or rounded grains also occur, while larger masses or nuggets are usually of a rounded mammillated form. The chief difference between the appearance of placer and vein gold lies in the fact that the former is always rounded, showing no sharp edges, even the crystals having their angles smoothed and rounded off. This has been pointed to by the advocates of the erosion theory of the origin of placer gold, as evidence in favour of their views, the roundness of the fragments being taken to prove that abrasion of the gold has been effected by attrition with water and grains of sand. The largest masses of gold yet discovered have been found in auriferous gravel. The "Blanch Barkley" nugget, found in South Australia, weighed 146 pounds, and only six ounces of it were gangue; and one still larger, the "Welcome" nugget, from Victoria, weighed 2,195 ounces, or 183 pounds, and yielded gold to the value of £8,376 10s. 6d. In Russia a mass was found in 1842 near Miask, weighing 96 pounds troy. The largest mass from California is given in the State Mineralogist's report as weighing 2,340 ounces, or 195 pounds, but no authentic cases seem to be on record of nuggets from this State weighing more than 20 pounds.

The minerals most common in auriferous quartz lodes or in the placer deposits are platinum, iridosmine, magnetite, iron pyrites, galena, ilmenite, copper pyrites, blende, tetradymite, zircon, garnets, rutile and barytes; wolfram, scheelite, brookite and diamonds are less common. Diamonds are associated with gold in Brazil, and also occasionally in the Urals and in the United States. The sulphides present in auriferous quartz frequently contain gold; the gold in such an ore is usually in part quite free, disseminated through the quartz, in which visible grains of the metal often occur, and in part locked up in the pyrites, whence but little can in general be extracted by mercury. It is, however, in all probability in the metallic state in pyrites, although this is not completely established. The subject is discussed in Section on "Gold in Pyrites," Chap. vii.

* *Loc. cit.*, and E. A. Smith on Bismuth, &c., *Journ. Soc. Chem. Ind.*, vol. xii. (1893), No. 4.

Among minerals other than sulphides which contain gold, the following may be mentioned, although none are of great importance in metallurgy:—**Calaverite** is a bronze-yellow gold telluride, usually containing a little silver, occurring in certain mines in California and Colorado. One analysis gave tellurium 55·5 per cent., gold 44·5 per cent., corresponding to the formula $AuTe_2$. **Sylvanite**, called also **graphic tellurium**, is a telluride of gold and silver, supposed to correspond to $(AuAg)Te_3$. It sometimes contains antimony and lead in addition. It is from white to brass-yellow in colour, and the arrangement of the crystals sometimes bears a resemblance to writing characters, whence the name graphic. It occurs in Transylvania, in Calaveras County, California, and in Colorado.

Petzite is a telluride of silver, Ag_2Te, in which the silver is partly replaced by gold. A specimen from the Golden Rule Mine, according to Genth, contained tellurium 32·68 per cent., silver 41·86 per cent., and gold 25·60 per cent. It occurs in Transylvania, Chili, California, Colorado, and Utah.

Nagyagite or **Foliated Tellurium** is remarkable for being foliated like graphite, which it also resembles in its colour, a blackish lead-grey, and in having a hardness of from 1 to 1·5 only Its density, however, is above 7. It occurs in Transylvania, and contains tellurium 32·2, lead 54·0, gold 9·0 to 13·0 per cent., sometimes with silver, copper and sulphur in addition. Other gold tellurides and some native gold amalgams are occasionally met with, but these minerals in which gold is an essential part are rarely of much importance as ores, as they seldom occur in sufficient abundance to be regarded as anything but specimens for collectors. In some few mines, however, notably at the Cripple Creek district, Colorado, in Transylvania, in Boulder County, Colorado, and at the Bassick mine in the same State, the value of the ore depends on the tellurides of gold contained in it.

One of the most striking differences between the ores of gold and those of all other metals lies in the extremely small proportion which the desired material bears to the worthless gangue with which it is accompanied. Occasionally hand specimens of vein stuff are found containing several per cent. of the precious metal, but these are of quite exceptional occurrence and have not the slightest economic importance. The greater part of the vein gold now being produced is derived from ores containing only about one part of gold in seventy or eighty thousand, whilst, under exceptional circumstances, a yield of one part in half a million parts of gangue may give handsome profits. Placer deposits are usually much less rich than this; the average amount of gold contained in those now worked does not exceed one part in one million, and in California deposits of gravel with only one part of gold in fifteen millions have proved susceptible of successful treatment by the "hydraulicking" method on a large scale.

Composition of Native Gold.—Native gold usually contains silver, which occurs in varying proportions, the colour becoming paler with the increase of silver. The finest native gold yet found is that from the Mount Morgan Mine, Queensland, which is 997 fine. The finest Russian gold was that formerly obtained at Katerineburg in the Urals, and yielded gold 989·6, silver 1·6, copper 3·5, and iron 0·5 (G. Rose). Gold dust from West Africa has been found to contain 978·1 of fine gold, the remainder being silver. The gold found in the British Isles varies from 800 to 900 fine, the remainder being silver; the specimens from the district around Dolgelly are sometimes a little over 900 fine. Particulars of the fineness of gold from Australia, California, &c., are given in the chapter on Refining, Chap. xviii. Gold is occasionally found alloyed with copper, and sometimes also with iron, bismuth, palladium, or rhodium. Rhodic-gold from Mexico was found to be of the specific gravity 15·5 to 16·8 and contained 34 to 43 per cent. of rhodium. Bismuthic-gold has been called *Maldonite*.*

Geographical Distribution of Gold.—Gold occurs in lodes in many districts composed of partially metamorphosed rocks such as slates or schists, while its occurrence in holocrystalline-metamorphic, or igneous rocks is comparativaly rare. Among sedimentary rocks, its occurrence is almost confined to the sands of rivers which run for a part of their course through crystalline formations, or more particularly through districts in which gold occurs in quartz veins. Such river sands are rarely quite free from gold. The beds of ancient rivers no longer existent are also frequently auriferous. In spite of the fact that the sea contains gold in solution, the aggregate amount perhaps exceeding that contained in the accessible portion of the earth's crust, nevertheless, unaltered marine deposits seldom or never contain a perceptible quantity of the metal, except in one or two cases of beach deposits formed by the erosion of auriferous land-formed gravels.

In the British Isles, gold is found in some of the streams of Cornwall and in lodes and river gravels near Dolgelly and in other parts of Wales, in Sutherlandshire and near Leadhills in Scotland, and in the County of Wicklow. The total amount which has been obtained from these localities is small, probably not exceeding 40,000 ounces, and little is now being produced. On the Continent of Europe, gold is most abundant in Hungary and Transylvania, where the gold occurs in quartz lodes contained in eruptive rocks of tertiary age, chiefly propylite, porphyry, diorite and granite. The minerals occurring with the gold are galena, blende and pyrites. In the German Empire, the gold obtained is chiefly derived from the smelting of argentiferous galena in which small quantities of the more precious metal are contained.

* Dana's *Mineralogy*, p. 108.

In Italy the only important mines are those of Pestarena and Val Toppa in North Piedmont, near Monte Rosa. The pyritic ores from these mines are treated by amalgamation. Gold is also found in the sands of the Rhine, the Reuss, the Aar, and other rivers, and in small quantities in Sweden.

The gold-bearing districts of Russia are (1) the Urals, (2) Siberia, Eastern and Western, whilst an insignificant amount is also derived from Finland and from the Caucasus. The gold was formerly derived chiefly from lodes both in the Urals and in Western Siberia. In the Urals, quartz-mining began in 1745, and the output from this source continually increased up to the year 1810, when it began to fall off, and has been trifling since 1838. The placers, which are mainly situated on the eastern slope of the range, were discovered in 1774, and the yield has continually increased since then up to the present time. The Siberian placers were discovered in 1829, although quartz-mining had been prosecuted since 1704.* The output from these continually increases, in spite of the exhaustion of the old placers and the reduction in the percentage of gold contained in those still worked; this increase of output is in consequence of the discovery of new placers further east. The auriferous gravels are all thin, shallow deposits, ranging from 3 to 20 feet in thickness, and as they are worked out, other gravels are opened-up further east, so that operations are being gradually transferred from west to east. When the exhaustion of these placers has proceeded further it may be expected that more attention will again be paid to the quartz lodes. The relative amounts produced by the different districts are given as follows:—†

	Ural, per Cent.	Western Siberia, per Cent.	Eastern Siberia, per Cent.	Finland, per Cent.
1876-80,	20·0	6·0	74·0	0·02
1890,	28·0	6·4	65·4	0·04

Almost the whole of the production is now derived from shallow alluvial deposits by means of sluicing operations, as described on pp. 58-61.

In India, almost all the gold now being produced is derived from the quartz lodes of the Colar gold-field, Mysore, in Southern India, in which work was begun in the year 1880. The ore is free milling, and is treated by amalgamation in the stamp battery. A little gold also comes from the Presidency of Madras. In China and Japan considerable quantities of gold are produced,

---

* *Mineral Industry* for 1892, p. 203.
† *Prod. of Gold and Silver in U.S.A.*, 1886, p. 81; *Mineral Ind.*, 1892, p. 203.

chiefly by various primitive methods; little is known of the methods used and of the amount produced in the former country.

From the United States a large percentage of the total gold production of the world is obtained. The chief producing States are California, Colorado, Dakota, Montana, Nevada, Idaho and Oregon, but smaller amounts come from many other States. The produce is now far more from lodes than from placer deposits, and in the treatment of auriferous quartz and pyritic ores almost all the known methods of treatment are applied in different localities.* Mexico produces considerable quantities of gold, and gold ores are also found in various parts of Canada, Colombia, Bolivia, Chili, Venezuela, Brazil, Peru, and the small States of Central America. The production of several of these countries was formerly much larger than it is at the present day, the reduction being especially marked in the cases of Brazil and Venezuela. On the other hand, the gold districts of British and French Guiana, of the Argentine Republic and of Uruquay are now being developed, and may eventually prove to be of considerable importance.

Gold is somewhat widely distributed in Africa, the chief sources of production in former times being the placer deposits of the Gold Coast and Abyssinia. The discoveries of auriferous quartz deposits in the Transvaal since 1884 have converted that region into one of the most important among gold producing countries. The richest deposits are in the Banket formation on the Witwatersrand, whence about nine-tenths of the gold obtained in the Transvaal is derived, the De Kaap field being next in importance. No placer deposits of value have been discovered, and the gold is mainly obtained by stamp-battery amalgamation, followed by treatment of the tailings by means of the cyanide process. The produce from both sources is being continually increased. The production by means of treatment with cyanide is now about 30 per cent. of the whole.

Gold is found in all the Colonies of Australia, together with Tasmania and New Zealand; the largest amount is now produced by Victoria, and Queensland, New South Wales, and New Zealand follow in the order named. The chief gold producing districts of Queensland are Charters-Towers, Rockhampton (where the Mount Morgan mine is situated), Croydon, and Gympie. Almost all the gold is produced from the quartz mines, the placers having been practically exhausted; thus in 1891, out of a total production of 576,439 ozs., 560,418 ozs. were produced from quartz, and only 16,021 ozs. from alluvial deposits, while in 1877 the amounts were—Placer gold, 164,778 ozs.; quartz gold, 188,488 ozs.—total, 353,266 ozs.†

* For full details as to the geographical and geological distribution and the nature of the gold ores found in the United States, the student is referred to the Annual Reports of the Director of the United States Mint, and of the Californian State Mineralogist.

† *Mineral Industry*, 1892, p. 192.

The production of gold in Victoria is now increasing, but is far less than formerly, owing to the exhaustion of the alluvial deposits, although the yield of the quartz mines has continuously increased.

In 1894 the production was divided as follows:—Alluvial gold, 254,308 ozs.; reef gold, 419,371 ozs.—total, 673,680 ozs. The chief producing districts are Ballarat, Sandhurst (Bendigo), Beechworth, Maryborough, Castlemaine, Gippsland and Ararat. In 1894 in Victoria the average yield per ton of quartz crushed was 8 dwts. 8 grains of gold, this being slightly less than the mean yield during the last decade.

In New South Wales about half the gold is obtained from quartz lodes; the pyritic ores are not yet effectively treated, and the river bank deposits have not up to the present been exploited on a large scale.* In New Zealand, only quartz lodes are worked in the North Island, alluvial workings being confined to the Middle Island. The most important districts are those of Kuaotuna, Thames, Coromandel, Waihi and Reefton in the North Island, and Ross and Kumara in the Middle Island.†

Origin of Gold Ores.—The origin of mineral veins, including those in which gold is contained, has long been discussed by geologists. The old theory that the quartz of veins was originally in a molten condition and was ejected from below into fissures is no longer maintained, although in 1860 H. Rosales brought forward evidence in its favour as far as the Victorian lodes are concerned. It is now admitted by all that the materials forming the veins have been transported in aqueous solution and precipitated where they occur. In a few exceptional cases, sublimation may have played a part. The view that the solutions found their way downwards from above has been abandoned, but the *ascensional* theory and the *lateral secretion* theory both have many adherents. The last-named theory has found its principal supporter during the last twenty years in Prof. F. von Sandberger, who pointed out that the gangue of many lodes varies in composition if the nature of the rocks through which they pass is changed, and claimed to have proved by analysis that the materials forming vein-stone are derived from the adjacent country rocks. He stated, moreover, that such minerals as augite, hornblende, mica, and olivine, which are essential constituents of crystalline rocks, contain small quantities of the heavy metals occurring in veins.‡ Although Sandberger did not try to detect gold in the silicates, this metal is not likely to be an exception. Prof. A. Stelzner objected to these conclusions, urging that small quantities of the sulphides of the heavy metals were probably mechanically mixed

---

* *Report of the Dept. of Mines and Agriculture, N.S.W.*, 1894.
† *Annual Report of the Mining Commissioner of New Zealand*, 1891.
‡ *Untersuchungen über Erzgänge.* Wiesbaden, 1882 and 1885. Useful abstracts are given in Phillips' *Ore Deposits* and in Le Neve Foster's *Ore and Stone Mining.*

with the crystals of minerals which Sandberger analysed in the belief that they were pure. Stelzner advocated the retention of the ascensional theory, which alone affords a satisfactory explanation of the difference in composition observable in neighbouring lodes passing through the same rocks, and apparently formed at different periods. The two theories are, however, not contradictory, and perhaps neither need be entirely rejected, the solutions being supposed to pass more or less freely in the plane of the lode, after they have been impregnated. For the origin of placer gold, see p. 67.

CHAPTER IV.

TREATMENT OF SHALLOW PLACER DEPOSITS.

The deposits grouped together under the name of "placers" comprise sands, gravels, or any loosely coherent or non-coherent alluvial beds containing gold. They have accumulated, owing to the action of running water, in the beds of rivers, or on the adjoining inundation plains, or on sea beaches. They fall naturally into two groups, between which no strict line of demarcation exists. These are—

(1) Shallow or modern placers, which are in or near existing rivers, and have not yet been covered by other deposits.

(2) Deep level or ancient placers, which now lie buried beneath an accumulation of débris or coherent rock, the rivers by which they were formed having often been deflected into other channels by more or less extensive changes in the physical geography of the district in which they existed. Beach deposits occur in each subdivision.

In this chapter, the first of these groups will be considered.

From the earliest times to the present day, shallow placer deposits have probably yielded more gold in the aggregate than has been derived from all other sources put together.

Shallow placers consist of loose aggregations of sand, gravel, loam and clay, accumulated by the action of existing rivers and streams, and not extending to a greater depth than 10 or 15 feet from the surface. They contain metallic gold in fragments of all sizes, ranging from the finest dust to nuggets weighing thousands of ounces. Auriferous sands are found in the beds of most rivers which flow during any part of their course through a region composed of crystalline rocks. If the rivers have rocky beds, gold may be found in the crevices, caught in natural riffles, and the whole may subsequently be covered by beds of sand.

Gold also occurs in river bars and banks, in river "flats," or inundation plains, in the dry beds of streams which only flow after heavy rains ("gulch diggings"), in terrace gravels on the sides of valleys high above the present level of the water ("bench diggings"), and on the sides and tops of hills ("hill diggings"). The last two subdivisions are evidently ancient rather than modern deposits. The gravels may contain boulders of any size, up to several feet in diameter, or may shade off into fine sand, while sandy clays, especially if on the bed-rock, are frequently very rich. In the Urals, the placer deposits often consist of heavy clays, while others are formed of waterworn fragments of auriferous quartz, talcose and chloritic schists, serpentine, greenstone, &c. Gold occurs under very various conditions in these deposits. It may occur in the grass roots on inundation plains, or near the surface of the gravels in river beds, or dispersed through the whole thickness of a stratum. More commonly, however, the lowest part of the superficial beds, just above "bed-rock" (the country rock of the district), is richest. In hollows, cracks, and crevices of the bed-rock, or, if it is soft and decomposed, in the substance of the upper part of the rock itself, to the depth of 1 or 2 feet, gold occurs in the greatest quantities. In pipeclays just above bed-rock, in Victoria, it was not uncommon to find 12 ozs. of gold or more in a single tubful of "dirt," and similar rich bed-rock deposits occur in California. The depth at which bed-rock is found varies greatly; it may crop out at the surface, or it may be buried beneath hundreds of feet of gravel, and great variations occur even in a single district. In the Urals, however, the thickness of the gravel is usually less than 3 feet thick, and is rarely more than 12 feet.

Methods of obtaining Gravel from Shallow Placers.—The gravel obtained from any placer deposit is, with exceptions, ultimately all treated alike, but the mode of winning the dirt varies with the necessities of the case. On flats and bars, the surface gravel, if rich enough, is loosened with pick and shovel, and then washed. If only the part just above the bed-rock will pay for treatment, it is reached by "stripping," or, if covered by too great a thickness of barren material, shafts are sunk, and short levels run from the bottoms of them in all directions. If the nature of the ground permits of it, tunnels are run without shafts, and the rich gravel is then followed from the surface, wherever it is found. This system was much practised in the early days in California, although now seldom to be seen in operation; it was called "coyoting," from the coyote, which lives in holes in the ground. When water was encountered in the shaft, it was drawn out by a bucket, until it came in too fast, when the claim was abandoned. In California, in somewhat later times, efforts were made to reach the gold in the river beds by deflecting the streams of water from their courses, and in other ways. These methods will be briefly described under the head "River Mining."

Methods of Washing the Gravel.—*The Pan.*—When the existence of gold in the placers of California and Australia first became known, the diggers were not acquainted with any apparatus which was well adapted to extract the metal. The household pan was used everywhere to wash the gravel, and though, in its original form, it was a difficult implement to use efficiently, it has retained its place in both countries for prospecting and also for washing small quantities of rich material, however they may have been obtained. The pan is usually made of stiff sheet-iron, is flat-bottomed and circular, and at bottom about 14 inches in diameter. The sides slope outwards at an angle of about $30°$ to the bottom, and are about 5 inches wide. A riffle is a useful addition, formed by the thickening or bulging inwards of the side, situated about half-way up the latter and running about half-way round the pan. The method of using this pan embraces several operations. First, it is filled to about two-thirds of its capacity with pay-dirt, of which it then contains from one-fifth to one-quarter of a cubic foot. It is then placed at the bottom of a water-hole or convenient stream, and the dirt is thoroughly broken up with both hands, care being taken not to leave any lumps of clay. As soon as the contents of the pan are reduced to the consistency of soft mud, the pan is grasped with both hands a little behind its greater diameter, inclined away from the operator, raised until the dirt is only just covered with water, and shaken from side to side, while a slight oscillatory circular motion is also imparted to it. The mud and fine sand is soon obtained in suspension in the water, and gradually passes over the far edge, which is lowered more and more, until little but the stones, coarse particles of sand, black sand and gold is left. The larger stones lie on the top and are removed by hand. The final stage consists in lifting the pan with a little water in it, and by a movement of the wrist, something like that used in vanning, causing the material to be spread out by the water, in a comet shape, in the angle of the pan. The "colours"—*i.e.*, yellow specks of gold—are seen at the extreme head of the comet, and also occur in the succeeding inch or two, mixed with the black sand, while the quartz-sand forms the remainder of the tail and is scraped or washed off. The gold is separated from the black sand by (*a*) amalgamation with mercury, or (*b*) drying and blowing away the black sand, a wasteful process. Liquid amalgam is readily separated from sand, and the mercury is then driven off by heat.

The Batea differs from the miner's pan in not having a flat bottom. It is of wood turned in a lathe, about 20 inches in diameter, conical, or more rarely basin shaped, and about 3 inches deep in the centre, so that the angle at the apex is about 160°. The gold collects at the lowest point and clings to the wooden surface under conditions when it would slide over iron. The batea consequently is more rapid and effective in obtaining a "prospect" than the pan, especially when the gold is fine, but is less frequently used in the United States and Australia. It had its origin in South America. It is usually now made of enamelled iron with a hole in the centre fitted with a cork, when it is for use in countries other than South America. It is considered by MM. Cumenge and Fuchs to be especially favoured by the negro race.

Prospecting Trough.—This instrument is used in the far east, especially by the Chinese, Malays, Annamites, &c. It is made of wood, and is shaped in the form of a very flat reversed rooftop, the angle between the long sides being about 150°. In place of a circular movement of the water an alternating rocking motion is used, the water flowing up and down. The instrument is easily handled, but is very slow.

Horn Spoons cut out of black ox-horns have been used by prospectors, especially to finish the work begun by the pan. The surface holds the gold well and shows "colour" very readily.

The Cradle or *Rocker* was introduced in California soon after the first rush to the diggings took place in 1849. It consists of a rectangular wooden box, about 3 feet long and 18 inches wide, resting on two rockers (D, Fig. 4) similar to those used for infants' cradles. The shape of the walls is shown in Fig. 4, which is a section of the apparatus. The method of using it is as follows:—

The gravel is shovelled into the riddle-box, A, the bottom of which consists of ½ inch mesh screen; the workman sits

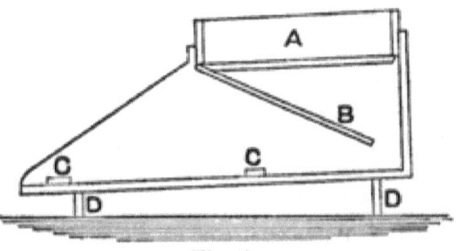

Fig. 4.
Scale 1 in. = 18 inches.

by the side of the machine and rocks it with one hand, while he pours on water by means of a dipper filled from a water-hole with the other. The dirt is disintegrated and carried through the riddle, and falls on the apron, B, which consists of blanketing. Here the fine gold is caught, and the dirt then passes out from back to front over the bottom, which is slightly inclined towards the front, and the coarse gold, black sand, &c., is caught in two or three riffles, C, each of about 1 inch in height, to which mercury is sometimes added to assist in retaining the gold. The rocking motion not only assists in the disintegration of the dirt, which is effected by the water, aided by the stones, but also prevents the sand from packing behind the riffles; in the event of this happening, gold would pass over the surface of the sand and be lost. Consequently the rocking should be quite continuous, since, after every pause, the sand in the riffles must be stirred up before recommencing. It is, therefore, desirable for two men to work together at the cradle, one to carry the gravel and charge it into the hopper, and to remove the large stones from the latter by hand, while the other man rocks the cradle and pours on water. It requires three or four parts of water to wash one part of gravel, and it is, therefore, better to carry the ore to water than to carry water to the ore. When a clean-up of the cradle is desirable, the riddle is removed, the apron is taken out and washed in a bucket, and the accumulations behind the riffles are scraped out and panned. Most of the fine gold in the dirt is lost by the cradle, and two men working together can only wash from 3 to 5 cubic yards per day, according to the nature of the dirt.

The Long Tom replaced the cradle in California after a short time, and was used there for some years, while it is still in operation in parts of Australia and Dutch Guiana. It consists of a sluice-box or trough (A, Fig. 5) about 12 feet long, 20 inches

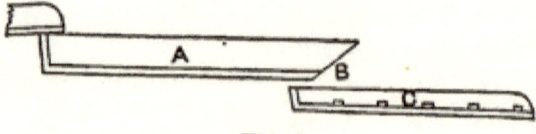

Fig. 5.

wide at the upper end, and 30 inches at the lower end, and 9 inches deep, with an inclination of about 1 inch to the foot. The lower end of the trough is cut off at an angle of about 45° and closed by a screen of punched sheet iron, B, which prevents large stones from passing through it. Below the screen is the upper end of the riffle-box, C, which is about 12 feet long, 3 feet wide, and at about the same inclination as the upper trough. It is fitted with several riffles, which are sometimes supplied with mercury. In working, a stream of water enters at the upper end of the sluice-box, into which gravel is continually shovelled, while a man breaks up the lumps with a fork, removes the large stones, and puddles the lumps of clay. Two to four men can work at one tom, and wash about five times as much in a day as can be done by one or two men with the cradle. Only the coarse gold is caught, and the machine is only suitable for washing small quantities of rich dirt where there is a plentiful supply of water. The material caught by the riffles is scraped out occasionally and panned, but the riffle-box is too short for close saving of the gold.

The Puddling-tub.—When water is scarce, as was the case in many places in Australia where rich gravels were found, the long-tom is inadmissible, and the puddling-tub is resorted to. This is particularly well adapted for washing clays, and is still used to disintegrate lumps of clay encountered in sluicing operations. It consists of one half of a barrel which has been sawn in two; into this the dirt is dumped and stirred up with water by means of a rake, until all the clay is held in suspension in the water, when a plug a few inches from the bottom is removed, and the slime run off. The operation is repeated until the tub is filled with gravel and sand to the level of the plug-hole, and this residue is then shovelled out and washed by the pan, the cradle, or by sluicing. Large boxes were used in Australia in this way in early days, the rakes being worked by horse- or steam-power; in 1860 no less than 3,958 boxes, worked by horses, were in use in Victoria alone.*

The cradle, long-tom and puddling-tub are now little used except by the Chinese, who gain a precarious livelihood with their help by washing over the heaps of tailings accumulated from sluicing or hydraulicking operations in Australia and California.

The Siberian Trough.†—In Siberia, in the Urals and in the valleys of the Obi, the Yenisei and the Lena, individual workers

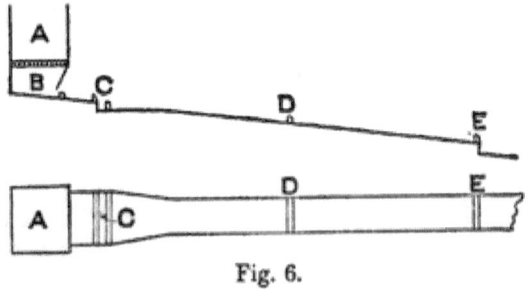

Fig. 6.

still exclusively use a trough, which differs from the long-tom mainly in requiring more constant attention on the part of the operator, and which resembles the old German buddle. The trough consists of a rectangular box open above and at one end. When sandy gravels are being treated, the bottom of the disintegration box (A, Fig. 6) which is about 40 inches square, is made of a perforated screen of wood or sheet-iron, having holes of from $\frac{1}{2}$ inch to 1 inch in diameter. The dirt is shovelled into this box, and, contrary to cradle-practice (see p. 45) if the gold is present in fine flakes, mercury is added here also, the amount depending on the richness of the auriferous material as determined by assay, the proportion used, however, being never more than 10 of mercury to 1 of gold. Water is directed upon the charge in the box, either by pipes from a

---

* Philips' *Met. of Gold and Silver*, 1867, p. 139.
† For further particulars see the account given by Cumenge and Fuchs in Frémy's *Ency. Chim.*, vol. v., L'or, 1st Section, p. 12.

reservoir or more often by pumping, and the fine material is carried through the screen and falls on to the table, B, while the pebbles are collected by hand and thrown away. If clay is being treated, no screen is used; the lumps are puddled in the box, and the mud carried over by an overflow of water. The table is slightly inclined, about 20 feet long, and, for the greater part of its length, is about 20 inches wide. It is furnished with five riffles, of which two (E, C) near the top are about 2 inches high, while the others (D, E) are of less height. The disintegration of the sand is completed on the table with the aid of a small rake continually used by the workman. When disintegration is complete, the stream of water is diminished in amount, and the workman continues to rabble the sands which have accumulated above the riffles, pushing the contents of the lower riffles up the table again, until the water runs clear, and little except pyrites is left behind the riffles where the so-called "grey concentrates" accumulate. These are either concentrated further on the same table, or removed and worked on a smaller table. In either case the stream of water is still further reduced, being graduated so as to carry away the last particles of quartz, together with all materials of moderate weight, such as garnets, rutile, tourmaline, &c., and even all the fine pyrites. If mercury has not been added previously, it is sprinkled on before this last operation, unless the gold is very coarse, when no mercury is added at any stage of the proceedings.

The "black concentrates," thus obtained, consist entirely of amalgam, magnetite, and the large grains of pyrites. The final operation, by which the amalgam is separated, is the most difficult, and requires the greatest amount of skill on the part of the operator. The material is worked on the same table with very little water, with the aid of a small rake, or more often with the hand of the workman, who kneels down by the trough for the purpose. Finally, all the pyrites having been washed away, the magnetite is removed with a magnet, and the amalgam collected. The tailings from the black concentrates are treated over again, together with the grey concentrates.

The apparatus just described treats about 500 lbs. of sand at one time, and can be worked by one man, but usually gives employment to four people (of whom three are frequently women), who can treat about 5 tons of sand per day. The degree of success attained depends largely on the skill of the workman; in Siberia and Russia the art is handed down from father to son, certain families devoting their whole lives to the work during many generations. These workmen attain such a degree of dexterity in the use of the trough, that practically the whole of the valuable contents of the gravels treated are extracted by them, but the work is only suited to those who are content with small earnings.

The Sluice.—The sluice has replaced all these implements for washing the gravel from shallow placers, where water is abundant. Sluices are constructed of "boxes," each of which resembles the upper part of the long-tom. The bottom of each box is made of rough boards, about 12 feet long and 1½ inches thick, cut 4 inches wider at one end than at the other; the total width is usually from 16 to 18 inches, while the sides are 8 or 10 inches high. It is considered by some New Zealand experts that sluice boxes are usually made too narrow, so that the current is too deep and fine gold is lost, while the flow is unduly checked by the sides, thus creating undesirable eddies. The box is held together with nails, and no attempt is made to render it watertight, as the swelling, caused by the absorption of water, and the filling up of chinks with sand and clay soon stops all leakages. The narrow end of the box fits into the wide end of that next below it, and so a sluice, made up of hundreds of boxes, can be rapidly put up or taken down and moved to another locality.

In all but the smallest and cheapest sluice boxes, extra strips of wood are affixed to the sides so as to protect them from the wear

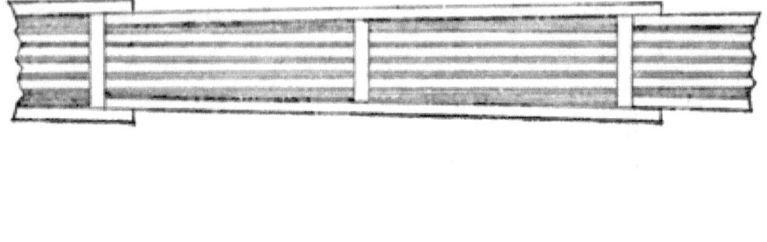

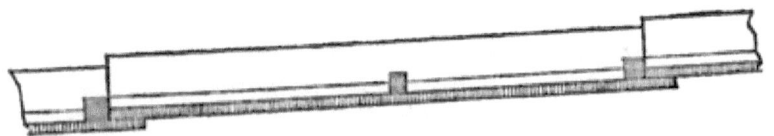

Fig. 7.
Scale = $\frac{1}{15}$.

caused by the attrition of the stones and gravel carried through by the current. When worn thin, these strips are replaced. The bottom is similarly protected by the riffle bars, whose main function is to catch the gold. These riffle bars are usually placed longitudinally, and are strips of rough wood from 2 to 4 inches thick, from 3 to 7 inches wide, and about 5½ feet long. They are wedged in the boxes at a distance of 1 or 2 inches apart by means of transverse bars, so that two sets of riffles are placed in each box in the manner shown in Fig. 7, which represents the whole of one box and parts of two others. The rectangular depressions thus formed between the bars are well adapted to intercept all heavy particles that pass down the sluice, such as gold, mercury, amalgam, pyrites, &c., which gravitate to the bottom of the stream. Sometimes the riffles are placed transversely, and sometimes for a short distance in zig-zag fashion. This arrangement does not retain anything, but affords a better chance of amalgamating the gold, which, together with the mercury (in this case fed in constantly), slides down the inclined riffles from side to side of the sluice, having sunk to the bottom by virtue of its high specific gravity.

When all is ready a stream of water is turned into the head of the sluice, where the dirt is shovelled in also. The amount of dirt shovelled in per man depends on the height of the lift and the nature of the soil, as well as on the labourer. It varies from 3 to 7 cubic yards per diem, and should average 5 or 6. The first gravel sluiced fills up most of the riffle depressions, leaving enough inequalities of surface, however, to intercept and retain the mercury, &c.

The length of the sluice varies with the consistency of the gravel, the fineness of the gold, the capital available, and the fall of the ground. It must be sufficient to complete the disintegration and then to catch the gold. The length may be adjusted by experiment; if, in the clean-up, the lowest boxes yield much amalgam, an addition to the length is necessary, while if they yield none, the sluice may be shortened by the removal of one or more boxes. Even in the latter case, however, the tailings would almost certainly contain some fine gold. The grade of the sluice is measured by inches per box, so that a grade of "12 inches" means one of 12 inches in 12 feet. The usual grades are from 8 inches to 20 inches per box, depending—

(1) *On the fall of the ground*, since the sluice cannot be raised far above it, nor sunk deep into it, owing to the increased expense thereby occasioned.

(2) *On the nature of the dirt to be washed.*—Tough, tenacious, clayey, or cemented gravels require higher grades to effect their disintegration than loose material. Instead of being disintegrated, clay sometimes becomes aggregated into balls, which roll down the sluices, picking up particles of gold previously caught in the riffles, and these lumps of clay must be removed by hand and puddled. There must be sufficient grade to enable the water to carry away all but the largest stones, so as to avoid unnecessary hand-picking, but on the other hand, while coarse gold is readily caught, fine particles are lost if the current is too rapid.

(3) *On the quantity of water available.*—The reduction of the grade lessens the duty of the water, so that if the supply of the latter is short or costly, the grade is made as steep as possible, consistent with saving a fair proportion of the gold. A steep grade reduces the necessary length of the sluice, as disintegration takes place sooner. Since a steep grade, a rapid flow, and deep

currents are best suited to effect speedy and thorough disintegration of the gravel, while a low grade, and slow and shallow currents are best adapted for saving the gold, the upper part of a sluice, for a sufficient distance to effect the complete disintegration of the gravel, is sometimes made of higher grades, or with narrower boxes than the lower part, which is occupied solely in catching the gold. When this is done additional supplies of water should be introduced at the point where the change is made, otherwise, the duty of the water being reduced, the sand packs in the angle where the grade is altered, and constant attention is required to prevent the stream from overflowing.

The opposite requirements of disintegration and gold-saving are more often supplied by "drops" and "undercurrents." A vertical fall of the pulp constitutes a *drop*, which is arranged as follows:—The sluice terminates in a "grizzly," or inclined grating made of parallel iron bars placed longitudinally to the stream and from 1 to 6 inches or more apart, according to the exigencies of the case. All the water and fine stuff pass through the grizzly and fall a distance of from 1 to 10 feet into a sluice below. The larger stones or boulders roll down the inclined bars, and are shot over a precipice (if possible) or on to a steep slope outside the sluice, as, unless some arrangement for removing these rocks is made, they will accumulate until they can no longer roll off the grizzly. The higher the fall, the more effectively it acts in causing disintegration. Sometimes, near the head of a sluice, the grizzly is omitted from a fall, and the boulders are retained to help in breaking up the gravel. The chief disadvantage in permitting them to remain with the rest of the gravel lies in the fact that they wear out the sluice, and that much water is required to wash them down.

The *undercurrent* is often used in conjunction with a drop. A grizzly with bars placed close together allows most of the water and fine material to pass through, while the coarse stuff is carried over and falls into the main sluice below. The fine material is carried off by a short sluice placed at right angles to the general direction of the main sluice, and is discharged into the upper end of a large broad box from three to ten times as wide as the sluice, and with its long diameter parallel to the main sluice. A number of check-boards help to distribute the stream evenly over the whole width of the box. This box, to which the name undercurrent is often given, although it properly belongs to the whole arrangement, is usually of lower grade than the sluice, and is in that case supplied with additional water. A broad, shallow stream is thus made to flow with reduced velocity over the surface of the box, and, as the latter is plentifully supplied with riffles and mercury, considerable quantities of fine and "rusty" gold and fine particles of amalgam, which would otherwise be swept away and lost, are retained by it. The tailings from the undercurrent are discharged into the main sluice below the drop.

Both grizzlies and undercurrents are used more frequently in hydraulicking than in shallow placer sluicing, in which the large stones are usually removed by a man with a blunt pronged fork, who also either breaks up the lumps of clay or removes them and puddles them in tubs.

The Use of Mercury in Sluicing.—Mercury is added at the head of the sluice after washing has been in progress for a sufficiently long time for all leakages to have been stopped, and for the lowest depressions to have been filled in with sand. The mercury is broken up into small particles either by pressing it through chamois leather above the sluice, or by letting a thin stream fall on to a man's hand, by which it is scattered, or by some similar method. The amount added varies with the richness of the dirt and the magnitude of the operations, enough being added to dissolve the amalgam formed. It is carried down the sluice and lodges in the riffles, the greater part being retained in the first few boxes. Fresh supplies are introduced every few hours at the head of the sluice, and sometimes at various points lower down the sluice also; in particular, mercury is added to the undercurrents, as it is especially valuable in catching the finer particles of gold which would otherwise be lost, whilst coarse gold can in great part be saved without mercury. Sometimes the latter is forced into the substance of the wooden riffles by driving an iron gas pipe into the wood, and filling it up with mercury, which is forced by the pressure of the column through the pores of the wood. The amalgam then forms on the surface and in the interstices of the blocks, and in cleaning-up this is scraped off. A better plan is to use amalgamated copper plates, which is now often done. These resemble the plates used in stamp batteries, described on p. 117.

The Clean-up.—The length of each "run," at the end of which the boxes are cleaned-up, varies, according to the richness of the gravel, from a day to a whole season, and is usually a week. The upper part of the sluice, which retains most of the gold, is usually cleaned-up more frequently than the remainder. A clean-up is begun by discontinuing the supply of gravel, and letting the water continue to flow until it passes through the sluice quite clear. The first six or eight sets of riffle bars are then taken up, and the sand, mercury and amalgam washed down, all the latter being caught by the first riffle left in. It is scooped out thence by a wooden ladle or iron spoon into a bucket, and the rich sand is collected and panned. The next few riffle bars are now taken up, and so on, or alternatively the work may be begun on several sections at the same time. Lastly, the whole sluice is carefully searched over, and particles of amalgam or mercury picked out of every crevice where they have lodged by spoons, penknives, &c.

The mercury thus collected is quite liquid, and is strained through chamois leather to separate the solid amalgam, which is then retorted. The retorts used in well conducted enterprises are similar to those in use in stamp mills, described on pp. 133-135. Amalgam obtained as the result of operations on a small scale, however, is often merely heated on a shovel over an ordinary fire, the mercury being driven off and lost.

Care must, of course, be taken to prevent partial clean-ups by unauthorised persons. A man with a shot-gun stationed by the sluice is a frequent preventive measure.

Tail Race.—The tailings from sluicing operations on low ground which has not much fall are removed through a covered-in wooden sluice, or, better still, through a large iron pipe. The work proceeds in the up-stream direction, and the worthless material stripped from above the pay-gravel is thrown on the top of the tail race, which thus passes through a mound of earth and discharges into the open air lower down the valley. As the digging and sluicing progresses up-stream, the tail race is lengthened and the sluice boxes proper are conveyed further up the valley so as always to be near the auriferous material last uncovered. This method originated with the Chinese.

Ground Sluice.—In some cases boarded sluice boxes are not used, but a stream of water is conducted to a little trench cut in the pay-dirt, which is soon enlarged by the action of the water, while the banks are at the same time shovelled or prised by the pick or crowbar into the sluice. The gold is caught in the natural riffles afforded by the uneven wearing of the bed, or rocks may be added to arrest the gold, no mercury being used. Ground sluicing is only adopted where the water supply is precarious, or the season very short, so that violent rains cause floods that would sweep away sluice boxes, and then are succeeded by dry intervals during which the boxes would warp and crack. Only the coarse gold is saved, while the duty of the water is usually much less than in wooden sluices. After a time, usually when the water gives out, the auriferous material is collected from the sluice and washed in a long-tom or cradle.

Booming.—This method of sluicing, which originated in the United States, is adopted when the water supply is insufficient for continuous operations. A dam with a light gate, capable of being easily lifted, is built just above the part of the valley where the auriferous gravel is situated. The water trickling down the valley accumulates behind the dam, and finally overflows at one point into a small rectangular box fastened to the end of a long lever. When full of water this box depresses its end of the lever and raises the dam-gate, so that all the accumulated water rushes out at once and scours the valley bottom. As the box falls it empties itself of water, and the dam-gate returns to its original position by its own weight.

This device is usually employed in connection with ground-sluicing, but a line of sluice-boxes might be used through which the sudden flood could carry gravel piled just above the head of the series.

Dry Diggings.—If no water can be obtained, it is sometimes profitable to concentrate very rich pay-dirt by winnowing, tossing it in a blanket until the lighter particles have been blown away, and finishing on a batea with or without mouth-blowing. It is of course a wasteful method of concentration. Several machines have been invented for dry concentration, some of which have attained a fair measure of success. These machines all use a blast of air by which the sand is kept partly in suspension, while it is moved by gravity down an inclined table which is furnished with riffles. The auriferous material must be quite dry or perfect disintegration cannot be accomplished. In a typical dry washer, the gravel is first made to pass through shaking screens, the mesh of which is adapted to the character of the material. The object of the screen is to eliminate the larger fragments, which are usually barren. The shaking screen delivers the material on to an inclined table formed of a wire-screen, covered with light canvas or some similar material through which are forced pulsating blasts of air. These sudden puffs throw up the sand and let it settle again alternately, and as a result the light material works down the table, while the gold is retained by the riffles, being too heavy to be tossed over them by the air.

Cement Gravels.—When gravels are cemented by iron oxides or clay so as to be too hard for disintegration in the sluice, except with the aid of a great number of drops, they are either sent to the stamp-mill or, with more advantage, treated by a cement-mill before being washed in the sluice. The treatment of cemented gravel is best preceded by exposure to the air for a time, frost, sun and rain having a very rapid effect on it. The cement-mill is a large iron pan fitted with coarse screens; it has revolving arms carrying plough-shares, which puddle the clay and break up lumps of gravel, but leave boulders untouched. Water, fed in with the dirt, carries the fine stuff through the screens, and at intervals the boulders are removed by various contrivances.

Tail Sluices are sometimes erected to intercept the tailings from one or several sluices with the object of collecting a further percentage of gold from the waste material. These tail sluices are made of much greater size than those described above, and in some cases pay for construction. The Kumara sludge channel, erected by the New Zealand Government to carry the tailings from the sluicing works of the district into the river, caught 957 ozs. of gold in the four years ending in 1891. This sluice is 3 feet 6 inches wide, and has a grade of 1 in 28, while the boxes which discharge into it are only 18 inches to 22 inches wide. Usually, in good work, the sluices are long enough to make the tailings too poor to be worked over again at a profit, except by the Chinese, until after they have been enriched by natural concentration in the rivers.

Fly Catchers were invented in Australia for the purpose of catching the fine particles of gold, which, successfully evading the riffles of all sluices, float down on the surface of the rivers. These devices consist of weirs constructed on piles driven into the river bed, and stretching across from bank to bank of the river. Boards covered with blanketing or coarse gunny-sacking are attached to the weirs and collect all particles floating on the surface of the water. At intervals the blankets are taken up and washed in a tank. These fly catchers soon pay for their cost of construction on many rivers, but are liable to be damaged by floods, and by being used as bridges by men and animals.

River Mining.—River mining consists in the working of auriferous gravels in the channels and beds of existing rivers, and may with convenience be made to include the exploitation of deep bars below the level of the water. It is conducted in three different ways:—

1. A portion of the river bed is laid dry by damming and fluming.
2. The gravel is raised by dredges or similar contrivances, operated from a boat.
3. Pits are sunk on the bank, and the gravels below water level excavated and raised to the surface by ordinary mining methods, or by the hydraulic elevator. The gravel won by either of these three methods is washed in the usual way by sluicing.

1. River Mining Proper.—This method is carried on chiefly in California. In the case of the larger rivers, or where only a small capital is available (from £100 to £1,000), *wing-dams* are built out from the bank, above and below the part of the river which it is desired to work, and a third dam, built parallel to the direction of the current, connects their mid-stream ends. The dams are usually constructed of boulders, with their interstices filled with gravel, and are only slightly higher than the river surface at low-water (summer) level. The water-flow is forced to the other side of the river channel, its level being little altered. The space thus cut off is pumped dry by a Chinese pump, worked by an undershot wheel placed in the current by the side of the mid-stream dam. The pay-dirt is usually covered by a mass of barren boulders and gravel, and is exposed by stripping or by sinking shafts and running drifts. The dirt is taken out down to the level of the bed-rock, which is cleaned thoroughly, all crevices being well searched. The pay-dirt is washed by sluicing, the water being supplied directly from the river, or, if necessary, raised to the required height by a dip-wheel (an undershot water wheel with small buckets placed near its periphery, which empty their contents into a flume). The boulders are removed by a derrick worked by water-power. As the construction of the dam can only be begun when the water is low, the season is usually very short; sometimes only a few days or hours are left, after the draining and stripping have been finished, for the actual collection and washing of the pay-dirt. Operations are stopped by the autumnal rise of the river, which overtops the dam and fills the pit, often coming so suddenly as to leave no time to remove any of the tools and machinery. Sometimes large returns of from £100 to £1,000 per day are obtained in this short time, and these may be sufficient to yield handsome profits on the undertaking, but the dams are always swept away in winter.

When enough capital is available, or the river to be operated on is small, head- and foot-dams are made, stretching across from bank to bank, and the water is carried off in a wooden flume and delivered into the river bed below the foot-dam. These dams are usually built of timber, faced with planks, and supported by earth and stones. The portion of the bed thus isolated is drained and worked as in wing-dam practice, the source of power being the water-race in the flume. The work is also in this case often cut short by the rising waters, which carry away the dams, flume, sluices, wheels, &c.

Of late years efforts have been made in two or three places in California to prevent the winter floods from stopping work. On the Feather River a permanent head-dam and flume has been constructed by an English company to drain a part of the bed, and tunnels have been made on the American and Feather Rivers to permanently drain large reaches and deliver the water at a lower point. None of these undertakings have as yet been strikingly successful, from various causes.

River mining is probably subject to more uncertainty than any other branch of gold mining. The whole capital invested may be lost, and all works and machinery swept away by a flood before the pay-dirt is sighted, while numerous instances are on record in which the alluvium on the river-bed, after having been laid bare at great expense, was not rich enough to pay for sluicing.

Between the years 1857 and 1880 the Californian river-beds were covered so deep by the tailings from hydraulicking that they could not be worked with advantage. Since the suspension of hydraulicking, however, and the gradual working down of the débris, some places have again become worthy of attention.*

* For a full account of river mining, with details of dam-construction, &c., the student is referred to the article on the subject by R. L. Dunn in the *Ninth Annual Report of the Californian State Mineralogist*, 1889, pp. 262-281. See also the *Eleventh Report*, 1892, pp. 150-153.

2. **Dredging.**—This method of winning the gravel has seldom met with any success, for the reasons that the gold, even when it exists in the river bed, cannot be got at in such a simple way. The bed-rock cannot by this means be cleaned and creviced, and if the poor gravels on the surface are thick, they cave in and slide down into the pit made by the dredge, so that the pay-dirt is never reached. Moreover, boulders impede or prevent the work. The favourite implements are various forms of vacuum lifts and pumps, the power being sometimes supplied by a hydraulic elevator. Water, gravel and stones are brought into the barge together, and the stuff is usually then and there washed and dumped into the river again. The system has made somewhat more progress in New Zealand than elsewhere, and the following details of recent work there are taken from a Government report on mines.* On the Molyneux River in New Zealand the centre bucket dredger has done good service, the difficulty of raising big stones being that most severely felt. Here it is found that little of the fine scaly gold of the rivers is caught in ordinary sluices, and the washing is done by separating all the coarser material by means of trommels, and passing the fine sand by itself over wide tables in a very shallow stream. The volume of water must be just enough to keep the tables free from sand, as if the latter begins to collect, it is believed that the fine gold is not being caught. The tables are covered with cocoa-nut matting, baize, blanketing, or even plush, each material having its advocates. On the Shotover River, the Sand Hills Dredging Company separates the stones by passing the gravel through a revolving cylinder 3 feet in diameter and 10 feet long, set with fine holes; 30 tons per hour are treated by this trommel. The washing is done on inclined tables which are 64 inches wide, and about 27 feet long, having a grade of 18 inches in 12 feet. These tables have a superficial area of 144 square feet only, which is insufficient for the treatment of 30 tons per hour. At the Waipapa Creek Dredging Company's works, which are situated on the sea coast, a Welman dredge is used, which acts on the centrifugal suction system. The pump is 3 feet 6 inches in diameter, and the delivery tube 13 inches in diameter, and the pump raises gravel from an area having a radius of 40 feet. All large stones are caught and separated from the fine stuff by a riddled hopper-plate. The gold is very finely divided, and is caught on plush mats which are washed every eight hours. The water for washing is supplied from a reservoir by means of an 18-inch pipe. Stones of 56 lbs. weight are lifted by this pump, which may be used with advantage on all low wet ground.

In Northern Italy a dredging plant has also been successfully at work for some years.

3. **Deep Bar Mining.**—Deep bar mining, as conducted by

* *New Zealand Mng. Comm. Report*, 1891.

the old methods, consisted in sinking shafts in the bank and running drifts below water level, either under the river bed itself or under deep bars, for the purpose of digging out the pay-dirt from the surface of the bed-rock and taking it to the surface to wash. It was mostly unsuccessful owing to the amount of water, quicksands, &c., encountered, although to-day these difficulties could be often overcome by the present improved practice in quicksand digging. As, however, this kind of digging is better suited to the efforts of a few individuals than to undertakings on a large scale, it is not likely to be largely reverted to. In particular, working in loose detrital matter under the river bed itself must always be hazardous to the workers, and the chances of success very dubious, although profits have been obtained where the pay-dirt was only worth 70 cents per cubic yard. Deep bars, however, are now being worked profitably by the hydraulic elevator, which has become of great importance in placer mining. This machine may be more conveniently described after hydraulicking has been considered.

Method of Working Siberian Placers.—The methods and apparatus employed in Siberia differ so markedly from those which have been adopted elsewhere, that they are well worth a special description, although they cannot usually be applied to placers found in other parts of the world, owing to the difference in economic conditions. In California the valleys are narrow and the grade steep, so that watercourses are usually close to the auriferous deposits, and the sluices can be made of almost any length, while still conforming to the general slope of the soil. In Siberia the slope of the valleys is so gradual that the flow of the water is almost imperceptible, and takes place through wide marshy tracts. The result of this is that the sluices must be short, being usually less than 100 feet long, and their upper ends must be raised on trestles. As a further consequence, also, the gravel must be excavated by hand and carried in waggons to the sluice, and the tailings removed in a similar way, the flow of water acting by gravity not being available for these purposes. The excavation is made in benches or terraces, working up the valley. The height of each bench above the lower one is about 5 feet, and the gravel is picked down from the face of each bank and shovelled into carts by which it is conveyed to the washing establishments. The additional expense entailed by these causes is balanced by the low cost of labour in the country, and an incidental advantage lies in the fact that the washing apparatus can be placed outside the limits of the river during flood time, and so may remain for a number of years undisturbed, while all the gravel in the district is being washed. Only surface deposits are exploited, no deep workings existing in the country.

There are three types of apparatus employed, each designed for washing a particular kind of deposit. They are :—

1. The Siberian sluice, which is used to wash light sand.
2. The Trommel, used for loamy sands.
3. The Pan, used for gravel which is cemented together by means of compact clay.

1. *The Siberian Sluice.*—The apparatus at Voltchanka, which may be taken as a type, consists of a head sluice and three secondary sluices, which are placed at right angles to the head sluice, and which leave it at different points and converge to a common centre, where the tailings are discharged. The head sluice begins at a height of 13 feet from the ground; it is about 90 feet long by 2 feet wide, and has a fall of about 1 in 14. The sands are dumped from the waggons on to a wooden platform situated above the sluice-head, and shovelled into the latter, a stream of water being turned in at the same time. After passing through a grizzly, the gravel runs over a series of cast-iron cross-bar riffles, which form a number of rectangular depressions (or *pigeon-holes*) in the bed of the sluice, by which the disintegration is favoured. The stream then flows over an iron screen, through which a part of it falls into the first secondary sluice, while the remainder continues its course over more pigeon-hole riffles. This arrangement resembles the Californian undercurrent. A second and a third screen open on to the other secondary sluices, and the part of the gravel (consisting chiefly of small stones) which has resisted disintegration, and has not passed through the screens, is then let fall into a hopper, whence it is removed to the tailings waggons.

The secondary sluices are wider than the head sluice, and have a steeper grade, and the amount of water and auriferous material passed is of course much less in each one than on the principal sluice. The sands first pass over a number of transverse riffles, and then over about 30 feet of blanketing, the sluice being widened at the same time, and subdivided by longitudinal wooden partitions, and the grade being as much as 1 in 6, while small drops are introduced at intervals of a few feet. The third sluice has a more gentle inclination than the others. The tailings fall into a shallow sump, and are instantly lifted out of it by a bucket elevator or "tailings-wheel" operated by water power, and stored in a hopper, whence they fall into waggons, by which they are removed to the dumping ground.

No mercury is used in these sluices, and only the production of "grey concentrates" is attempted, this work being continued from 6 a.m. to 7.30 p.m. every day, after which the concentrates are collected from the riffles. They consist of gold in scales and plates, magnetic iron oxide, pyrites, rutile, together with some quartz, &c. They are treated with mercury in the Siberian trough or on inclined tables, the method being that described on p. 47.

The apparatus described above treats 500 tons of gravel per

day, the labour required being furnished by twenty men and ten horses. The gravel treated contains an average of from 12 to 15 grains of gold per ton, rarely falling below 6 grains per ton; the exceptional richness of 3½ dwts. per ton has been observed. The gold is chiefly found in the head sluice, where 70 per cent. is caught, 30 per cent. being caught on the secondary sluices. At a similar establishment at Tchernaia-Retchka, however, where the gold is less finely divided, 97 per cent. was caught on the head sluice, and only 3 per cent. on the secondary sluices. The amount of water used at Voltchanka is about six times the weight of the gravel. The cost of construction of the works was 70,000 roubles, or about £7,000.

2. *The Trommel.*—Gravels which are too compact for satisfactory disintegration in the short sluices described above are subjected to a preliminary treatment by a trommel. At Berezovsk the trommel is of sheet iron of 9 mm. thick, having holes in it of about 1 mm. in diameter. The trommel is about 12 feet long, 3½ feet in diameter at one end, and 4½ feet at the other, and is set inside with denticulated plates of iron to assist in the disintegration effected by the water. The machine is driven by a water wheel, and is sufficient for the disintegration of from 400 to 500 tons of gravel per day, requiring the expenditure of about 3 horse-power to drive it. The amount of water used in the trommel and on the tables is 67·5 litres per second, or about seven and a-half times the weight of the ore. The washing is effected on inclined tables only 30 feet long and 12 feet wide, and with a grade of about 1 in 4, placed with the incline at right angles to the length of the trommel. Near the head of the table, and stretching across it, is a deep trough-like depression, and below this there are a number of transverse riffles, in which grey concentrates are caught and treated as usual. The Berezovsk establishment employs twenty-five men and fourteen horses constantly; the trommel usually lasts for two seasons.

3. *Pan Washings.*—Sandy clays cannot be economically disintegrated in a trommel, and are, therefore, treated in a washing pan, which bears a strong resemblance to the cement pans employed in California. The pan usually consists of cast iron, and is from 8 to 16 feet in diameter, with vertical sides from 1 to 5 feet high. The bottom is of cast iron or sheet iron, and has numerous holes in it of about $\frac{1}{15}$ inch in diameter, widening downwards. The bottom is divided into 25 sectors, between which are deep grooves for the collection of the pebbles. Through a circular opening in the centre of the pan there passes a revolving axis to which are suspended eight horizontal arms studded with vertical iron teeth, some of these being shaped like plough-shares. The revolution of these arms effects the disintegration of the sandy clays, which are fed into the pan together with water and puddled until fine enough to pass

through the holes in the bottom, and the stones are removed at intervals by opening little gates placed opposite the radial grooves. The disintegrated gravel falls from the pan on to concentration tables, similar to those used after disintegration in the trommel. At Berezovsk the pan is $11\frac{1}{2}$ feet in diameter and 5 feet deep, and the arms revolve at the rate of 25 turns per minute. From 50 to 55 tons of material are treated in twelve hours, the water consumed, including that required for power, being ten times the volume of the sand.

Beach Mining.—Beach mining is a comparatively unimportant form of shallow placer mining. The sea beaches on parts of the coasts of California, Australia and New Zealand contain small quantities of gold, which has been proved, in all cases in which the matter has been investigated, to be derived from the cliffs, which mostly contain a still smaller quantity. Some streaks of black sand, however, in the "Gold Bluff," California, have yielded $135 or $6\frac{3}{4}$ ozs. per ton by actual working.* The waves of the sea wash down and partially concentrate the poor sands, and, under certain rather exceptional circumstances, as the tide goes out the surface of the beach is left covered with black sand, in which numerous specks of gold occur. This is carefully scraped up and transported inland to be washed, as sea water is not well adapted for the purpose, although it is used by one Californian company. The next tide usually washes away all the valuable material which has not been collected, or else covers it with barren sand. There is great difficulty in washing the black sand in California, as it consists largely of rounded grains of magnetite, the density of which is about 5·0, while the gold is in minute flakes and scales, which can be seen under the microscope to be oblong in shape, and thicker at the sides than in the middle (a shape due to continued pounding of a malleable material). This form is so easily moved and buoyed up by water that it is difficult to get a "colour" with the pan, and the amount caught by the mercury in sluices or long-toms is usually an insignificant proportion of the total assay value of the sand. The industry is generally a languishing one.

Treatment of Shallow Placer Gravels by Steam Shovels and Amalgamated Plates.—A plant for the treatment of placer deposits, which are similar in nature to those worked in Siberia, has been devised in America, where labour-saving contrivances are indispensable in all such cases, owing to the high rate of wages. The apparatus consists essentially of a combination of a steam shovel or navvy, and an amalgamator consisting of a wide waggon-shaped wrought-iron trough loosely lined with silver-plated amalgamated copper plates, which form a series of steps or riffles at the sides of the trough. The material is elevated into a hopper by the excavator, and thence is charged into a revolving trommel

* *Prod. Prec. Met., U.S.A.*, 1884, p. 557.

placed inside the trough. Here the disintegration of the gravel is effected, and the fine material falls through into the trough, while the stones are discharged outside at the end of the trommel. At the bottom of the trough there is a water pipe, carrying water at high pressure, and in that pipe a series of jets pointing alternately forwards and backwards. The result is to give a series of eddies or whirlpools in the water in the trough, and the sand and fine gold is continually carried up to the surface near the middle line of the trough, and in descending again near the sides it comes in successive contact with several of the plates, which form a series of steps. The fine sand is eventually discharged at the end of the trough. It is stated by the Bucyrus Steam Shovel and Dredge Company, by which the machinery is manufactured, that such a machine erected in Montana has a capacity of from 600 to 800 cubic yards of gravel per day, all the water required being supplied by a pump raising 500 gallons per minute, the proportion being only three of water to one of gravel, which seems much too small. It is stated that gravel containing only 12 cents of gold (*i.e.*, about 3 grains) to the ton has been treated successfully by this machine, but exact records of continuous work done by it are wanting. It will be noted that special means must be adopted in each case for the handling of the tailings.

CHAPTER V.

DEEP PLACER DEPOSITS.

Nature and Mode of Origin of Deposits.—This discussion is necessary in order that the description of the methods of treating the deep placer gravels may be intelligible. Both in Australia and California, besides the superficial placer deposits situated in or near the existing rivers, which in the deep cañons of the Klamath and other rivers in the extreme north of California attain a thickness of 250 feet, there exist auriferous gravels which bear no apparent relation to the present drainage of the country. These gravels often attain enormous thickness, and are in many places covered by volcanic rocks, consisting of basaltic lavas and tuffs, which are sometimes interbedded with gravel and loam. This latter circumstance shows that intermittent action of the volcanic vents, with long intervals of repose, has taken place. There has been some difficulty in accounting for the origin of these deep placers, and it has been ascribed in succession to the agency of the sea, of ice, and (for California) of a huge river flowing from north to south at right angles to the direction of flow of the existing rivers. None of these views are now entertained, and the "fluviatile" theory is generally accepted, the origin of the gravel being ascribed to the depositions of ancient rivers flowing in courses roughly parallel to those of existing rivers. The geological age of these ancient rivers has not yet been determined with certainty, but though they may be Pleistocene, the balance of palæontological evidence is perhaps in favour of Whitney's view that the deposits were formed in the Pliocene period.

The ancient Californian rivers probably had their sources at somewhat higher altitudes than those now existing, and had more uniform general grades, the slope of their beds corresponding more nearly to the general slope of the country. The existing rivers, on the other hand, have steep grades in the upper parts of their courses, followed by comparatively level stretches below. The old rivers, however, like their successors, had rapids, falls and level stretches, the grade varying from 5 feet to 250 feet or more per mile. The Pliocene rivers ran in valleys which were broad and shallow in comparison with the present deep precipitous cañons, and the volume of water was in general much greater than that delivered by their representatives of to-day. The width of the valleys varied from 100 feet to fully 1½ miles (which is the width at Columbia Hill), and the depth must have been often over 1,000 feet. These valleys were already partly filled up by accumulations of gravel, when the outbreak of volcanic activity in many cases filled up the remainder, and the streams were deflected into other channels, which often lie close alongside the old cañons. These new channels have been excavated by the running water until they now lie much below the level of the beds of the Pliocene rivers, and consequently the gravels which were deposited in the old valleys now sometimes crown the highest ground in the district, the general level of the country having been greatly reduced in height. The new channels have been cut partly in the old country rock and partly in the Pliocene auriferous gravels and their covering of volcanic rocks. Sometimes the course of the present cañons cuts that of the old at several points owing to the sinuosity of both, see Fig. 8, in which A represents the modern river, and B the ancient one. The result is that sections of the old valley from bed-rock to surface are exposed in the sides of the cañons, usually at some height above the present level of the water, and it was at such points as these that the discovery of the existence of the deep placers was first made. The hard covering of basalt has served to protect the more friable gravels, which have been for the most part removed in those places where the lava has been worn away or has never existed, so that

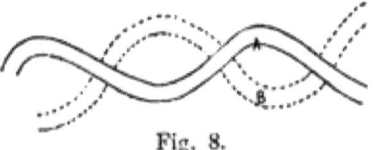

Fig. 8.

the largest tracts of gravels still existent lie beneath the volcanic rocks.

Fig. 9 represents a section across two ancient channels (B, B) and a modern cañon, that of the American river. Here, A is the volcanic capping, which is 800 feet thick above the Red Point channel; B, B are the auriferous gravel channels; C, C are deposits of gravel on the "rims," containing gold in places; D is the bed-rock, consisting of dark-blue slates; E is a barren deposit of angular débris and boulders; F F are prospecting tunnels, which were put in at too high an altitude; F' is the tunnel bored with the object of reaching the bottom of the gravel deposit; H are prospecting winzes sunk in order to discover the position of the gravel. The space included within the dotted lines N M Y M' N' has been obviously denuded since the deposition of the volcanic cappings, the soft slate rims N M R and N' M'

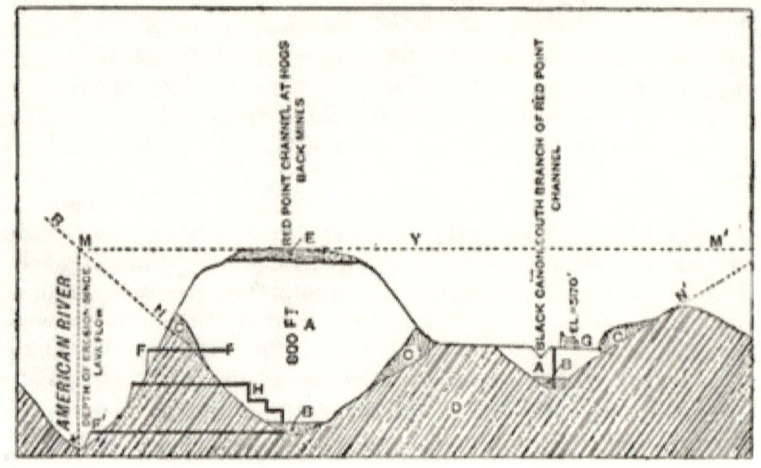

Fig. 9.

having been worn away, while the hard lava has resisted erosion. The vertical depth from M to the American river is about 1,800 or 2,000 feet.

This condition of things is that prevailing in California, but in Victoria the structure closely resembles that just described, with the exceptions that the old valleys were smaller and that the erosive action of the rivers since the deposition of the basalt has been comparatively slight, owing to the slight grades of the streams caused by the low elevation of the country and to the small amount of the rainfall. In consequence of this the basalt has usually not been worn through, and the "deep leads" or old river bottoms are often below the level of the present streams, so that although a larger proportion of the Pliocene gravel remains, it is more difficult and expensive to mine.

The shallow placers, at any rate in California, have resulted in the main from the erosion of these deep placers, the materials of which, having undergone a natural concentration in the ground sluices afforded by the river beds, furnished the wonderfully rich river-bed and bar deposits, which yielded so much gold between 1848 and 1860. The deep level gravels vary greatly in thickness, as has already been stated, being only 2 feet thick at Table Mountain, Tuolumne County, and over 600 feet thick at Columbia Hill, Nevada County, California, and averaging from 100 to 300 feet, the thickness varying with the nature of the river bed, and the subsequent erosion. The gravel consists in slaty districts chiefly of quartzose sand, the fine materials furnished by the disintegration of the slate having been for the most part swept away, and the products of the quartz veins contained in the slate being left. Near bed-rock, but at no higher level, there is often a collection of large boulders, varying in size up to 10 feet in diameter, consisting mainly of quartz, but subordinate to these are others usually similar in character to the bed-rock. These boulders, though rounded, are too large to have been transported far by running water, and have probably been polished by the attrition of the sand carried over them. At the Forest Hill Divide, Placer County, California, the gravels consist almost entirely of pure white quartz, but at other places such quartz is rare. It is to be noted that it is only in slaty districts, where the gravels are mainly quartzose, that rich auriferous deposits occur. In granite districts, where the gravels are composed of more heterogeneous materials, and in cases where they consist of volcanic boulders and detritus, little or no gold is found. The lower parts of the gravels are often cemented into a conglomerate, called "cement," by infiltration of silica, oxides or sulphides of iron, or, rarely, carbonate of lime; when the gravels are covered with lava, the whole thickness is in some cases converted into cement. The upper parts of the gravels often contain pipe-clay in greater or less quantity, either in pure beds or mixed with sand. Fossil leaves occur in the clays, and drift wood occurs throughout the whole of the deposits in extraordinary abundance, particularly in Australia; this wood is for the most part silicified or replaced by sulphide of iron. The higher portions of the gravels are often altered by the action of air and water on the iron sulphide, thus forming ferrous salts, hæmatite and hydrous sesquioxides, which colour the gravels red and brown respectively. The upper gravels are hence called "red gravels." The lowest layers, being protected from alteration from above, are coloured dark blue-grey by the ferrous sulphide contained in them, and are hence called "blue gravels," their occurrence giving origin to the old "blue lead" theory, owing to their uniformity of colour over wide areas in California. The sulphide of iron which incrusts fossil bones and teeth found in the gravel, and replaces the substance of drift wood, was formerly believed to be derived from below, as the result of metamorphic action going on in the country rock.

Distribution of Gold in the Gravels.—The gold is found chiefly either in contact with or just above bed-rock. If this consists of soft slate, and especially if the planes of cleavage are at a high angle to the horizon, particles of gold are often found in the natural riffles thus formed, and are disseminated through the rock to the depth of a foot or two. If depressions, pot-holes, or fissures exist in the old river bottom, they are usually very rich in gold. Where, as often happens, there is a channel, or "gutter," to adopt the Australian expression, cut by the stream in the lowest part of the valley, the gravel filling it is usually much richer than that found elsewhere. Such rich portions, often only a few feet wide, and of insignificant depth, but extending to considerable distances in the direction of the stream, are called "leads." As a result of these circumstances, the "blue gravels" happen to be richer in general than the "red gravels," from which arose the old theory that only blue gravel pays to work. Coarse gold and nuggets chiefly occur near bed-rock in the deeper parts of the channel, but the "rim-rock" gravels are also often rich. The "rim-rock" is that portion of the bed-rock which forms the sides of the old valley, thus lying considerably higher than the central channel. The richness of gravels here doubtless arises from the existence of old bench or terrace gravels, which are consequently the oldest of the whole series, being formed before even the gutter gravels. Rich streaks also occur at various levels in the gravels, often resting on "false bottoms," which consist of impermeable beds of clay or some similar material. Sometimes these streaks are richer than those encountered at bed-rock, as, for example, at the Paragon Mine, Placer County, California. Although it is concentrated in this manner at various points, gold nevertheless occurs disseminated through the greater portion of the red gravel, where, however, it is in a finer state of division and less abundant. Besides existing as free particles, gold may occur in quartz boulders, although this is rare. For instance at the Polar Star Mine, Dutch Flat, California, a white quartz boulder was found, which contained 288 ozs. of gold. Gold may also occur, together with pyrites, replacing the substance of drift wood.

The amount and position of the gold varies, as in the case of the present rivers, with the grade, the shape of the valley, the volume of water, the amount of gravel being carried down, &c. "An underloaded current—*i.e.*, a current charged with less detritus than it is well able to carry—is apt to cut its bed, and prevent the accumulation of gravel. A greatly overloaded current will deposit too rapidly to admit of the concentration of the gold dust."* Under conditions intermediate between these

---
* Ross E. Browne, *Tenth Report of the Cal. State Mineralogist*, p. 448.

extreme states, the current may be just strong enough to keep its bed clear from all accumulations except a small quantity of coarse gravel and the coarse gold, which is caught in the natural riffles, and thus all the conditions necessary to form a rich bed of pay-dirt may be present. If, however, the bed consist of granite or other rock which wears in smooth and rounded shapes, little gold will be caught. Slates, consisting of layers of uneven hardness, wear irregularly, and afford a good gold catching surface. The conditions noted above as necessary to form rich gravels cannot be expected to have been prevalent over great distances. "An increase of grade or narrowing of the channel will cause an increase of velocity, and the same stream may be underloaded in a narrow steep section, and overloaded in a broad flat section."* The difference of velocity between the middle and sides of a stream, and between the inside and outside of a bend, may give the right conditions in one part of a river bed and not in another. Thus with high grades, rich gravels should occur in the less rapid, and with low grades, in the more rapid parts of a stream. Having regard to such considerations, the richer parts of existing rivers can be pointed out with little trouble. When, as in the Pliocene rivers, the beds are buried to a depth of hundreds of feet, the richer parts are more difficult to find.

The history of the Pliocene rivers began with a period when excavation exceeded deposition, when the rivers were underloaded for at least a portion of each year, and the channel was constantly being deepened. Some bench or terrace gravels were formed at this time, and being at the sides of an underloaded river tended to be rich. The river bed, although rocky and comparatively free from sand, would perhaps accumulate some coarse gold, which as the channel deepened was no doubt in part ground up into fine particles and carried off, but at the time when the excavation had reached its lowest point, some of this coarse gold would certainly be present. When the underloading of the stream ceased, whatever caused the cessation, a pause must in many cases have occurred before the gravel proved too much for the stream to carry. During this pause the conditions for gold catching were favourable, and hence rich gravels were formed on bed-rock in the gutters or channels. Then, as the streams became overloaded, sand and gravel accumulated rapidly, so that little concentration of the gold in them could take place. The rivers flowed over thick sand banks and, in consequence, frequently changed their courses. The sands, being deposited by overloaded rivers, of course contained fewer and smaller boulders, and the thick masses of poor sand thus went on accumulating until the volcanic outbursts put an end to the process.

Origin of the Gold in the Placers.—The origin of the gold in deep placers has long been a vexed question. It was

* Ross E. Browne, *Tenth Report of the Cal. State Mineralogist*, p. 448.

formerly accepted without question that the erosion of auriferous quartz lodes existing at higher altitudes furnished both gravel and gold. In support of this it was urged that the same districts which furnished auriferous gravels abounded in quartz veins at higher levels, while Whitney pointed out that numerous lodes were intersected by the valleys and were still to be seen in the bed-rock. On the other hand, the fact that the nuggets found in the drift are much larger than any masses of gold encountered in veins, and that the placer gold is of superior fineness, are difficulties in the way of accepting this theory. Moreover, Egleston states that nuggets as large as a man's fist have been found embedded in the midst of fine sand, whither they could not have been carried by the action of running water, but authentic instances of such finds seem to be lacking, nuggets usually occurring in coarse gravel, among boulders. It is further declared by the opponents of the erosion theory that if a small quantity of soft material like gold, mixed with lumps of hard quartz, were washed down by water, then, long before the quartz could be reduced by grinding to the condition of grains of sand, the gold would be worn down to such a fine state of division that none of it could lodge in the river bed at all. In opposition to this contention, it may be urged that the extreme malleability of fine gold would make this comminution very slow, and scales of the metal are said to have their edges blunted and thickened by the pounding action of dry sand moved by the wind, instead of having them worn away. However, it is certain that the form of placer gold is different from what might have been expected if it consisted of water-worn vein gold. Nuggets are mammillary in form, and generally appear more like concretions than water-worn fragments, in spite of the fact that they often include more or less quartz.

In 1864, in order to account for these and other facts, Mr. A. C. Selwyn, of Victoria, suggested a theory of solution in which it is supposed that the gold disseminated through the rocks and drifts is dissolved by percolating waters which contain acids and salts in solution, and is reprecipitated around certain centres. Selwyn considered that the waters capable of dissolving gold must have acquired this property by passing through the beds of basalt, &c., overlying the drifts, inasmuch as large nuggets occur in districts where basaltic eruptions have taken place, while, where these are absent, the gold is very fine, and nuggets can scarcely be said to exist. The fact has long been known that gold is soluble in certain dilute solutions of salts, likely to be met with in nature, such as a mixture of nitrates with chlorides, bromides or iodides, or as the haloid ferric salts. This has been firmly established by the researches of Skey, Daintree, Egleston and others. Also, the precipitation of gold from these solutions around nuclei consisting of particles of gold, pyrites, &c., by organic matter present in the liquid, has been studied, and efforts made to form nuggets similar to those found in nature, without much success. This, however, is not surprising since the conditions in nature, including almost unlimited time and immense quantities of exceedingly dilute solutions, cannot be reproduced in the laboratory. Among other pieces of evidence against the erosion theory which have been cited, may be mentioned the fact that some gold placers occur at higher levels than any quartz veins yet discovered or likely to be discovered; also that nuggets have been found embedded in decomposed rocks in positions to which they could not possibly have been carried by running water, so that these nuggets at least must have been formed by accretion. Some regard must also be paid to the prevalent belief among diggers that if a little "seed gold" is left in the tailings from sluicing operations, the deposit will grow in richness so as to be worth working over again after a few years.

Some of these arguments have been met by the exponents of the erosion theory. The fineness of the placer gold has been accounted for by supposing that the impurities (silver, copper, &c.) formerly present in the native gold have been dissolved away by meteoric water, in which they are much more soluble than gold is. The existence of large masses of gold in placer deposits was accounted for by Whitney by assuming that the upper portions of the lodes, now washed away, were richer, and contained larger masses of gold than the remains of the lodes now left, but Liversedge has shown[*] that this assumption is not necessary. Some nuggets too have been found showing undoubted signs of erosion by water, but these are rare. Liversedge has recently adduced evidence (*loc. cit.*) that, even if the small particles of gold found in placers have grown by accretion, nuggets cannot have appreciably increased in size. The suggestions made to account for the great richness at bed-rock— viz., that gold has "settled" through the quicksands, or that the gold solution has remained longest in contact with the sand nearest bed-rock—are not wholly satisfactory, and must be supplemented by some such explanation as that given above, p. 67.

Minerals occurring in the Placer Deposits.—In California, if quartz grains and silicified wood are excepted, the most abundant mineral is black iron-sand, which usually consists of magnetite, although menaccanite, a form of hematite in which part of the iron is replaced by titanium, also occurs. These minerals must have been derived from the lavas, as neither of them are known to occur in the quartz veins of the country. Platinum and its allies are usually present in more or less abundance; thus iridosmine occurred to the extent of 1 in 100,000 of the gold in the early days, and increased afterwards to 15 or 16 times that pro-

---

[*] *Proc. of the Royal Soc. of N.S. Wales*, Sept., 1893.

portion. Grains of native copper, nickel, and perhaps lead have also been detected, and a few diamonds occur, while garnets, small crystals of zircon and cinnabar are very abundant.

METHODS OF TREATING DEEP PLACER GRAVELS.

Both in California and Australia, when the first gold discoveries were made, the river beds and bars were at once explored, and soon afterwards the flats closely adjoining them. Subsequently, the bench gravels situated in the same valleys, and the side ravines and gulches, which remained dry during most of the year, were prospected and worked, owing to the rapid growth of the mining population and to the fact that the exhaustion of the shallow placers was already beginning to make itself felt. The result was that the exposed edges of the outcrop of some of the deep leads were found, and the pay-dirt followed into the hill-side by drifting. Then, as in many cases it was found that the gravels overlying the pay-dirt on hill-sides, although poor by comparison with the earth below them, nevertheless contained a small quantity of gold, the idea was evolved of breaking down the whole bank by jets of water, and passing all the material through the sluices. Hence arose the practices of drift mining and hydraulicking,[*] of which the former is largely used in California, while the latter, though much used in California prior to 1884 and in New Zealand, cannot be applied in Australia or Siberia owing to the general flatness of the country. In Siberia, only shallow placer deposits are worked. In Australia, the deep leads are usually reached by shafts, since the surface of the country is not intersected by deep cañons, as in California.

Hydraulicking.—This method of working consists, as has been already stated, in breaking down banks of gravel by the impact of powerful jets of water, and passing the disintegrated material through a line of sluices, without the agency of hand labour. The chief requisites for the successful application of hydraulicking are—

1. Large quantities of auriferous gravel, not less than 30 feet in thickness, and not overlaid by any appreciable thickness of barren material, which would necessarily be passed through the sluices with the pay-dirt. The gravel treated need not be rich, a mean yield of less than 1 grain of gold per cubic yard being often enough to furnish profits if the operations are on a sufficiently large scale.

2. A plentiful and uninterrupted supply of water throughout considerable portions of the year.

---

[*] The invention of hydraulicking is ascribed to Edward Mattison, of Sterling, Connecticut, who used the method in 1851, on a small scale.

3. Sufficient fall in the ground so that (a) the water may be delivered under the pressure of a head of from 100 to 300 feet, (b) the tailings can be easily carried away to a large dumping ground, which is most conveniently either the sea, or a large and rapid river.

Commencement of Operations. — In California, the naturally occurring banks or cliffs in the gravels in the sides of the gulches were first selected for attack. Later, some of the deposits occurring in those channels which are not intersected at favourable points by the present system of drainage were operated on. It is necessary in such cases to run a tunnel from the nearest cañon in the bed-rock to the lowest point in the gravel, this point being found or guessed at by prospecting operations. The tunnels are often of great length, that at the North Bloomfield mine, Nevada Co., Cal., for example, being 7,874 feet or $1\frac{1}{2}$ miles long. One or more shafts are then sunk from the surface through the gravel to the tunnel, and washing operations are begun by ground-sluicing, letting the water and gravel fall down the shaft and run through the tunnel, in which the sluices are sometimes laid. The surface near the shaft is thus gradually lowered, or it may be terraced by hand labour, until an excavation is made of sufficient size to enable the ground to be attacked with the hose. Washing then proceeds regularly in this manner, the bank being broken down by jets of water, and the products being allowed to fall down the shaft and pass through the tunnel. Care must, of course, be taken in the initial stages of the work to prevent the blocking of the shaft, either by the caving in of its walls, which are held apart by heavy timbering, or by runs of ground, which may occur when the upper terrace is undermined by the jets. After work has proceeded for some time, however, the gravel immediately round the shaft is all washed away down to bed-rock, and the shaft is then almost or entirely obliterated (according to the location of the tunnel), giving place to a huge open cut or funnel-shaped excavation, at the bottom of which is the upper end of the tunnel. Having thus briefly indicated the usual sequence of events in opening a hydraulic mine, the plant and the ordinary method of working may be considered under four heads.

1. The supply of water.
2. Breaking down the banks.
3. The washing proper, or sluicing.
4. Disposal of the tailings.

1. *The Supply of Water.*—The amount of water required by large undertakings is far more than can be obtained from the rainfall on the hills immediately round the mine. Thus the North Bloomfield Mine, in the season of 1877-8, used between sixty and seventy millions of gallons per day, or enough for the total supply of a city of three million inhabitants. The workings,

moreover, must necessarily be considerably above the level of any large rivers in the neighbourhood, so that it is often necessary to build huge reservoirs at convenient spots to store up the rainfall and melted snows of large districts, and to convey the water thence to the mine by ditches, flumes, or pipes. In California the reservoirs have an aggregate capacity of 8,000,000,000 cubic feet, the largest, that of the South Yuba Water Company, having a capacity of 1,800,000,000 cubic feet.

The ditches conveying the water from these reservoirs to the mines pursue a course determined by the contour of the country. They are usually cut in the sides of hills, and are given as far as possible a uniform grade, which in different ditches varies from 5 feet to 40 feet per mile. With the higher grades more care in the lining of the ditch is necessary to prevent erosion of the banks. The ditch may be lined by stones or boards, or left unlined, but the loss from leakage amounts to several per cent. of the delivery of unlined ditches. The ditch is sometimes buried in the ground to prevent damage by avalanches, to keep the water from freezing, and to reduce the loss by evaporation, which, in the case of one ditch in California, was found to be about 12 per cent. of the amount delivered from the reservoir. Wooden flumes are also used instead of ditches, but though cheaper, are more liable to be damaged. The dimensions of the ditches vary enormously; they range up to 100 miles in length in California, and are from 2 to 15 feet wide, and from 1 to 5 feet deep. In New Zealand ditches have been used less than wooden flumes, but latterly, the use of sheet-iron pipes has been greatly extended there, and many flumes have already been replaced by them. These pipes are made as much as 30 inches in diameter, and are cheap, durable and easily repaired; they entirely prevent losses by leakage and evaporation, and deliver more water under similar circumstances than flumes, as they offer less frictional resistance to the flow.

Where it is necessary to cross side cañons or gulches the flume or iron pipe is carried over on a light tressle bridge, or the water is passed through an inverted siphon formed by a wrought-iron pipe which passes down one side of the valley and up the other, being filled from a head-box or reservoir, and delivering the water at a lower level on the other side. At Cherokee, Butte County, an inverted siphon pipe is used to carry the water across a ravine 873 feet deep. The diameter of the pipe is from 30 to 34 inches, and its greatest thickness (where there is a pressure of 384 lbs. per square inch) is 0·375 inch.* The Miocene Ditch Company, operating in the same County, carried their flume, of 4 feet wide and 3 feet deep, round the face of a bluff 350 feet high, supported on L-shaped, iron brackets made of bent T rails, soldered into holes previously drilled by

* *Ninth Report, Cal. State Min.* (1889), p. 125.

men let down the face of the cliff by ropes. Successful ditch construction often depends on such engineering feats.*

The water is delivered at a convenient height above the workings into a head-box consisting of a small wooden reservoir from which the pipes take their origin. In many cases the reservoirs and ditches are owned by separate companies, or, as in New Zealand, by Government, and the water is sold to the miners by measure. The unit in New Zealand is a "government head," and in the United States a "miner's inch," following the system in vogue in Spain and Italy. The amount of water that will flow through an orifice 1 inch square, cut in a board of 1 inch thick, under a head of water that varies with the custom of the locality but is usually from 4 to 8 inches, is called a miner's inch. The amount of flow in twenty-four hours is called a "twenty-four hour inch," and similarly there are ten-hour and twelve-hour inches. The quantity of water in a miner's inch varies with the head of water used and the form and size of the orifice for delivery. Thus the amount delivered from an orifice 25 inches long and 2 inches wide is reckoned as 50 inches, although it will be more than fifty times as much as the delivery from an orifice 1 inch square. The twenty-four hour inch under a head of 7 inches amounts to about 2,230 cubic feet.†

The water is conveyed from the head-box by pipes, which were formerly made of canvas hose, to which iron rings 3 inches apart were added for pressures of over 100-feet head. These latter are called "crinoline hose," and were generally made of from 6 to 8 inches in diameter. They were used for some time, but were found to suffer from rapid rotting, and to be liable to burst at pressures of above a 200-feet head, and are now completely replaced in the United States by sheet iron. In New Zealand the canvas hose still lingers, but is now being rapidly replaced. The iron feed-pipes are made from 10 to 15 inches in diameter, and the thickness varies according to the pressure which they may be called on to withstand. Sharp bends in them are avoided as the flow of water is checked thereby. They are liable to collapse if the level of the water in them is reduced, and a partial vacuum formed inside; hence, as in the case of all other sheet-iron pipes used in hydraulicking, they are fitted with valves, which are constructed so as to freely admit air from without. The best and cheapest form of these is that used in New Zealand, which has a 2-inch hole and a rubber clack like that used in pumps. The valve is always open until the rising water lifts it up on its surface, and closes the orifice. The water is discharged

---

* For details of the cost of construction, &c., of ditches, the student is referred to Egleston's *Mining and Metallurgy of Silver, Gold and Mercury in the United States*, vol. ii., pp. 120-180.

† For further details concerning miner's inch *vide* Art. by P. M. Randall in *Precious Metals in the United States*, 1884, pp. 558-572.

through a nozzle called a "giant" or "monitor." The nozzle was at first a sheet-iron tube, having an aperture 1 inch in diameter, and was held in the hand. The size of the nozzle was gradually increased, until it has now reached a diameter of 11 inches. Such a stream, under a head of 200 feet, requires special appliances to control it, deflect it at will, and prevent the nozzle from "bucking." A number of ingenious contrivances have been devised both in America and New Zealand, but all the monitors bear a general resemblance to the one shown in Fig. 10.

2. *Breaking Down the Bank.*—When the jet is first directed against the bank, the water spatters in all directions, then buries itself a little, and after a time, in loose ground, a "cave" takes place, the undermined bank falling down. By the method of undermining, the power of the giant is much increased, especially where hard and soft layers alternate. When large caves are about to take place the water is turned off, as otherwise the

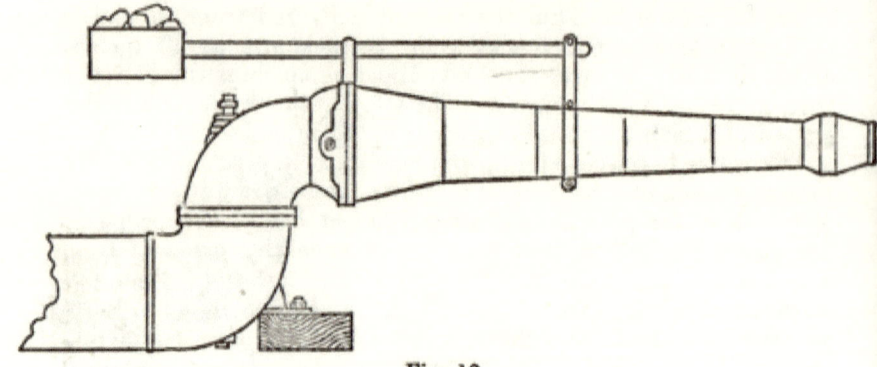

Fig. 10.
Scale, 1 in. = 4 ft.

ground may run so far as to overwhelm the monitor and the workman directing it. The nozzle is placed as near to the bank as possible, consistent with the safety of the workers, so as not to waste too much of the initial velocity of the stream of water. Consequently, lofty banks are not advantageous, and if they exceed 200 feet they are usually worked in terraces of 100 feet or so in height. In some parts of the Spring Valley Mine, however (see Fig. 10a), a bank of 450 feet high was worked in a single bench, and it was then not unusual for the runs of ground to bury pipes which were throwing 7-inch streams from a distance of 400 feet from the face of the bank.

The jet is, if possible, delivered unbroken against the face of the bank, as its disintegrating power is thus kept at its maximum. However, in some cases, where it is cemented, the gravel is too hard to be economically broken down by the water alone, and blasting is then resorted to, a drift being run into the bank, and cross-cuts made at the end in which the powder is placed; the

DEEP PLACER DEPOSITS. 75

Fig. 10a.

drift is then filled up, and the charge exploded by electricity. It is more economical to blow out the base of the bank, as the upper part then falls by its own weight and can be broken up by the water. Sometimes arrangements are made to explode very large blasts: thus, at the Blue Point Gravel Mine, a charge of 50,000 lbs. of powder was exploded at the end of a drift 275 feet long in the year 1870, and 150,000 cubic yards of gravel were brought down, while at another mine, 3,500 lbs. of dynamite were exploded in 1872, and 200,000 cubic yards of gravel disintegrated at once.

3. *Washing the Gravel in the Sluices.*—The sluices in which the gold is caught are constructed on exactly the same principles as those already described, but are larger and, though usually made of wood, are of more massive construction, in accordance with the great quantities of gravel to be handled and the continuous nature of the work. The sluices are commonly called "flumes," but it is better to restrict the use of this word to a conduit for carrying water only. The sluice boxes used in hydraulicking, though, as usual, only 12 feet long, are as much as from 3 to 6 feet wide and from 2 to 3 feet deep; they are lined with heavy planks on the sides, and the pavements are made of more durable materials than is usual in shallow placer sluicing, wooden blocks, rocks, or T railroad iron being most usually employed. The wooden blocks are from 12 to 30 inches square, and from 8 to 18 inches deep. They are usually made of one of the softer varieties of pine (*e.g.*, the "digger" pine, Pinus sabiniana, and others), which "brooms up" under friction, and thus presents a better catching surface. The blocks are cut across the grain of the wood, and are set side by side across the sluice, each row separated from the next by strips of wood to which they are nailed, while they are also kept in position by the side lining which is placed upon them. The interstices in the block pavement act as gold catchers, and are filled with small stones, or, with less advantage, allowed to fill up with gravel when washing begins. On account of the rapid wearing away of the wood, much of the gold and amalgam caught is scooped out and carried off again. Wooden block riffles only last from two to four weeks when in heavy work, but are easy to take up and put down again in cleaning up; they are discarded when worn so as to be only 4 to 6 inches thick.

Rock pavements are made of those boulders which are most easily obtained in the particular district. Basalt is generally used, oval stones of 15 or 18 inches long and from 9 to 12 inches thick being selected and placed on end, with a slight slant in the direction in which the current flows. They are held in place by wooden planks which divide the sluice into compartments, so that if one stone works loose the pavement as a whole is not affected. The interstices as before are filled with gravel. Rock pavements are very durable, lasting from three to six months, but require more grade to the sluice, and occasion loss of time in cleaning-up and re-paving the sluices. Consequently, they are never used near the head of a sluice, where cleaning-up is a frequent operation, but are often used for the lower parts of sluices, where they sometimes alternate with block riffles, and are especially suited for tail-sluices which are only cleaned-up once a year. Rock pavements cost less than other forms of riffles.

Iron riffles, which usually consist of T-iron rails, are placed longitudinally in the sluice, closely packed side by side. They present a large amount of space available for catching the gold and amalgam, last well, present little resistance to the current (so that the grade may be low while the duty of the water remains high), and are easily taken up and put down. They are, therefore, generally used at the head of the sluices. Though their first cost is higher than that of wooden blocks, they are more economical in the end, owing to the saving of time in cleaning-up and to their longer life. Egleston * instances the results of experiments made at the Morning Star Claim, California, where three sections of sluice, each 65 feet long, were laid at a distance of 300 feet from the face of the bank which was being worked. The first section was as usual laid with wooden blocks, the second with old rails, and the third with rocks. When the clean-up was made the middle section gave 9 ozs. more of gold bullion than both the others combined. If old rails cannot be had, strips of wood bound with iron are used, but are less durable and satisfactory.

At the Blue Spur Consolidated Gold Company's plant, Gabriel's Gully, New Zealand,† where the sluice is necessarily very short, most of the stones are first separated from the gravel, and the finer material is then passed over a sluice paved with trans-

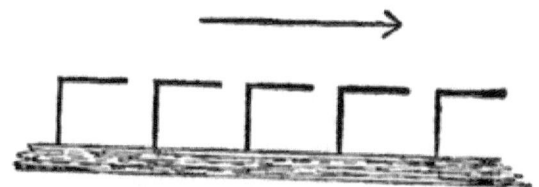

Fig. 11.
Scale = ¼.

verse angle-iron riffles, placed with the hollow side facing down stream (Fig. 11, in which the arrow shows the direction of the stream); these iron riffles are placed 2 or 3 inches apart. Below the section containing these riffles, there is a false bottom to the sluice, formed by an iron plate perforated with small round holes,

* *Gold, Silver and Mercury in the United States*, vol. ii., p. 218.
† *Report of the Mining Commissioner of New Zealand*, 1891.

through which some of the water and the finest particles of the gravel fall on to cocoa-nut fibre matting, laid on the true bottom of the sluice. Here the fine gold is caught, the principle being similar to that used in undercurrents.

The sluice is often divided into two by a median longitudinal partition, so that one side may be at work while the other is being cleaned-up or repaired, both sides being sometimes worked when water is very plentiful. There are usually unpaved rock-cuts above the sluice, leading to it from the places undergoing the process of piping. These rock-cuts are rarely supplied with mercury, and very little gold is usually caught there. The sluice may be placed above the tunnel, or in the tunnel itself (one way of preventing unlawful cleaning-up), or below it. In the case of the North Bloomfield Mine, the irregular slates, dipping at a high angle, forming the floor of the tunnel, were used as natural riffles for the lower part of the sluice, and thus all cost of wooden frames, pavements, &c., was saved, but the floor of the tunnel was lowered by 3 feet, and deep holes worn in it, after 22,000,000 cubic yards of gravel had passed through it.

The length of the sluice, if capital is not lacking, depends on the cost of construction and of the maintenance, as compared with the value of the gold saved owing to the increased length of the system. The length may be diminished by a plentiful use of drops, grizzlies and undercurrents, all of which are described above under the head of shallow placer sluicing; they are made of proportionately large size in hydraulicking. Coarse gold is of course soon caught, but fine gold may successfully evade all the riffles of a long sluice. The Spring Valley Mine has three parallel lines of sluices, each 2½ miles in length, and it is estimated that 95 per cent. of the gold contents of the gravel is caught.* This length is unusual, the average not exceeding about 1,000 feet. The cost of construction is from $25 to $35 per box (of 12 feet long) at the larger Californian mines, and less in New Zealand. Here, by a number of contrivances, especially by eliminating all but the smallest stones by means of grizzlies or trommels, the length of the sluices has been greatly diminished, but American engineers regard the retention of the stones as desirable, owing to their action in assisting disintegration.

Grade of the Sluices.—The grade depends on the available fall of the ground and on the character of the material to be washed. The minimum is from 2 to 4 inches per box, such low grades being sometimes enforced by the nature of the ground, sometimes adopted from choice if the gravel is light, the gold fine, and water plentiful. With these low grades, however, disintegration is slow and incomplete; stones, unless they are small, cannot be sluiced; large ones block the sluices and must be removed by hand, and the "duty" of the water, as regards sand, is

* *Ninth Report Cal. State Min.* (1889), p. 129.

greatly decreased. The 6-inch grade is that most generally used, but as much as 12 inches per box, where it can be obtained, is regarded as desirable in good American practice. Higher grades than this are unusual. Steep grades effect disintegration rapidly, thus shortening the length of the sluice, and enable all but the largest rocks to be sluiced, but less gold is then caught and a more plentiful use of undercurrents is necessary. In New Zealand an idea is gaining ground that sluices have been made too narrow, and that if the width is greatly increased and the grades diminished, and the depth of the current thus reduced to a minimum, the losses of gold will be reduced. Under such conditions all rocks must be removed by grizzlies, and in any case the cemented gravels, so common in the Californian blue leads, could not be treated in this way. It is considered necessary to have a sufficient depth of water to cover the largest boulders to be sluiced, but it undoubtedly diminishes the amount of gold saved if the water is more than two inches deep. Nevertheless, a depth of water of from 6 to 9 inches is not unusual. Where poor or top gravel is being "piped," it is worked off as rapidly as possible, and with less regard to the percentage of gold saved than when rich stuff is treated, but with more regard to the number of cubic yards treated.

The Use of Mercury.—Mercury is added in great quantities several times a day in America on the principle that "the more quicksilver added the greater are the chances of catching the gold," but less frequently and in less quantities in New Zealand. In some Californian sluices from 2 to 4 tons of mercury are in use at once. The feeding is regulated by the appearance of the amalgam in the sluice, the additions being made near the head-box and in the undercurrents. The loss of mercury is usually from 10 to 15 per cent. of the amount used per run. When cemented gravels are being treated, owing to the extra amount of trituration required, the loss may be as high as 30 per cent. These losses are the more serious, for the reason that amalgam is more easily lost than pure mercury, so that a heavy loss of mercury denotes a heavy loss of gold.

Cleaning-Up.—The process does not differ from that described under the heading of shallow placer mining. It is advisable not to defer the clean-up too long as losses of amalgam are caused by the wearing of the riffles. Usually from 50 to 95 per cent. of the total yield of amalgam is caught in the first twenty or thirty boxes, which are cleaned-up frequently. The following table[*] shows the percentage yield of the various sections of the sluices, &c., at the North Bloomfield mine, California, for the year 1877-8:—

Total yield, ... $311,276.20.

[*] *Ninth Report Cal. State Min.*, p. 131.

Near bank, from rock cuts in mine (all in gold dust, no quicksilver being added in the rock cuts),	4·57	per cent.
Sluice in tunnel (1,800 feet),	86·26	,,
Tunnel below sluice (6,000 feet), see p. 78,	4·50	,,
Cut below tunnel (200 feet),	0·81	,,
Tail sluices (300 feet),	1·21	,,
From seven undercurrents,	2·65	,,
	100·00	

The first undercurrent caught five times as much as the sixth, and nearly three times as much as the seventh, which was of double size. The yield of the seventh ($947) induced the Company to add another undercurrent. This mine affords an example of the difficulty of catching fine gold. The gold loss was unknown, but was believed not to exceed 5 per cent. of the contents of the gravel.

The bullion obtained by retorting the amalgam from the sluices is finer than that from quartz mills, and is sometimes 990 fine in Australia, although Californian placer gold is often as low as 850 fine. The remainder is mainly silver, but copper, lead, iron, and some of the minerals existing in the gravel also occur. The amalgam from the head of the sluices yields finer gold than that caught lower down and in the undercurrents.

Tailings.—The disposal of the tailings is one of the most important points to be considered in hydraulicking. Where there is not sufficient fall to enable the tailings to be removed from the lower end of the sluice without pumping, hydraulicking is impossible. The tail-sluices usually terminate on the side of a cañon, in a river, or in the sea. The enormous amount of loose sand and gravel, delivered from the hydraulic mines of Placer County, California, and the neighbouring counties into the Yuba and Feather rivers prior to 1880, filled up their beds to such an extent that in rainy weather disastrous floods ensued, and much valuable agricultural land was buried beneath sterile drift deposits and rendered worthless. The farmers thereupon took action against the Mining Companies and obtained a perpetual injunction forbidding them to discharge their tailings into these rivers. The result has been to stop hydraulicking in these districts, and the efforts to work the deep leads more extensively by drifting, or on the other hand, to impound the tailings by dams made of brushwood, or to return them to their original position, have not resulted in unqualified success. Consequently the gold winning industry has not been maintained on the extensive scale it had assumed prior to the action of the courts.

The Yuba and Feather rivers, in which it was estimated that from 750,000,000 to 800,000,000 cubic yards of gravel were contained, have continued during the last ten years to carry this material lower down and to distribute it over the level ground of the plains. The result has been that ground-sluicing on a tremendous scale has been carried on, and the tailings in the upper parts, enriched by the removal of the valueless sands, will probably pay to work over again in the immediate future. The impounding of tailings behind brushwood dams is now believed to afford a solution of the difficulty in California, and hydraulicking will probably be recommenced on an extensive scale before long.

Drift and Shaft Mining.—The problem of reaching the rich leads lying in the gutters of the old river channels, without removing the superincumbent masses of poor material, was attacked in California before hydraulicking had been invented. The method of following rich beds which passed into the hill-side by means of timbered tunnels was practised before the upper beds were suspected of containing any gold. The rise and rapid progress of hydraulicking probably checked the development of this tunnel or drift method between the years 1855 and 1880, but it is now the most important branch of the placer mining industry in California, while, as already stated, the analogous method of shaft mining has always been almost exclusively in use in Australia. Although hydraulicking was used wherever practicable after its introduction, many deposits occur where it cannot be employed. Where the upper gravels are thick and almost sterile, and in particular where the pay-gravel is covered by a great depth of lava (as for example at Table Mountain, California), drift mining must be resorted to. Want of sufficient grade, or of dumping ground, or of water supply, may also render hydraulicking impossible.

At first work was done in California only on those leads which were intersected by one of the existing cañons, so that the drift could be carried in paying ground from the start. In later times, when the country had been more carefully surveyed, and the probable courses of the Pliocene rivers roughly indicated, efforts were made to reach leads at points far from any outcrop. In order to do this, a tunnel is driven in the bed-rock from the nearest cañon to a point just below the gravel which it is proposed to mine (Fig. 9, p. 64). Formerly tunnels were often driven when the proofs of the position, depth and value of the channel rested on evidence which would now be deemed insufficient. Consequently they sometimes completely failed in their object. Instances are on record of tunnels being driven hundreds of yards in directions in which there was absolutely no chance of encountering a river channel. The tunnel was frequently put in too high, so that it was necessary to sink a shaft at the inner end of the drift in order to reach the pay-gravel, and thus the extra expense of raising water and gravel vertically for some distance was incurred. During the past few years, the location of the ancient channels, even at distances of several miles from the nearest outcrop, has been determined with some precision by engineers who have made a special study of the subject.

If the bed-rock tunnel is successfully placed, so that, while there is a fair grade outwards throughout its length, its inner end is a few feet below the lowest point of the channel, the succeeding operations are much cheapened and simplified, as "upraises" to the gravel then serve to drain the workings and to deliver the pay-dirt into cars, which convey it to the tunnel mouth. The methods of cutting out the gravel bear a strong resemblance to those used in coal mining, if the channel is wide and of fairly uniform value, but the gravel is of course frequently tested by panning. A detailed description of these methods of mining the gravel, however, may with more propriety be given in treatises on mining, than in this volume.*

The pay-gravel is carried to the tunnel mouth in cars, which are propelled by hand, by horses, or by steam power, according to the magnitude of the work. At Bald Mountain, California,† the gravel is brought through a tunnel 7,000 feet long in a train of eighteen cars, each holding 2 tons, by a locomotive which performs the trip in five minutes. The cost of transportation at this mine is stated to be, by man-power 21 cents per carload (of 2 tons), by mule-power 9 cents per carload, by steam $4\frac{3}{4}$ cents per carload. Outside the tunnel the gravel is dumped into bins, whence it is delivered, by gravity if possible, either to the sluices or to the batteries. In many cases the gravel is cemented into a conglomerate which is too tough to be easily disintegrated in the sluices. It is then passed through "cement mills," which closely resemble the stamp battery to be described in the next chapter, the chief differences to be noted being in the facilities for delivery. Double discharge mortars are used, and the screens are very coarse, the mesh being usually about $\frac{3}{16}$-inch, but varying up to $\frac{1}{2}$-inch in diameter. One battery of ten stamps, each weighing 950 lbs., making 94 drops of 9 inches in height per minute, will crush about 40 or 50 tons of gravel in ten hours so that it will pass through a $\frac{3}{16}$-inch mesh screen. Mercury is put into the mortar, and most of the gold is usually caught there on amalgamated copper plates, but copper plates outside the mortar are also used as in quartz-milling, and rubbers are employed to brighten the gold. If well-arranged plates are laid down, the number of sluice boxes which can be added with advantage is very small, a length of from 50 to 300 feet being used, the former limit being most common. No attempt is made to save the auriferous magnetic sands and sulphides which these conglomerates usually contain.

In cases where cement mills are not required, the gravel is washed in sluices which differ little from those already described. The boxes are not more than from 18 inches to 24 inches wide

---

* A detailed description is given by Russell L. Dunn, in *Eighth Report of Cal. State Min.*, 1888, p. 736.
† *Ninth Report of Cal. State Min.*, 1889, p. 118.

and deep, and the series is seldom more than 300 or 400 feet long Iron riffles are most in favour. Where the amount of gravel to be washed is small, or the water is scarce, the gravel is allowed to accumulate for some time and the water stored in a tank or reservoir. It is in some cases a great advantage to keep compacted gravels exposed to the air during a few months before washing them, as they "slack" and disintegrate under the influence of the weather, and subsequently are more easily treated, while for a similar reason, tailings are sometimes impounded, and re-washed after some time has elapsed. The disintegration of cemented material, which has been "slacked" by exposure to the weather, is usually completed in a cement-pan. This is a cast-iron pan with perforated bottom, and with a gate in the side for the removal of boulders, which are mostly barren and are separated from the auriferous material by this system, instead of being crushed and mixed with it, as is the case when stamp-mills are used. In the pan, four revolving arms, furnished with plough-shares, break up the gravel, which is carried through the apertures in the bottom by a stream of water, and falls into the sluice. A pan of 5 feet in diameter and 2 feet in depth will treat from 40 to 120 tons per day, according to the nature of the gravel. (See also p. 60.)

The Hydraulic Elevator.—In this machine a jet of water under high pressure forces water, gravel, and boulders up an inclined plane, and delivers them all at the head of the sluice, which may be as much as 100 feet above bed-rock. The differences in construction between the machines made in Australia, New Zealand, and the United States are only matters of detail. They consist essentially of an upraise pipe, usually of wrought iron, having a diameter of from 12 to 24 inches, which terminates below in an open conical funnel; a hydraulic nozzle, delivering water under the pressure given by a head of from 100 to 500 feet, projects into this funnel, and sand and gravel can also enter round the sides or through a special orifice. The inclination of the upraise pipe is usually from 45° to 65°. The top of the upraise pipe is turned over and terminates above a sluice, into which the gravel falls and is washed in the ordinary way. The subjoined figure shows the arrangements at the base of the upraise pipe of the elevator manufactured by Mr. J. Henry of San Francisco. In Fig. 12 Nos. 1 to 13 are castings, No. 14 consisting of wrought iron; a ball joint is formed by Nos. 3, 4, and 5, enabling the pipe bringing the water to be moved. The nozzle and the lower part of the upraise pump are sunk in a sump excavated in the bed-rock, and the gravel is washed down by any means (usually by a jet from an ordinary hydraulic nozzle) into this sump. The entrance to the upraise pipe is protected by a coarse grating which prevents large stones, pieces of wood, &c., from entering it. The force of water is enough to complete the disintegration of the gravel during its passage through the upraise pipe, so that a short sluice is enough to effect the washing proper. If the excavation is carefully arranged, it may be kept funnel-shaped, so that the elevator, once placed in a sump, may be worked there permanently without being moved. When the pit is large enough, the washing may be done inside it, only the tailings being raised to the surface by the hydraulic elevator. This was done at Oroville in 1880,* the flume to deliver the tailings into the river being about 500 feet long. The head of water required varies according to the vertical height through which the gravel must be raised; a head of about 70 feet is required for every 10 feet of vertical upraise.

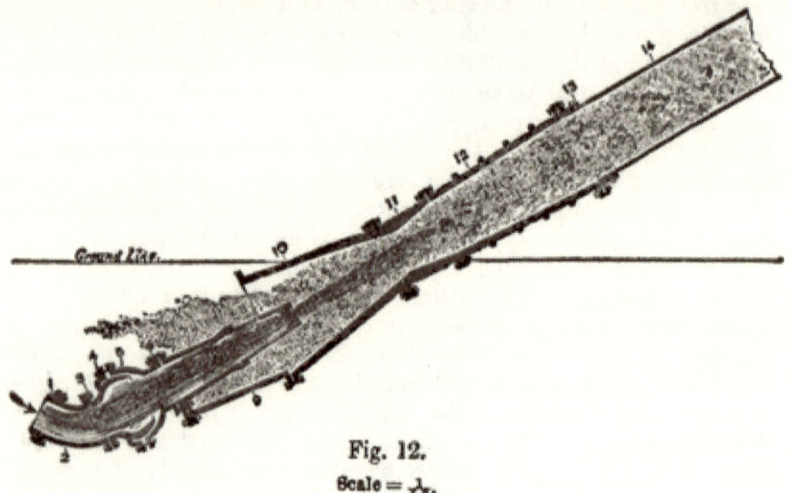

Fig. 12.
Scale = $\frac{1}{15}$.

Since it is just as expensive to raise water as gravel, arrangements must be made to deliver as much gravel into the sump as can possibly be raised by the jet, otherwise the expense per cubic yard will be greater, and there will be too much water with the gravel for satisfactory treatment in the sluices.

Wherever the necessary head of water is available, the hydraulic elevator is now recognised as the best method of working flat placers, river-bars, &c., or any deposits which are either below the water level of the district, or which have not sufficient fall for the disposal of the tailings by gravity. It is in wide use in California and New Zealand. The following instances of work in both countries may be given:—At the Blue Spur Consolidated Gold Mining Company, Gabriel's Gully, New Zealand,† tailings, which have accumulated close to the sea on the foreshore, are sluiced in this manner. The vertical upraise is 60 feet, the angle of inclination of the upraise pipe

---

* *Prod. Gold and Silver in the U.S.A.*, 1880, p. 15.
† *New Zealand Mining Commissioners' Report*, 1891, p. 65.

being $63°\cdot 5$; about 480 tons of gravel are raised per shift, the head of water used being 400 feet, while the amount of water used in each elevator is seventeen government heads. The sluice is short, and has an inclination of only $3\frac{1}{2}$ inches in 12 feet; the upper parts are fitted with transverse, patent ⌐-shaped, angle iron riffles, in which the angle faces up stream (see Fig. 11). The lower parts of the sluice have a false bottom of wrought-iron plates, perforated with round holes; beneath these plates is the true bottom of the sluice, covered with cocoanut matting in which fine gold is caught. The tailings are discharged into the sea.

At Quartz Valley, Siskiyou County, California,* on hard ground, where the elevator was first used, it took forty-three days to work out a piece of ground 300 feet by 250 feet, which was of an average depth of 18 feet. The bank was washed down with 600 miner's inches of water, and went to the elevator through a 30-inch bed-rock flume, which had a grade of 5 inches in 12 feet. The water and gravel were raised through a 20-inch elevator pipe without any contraction at the throat. It was set at an angle of 40°, and the pipe was 42 feet long, the vertical upraise being thus 28 feet. The force used was 1,000 inches of water with a head of 230 feet, delivered through a 6-inch nozzle, and the gravel was emptied into a sluice 6 feet by 3 feet, with a grade of $1\frac{1}{2}$ inches in 12 feet. When 3,500 inches of water were running in this sluice, they could not carry off all the gravel raised by the elevator. The work was done without any delay from stoppage of the machine, and there were no repairs, the wear of the elevator being very little.

Economic Conditions in Placer Working—*Shallow Placer Deposits.*—In work by individuals the results differ greatly, both according to the strength and skill of the worker and to the contents of the gravel. Under the best conditions of climate, a strong, well-nourished, American digger may be able to raise by the shovel from 10 to 12 cubic yards of gravel per day and throw it into a receptacle 3 feet above the ground. Native labour cannot be expected to effect so much, and in French Guiana it is reckoned that only about half a cubic yard of earth per man per day can be shovelled into the sluice. If the workman must wash the gravel, as well as raise it, much less can be accomplished. For an active man, it is a fair day's work to dig and wash from fifteen to twenty pans of dirt, the amount treated thus not exceeding about 10 cubic feet. On the other hand, with the cradle, the output may be from $1\frac{1}{2}$ to 2 cubic yards per man in a day, while with the long-tom it may rise to 3 or 4 cubic yards per man. In the Siberian trough, only from 1 to $1\frac{1}{4}$ cubic yards can be treated by one worker per day, but a large percentage of the gold is believed to be saved in this

* Egleston's *Silver, Gold and Mercury in the United States*, 1890, vol. ii., p. 307.

apparatus. The minimum contents in gold, which will make the gravel worth treating, depends, of course, on the cost of labour in the country in which the deposit exists, since few men will continue to work for less reward than they could obtain in other employments. The wages of placer miners in various countries is as follows:—In California, from $2½ to $4 per day; in Colorado, Montana, Dakota, and the neighbouring States, from $2 to $3½ per day; in Australia, from 6s. to 10s. per day; in French Guiana, the negroes and coolies are paid about 3s. 4d. per day, exclusive of rations. The wages in Siberia are given below. In California, the Chinese are content to earn about 75 cents or 3s. per day. In early times, in California and Australia, when the virgin shallow deposits were being worked, large sums were often realised by individual diggers, cases being on record in which 5 ozs. of gold were obtained from one pan of bedrock-scrapings lying under heavy gravel, and earnings of several hundred dollars per day were not uncommon. The rich gravels, from which these results were obtained, are now all worked out, and instances are rare in which more than $1 or $2 per day can be earned with the pan.

Concerted work, with the aid of the sluice, is much more effective; in California gravel containing about 1 pennyweight of gold per cubic yard is worked at a profit, the dirt being lifted into the sluice by hand-labour, and the tailings removed by sluicing with water; at Ballarat, in Australia, where the gravel is raised to the surface from underground workings through a vertical shaft several hundred feet deep, and subsequently washed, 12 grains of gold per cubic yard of material pay for the treatment, while in Siberia, as stated below, the cost is even less. In French Guiana, the unhealthiness of the climate and the cost of supplies render it impossible to work gravel containing less than about 3 pennyweights of gold per cubic yard.*

In Siberia,† the distance of the workings from the nearest town, and the traditions of the industry, require workmen to be hired by the year, or, in cases where no work is attempted in winter, for the season. The best workmen, employed in actual washing, receive 1 rouble (about 2s.) per day; the diggers and carters 75 copecks (1s. 6d.), and the women and boys from 40 to 50 copecks (9d. to 1s.) per day: rations are also distributed in mines far from towns. Sometimes piece-work is done, the price paid being from 2¼ to 2¾ roubles per cubic *sagen*, or from 4d. to 5½d. per cubic yard (!); the horses and men are then both supplied by the workman. The total cost of treatment of the gravel varies greatly with its geographical position. On the banks of the Lena, where the season only lasts for five months, it is estimated that the gravel must contain 2 *zollatniks* of gold per 100 *pouds*, or 4⅓ dwts. per cubic yard; but in the neighbourhood of Ekater-

* Frémy, *Ency. Chim.*, vol. v., L'or, p. 48. † *Ibid*, p. 47.

ineberg, the deposits in the bed of the Pechma, a tributary of the Obi, are worked when they only contain from 8 to 9 grains per cubic yard. This is less than that required in other countries in order that a profit may be made from similar deposits, viz., those in which the gravel must be handled at least twice.

Deep Placer Deposits.—The cost of hydraulicking and drift mining necessarily varies enormously with the conditions. In hydraulicking, it depends largely on the magnitude of the operations. With large quantities of water available at a cheap rate, big banks of soft gravel, and a well-constructed and convenient tailings-tunnel, the cost has been reduced to from 2 to 3 cents per cubic yard in California, while the average cost there in 1884 was about 10 cents, and only in exceptional cases amounted to as much as 50 cents per cubic yard. This is without reckoning interest on capital expended. As already stated, hydraulicking is rarely possible in Australia, and then only on a small scale. A case in Australia is cited, however, by Mr. G. Warnford Lock, in which three workmen washed 150 cubic yards of gravel per day with 1,875 cubic yards of water, the jet being directed against a bank about 30 feet high. The cost of working was £2 per day, being made up as follows:—

	£	s.	d.
Three men at 8s.,	1	4	0
Water,	0	10	0
Maintenance, repairs, &c.,	0	6	0
	£2	0	0

The cost was, therefore, a little more than 3d. per cubic yard, and a saving of 1½ grains of gold per cubic yard would, in this case, pay expenses. The cost of production in the year 1875 of 1 ounce troy of gold in hydraulicking is given opposite for two Californian mines,[*] viz., the La Grange Company, which worked on a thin bed of gravel, and the North Bloomfield Mine, in which the bank was 250 feet high.

In drift mining, the cost of treating uncemented gravel is less than that of treating conglomerate. The cost at the Hidden Treasure Mine, on the Damascus Channel, Placer County, California, was $0·9236 per carload of 1 ton in the spring of 1888, the yield being $1·2347 per ton. Here the ground could all be "picked" down, requiring no blasting, and the gravel could be sluiced without crushing, but the timbering was costly. The working of this mine was exceptionally cheap, as, under ordinary conditions, a cost of from $1·50 to $1·75 per carload is as low as can be expected, and in many mines the cost rises to $3 per carload. The cost of milling cemented gravel is from 20 to 40 cents per ton, the capacity of the mills being from 5 to 12 tons per stamp in twenty-four hours.

[*] *Ninth Report Cal. State Min.*, 1889, p. 133.

	La Grange		North Bloomfield	
	Per cubic yard of Gravel.	Per ounce of Gold.	Per cubic yard of Gravel.	Per ounce of Gold.
Water,	0·8 cents.	$1·43	0·75 cents.	$2·09
Labour,	3·6 ,,	6·85	1·40 ,,	3·93
Materials,	1·0 ,,	1·81	0·32 ,,	0·88
Explosives,	...	...	0·35 ,,	0·98
Blocks,	...	...	0·18 ,,	0·50
General expenses,	...	...	0·25 ,,	0·70
Contingent,	0·6 ,,	0·26	...	..
Salaries of officials,		0·94	...	...
Taxes,		0·09	...	...
	6·0 cents.	$11·38	3·25 cents.	$9·08

One cubic yard usually contains from 1¼ to 1⅝ tons of gravel.

The relative costs of working the various classes of gold deposits in California, by methods applicable to the respective classes, are given by John Hays Hammond,[*] as follows:—

Cost per Ton of Material treated.

1. Auriferous vein deposits (mining and milling), . $3 to $10.
2. Drift mining, ... 75 cents to $4.
3. Miner's pan, ... $5 to $8.
4. Rocker or cradle, ... $2 to $3.
5. Sluicing, ... 75 cents to $1.
6. Hydraulic method, - . . . . . 1¼ to 8 cents.

CHAPTER VI.

QUARTZ CRUSHING IN THE STAMP BATTERY.

Primitive Methods of Crushing and Amalgamation.—In all countries, when the richest alluvial deposits have been worked out or have all been taken possession of, efforts have been made to extract the gold from the various hard, auriferous materials met with in veins. The earliest machines used for the purpose seem to have been stone mortars, in which the gold quartz was crushed by stone-hammers, or by large rocks raised by levers and let fall, whilst the fine material was subsequently washed out with water. The heavy particles of gold then settled to the bottom of the mortar, the lighter, worthless material being washed away. Hollowed-out stone mortars, in which this

[*] *Ninth Report Cal. State Mine*, 1889, p. 105.

operation was conducted, have been found in Wales, in Central America, in the Pyrenees, and in Transylvania. Diodorus Siculus, the Greek historian, has given a detailed description of the method of gold-quartz reduction employed in the mines of Upper Egypt 1,900 years ago. The ore was first reduced to coarse powder in mortars, then finely crushed in hand-mills resembling the flour-mills of the present day, and finally washed. In order to separate the pulp from the uncrushed lumps, the Egyptians, in common with other races in ancient times, employed sieves, but in extracting the gold from auriferous sands they used raw hides, on which the flakes of gold were entangled.* These devices closely resemble those still in use in many parts of the world.†

The date of the first use of mercury for amalgamation is unknown. Pliny mentions the fact that mercury will extract gold from its ores,‡ and it has no doubt been used for this purpose for the last 2,000 years. It is mentioned by Biringuccio in his treatise (which was published in Italian in 1540), and had apparently been a secret art for some time previously. Mercury was introduced into Mexico as a means of extracting the precious metals by Bartolomé Medina in the year 1557,§ and its use doubtless soon spread to Peru and the neighbouring countries. When Barba wrote in 1639 there were three amalgamation machines in use in Peru,‖ viz., the *mortar* (used in the Tintin process), the *trapiche* or Chilian mill, and the *maray*. The *mortar* was hollowed out in a hard stone, the cavity being about 9 inches both in diameter and in depth. The ore was triturated with water and mercury in this mortar with the aid of an iron pestle, while a stream of water, flowing through, carried away the crushed particles. The slimes were roughly concentrated, but most of the gold remained in the mortar with the mercury.

The *maray*, although equally primitive, was probably of greater capacity; it makes use of the principle employed in the bucking hammer. About the year 1825 Miers¶ saw it in Chili, where it is probably still in use. It consists of a flat or slightly concave stone, 3 feet in diameter, on which lies a spherical boulder of granite about 2 feet in diameter. This is rolled to and fro by two men seated on the ends of a long, wooden pole, which is

---

* Beckmann's *Hist. of Inventions*, vol. ii., p. 334.
† An interesting account of some primitive methods of treating gold quartz now employed by the Chinese is given by Henry Louis in the *Eng. and Mng. Journ.*, Dec. 31, 1892, p. 629, from which it appears that these methods bear points of resemblance to those of the Egyptians.
‡ *Nat. Hist.*, xxx., cap. vi., sec. 32.
§ Georg Agricola's *Bermannus* . . . übersetzt von Friedrich August Schmid, 1806, p. 49.
‖ *Arte de los metales*, lib. iii, cap. 16.
¶ *Travels in Chile and La Plata*, by John Miers, vol. ii., p. 390.

firmly fixed to the boulder. Ore, water and mercury are ground together in this machine, and then washed down.

The *Chilian mill* closely resembles the edge-runner mill of the present day, which is used for grinding and mixing mortar, &c. The Peruvian *trapiche* had a similar circular bed of hard stone, but only one stone-runner, which was driven by mules. The Chilian mill is still used to prepare ores for treatment in the *arrastra*, which was not mentioned by Barba, and may perhaps be regarded as an outcome of the *trapiche.*

The **Arrastra** was also one of the earliest crushing machines in use in America, being introduced at the same time as the Patio process—*i.e.*, about 1557—and is still in wide use in Mexico, although chiefly in the treatment of silver ores by the Patio process. It is a circular, shallow, flat-bottomed pit, 10 to 20 feet in diameter, and paved with hard, uncut stones. Granite, basalt,

Fig. 13.

and compact quartz are all used for the rock pavement, which is made 12 inches thick, and is either placed on a bed of well-puddled clay from 3 to 6 inches thick, or set in hydraulic cement, so that no chink or cranny remains, into which the mercury or amalgam can find its way. In the centre a vertical shaft revolves, carrying two or four horizontal arms, to each of which is attached a heavy stone by thongs of bullock hide, or by chains. These grinding stones weigh from 400 to 1,000 lbs. each, their forward ends being about 2 inches above the floor, whilst their other ends drag on it. They are moved by mules walking round outside the arrastra, or by water- or steam-power, the speed varying from four to eighteen turns per minute. Fig. 13 represents an arrastra of the simplest description; at the front the stones forming the edge have been removed, so as to expose a section of the rock pavement.

Ore of about the size of pigeons' eggs is introduced, enough water being added to make the pulp of the consistency of cream, and mercury is sprinkled over it after most of the grinding has been done. When the ore cannot be ground any finer, more water is added to dilute the pulp, the mule is driven slowly for half an hour to collect the mercury, and the pulp is then run out and another charge shovelled in. An arrastra of 10 feet in diameter takes a charge of 500 to 600 lbs. of ore, and treats about 1 ton per day of twenty-four hours. The amount of wear sustained by the grinding stones is equal to from 6 to 10 per cent. of the ore crushed. The output is so limited that the use of the arrastra has never been general outside Mexico, although it has been used in almost every district in the United States for a short time after the commencement of quartz mining in the particular locality.

It is often stated that the results obtained, so far as the percentage of gold extracted from refractory ores is concerned, cannot be equalled by any other amalgamating appliance, and that the Mexicans, using the arrastra, formerly treated at a profit ores which hardly yield any gold to the stamp mill, or even to the amalgamating pan. In consequence of its power of saving gold and the cheapness with which it can be erected and worked, the arrastra is still valuable for prospecting. In preparing the bed for this purpose, every care must be taken that the surface is even, and all joints properly cemented, or else the mercury and amalgam will in great part find their way into the foundations.

The reasons for the high extractive power of the arrastra, when treating certain ores, are no doubt to be found in the extreme fineness to which the ore is reduced, and in the prolonged contact between the ore and mercury, which is maintained while they are being ground together. Moreover, the grinding action of the dragged stones keeps the particles of gold bright and in a suitable condition for amalgamation, without exercising force enough to flatten and harden them. The relative advantages of grinding surfaces consisting of iron and stone are less certain, and probably vary with the nature of the ore in course of treatment. An instance is given by Colonel Harris,[*] in which stone was better than iron. In this case, at Cerro de Pasco, Peru, the old method of grinding ores in circular pans, by edge-runners of stone or granite, was found to entail a rapid wearing of the edge-runners, and, in order to remedy this, the runners were shod with iron. The returns at once fell off, and on careful trial it was found that the yield in the old machines was 15 per cent. more than in the new ones. Other similar instances are on record, but it must nevertheless be conceded that, with some ores, the presence of iron is necessary for good work, chlorides of silver and mercury being reduced by it, which would otherwise be lost in the tailings. The somewhat extravagant views often expressed, upholding the great superiority of the arrastra over its more modern rivals as an amalgamating machine for gold ores, are

[*] *Min. Journ.*, December 24, 1892, p. 1466.

perhaps hardly justified, since a direct comparison with stamps or pans has been possible in only a few instances.

One of the most remarkable of these instances is that afforded by the experience at the Pestarena Mines, Val Anzasca, Italy. The ore of this mine contains about 20 per cent. of pyrites, and is of somewhat low grade, rarely containing as much as 15 dwts. of gold per ton. Efforts were at first made to treat it by means of stamps and amalgamated plates, with the result that only 65 per cent. of the gold was extracted. Better results attended the introduction of the Francfort mill, a modified arrastra, driven in this instance by steam; this mill is substantially a wooden pan, with dies and shoes of stone. The mercury was added to the pulp after it had been finely ground, and the amalgamation and grinding of the pulp subsequently kept up for seven hours. From ore containing 12·3 dwts. per ton, the mill extracted 10·2 dwts., or 83 per cent., for a considerable period of time, while in the year 1890 the average extraction was 79·4 per cent. There were twenty-eight mills, each of which treated 1,200 lbs. of ore per day. The stone bed is said to last ten months, and the shoes from six to eight weeks, the total cost of treatment being very low.

Successful work by the arrastra on gold ores is also being done at the present day at the Bote Mine, Zacatecas. The material treated is an auriferous silver ore, the gold being extracted in the arrastra, and the silver subsequently obtained by means of the Patio process. Here the closely-fitting cut-stones forming the bottom of the arrastra are first coated with an amalgam of silver, and the ore then ground for twelve hours, mercury being added to it to the extent of one and a-half times the weight of the gold present, so as to form a "dry" amalgam. The extraction of gold in this case, however, is only 60 per cent., a result which is probably not better than that which would be obtained in pans and settlers, if economical conditions admitted of their employment. The necessity of keeping the amount of mercury low so as to obtain "dry" amalgam, and thus to prevent loss by leakage into the foundations, is one of the objections to the use of a roughly made stone-floor for amalgamation, the percentage of extraction being necessarily reduced by this precaution. There were about 100 arrastras at work in California in 1889, each treating from 1 to 3 tons of ore per day. They are used where only small quantities of high-grade ore are available for treatment.

At Smartsville, Nevada County, the arrastra has lately been applied to a new purpose, that of crushing and amalgamating hard cement-gravel obtained from a drift mine.* The cement is coarsely crushed by being passed through a Gates rock-breaker, and is then charged into four arrastras, each of which is 12 feet in diameter and 3 feet deep, and capable of containing from 5 to

* *Eleventh Report Cal. State Min.*, 1892, p. 315.

9 tons of gravel. The grinding is effected by four blocks of diabase, each weighing from 600 to 1,000 lbs.; the rate of revolution is 14 times per minute, the time of grinding being one hour. A tablespoonful of mercury is fed in with each charge, and the total loss is only 10 per cent. of this. The pavement costs $40 and lasts for six months. At the end of the hour, a gate is opened in the side of the arrastra and the charge run out into a sluice, 200 feet long, where the mercury and amalgam is caught by means of riffles. The capacity of one arrastra is 50 tons of hard cement per day, or 75 to 90 tons of soft surface-gravel per day. The cost is from 6 to 8 cents per ton, and more gold is extracted than when a stamp battery was used at a cost of from 20 to 40 cents per ton. One man attends to the four arrastras, and another to the Gates crusher.

Iron Prospecting Arrastra.—The difficulties of getting suitable stone for the arrastra in some localities, and of constructing the pavement so as to prevent loss of mercury, have led to the introduction, for prospecting purposes, of wrought-iron pans with steel bottoms on which stone drags are used. One of these so-called arrastras is 3 feet in diameter and 12 inches deep, with an axis revolved by bevel gear, which is placed below the bottom. The upper part of the axis carries a yoke to which stone drags are attached by chains. It is doubtful if the steel bottom is as good for amalgamation as the stone pavements, but the losses of mercury by leakage are certainly avoided. Perhaps a thin pavement of flat stones might be put inside such a prospecting pan with advantage, when working on some ores.

The Stamp Battery.—The stamp battery was, no doubt, evolved from the pestle and mortar, but was not introduced until a comparatively recent date. Beckmann[*] states that mortars, mills and sieves were used exclusively in Germany throughout the whole of the 15th century, and in France stamps were unknown as late as the year 1579. It has been suggested that the origin of stamp mills was probably due to the gunpowder manufacture, and it seems certain that in 1340 stamp mill, used in connection with this industry, existed in Augsburg, and that Conrad Harscher, of Nuremburg, owned one in 1435. They were first applied to the gold industry at the beginning of the sixteenth century, a doubtful record stating that they were introduced into Saxony by Count von Maltitz in 1505, whilst in 1519 the processes of wet-stamping and sifting were established in Joachimsthal by Paul Grommestetter, who had some time previously introduced them at Schneeberg. The improvements gradually spread through Germany, and detailed descriptions and drawings of the apparatus are given by Agricola[†] in 1556, from which it appears that the earliest battery consisted of a single stamp,

---

[*] *Hist. of Inventions*, vol. ii., p. 334.
[†] *De re Metallica*, vol. viii., p. 247.

raised by means of two levers fixed to the axis of a wheel. Dry crushing was at first employed, but the great production of dust soon led to the use of water in the mortar. In some Hungarian mines, Bennett H. Brough * recently saw some primitive stamps in use, resembling those drawn by Agricola, weighing only 100 lbs. each, and having stamp heads made in some cases of hard blocks of quartzite. At that time, in cases where the conditions of water supply were favourable, these stamps were able to treat with profit an ore containing as little as $2\frac{1}{2}$ ozs. of gold to 50 tons of ore, and, at Zell, in the Tyrol, they were able to treat a slaty material containing 1 oz. of gold to 50 tons of ore. Such economical work is seldom possible with the heavy modern Californian stamp under the most favourable circumstances.

The German stamp has a rectangular stem made of wood or, latterly, of iron, with an iron head, the total weight never exceeding 300 to 400 lbs. It was introduced almost unchanged into France, Cornwall, and, after the discovery of gold in 1849, into the United States, but has given place in all new districts to the Californian stamp, and need not be fully described here. In 1850, the first Californian stamp-mill was erected at Boston Ravine, Grass Valley. The stamps consisted of tree-trunks shod with iron, and the framework was constructed of logs.

The ordinary method of reduction and amalgamation of gold quartz in a stamp battery now consists of the following operations :—

1. The ore is broken down to a moderate size, varying from that of a man's fist to a nut, by passing through the jaws of a stone-breaker, or by hand hammers.

2. The ore is then fed into the mortar-box of a stamp mill, where it is pulverised to any required degree of fineness. In wet crushing, a stream of water is introduced also, and the blows of the stamps splash the water and pulp against screens set in the side of the mortar, the finely-divided ore being ejected in this way. In some cases the mortar-box is partly lined with amalgamated copper plates, by which some of the gold is caught and retained, mercury being in this case usually fed into the mortar-box with the ore and water.

3. On issuing from the battery, the pulp is allowed to run slowly over a series of inclined, amalgamated, copper plates, by which a further percentage of the gold is amalgamated and retained.

4. The tailings are sometimes further treated by running over rough hides or blankets, by which some particles of gold and pyrites are retained, or the pyrites are separated from the valueless sands by concentration on some form of vanner or jig. These concentrates are subjected to further treatment, usually either by smelting or by chlorination.

* *Proc. Inst. Civil Eng.*, Session 1891-1892, part ii.

5. At intervals the gold amalgam is wiped off the copper plates, the excess of mercury separated by squeezing through filter bags of chamois leather, buckskin, or canvas, and the solid amalgam thus obtained is retorted so as to distil off the mercury, and the gold is then melted.

In the following pages much information has been derived from (1) "The Milling of Gold Ores in California," by J. H. Hammond, published in the *Eighth Report of the Californian State Mineralogist*, 1888; (2) the series of articles on "The Variations in the Milling of Gold Ores," by T. A. Rickard, published in the *Engineering and Mining Journal* in the years 1892 to 1895; and (3) "The Amalgamation of Free-Milling Gold Ores," by Louis Janin, Jun., published in the *Mineral Industry for* 1894.

The stamp battery must be regarded from two different points of view—viz., (a) as a crushing machine, (b) as an amalgamating machine, and it should be remembered that the modifications designed to make it a more efficient crusher often reduce its power as an amalgamator, and *vice versâ*. Rickard has pointed out that two typical methods are the Californian practice, for the treatment of "free-milling" gold ores, and the Colorado practice, for the treatment of ores, sometimes called "refractory," which, however, yield most of their gold when carefully treated in the stamp battery. The word "refractory" is better reserved for those ores which cannot be satisfactorily treated by direct amalgamation, whatever be the method adopted. The stamp battery is also used purely as a crushing machine in preparing certain silver ores (whether gold is present in addition or not) for treatment by other processes (*e.g.*, the Washoe, Reese River, and lixiviation processes), which properly appertain to the metallurgy of silver and will not be described in this volume, although a short account of pan-amalgamation is given on pp. 158-163.

The Californian practice consists briefly in crushing the ore and effecting its discharge from the battery as rapidly as possible. With this object in view heavy stamps are used, running very fast, with a small drop; the screen area is large and placed as low down as possible, and the mortar is made narrow, with nearly vertical sides. These arrangements all increase the output of the battery. There are usually amalgamated copper plates inside the mortar on the discharge side only. The method is suitable for ores containing coarse, free gold which is easily amalgamated, and which is caught largely on the inside plates, in spite of the short time during which the ore remains in the mortar. A small amount (1 to 5 per cent.) of pyrites, especially of clean iron pyrites, does not interfere with rapid amalgamation; but if the percentage of this mineral is high (10 to 20 per cent.), and especially if the easily decomposable variety of sulphide of iron (marcasite), or some sulphides of lead and zinc, or if compounds of arsenic or antimony or other base minerals are present, the amalgamation is greatly retarded or prevented, and the Colorado practice is resorted to. If the gold is fine, or if, for any other reason whatever, amalgamation is difficult, the Californian practice must be modified in the same direction.

The Colorado practice was first devised in Gilpin County, Colorado. In the early days of this gold-field, about the year 1860, after the arrastra had been discarded as being too slow, the fast-drop and shallow-discharge batteries, like those used in California, were introduced, and gave good results in working on the oxidised surface-quartz, 60 per cent. to 75 per cent. of the gold being extracted. As the mine workings got lower, however, the percentage of pyrites steadily increased, and the mills gave poorer and poorer results, until a return of only 30 per cent. to 40 per cent. was obtained. A return to efficient working was only made after a long series of costly experiments, which resulted in the present slow-working, long-drop stamps with a wide, roomy mortar and a very deep discharge, the lowest part of the screen being more than a foot above the dies, instead of only 6 inches as in California. Amalgamated plates were put inside the mortar on both feed and discharge sides. The object of these arrangements was to keep the ore in the mortar for a long time, so as to increase the chance of catching the gold on the inside plates. The duty of the stamps was of course greatly diminished, the output of a typical Colorado battery being only about 1 ton per stamp per day of twenty-four hours, while with Californian practice it is from 2 to 5 tons. Nevertheless the ultimate object is equally well attained by both types of battery—viz., the extraction of from 75 per cent. to 95 per cent. of the gold present, including that saved in the concentrates.

Ores containing more than 20 per cent. of sulphides, unless they are clean, cubical iron pyrites, usually come under the heading of true refractory ores, and cannot be treated by any kind of stamp-battery amalgamation. Their treatment is considered in succeeding chapters. The following is a general description of the machinery employed in stamp-battery practice. In the sequel the modifications adopted in different districts will be discussed, and the conditions for successful amalgamation of different classes of ore referred to.

1. **Rock-breakers.**—There are two classes of these machines in general use, viz.:—(a) Those constructed on the reciprocating-jaw principle, and (b) gyratory crushers. The Blake and the Dodge crushers are representative of the former class, and the Gates crusher of the latter.

The *Blake Crusher* is shown in section in Fig. 14. The rock is crushed between the stationary jaw, BC^1, and the swinging jaw, D, which is pivoted at E, and moved by the eccentric, F, through the toggles, J K. The swinging jaw, in order to be as light as possible, should be made of steel, cast hollow and braced by ribs, but is usually composed of cast iron. The machine works at about 250 revolutions per minute. At each revolution the moving jaw is advanced about ½-inch towards the other, and the lumps of rock which have dropped down between the jaws are broken; as the moving jaw recedes, the fragments slip lower down and are further crushed at the next advance, and this process is repeated until the ore is small enough to pass out at the opening at the bottom. The distance between the jaws at the bottom limits the size of the fragments, and this distance may be regulated at will by moving the wedge, L, or by changing the length of the toggles, J K. The capacity of the machine is great, being about 300 tons of ordinary rock per day of twenty-four hours in the case of the machine whose

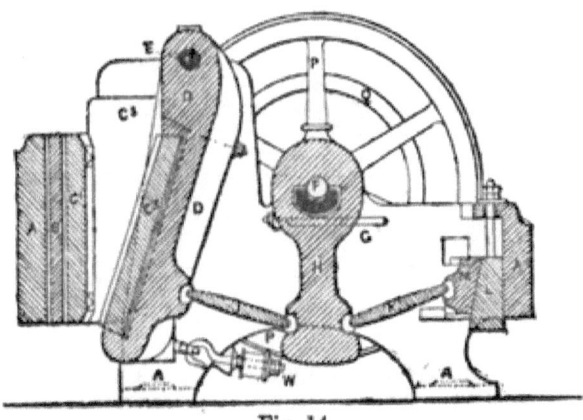

Fig. 14.

dimensions at the mouth are 20 inches by 10 inches, when the lower edge of the jaws are set to approach within 1½ inches of each other. The power required for this is stated to be 14 H.P.

Many modifications have been made in this machine by various makers since the patent expired, and of these the Blake-Marsden machine has come into more extended use than any other reciprocating-jaw crusher. The most important alteration is the pivoting of the moving jaw below instead of above, which is adopted in the *Dodge crusher* shown in Fig. 15. The effect of this arrangement is to make the product more uniform in size, and as there is little or no motion of the movable jaw at the delivery aperture, this may be made as narrow as desired, so that a finer product can be obtained, although it is at the expense of capacity. The Dodge crusher is more particularly recommended for fine crushing in concentration works, or where the product is to be subsequently passed through rolls. A preliminary treatment of the ore in a Blake crusher is desirable, where fine crushing by a Dodge is resorted to.

Authorities differ as to the relative advantages of the two positions of the pivot. A. Sahlin decides * in favour of placing the pivot below, assigning as a reason that more work is necessary to crush comparatively fine material than to break down large pieces of rock in the upper part, where the points of contact between the crushing jaws and the rock are few. The shortest stroke and great leverage should, therefore, he considers, be at the bottom, where the work is heaviest. On the other hand, W. P. Blake dissents from all these views.

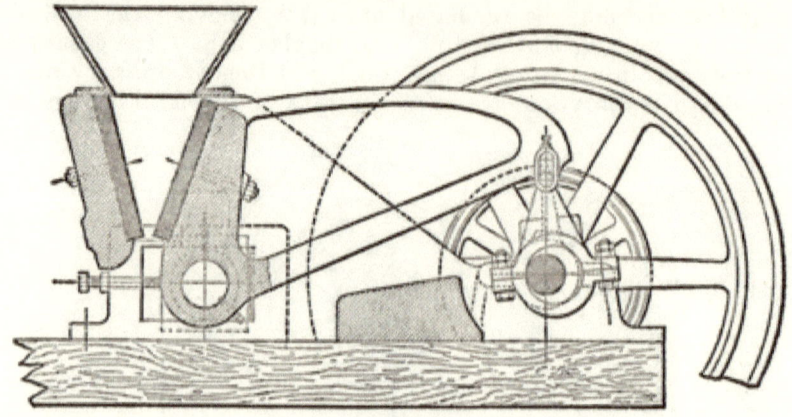

Fig. 15.

Gates Crusher.—This machine has been in use for fifteen years in crushing macadam, ballast, and iron ore, chiefly in the United States, but has not long been applied to crush gold ores. It is now largely used, however, both in America and in South Africa, being probably the most economical rock-breaker where large quantities of ore are being treated. It is shown in Fig. 16, and consists of a vertical shaft of forged steel, G, rotated at the bottom by a bevelled wheel, L, placed $\frac{1}{2}$ inch out of centre. At the top of the shaft is a chilled-iron breaking head, F, and the shell surrounding this is lined with twelve chilled-iron, concave pieces, E. These form the crushing faces. The shaft, G, has a gyratory motion imparted to it by the eccentric box, D, attached to L, and the rock is thus crushed, without grinding, between the head and liners. The distance between the crushing surfaces at the bottom may be regulated by set-screws. With dry ore this distance may be as low as $\frac{3}{8}$ inch, no pieces larger than this being allowed to pass. It is stated that this machine works with less expenditure of power than the Blake crusher, and that its product is more uniform and can be made finer. Its first cost, however, is higher, and, what is of more importance, the

* *Trans. Am. Inst. Mng. Eng.*, 1892.

repairs are troublesome and expensive, but it certainly works well in many places where it has been adopted.

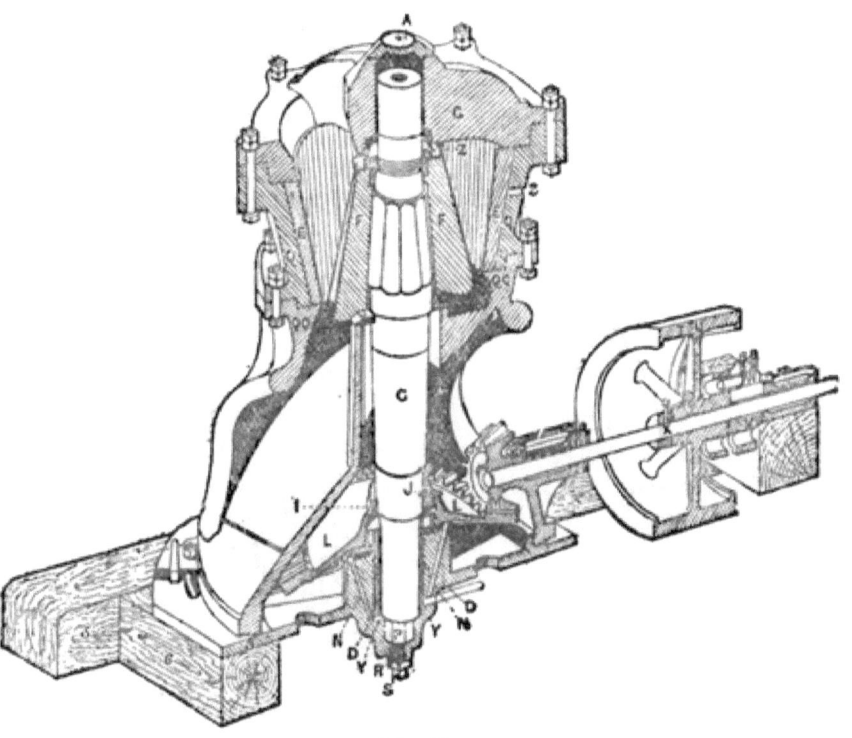

Fig. 16.

The Schranz Stone-breaker.—This machine (Fig. 17) is of interest, as it forms a link between the ordinary stone-breaker, with reciprocating jaws, and the crushing rolls. The movable jaw, instead of having a reciprocating motion, has a rocking one, somewhat similar to the motion of a circular blotting-pad. The jaws are of cast steel, and the space between them is regulated by means of the vertical screw, I, which adjusts the wedge, c. The short connecting-rod, b, made of cast iron is made with a narrow section, so that it is broken before any other part of the stone-breaker, if some very hard material should find its way between the jaws. The machine has the advantage, in common with the Dodge crusher, that the opening at the bottom does not vary, so that a uniform-sized product is secured, the maximum diameter of which can readily be reduced to 8 mm. (0·3 inch). The machine can thus take the place of roughing-rolls. The large sized machine, working at 250 revolutions per minute, and with an expenditure of 10 to 12 horse-power, is stated to crush from

4 to 5 tons of rock per hour to the size given above, and thus compares favourably with the Dodge crusher. It is in use at Laurenburg, on the Lahn, and at other places in Germany, and gives great satisfaction.

Material Employed for the Crushing Surfaces.—The selection of the most suitable material for the working parts, and especially for the crushing surfaces, of reduction machinery is a matter of the greatest possible importance, as the economy effected by using a durable material, which seldom requires renewal, is very great. When rock is crushed by repeated blows, as in the case of ordinary rock-breakers, stamps, and, perhaps,

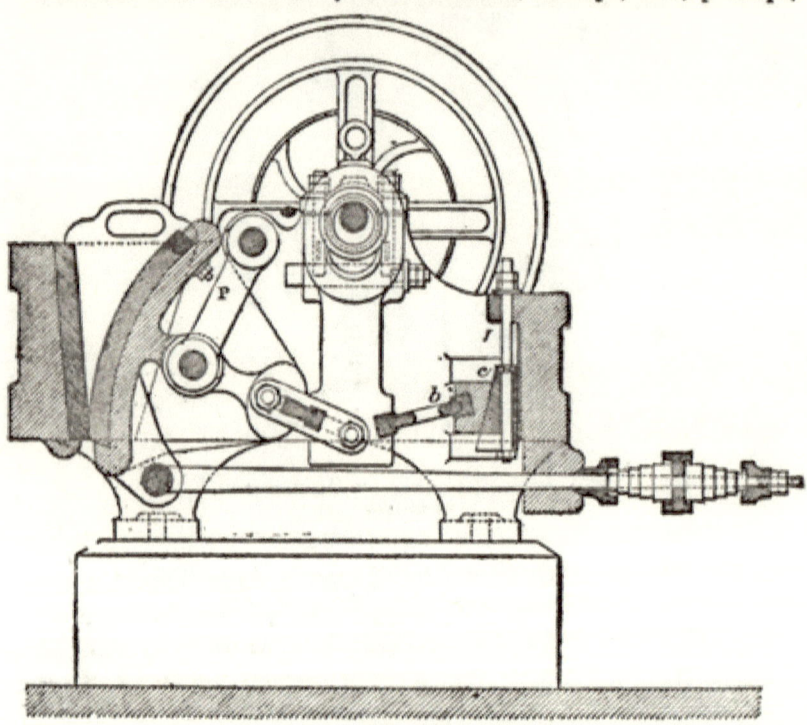

Fig. 17.

rolls, it does not follow that the hardest material is always the most durable. A substance is wanted that will not be broken or caused to crystallise unduly by a blow, but which, at the same time, will show neither signs of deformation nor rapid attrition by the impact of fragments of quartz. Both chilled iron and hard chromium-steel have been proved to be very useful for the surfaces of impact machines, but the softer and tougher open-hearth mild steel has also been found advantageous for the surfaces of rolls. The quality of toughness seems on the whole to be of even more importance than mere hardness.

No doubt different qualities are required for the crushing surfaces in different machines, and even in the same machine when operating on different classes of ore. Thus chromium-steel, which has been found so useful for the manufacture of stamp heads and dies, was not a success when used in the Huntington mill, where the action is rather one of grinding than of impact.

Perhaps Hadfield's manganese steel is the best material for the jaws of stone-breakers. At the Penmaenmawr granite quarries, the plates inserted at the sides of the jaws to protect the outer casing of a Blake crusher were worn out in two months, when ordinary steel was used, while similar plates made of cast manganese steel, after being in service for more than twelve months, had only been worn to the extent of $\frac{1}{4}$ inch, and were still at work. In stamp-mills, the crushing faces, if made of chilled iron, last for six or eight months, while steel faces last twice as long. The whole question of the nature of the materials to be used for different purposes is one for the metallurgist, and too much attention can hardly be paid to it in the future.

Position and Use of Rock-Breakers.—The aperture of a rock-breaker should be placed on a level with the floor, so that the ore can be dumped down by the side and shovelled into the jaws. As noted below (p. 121), it is now becoming customary to place the rock-breaker in a separate building distinct from the battery house.

A grizzly is employed to separate the fine material, which is passed straight to the stamps, while in some cases the material from the rock-breaker is also screened and part returned to be put through again. Revolving trommels and flat shaking screens have both been used for this purpose.

The efficiency and economy in crushing, attained by the machines on the reciprocating-jaw principle, are so fully recognised that there has been a tendency of late years to use them to reduce gold-quartz to a very small size before feeding it into the stamp batteries, the capacity of which is greatly increased in this way. For fine crushing, multiple-jaw crushers, on the same principle as the Blake, have been constructed, but have not passed into general use; the use of a pair of rolls between the rock-breaker and the stamp battery has also been advocated. This is done by the Huanchaca Mining Company at Antofagasta, Chili, and gives an increase of capacity to the stamps of over 20 per cent., while the cost is trifling. The size of the ore best suited for feeding into a stamp battery may be roughly put down as about $\frac{1}{4}$ inch for light stamps and $\frac{1}{2}$ inch for heavy stamps. If the size of the material is much smaller than this, no advantage in speed is gained, while the jar given to the stamps and framework is greatly increased. At the present day few large mills are erected without rock-breakers, which have also been successfully added to many old mills. Nevertheless, they

102　　　　　THE METALLURGY OF GOLD.

are absent or rarely seen in several of the richest gold-fields of the world. In Gilpin County, Colorado, they are not used; in Victoria the extent of their use may be judged from the fact that there were 5,901 stamp-heads in operation in 1891, and only twelve stone-breakers, and in other parts of Australia they have been similarly neglected.

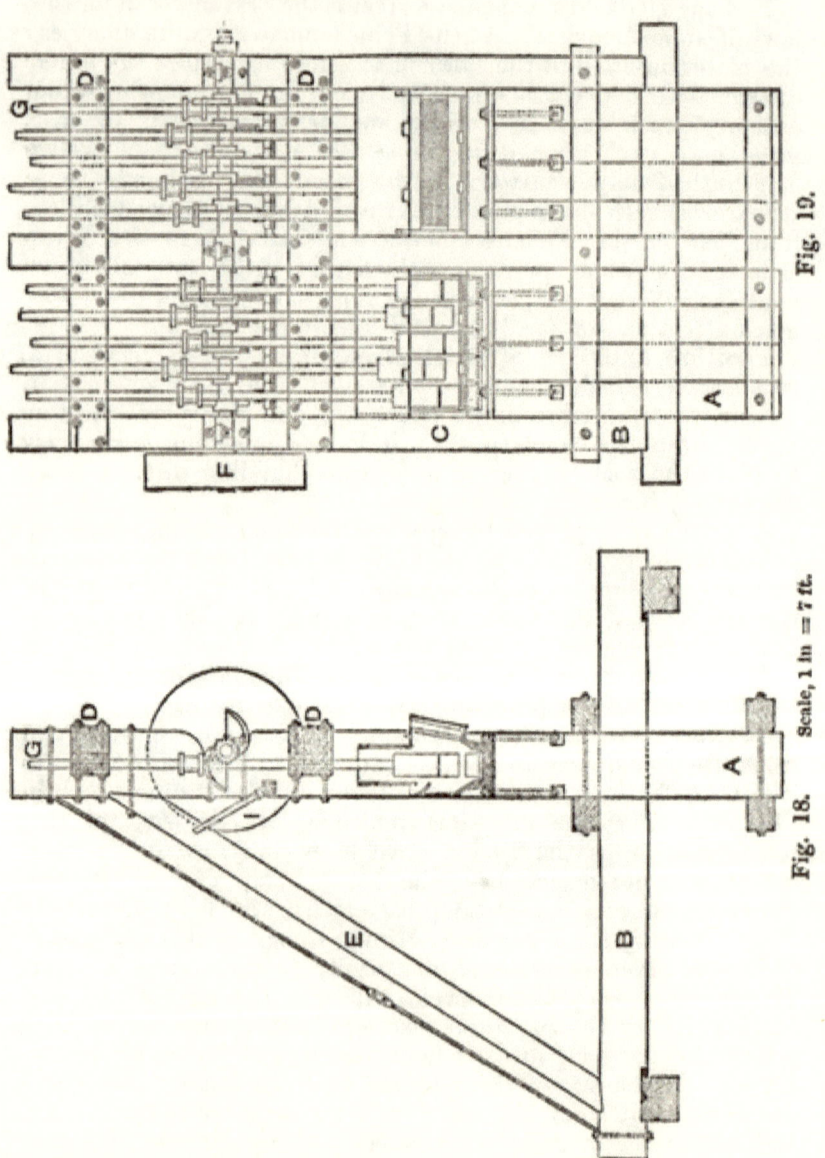

The Stamp Battery.—Californian "gravitation" stamps are

in general use at the present day for crushing gold ores. A stamp is a heavy iron, or iron and steel, pestle, raised by a cam keyed on to a horizontal revolving shaft, and let fall by its own weight. Stamps are ranged in line in groups of five stamps each, which have a mortar-box in common. Fig. 18 represents the side view, and Fig. 19 the front view, of a ten-stamp battery, with the amalgamating tables removed to show the foundation timbers or mortar-blocks, A.

The foundations are of the highest importance, as, if they are badly made through carelessness or false economy, the efficiency of the battery is greatly decreased, and it soon shakes itself to pieces. The blow of a stamp is partly employed in crushing the ore, and is partly expended in producing a concussion or jar acting on the framework and foundations. The amount of energy used up in the latter way depends largely on construction, for details of which the student is referred to the volume on *Metallurgical Machinery*. In preparing the ground for the foundations, the earth is removed until bed-rock is reached if possible, and the latter is then carefully smoothed and sometimes covered with a layer of cement. The wooden mortar-blocks of from 6 to 14 feet long are placed upright in this trench, and the space round filled up with sand, or, as in the Transvaal, solid masonry is built round the blocks. The *framework* is now usually made of wood, which is a far more satisfactory material for the purpose than iron. It consists of the massive battery sills, B, on which rest the battery posts, C, and the braces, E. The posts are held together by the stamp-guides or tie-timbers, D. Wooden braces have completely replaced iron rods, which allow the battery to spring. It is better to place the braces on the discharge side alone, thus leaving more room to work on the feed side.

The Mortar.—The mortars are made of cast iron, but differ in shape according to the nature of the ore and the corresponding modifications made in the course of treatment. They weigh from $1\frac{1}{2}$ to 3 tons, being especially thick at the bottom where there is the greatest strain. An ordinary mortar is about 4 feet 7 inches long, 50 inches high, and 12 inches wide on the inside at the level at which the dies are set. The bottom is from 3 to 8 inches in thickness, and has a heavy flange cast on it, by which it is bolted to the mortar-blocks. These are tarred over, all cracks in them having been filled with sulphur, and are then covered with three thicknesses of blanket, carefully coated with tar on both sides. The mortar is placed on these blankets and securely bolted down. This arrangement lessens the chance of the mortar working loose, the jar being diminished. A sheet of rubber, $\frac{1}{4}$ inch thick, may be used instead of the blankets. Figs. 20 and 21 represent sectional elevations of two forms of mortars in use in the United States, Fig. 20 showing a single-discharge narrow mortar for wet crushing, and Fig. 21 a wide double-discharge mortar for dry crushing. In both figures, *b* is the feed-opening through which the ore is introduced into the mortar; *c* is the bed on which the die is placed; *d* is the screen-opening. Single-discharge mortars vary in shape, according as they are intended to contain amalgamated copper lining plates both at the front and back, or on the screen side only. The chief difference between them is in the feeding arrangement; in the former case the back plate is put in a recess, and is protected from the falling rock fed into the battery. The mortar shown in position in Fig. 18 is one of this kind. The plates catch the coarse gold inside the mortar when the pulp is flung against them. Occasionally copper plates are also put at the ends of the mortar. All the plates, which are bolted through the mortar itself or to its lining plates, must be so arranged that they can easily be taken out and cleaned, as, when

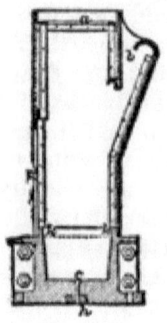 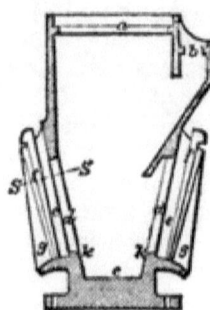

Fig. 20.   Scale, 1 in. = 3 ft.   Fig. 21.

very rich ores are being treated, precious metal accumulates on them very fast.

The width of the mortar varies from 10 to 14 inches at the level of the bottom of the screens. As has been already mentioned, narrow mortars are best fitted for rapid discharge, but, if hard flinty ores are to be crushed, a narrow mortar causes frequent breakage of the screens, unless the discharge is deep — *i.e.*, unless the bottom of the screens is a considerable distance above the surface of the dies. By this latter arrangement the output is reduced, since, the nearer the screens are to the dies, the more rapid is the discharge. The depth of the discharge is only about 6 or 7 inches in California, where an adjustable battery-screen has been introduced to keep this depth constant, in spite of the wearing of the dies. In the battery thus modified, the screen-frame is supported on a wooden block, which is easily removable, and to which the copper plate is bolted. When the dies wear down, this block is replaced by one of less height, to which a suitable plate has already been fixed.

Mortars often have a lining of cast-iron plates, bolted to them near the bottom, to protect them from the rapid wear due to the splashing of the pulp. These plates last from six to nine months, and can be replaced when worn out.

The *splash-box*, not shown in the figures, and now often omitted, is bolted to the outside of the mortar just below the screens. It is rectangular, consists of wood or iron, and is of the same length as the mortar. It receives the pulp as it passes through the screens, and distributes it evenly over the amalgamating tables by a number of spouts, usually three. Instead of the splash-box, a splash-board is now almost universally employed, the usual material for it being heavy canvas. The old form of mortar had its upper part, or housing, of wood (see Fig. 44, p. 195), but, as mercury is lost through the smallest aperture, and it was difficult to make these wooden housings quite tight, mortars are now cast in one piece, including the housings. The roof of the mortar is made of 2-inch planking, through which holes are cut to admit the stems of the stamps and the water pipes.

When the mortar is in place, the dies are put into it, a layer of sand being often introduced first. The dies consist of two parts, the *footplate* and the *die* proper, or *boss*.

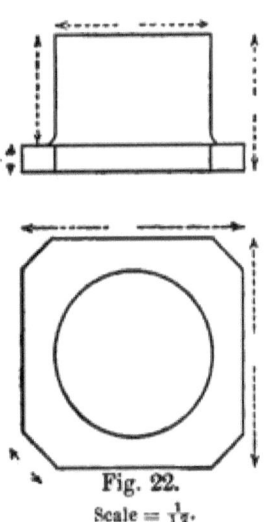

Fig. 22.
Scale = $\frac{1}{12}$.

Fig. 22 shows, in plan and elevation, one of the many forms of dies in use; here the footplate is almost square, so as to fit the mortar; it is 1 or 2 inches thick, and 10 or 12 inches square. The boss is cylindrical, from 3 to 6 inches high, and of the same diameter as the shoe. Shoes and dies are made either of iron or steel, hard chilled-iron being used for wet crushing, and soft iron for dry crushing. Cast-steel dies and shoes have been often tried, but owing to their tendency to chip and to their uneven wear, their introduction formerly met with little success. Nevertheless, in most mills remote from foundries, where transportation is an important item in the cost of supplies, steel shoes and dies have now replaced those of iron, as the life of steel is from two and a-half to three times that of iron, and the cost only about twice as great.* About seven years ago, chrome-steel shoes and dies were introduced into California, and proved their superiority over those manufactured from most other kinds of steel; forged steel and, according to Janin, manganese steel have also been used with success. Sometimes steel shoes and iron dies are used, the wear being more even than when both consist of steel. At the Patterson mine in 1888, a set of iron shoes and dies lasted

* *Eighth Report Cal. State Min.*, 1888, p. 708.

on an average six weeks, crushing 1,680 tons of ore, while one set of chrome-steel shoes is said to have lasted fourteen months, crushing 16,800 tons, or ten times the amount crushed by iron.* The wear of iron shoes and dies is stated to be about 2 or 3 lbs. per ton of ore crushed in California. In Colorado, at the New California Mine, the wear of shoes was 11·3 ozs. of iron per ton of ore, and that of the dies 4·5 ozs. At the Robinson Mine, South Africa, the wear of steel shoes and dies is, according to Mr. Harland, 0·45 lb. per ton crushed for shoes and 0·30 lb. per ton for dies. When the boss is worn down to within from $\frac{1}{2}$ inch to 1 inch of the footplate, the die is replaced. Dies wear more slowly than shoes since they are protected by a layer of pulp, which is from $1\frac{1}{2}$ to 3 inches thick. The dies should all be renewed together, as it is important that those in the same battery should be of equal height, otherwise one or more will become almost bare of ore, and a disastrous pounding result. If a die breaks, it should not be replaced by a new one, but by one worn to the same extent as the others in the battery. Iron false-bottoms or chuck-blocks are placed beneath partially-worn dies, so as to keep the depth of discharge constant.

The *cam-shaft*, H, Fig. 19, is of wrought iron, and about 5 inches in diameter (A, Fig. 23). It is now usual to have a separate cam-shaft for each five or ten stamps, which have thus a separate driving wheel. The advantage of this arrangement is that repairs can be done to one or more stamps without necessitating the stoppage of the whole mill, as used to be the case when there was only one cam-shaft. The cam-shaft is placed at a distance of from 5 to 10 inches from the stem-centre, and is 9 to 10 feet above the mortar bed. The bearings rest on supports attached to the battery posts on the discharge side.

The *cams* are made of cast iron, with chilled faces, which are 2 to 3 inches wide (B, Fig. 23). The double cam, two views of which are shown in Fig. 23, is now in almost universal use, though single and treble forms have been employed. Sometimes cams are cast in two pieces which are bolted together, so that when one is worn out, it can be replaced without first removing the other cams on the shaft. It is stated, however, that these sectional cams work loose, and are not much used in consequence. The hub is always strengthened by a band of wrought iron shrunk on it. The shape of the cam face is the involute of a circle slightly modified at the end so as to stop the upward motion gradually. The radius of this circle is equal to the distance between the centres of the cam-shaft and stem, which depends on the height to which the stamp is to be lifted, so that the curve of the cam varies with the drop. A cam should last several years unless broken through being a faulty casting, or through carelessness in letting the stamp fall when hung up.

The cam-face works against the iron collar or *tappet*, shown in

* *Eighth Report Cal. State Min.*, 1883, p. 656.

plan and section in Fig. 24, which is bored out at A to fit the stem of the stamp. The tappet is fitted with a wrought-iron gib, which is pressed against the stem by two or three keys behind it, thus binding the tappet firmly on the stem while, at the same time, admitting of rapid adjustment to another position. The entire end surface of the tappet comes in contact with the cam-face, by which the stamp is raised and, at the same time, rotated, all its parts being round. The effect of this is, that the shoe does not strike the ore in the mortar in exactly the same place twice in succession, and the wear of its face is made more uniform. The greater part of the revolution takes place during the raising of the stamp, but the latter does not quite cease to rotate as it falls, and a slight grind-

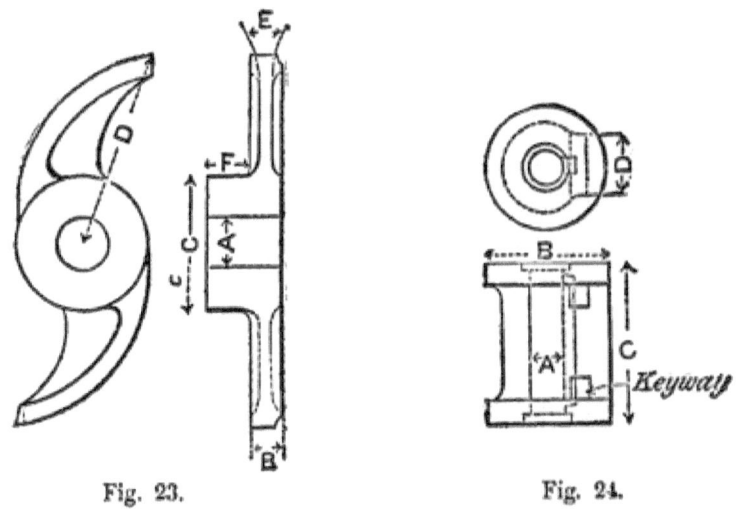

Fig. 23.　　　　　Fig. 24.

ing action on the ore has been noticed by many observers. The amount of rotation varies with the fall, the extent to which the cam and tappet are greased, and the state of wear of their surfaces. A little grease is always added to reduce wear, but, if too much is present, the stamp does not revolve at all, while, according to J. H. Hammond, when the tappet is in the right condition, one revolution is effected in from four to eight blows, with a 6- or 8-inch drop. Other observers find the usual rate of rotation more rapid, and in Gilpin Co., Colorado, where the average drop is from 16 to 18 inches, the stamp makes from $1\frac{1}{4}$ to $1\frac{1}{2}$ revolutions at each blow. Tappets should last for four or five years; and, having both ends alike, they can be reversed when one end is worn out, and their worn and grooved faces can be planed down when necessary. Some millmen assert that tappets may be broken by the cam if keyed too tightly to the stem.

As the cam-thrust is not applied at the centre of the stamp, there is always a considerable side pressure, which greatly increases the friction in the guides and wears the latter out, besides causing a loss of power. Moreover, another result of this is that the stamp tends to be inclined (not vertical) when it is released, and so the blow on the die is given slightly to one side—*i.e.*, the side of the dye on which the cam works. Consequently, there is a tendency for this side to wear down more quickly than the other. To obviate these disadvantages, cams have been introduced at Johannesburg with a wide hub, and the two blades set one at each end of the hub, so that they work on opposite sides of the stamp and cause it to revolve in different directions at each successive uplift. The Blanton cam, now generally used on the Rand, is fastened to the shaft by a semi-circular wedge pinned to the cam-shaft, no keys being necessary.

The pulley on the cam-shaft (F, Fig. 19) is made of wood on cast-iron flanges; if iron alone were used, the rapid succession of jars, caused by the dropping of the stamps, would soon cause the material to crystallise and break. A tightener pulley on the belt driving the cam-shaft is often used, by which the stamps can be put in motion or stopped without interfering with the driving power.

The *stamp* itself consists of three parts, the stem, the head, and the shoe. The *stem* (G, Fig. 19) is from 12 to 15 feet

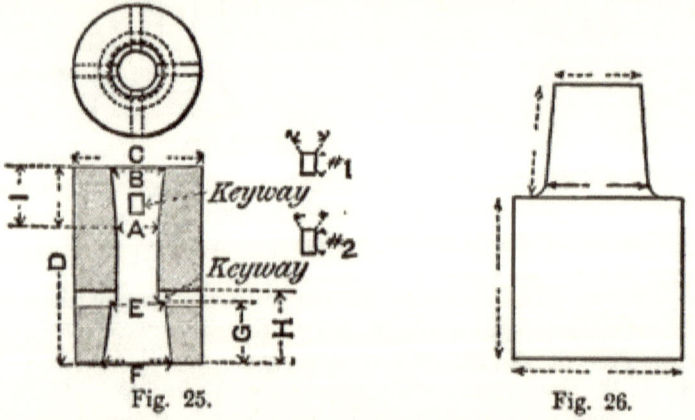

Fig. 25. Fig. 26.

long, and from 3 to 3½ inches in diameter; it is made of wrought iron, and has both ends tapered for a length of 6 or 8 inches to fit the heads, so that, if one end is broken off, the stem can be inverted and the other end used. The *head* (Fig. 25) and shoe (Fig. 26) are made of equal diameter—viz., about 8 to 10 inches. The head is of tough cast-iron, about 15 to 20 inches high, and has a tapered socket at each end, the upper one (A B) for the stem and the lower (E F) for the tapered shank of the shoe. When these are driven into their respective sockets, into which a few strips of wood are inserted to keep the two metal surfaces from touching each other, a few blows by the stamp bind them securely together, no other fastening being necessary. Slots are provided (shown in the figure), at the base of the two sockets, through which wedges may be driven to force out the shoe or stem when necessary. The head is often strengthened by bands of wrought iron, shrunk on at each end, to prevent splitting by the wedge-like action of the tapering stem and shoe. The head lasts several years, being rarely ruptured. The *shoe* (Fig. 26, which is on a larger scale than Fig. 25) consists of two parts, the shank, which fits into the head, and the shoe proper or butt. The latter is made of very hard white iron, and the shank of softer iron; steel is also used, as has been already mentioned. The diameter of the shank is about half that of the butt. The shoe is replaced when the butt, which is from 6 to 12 inches in length when new, has been worn down to about 1 inch in length. To keep the total weight of the stamp constant, several sizes of heads are sometimes used in one mill, the heavier heads taking partly-worn shoes. "Chuck-shoes" are inserted between heads and shoes with the same object.

The relative weights of tappet, stem, head and shoe, which together make up the stamp, vary considerably. There is an advantage in increasing the weight of the stem, as one of small diameter tends to spring and bend from the blow of the cam, or when the stamp falls, and to wear the guides rapidly. The stem weighs from 250 to 475 lbs., the tappet from 80 to 130 lbs., the head from 175 to 370 lbs., and the shoe from 100 to 230 lbs. The total weight of the stamp is usually from 650 to 1,150 lbs., but is sometimes as low as 450 lbs., and, for prospecting purposes, the weight is only from 100 to 300 lbs. Old dies weigh from 20 to 50 lbs. when they are discarded, and old shoes from 25 to 40 lbs. A steel tappet on a 900-lb. stamp weighs 112 lbs.

The stamp stems are guided in boxes bolted to the wooden cross-ties, which also serve to hold the battery posts together. There are two of these *guides* (D, Fig. 19), one within a foot or so of the top of the battery posts, and the other as low as the raising of the stamp head will allow. The depth of each guide is about 15 inches, and the stems are fitted closely to the guides, metal boxes being used occasionally, although wood is much more general. The guide-boards are sometimes pierced with large square holes in which bushes of wood, with the grain parallel to the length of the stamp, are placed fitting the stem exactly. In this way, the guide-boards themselves are preserved from wearing out. Sectional guides, consisting of a series of iron keys enclosing wooden bushings, are also used. In this case each stem has a guide to itself, and the bushings can be renewed by hanging-up the one stamp without stopping the other stamps in the battery.

Each stamp is provided with a *lever* or *jack* (I, Fig. 18) made of wrought iron, or of wood protected by iron. The jack is for the purpose of raising the stamp and hanging it up out of reach of the cam. When this is to be done, a strip of wood, an inch or more thick, is laid with one hand on the cam as it rises, and the stamp is thus raised an inch higher than usual, so that the jack can be slipped in under the tappet with the other hand. The stamp is thus suspended above the cam and can be repaired without stopping the others, while it can only be released in a manner similar to that in which it was hung up. Above the stamps there is a double rail, on which is a movable pulley; by this the stamps, &c., can be lifted up for repairs. When the stamps are to be set up, the head is put on the die and the stem dropped into it, iron being left sometimes on iron, but more usually some canvas or other packing being put into the head socket and the stem dropped into that. The stamp is then raised and dropped into the shoe, the shank of which is often surrounded by strips of wood for packing. As already stated, the parts are soon wedged firmly together by raising and letting fall the stamp a few times.

The height of the "drop" of the stamps varies from 4 to 18 inches, and the number of drops per minute varies from 30 to over 100. These depend on one another to a great extent, an increase in the height of the drop being necessarily accompanied by a diminution in the number of drops per minute. With a drop of $8\frac{1}{2}$ inches, about 95 blows can be obtained, the tappet then just having time to fall after leaving one face of the cam, before the other begins to raise it. As, within certain limits and under certain conditions, an increase of speed results in an increase of yield of pulverised ore, efforts have been made to raise the number of blows per minute. It has been proposed to use two cam-shafts, one above and one below the gib-tappets, a speed of 250 blows per minute being thus attained, according to the statements of the manufacturer, but this device has apparently not passed into use. The subject will be returned to when the conditions for successful amalgamation are discussed.

Screens.—The screens are set in iron frames, which now usually slide in grooves cast in the mortar, and are keyed to it, but were formerly fitted into recesses and bolted. They are made either of wire-cloth, or of Russian sheet-iron or steel, in which holes are punched. The sheet-iron is about $\frac{1}{35}$ inch thick and weighs about 1 lb. per square foot. The rough ("burred") side of the punched plate is put on the inside, so that the holes widen outwards, and are thus prevented from becoming clogged. The holes are round, oval, or consist of long slots (from $\frac{1}{4}$ to $\frac{1}{2}$ inch long) ranged parallel or inclined to each other. The shortest diameter of these holes ranges from about $\frac{1}{80}$ to $\frac{1}{12}$ inch or more, according to the nature of the ore and the method employed in its treatment: the usual size is from about $\frac{1}{24}$ to $\frac{1}{16}$ inch.

The relative advantages of wire-cloth and sheet-iron are not yet beyond dispute, and vary with the nature of the ore. Slots appear to be better suited for discharge than meshes, but, on the other hand, there is a great loss of discharge area in the use of punched iron. Thus a wire mesh screen, containing 18 holes to the linear inch, has 324 holes to the square inch, while a round-punched sheet-iron screen has only 140 holes of the same size per square inch. At Otago, N.Z., where friable quartzose ores are treated, both kinds are used, but the wire-cloth (No. 18) lasts a little longer, and costs nearly 25 per cent. less than the round-punched Russia iron. In Gilpin County, Colorado, the "burr-slot" screen is used, having horizontal slots placed alternately. The screens are inverted after a time, as the lower part wears out soonest.

In California, both steel and brass wire, and slot- and needle-punched tinned- and sheet-iron are used as screen materials, steel wire screens, however, being seldom used, owing to their rapid corrosion by rusting. The sheet-iron consists of the best soft Russia iron and has a smooth glossy surface, and the slots and needle-holes are almost universally burred. The size is named according to the number of the sewing-machine needle by which the holes were punched, the sizes numbered 5, 6, 7, 8, and 9 being those most used. The following table gives details of these screens:—

No. of Needle.	Corresponding Mesh.	Width of Slot.	Weight per Sq. Foot.
5	20	0·029 in.	1·15 lbs.
6	25	·027 ,,	1·08 ,,
7	30	·024 ,,	0·99 ,,
8	35	·022 ,,	0·92 ,,
9	40	·020 ,,	0·83 ,,

The burred horizontal or horizontal-alternate slots are those most in use. Tinned-iron screens are made thinner and, therefore, have the advantage of allowing a more rapid discharge. The brass-wire screens used vary from 16 to 60 mesh, 30 to 40 being the most common; with their use the pulp discharged is more uniform in size than if slots are employed. Although sheet-iron is the material usually employed for screens, it is often preferable to use copper, as pyritic ores, if kept for any length of time after being mined, soon become oxidised and acidified, and the ferrous sulphate thus formed corrodes iron rapidly, whilst the water used is often more or less acid if it comes from mines. Copper is not attacked in the same way. T. A. Rickard has shown[*] that the life of screens in the mills at Blackhawk, Colorado, diminishes as the creek is descended, the water becoming more and more impure. At the South Clunes United

[*] *Eng. and Mng. Journ.*, Sept. 3, 1892.

Mill, Victoria, iron punched gratings lasted less than a week, but, on introducing copper plates containing 100 holes per square inch, the life of the screen was increased to a month, and 275 tons of ore were passed through it. At Blackhawk, the iron screens last while from 80 to 430 tons are passed through them, according to the position of the mill. At Grass Valley, the average is 200 tons, at Bendigo, 134 tons, and at Otago, N.Z., only 40 tons. In California the brass-wire screens last from 10 to 14 days, corresponding to a passage of 120 to 140 tons, and the Russia-plate lasts from 15 to 40 days, the average being 30 days, corresponding to the passage of about 330 tons. The rate of wear of the screens depends greatly on their position, being more rapid with a shallow than a deep discharge, and more rapid in a narrow than a wide mortar. Pieces of iron or wood in the ore may cause the screen to break, and these should, consequently, be removed from the ore as far as possible. The battery is hung up now and then, so that a thorough inspection of all the screens may be made, and those that are broken replaced.

The screens were formerly set vertical, and this system still prevails generally in Australia. In the United States, however, they are now placed at an angle which varies somewhat but is never far from 10°, and this has been found to facilitate discharge. In wet stamping, screens are usually placed on one side of the mortar only — viz., that opposite the feeding side. In cases where the discharge is required to be as rapid as possible, the screen area is increased, and double discharge (front and back) mortars have been made, but have not been used much, except for dry crushing, and screens at the ends of the mortars are used at Harriettville, Victoria (Rickard). The area of the screen is from 3 to 4 square feet per battery in California, the height from the bottom to the top of the screen being from 8 to 10 inches; in Gilpin County, the screens are usually 8 inches high and $4\frac{1}{2}$ feet long. Opinions differ as to the necessary amount of screen area to be used. By some experts it is contended that the capacity of a battery is really limited, not by its crushing, but by its discharging power. Thus, by a number of experiments conducted some years ago at the Metacom Mill, California, it was shown that when crushed pulp instead of the unbroken ore was fed into the mortar, the rate of discharge was not increased. This was taken as a convincing proof that the discharge area is usually far too low. Nevertheless, double discharge is hardly ever used for wet-crushing mills, the various objections that are made to it being summarised as follows:—

1. Inconvenience is caused in the arrangement of the copper plates, both inside and outside the battery.

2. So great a quantity of battery water must be used, that the pulp is too thin for efficient amalgamation on the plates.

3. The ore does not stay long enough in the battery to be effectively amalgamated.

When, however, battery amalgamation is not attempted, double discharge might be advantageous, and, in particular, it is to be recommended with heavily sulphuretted ores, containing brittle pyrites. The object to be attained in this case is to leave these sulphides unbroken as far as possible, so as to facilitate concentration, since great losses from sliming will inevitably result if they are subjected to repeated blows in the battery before being discharged.

Order of Fall of Stamps.—The order in which the stamps drop is of considerable importance. If they were let fall in succession from one end to the other of the mortar, the pulp would be driven before them, so that the stamp which fell last would have its die covered by too deep a cushion of ore, while that at the other end would be almost bare. The result to be obtained is to keep the ore equally distributed through the mortar, so that each stamp shall do the same amount of crushing, although it is inevitable that the middle stamps should be more efficient than the end ones in discharging the ore. The order most favoured in California is 1, 4, 2, 5, 3, and that on the Rand is 1, 3, 5, 2, 4, whilst the orders 1, 5, 2, 4, 3 and 1, 5, 3, 2, 4 are also often used. Several other orders have their advocates, and are probably little inferior to the above for the particular ores on which they are employed. Since the end stamps are of less efficiency than the others, it has been argued that a larger number of stamps in one mortar would be advantageous, and at Clausthal, in the Hartz Mountains, there are usually from nine to eleven stamps in a battery,* placed close together, space being greatly economised in this way. Long and wide experience has, however, proved that the best number is five.

Feeding.—Ore is fed into the battery either by hand or by automatic machines. It is often asserted that really intelligent hand-feeding is better than the automatic method, since the stamps are not all equally efficient. At any rate, hand-feeding is still persisted in on some Australian gold-fields, although, perhaps, not always performed with intelligence. The feeder on small mills is often expected to break down the big pieces of ore with a sledge hammer, a rock-breaker not being used, but this method of working may be safely set down as irrational and uneconomical, and the result usually is that large and small pieces go into the mortar together. In the United States and in the Transvaal, self-feeders are almost universally employed in modern mills. The art of feeding consists in keeping the depth of pulp on the dies constant throughout the battery, as long as the work is carried on. This is much better done by automatic machinery than by hand, and it was found that by the introduction of the

* Meinecke, *Proc. Inst. C. Eng.*, Session 1891-1892, part ii.

former in California the capacity of the stamps was increased by 15 to 20 per cent., while the wear of shoes and dies was decreased by 25 per cent., and that of the screens by 50 per cent. It is not difficult to discern the cause of the advantages, for, if the dies are insufficiently covered with ore, less crushing is done, while a greater concussion must be taken up by the stamp and by the die, mortar, &c. If the die is quite bare this concussion is so great that the stem may be bent or broken, and the shoe and die battered. On the other hand, if the ore is too deep in the mortar, there is so thick a cushion that much of the force is taken up in compressing without crushing it; whilst, besides the reduction of output, the head, under these circumstances, sometimes becomes detached from the stem, which is broken or

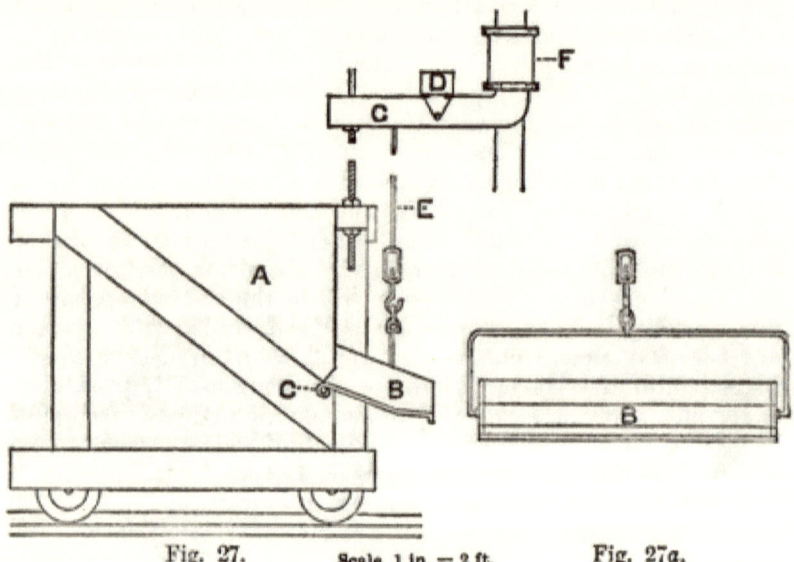

Fig. 27.   Scale, 1 in. = 2 ft.   Fig. 27a.

battered by the next blow. The maximum capacity is obtained with "low feeding," the depth of pulp on the dies being about 2 inches or less.

One advantage of even feeding is that a larger proportion of gold is caught, owing to the more regular and even flow of the pulp over the plates, the danger of scouring being diminished.

There are two classes of automatic feeders, each consisting of a pyramidal hopper with inclined floor, connected with the feed-opening of the mortar-box by a spout or shoot. Stanford's automatic feeder, which was the first invented (in California, about 1870), and is still used, is a typical example of the one class, and Hendy's Challenge feeder, the best machine devised, and the most largely used, is a good example of the other class.

Stanford's feeder, shown in Fig. 27, consists of a hopper, A, with an adjustable spout, B, which is hinged at C. Fig. 27a is a front view of B, showing how it is suspended. The spout is attached to the vertical rod, E, which is hung to the lever, G, and this has its fulcrum at D. Near the top of a stamp-stem (usually the middle one in the battery), the feeding-tappet, F, is keyed. When the battery is full of ore, the tappet does not come down far enough to strike the lever, G, but, when the ore gets low, the lever is struck, and the result is that the spout, B, is jerked up and down again, the ore being thus thrown forward.

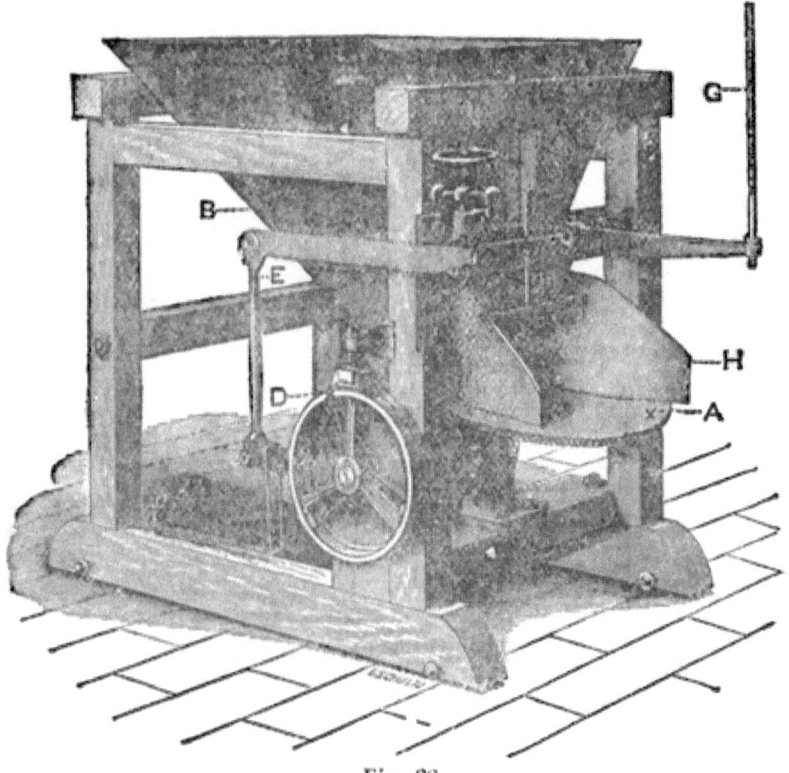

Fig. 28.

This machine works well with dry ores which are moderately fine, but, if the ore is wet, and especially if it is argillaceous, it sticks in the hopper until at last a powerful jerk brings it down with a run, the mortar-box is filled up, and all the evils of over-feeding result. The consequence is that the battery is fed as irregularly as by the worst hand-feeding.

The Challenge feeder (Fig. 28) is constructed so that the tray, A, below the sheet-iron hopper, B, is revolved in a horizontal plane by means of a gear-wheel below it, shown in the figure, and this gears with teeth set in the bottom of the tray, A.

The gear-wheel is set in motion by a friction grip, D, placed on the outside of the frame, and actuated through the lever, E, by the bumper-rod, G, against which the tappet of the stamp strikes. At each partial rotation a given quantity of ore is scraped off by the stationary wings or side plates, H, resting on the tray, A. This amount of ore is regulated by the condition of the mortar. The machine is especially adapted for very wet or sticky ores. It weighs, with the frame, about 900 lbs., and is the most expensive automatic feeder in the market. Other automatic feeders, such as the Tulloch and the Roller machines, constructed on a similar principle, the ore being scraped and not shaken into the mortar, are cheaper than the Challenge, and are doing excellent work on certain kinds of ore. In all cases one feeder is sufficient for a battery of five stamps, more ore being fed to the middle stamps, where the most work is done, than to the end ones. The feeders are now sometimes suspended from above instead of being supported from below.

Water Supply.—The water is supplied to the stamps by horizontal pipes passing just above the top of the housing of the mortar-box, with an opening supplied with a stop-cock opposite each stamp. The water is commonly made to impinge against the stem of the stamp, either just above or below the wooden casing, and thence run down into the mortar, thus keeping the stem clean. In front of the battery there is also another pipe of about half the size, to supply water to the tables to help carry off the pulp. This pipe is often pierced with pin-holes, so that the water is supplied as a number of fine jets. The water is warmed by steam in winter in many mills, from a belief that amalgamation is promoted by warmth. This belief is founded on general experience rather than on exact experiments, and some results appear to point to a different conclusion.

Thus, Professor Le Neve Foster obtained the following results* at the Pestarena Mill in Italy, during the years 1869-70:—The average temperature of the water supplied to the mill during the six summer months was $52°$ F., and the average temperature of the water supplied in the winter months was $39°·4$ F., and yet, in spite of that, and of the fact that the average temperature, for instance, of the month of January, 1870, was as low as $33°·6$ F., he extracted $3·1$ per cent. more gold with the cold water than with the warm. These figures do not necessarily prove that cold water is better for amalgamation, as there were in this instance other matters to be taken into consideration, but they show that amalgamation is possible even when the temperature of the water is on an average only $39°$. The difference in the results in this case might have been due to the turbidity of the water (which was derived from the glaciers at Monte Rosa) in the summer, and its clearness in winter, or to the fact that the pyrites were more liable to decompose in warm weather, and so additional sickening of the mercury was caused in summer.

* *Loc. cit.*

The amount of water used varies from $1\frac{1}{4}$ to $6\frac{1}{2}$ gallons per stamp per minute, the average in California being about $2\frac{1}{2}$ gallons, on the Rand about $5\frac{1}{2}$ gallons, and in Colorado only about $1\frac{3}{4}$ gallons. In California, with fast-running rapid-discharge batteries, the amount of water per ton of rock crushed varies from 1,000 to 2,400 gallons, the mean being about 1,700 gallons, while in Colorado the average amount is as high as 2,500 gallons. Besides varying with the method of crushing adopted, the amount of water varies with the nature of the gangue, clayey ores requiring more, while the large quantity required by sulphide ores is due to the deep discharge necessitated by the difficulty of catching the gold, as well as to the high density of the pulverised material, which renders it more difficult to convey in suspension over the plates. As a rule, the more rapid the output, the less water per ton of ore is required in the battery. Coarse crushing requires less water in the battery, but, on the other hand, more has to be added on the plates. The amount of battery water per ton is increased by over 20 per cent. by the employment of double-discharge mortars. The amount of water to be added on the plates varies with their grade, as well as with the density and size of the particles of crushed ore. It should be only just enough to prevent the pulp from accumulating on the plates, as any excess over this tends to check amalgamation and to scour the plates. The average duty of a miner's inch in a gold stamp-mill is given by P. M. Randall[*] as 12 tons of quartz, if the head under which the water is supplied is 4 inches, and 15·88 tons, if the head is 7 inches. This gives the proportion of the volume of water to that of ore as 11·1 to 1. This may be compared with the proportion of between 7 and 10 to 1 in Siberian placer working, and as much as 30, or even 50 to 1 in hydraulicking. It may be mentioned that a ton of 2,000 lbs. of quartz occupies about 13 cubic feet when unbroken, and about 20 cubic feet after having been broken up, so that in a lode a cubic yard contains about 2 tons, and in a tailings heap only about $1\frac{1}{3}$ tons.

Amalgamated Plates.—These plates are usually made of copper, and are as much as $\frac{3}{8}$ inch thick for the inside of the battery and $\frac{1}{16}$ to $\frac{1}{8}$ inch thick for the outside. The average weight used in California is 3 lbs. per square foot. It was formerly laid down as a general rule that the heaviest plates were the best, as they last longer and are not so easily dented, but comparatively light plates are now used. The copper should be of the best quality, and, if it is hard, it must be annealed before applying the mercury, so as to make it absorbent. This was formerly done by heating it from below as uniformly as possible until sawdust laid on the upper surface was ignited. The plate is then straightened by blows of a wooden mallet, striking

[*] Article on Practical Hydraulics in *Sixth Report Cal. State Min.*, 1886.

a block of wood laid on it, and the surface is carefully cleaned by scouring with sand or fine emery paper until quite bright, and washed with strong soda to remove all traces of grease. Cleaning may also be effected by nitric acid diluted with 9 parts of water, or by a $2\frac{1}{2}$ per cent. solution of potassium cyanide, rubbed on with a woollen rag and carefully washed off with water. As soon as the plate is clean, it is rubbed with a mixture of fine sand, sal-ammoniac and mercury by means of a brush, the sal-ammoniac preventing the recommencement of oxidation. More mercury is sprinkled on and wiped over with a piece of rubber until no more can be absorbed, the whole surface being now thoroughly coated and bright. After having been left for an hour, the plate is finally washed with clean water.

If the plates were used in this condition they would not catch the gold very well at first, but would continually improve until they had become coated with gold amalgam. In order to make them efficient from the start, they are usually coated with gold amalgam or silver amalgam before being laid down in the mill. Gold amalgam is most effective but is seldom used, as it is so much more expensive than silver. The amalgam is rubbed on with a piece of india-rubber, the plate being wetted with a solution of sal-ammoniac to keep it bright, and such plates last for years without further treatment. A more usual method of preparing the plates in California is to coat them with electro-deposited silver. This plating is done by certain establishments in California for most of the mills, but it can be done on the spot without much difficulty, the plant required being inexpensive. After being silvered, the plates have the mercury applied to them. They absorb a large amount of mercury, catch gold well, and are little trouble to keep clean. The plates need not be re-silvered, except after scraping and sweating (see p. 133), as they become coated with gold amalgam in the course of time. About 1 oz. of electro-deposited silver is required per square foot of copper plate. Silvered plates are not used inside the mortar.

The position of the plates is as follows:—The lower edges of the inside plates are level with the upper surface of the pulp, when the battery is working properly—*i.e.*, they are usually at $1\frac{1}{2}$ or 2 inches above the surface of the dies. The plate on the feed side is generally about 9 to 12 inches wide, and is of the same length as the battery; it is bolted to the mortar itself, and its angle of inclination varies with the shape of the latter, so that the angle of inclination is sometimes 40°, and it is sometimes nearly vertical. The plate on the discharge side is inclined at an angle of 10° or 20° to the vertical, and is as wide as the space below the screen permits, being usually from 3 to 6 inches wide. It is fixed to a wooden chuck-block, which has its top bevelled off so as not to obstruct the screen opening. The block is bolted to the mortar with some thicknesses of blanketing between, in order to make a tight joint. Several sizes of these chuck-blocks, with their copper plates attached, are kept in one and the same mill, a wider block being substituted for a narrower one when the wear of the dies has proceeded to a certain extent.

At the Alaska Treadwell Mill, and at one or two mills on the Rand, the cast-steel linings of the mortars are furnished with several horizontal slots or recesses for the collection of amalgam, and these take the place of the back copper plates. The front or chuck-plate is, however, retained in these mortars. This is shown in Fig. 45a, Chap. X.

The outside plates are fastened to a wooden table with copper nails, or wooden clamps, or by wedges driven into the raised edges of the table. The table is as wide as the battery (4 feet 7 inches) and usually from 6 feet to 8 feet long. In California a length of 2 or 3 feet of plates of the same width, the *apron plates*, are interposed between the battery and the tables proper, on to which there is a drop of 2 or 3 inches. Below the tables there is a succession of four or five *sluice plates*, each about 30 inches long, with drops of 1 to 3 inches between them. They are usually made narrower than the others, and are frequently only 12 or 18 inches wide, but this practice is not to be commended, as the stream of ore and water, forced into a narrower channel, becomes deeper and flows more rapidly and tumultuously, with the result that the contact between the ore and amalgamated plates is much reduced, and very little gold is caught. The use of the drops is to assist in catching the float gold and to separate the amalgam which has become floured and mixed with the pulp. In place of the composite arrangement described above, a single length of 12 or 14 feet of plates is often used, especially in the Transvaal.

To aid in catching the float gold, swinging amalgamated plates have been introduced, and are in use in the sluices below the batteries of many Californian mills. They are also used in hydraulicking. The swinging plate consists of a curved strip of silver-plated amalgamated plate about 3 inches deep, and of the same width as the sluice in which it is hung; it is suspended on eyes through which wires pass. The plate thus hangs, half submerged, with its concave side up stream, and is kept swinging by the current, so that all floating particles of gold must come in contact with it. It is found in practice that, immediately under each plate, across the sluice, a line of amalgam which has dropped from the plate accumulates. The plates are placed a few feet apart. They cost little and are very effective.

Mill Site.—This should be easily accessible by road, rail, and water, if possible; moreover, it should be near both wood and water, and there should be a good fall of the ground. The least fall that is considered sufficient in California is 33 feet from the mouth of the rock-breaker to the floor on which the concentrators are placed, when rock-breakers are used, followed by stamps, copper-plated tables, sluice plates, and two successive concentration tables. If a second concentrator is dispensed with, however, and space otherwise economised as far as possible, 29½ feet may be enough.

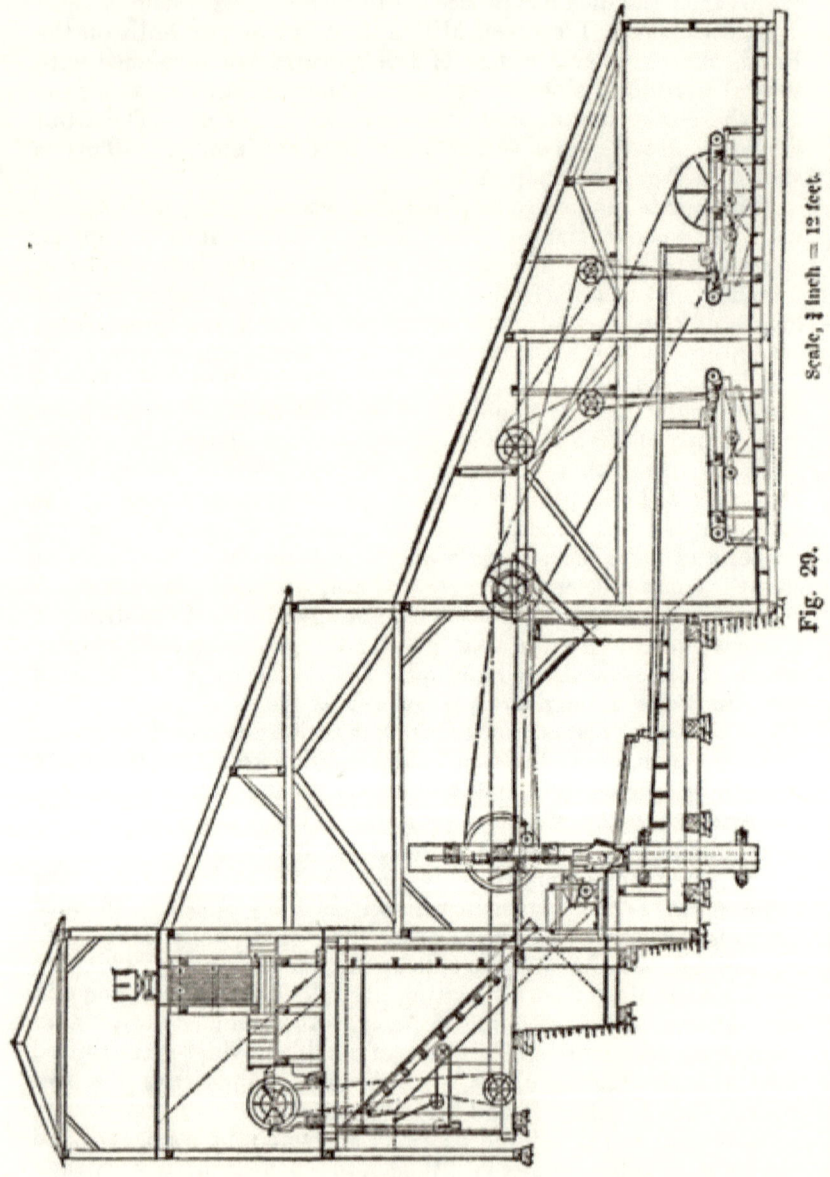

Fig. 29. Scale, ⅜ inch = 12 feet.

Arrangement of the Mill.—The general disposition of the machinery is shown in section in Fig. 29. This represents a mill in which the ore is delivered from the ore-cars through a grizzly on to the rock-breaker floor, and thence by a shoot to the automatic feeders of the stamp battery; the pulp, after passing over the plates, is conveyed by sluices to the double row of "frue-vanners" (described on p. 176), which are shown standing back to back on the lowest floor. The ore-bins should be of sufficient capacity to hold several weeks' supply for the stamps, in case of breakdowns or other delays in the mine. Their floors are made of planking, which is laid with the lengths in the direction of the slope, for, if placed transversely, the boards wear fast, and the ore packs at the edge of each one, with the result that its movement is impeded and must be assisted by shovelling. The slope should be at least 45° in order to enable the ore to move downwards by gravity, when the lowest portion is drawn from the shoot. Ore-bins with flat bottoms have greater capacity, but necessitate an additional handling of the ore. The sills of the bins should be placed horizontally on terraced ground, not on the slope of the hill. A shoot from the ore-bin door leads to a grizzly, through which the fine ore drops into the main battery ore-bin. The larger pieces of rock are discharged either into a coarse ore-bin or else upon the platform by the side of the rock-breaker and on a level with its mouth, into which it is shovelled by hand. The former course is preferable, as in that case the rock-breaker can be fed continuously by a gate in the coarse ore-bin, which is opened and shut by a rack and pinion. By this arrangement there is a saving of labour, but the chief advantage is that the rock-breaker is thereby kept constantly at work. At the North Star Mill, California, it was found that when, by arranging for a continuous feed from the coarse ore-bin down a shoot leading direct to the rock-breaker, the latter was in constant work, it absorbed 12 horse-power, as against 8 horse-power when in intermittent work, but its output was over 50 per cent. more. When the stamp battery is used only to crush the ore, which is subsequently treated in pans or other amalgamators, or by concentration, it is of great advantage to separate the fine product of the rock-breaker by sieving, instead of passing the whole through the stamps. This arrangement increases the output and prevents unnecessary sliming of the ore, thus greatly reducing the loss of sulphides when an attempt is made to save these by concentration. As observed on p. 101, it is becoming customary in large mills to place the rock-breakers in separate buildings.

The product of the rock-breaker is mixed as thoroughly as possible with the ore, which originally passed through the grizzly, and led by means of a shoot direct to the automatic feeders, which should be mounted on wheels, so as to be readily movable. Plenty of space must be left behind the feeders for convenience in repairing and in the exercise of supervision. The amalgamating tables should also be easily accessible, space being left to pass between them, and the same remark applies to the sluices and the tables or other appliances for concentration. All shafts, bearings, &c., should also be easily accessible, so that oiling, re-lining, and repairs may be readily done.

The tailings are discharged into a large sluice by which they are carried into a river, or into the sea, or run into settling pits, or impounded behind dams. One of the two latter courses is adopted, either if water is scarce, so that it is necessary to use it over again, or if the discharge of tailings is forbidden by law, or if the tailings are rich enough to be subjected to further treatment, at once or at some future time.

The whole of the machinery is contained in a strong building of timber or corrugated iron, to protect it and the workmen from the weather, and to prevent theft of amalgam. The interior should be lighted by as many windows as possible, in order to facilitate superintendence and repairs. In cold climates, the mill buildings are sometimes warmed by passing the flues from the boiler fires through them from end to end, before leading the products of combustion to the stack, or by steam-pipes.

CHAPTER VII.

AMALGAMATION IN THE STAMP BATTERY.

Treatment of the Plates.—In order to keep the plates in proper condition so that successful amalgamation may be maintained, they must be prepared carefully, and the closest watch kept over them. The silver-plated copper table is preferred in California from the ease with which it is kept clean, but is not used in the Transvaal. It is not considered desirable to put on it as much mercury as it will hold, since, if the amalgam is too fluid, losses are sustained by scouring, but, on the other hand, if the amalgam becomes too hard and dry from absorption of gold and silver, further amalgamation is checked and fresh mercury must be added.

The condition of the inside plates is regulated by the amount of mercury supplied to the mortar. In Colorado there is an opening at the front of the battery and above the screen frame, ordinarily covered by canvas, which can be lifted up by the millman, who introduces his arm, and determines by passing his hand over the front plate whether the right amount of mercury is being added by the feeder. The regulation of the addition of mercury is thus effected without removing the screen frame. The amount added varies with the conditions of crushing and the richness of the ore, but in general from 1 to 2 ounces of mercury are fed in for every ounce of gold contained in the ore. The finer the state of division of the gold and the more sulphides there are contained in the ore, the more mercury is required. It is fed into the battery at stated intervals of from half an hour to an hour. In some mills, particularly in Australia, amalgamation in the battery is not attempted, no inside copper plates being provided, and under these circumstances it is not usual to feed mercury into the battery.

The practice of feeding mercury into the battery, although almost universally pursued in the best mills in the United States and South Africa, still meets with opposition from certain experienced millmen. The objections urged against it are mainly that the mercury so introduced, and the amalgam formed through its agency, tend to become so excessively subdivided that a high percentage is lost through flouring; moreover the mercury is liable to sicken when the ores contain sulphurets. These evils, no doubt, exist, and tend to increase with the percentage of sulphides present, while arsenic and antimony in particular cause heavy losses of both mercury and amalgam if battery amalgamation is attempted, but with ordinary free-milling ores such losses are not serious.

The amalgamated plates are dressed as frequently as is necessary, the length of time allowed to elapse between two operations depending partly on the richness of the ore. To dress the plates, the battery is stopped, and the amalgam wiped off the plates with a brush or piece of indiarubber held between two pieces of wood, so that only about half an inch of it projects; and, if necessary, fresh mercury is added. The amount of mercury put on the plates should be enough to keep their surfaces in a pasty condition, but not enough to gather into liquid drops or to run off. The plates are wiped three or four times a day, or as often as every two hours when rich ores are being treated, whilst, with exceptionally valuable materials, the operation may be necessary at intervals of a few minutes only.

Discoloration of the Copper Plates.—The plates often become stained by the formation on them of oxides, carbonates, or other compounds of copper through the corrosive action of the water and pulp. Ores containing decomposing sulphides acidify the water and thus cause the corrosion of the plates, a yellow film being formed on the surface of the metal. The presence of carbonic acid in the water is equally harmful, but Aaron points out that the addition of slaked lime to the water neutralises the acid substances and diminishes the tarnishing. The yellow, brownish, or greenish discoloration, the so-called "verdigris," appears in spots and spreads quickly, especially on new plates, those which have been silver-plated being less liable to become dirty than the others; whilst when a plate has become covered with a thick layer of amalgam it is not readily discoloured. When these stains appear the plate must be at once cleaned, as the stained part catches little or no gold. The chemicals used for the purpose are generally sal-ammoniac and potassic cyanide, the operation being conducted as follows:— The battery is stopped, the plates rinsed with clean water, and a solution of sal-ammoniac applied to the stained parts with a scrubbing brush, and left covering them for a few minutes in order to dissolve the oxides. It is then washed off, a solution of potassic cyanide rubbed on to brighten the plate, and almost instantly washed off, fresh mercury being then added if necessary. Janin states (*Mineral Industry*, 1894, p. 332) that long brushing with potassium cyanide is necessary, as otherwise the spots reappear when the water is turned on.

Whisk brooms are perhaps better than indiarubber for brushing the plates; these brooms are cut down to a short length so as to be stiff enough. The plate is brushed all over and the amalgam thus thoroughly loosened from it, after which, commencing at the top where the amalgam is thickest, it is subjected to a systematic stiff brushing, each stroke being directed longitudinally down the table, and not towards the centre. The surplus amalgam is thus brushed to the lower end of the plate, whence it is removed, and a thin coating of amalgam is left over the whole surface of the plate, excluding the air and preventing the formation of "verdigris."

Chemicals Used to Promote Amalgamation.—The use of chemicals to aid amalgamation was formerly much more general than at present, although battery-men were less afflicted by the rage for nostrums than pan-amalgamators. Almost the only chemical now used on a battery in California or Colorado, both to promote amalgamation and to clean the plates, is cyanide of potassium, and the use even of this reagent is becoming less general every year. A dilute solution is believed to promote amalgamation, but probably its action consists merely in thoroughly cleaning the plates and the mercury from all trace of oil, grease, and base metallic oxides. 1 or 2 lbs. of potassic cyanide should be enough to supply a 40-stamp battery for twelve months when treating free-milling ores, by which the mercury is not much affected. The difference between such a mill and one running on base ores may be judged from the fact that at the Hidden Treasure Mill, Colorado, where there are seventy-five stamps, 260 lbs. of cyanide were used in a year. Here the plates are dressed every twelve hours with a weak solution containing 2 oz. of cyanide in 3 gallons of water, the operation being necessitated by the acidity of the water which comes from the mine, and is further contaminated by sulphates formed in the ore. Sodium is now used mainly to clean the mercury, after it has been retorted, before again using it in the battery. A weak solution of caustic soda is used to remove grease from the plates.

The merest trace of any kind of grease or oil is very prejudicial to successful amalgamation, forming a film over the plates and over the little globules of mercury, and thus preventing contact between them and the particles of gold. Ammonia or potassic cyanide removes the film, but a saturated solution of wood-ashes, or of soda carbonate or caustic soda is more generally used than these reagents. Grease may consist of the tallow dropped from the miner's candles, or the oil from the loose steam which is sometimes used to warm the battery feed-water, or from the bearings of the machinery.

Mercury.—The mercury used for the battery and the plates should be quite free from base metals, such as copper, zinc, lead, or tin. If these are dissolved in the mercury they become rapidly oxidised and soon cause it to "sicken," so that it breaks up into a number of minute globules, each coated with a layer of base metallic oxide. These globules are not only useless for amalgamation, but as they are very minute and refuse to coalesce, they are carried away in the tailings, together with such gold as may have been taken up before the oxidation had proceeded far. Such impure mercury, when used to coat the plates, also causes their discoloration by oxidation of the dissolved base metals. Mercury acts best when it already contains gold and silver, and it is customary to dissolve some silver in the new mercury applied at the starting of a mill, if no old mercury can be had to mix with it. When work has been some time in progress there is no difficulty on this account, as the quicksilver, which has been squeezed through a filter to separate the amalgam, carries some dissolved silver and gold with it.

Grade of Plates.—The grade or inclination of the plates varies with the nature of the ore to be treated, heavy pyritic ores requiring a higher grade than light quartz, while the coarser the crushing the steeper must be the grade. In California the copper tables have an inclination of from $1\frac{1}{2}$ to 2 inches per foot, the apron plates from $\frac{1}{2}$ to $1\frac{3}{4}$ inches per foot, with an average of $1\frac{1}{2}$ inches, and the sluice plates from $1\frac{1}{4}$ to $1\frac{1}{2}$ inches. With heavily sulphuretted ores a grade of from 2 to $2\frac{1}{2}$ inches per foot is used. The steepness of the grade is of great importance, as on it, and on the amount of water supplied, the attainment of the necessary contact between the ore and the plate depends. When the pulp is flowing properly, it travels down in a series of little waves and ripples, and, in consequence of the friction between the plate and the film of water in contact with it, the upper portions of these little waves travel faster than the lower parts, so that the motion becomes one of tumbling over and over. As a result of this, if the plate is long enough, every particle of pulp comes in contact with the amalgamated surface, and the perfect extraction of amalgamable gold, mercury and amalgam is obtained.

Muntz Metal Plates.—The use of Muntz metal (which consists of copper 60 per cent., zinc 40 per cent.) for amalgamated plates is of great interest. It differs from copper in catching gold well as soon as the plate is amalgamated, not requiring to be covered with gold- or silver-amalgam before it begins to do good work. Moreover, the amalgamated surface is very superficial, since the mercury does not sink in so far as it does into a plate composed of pure copper, so that only a small quantity of mercury is required to cover it. The result is that cleaning up is easy and rapid, no iron instrument being necessary, but rubber being always sufficient. These properties make it particularly valuable for custom mills, where it is desirable to catch as much as possible without mixing the amalgam obtained from two parcels of ore crushed in succession. On the other hand, as it holds little mercury, it cannot absorb much gold, and must be cleaned-up at frequent intervals.

The mercury on Muntz metal plates does not suffer so easily from "sickening" as that on copper plates; it has been suggested that this is due to the electrolytic action of the copper-zinc couple, which sets free nascent hydrogen, and so reduces the compounds of mercury and other metals which have been formed. It follows that Muntz metal plates are preferable for ores containing large amounts of heavy sulphides or arsenides. The greenish-yellow stains (called "verdigris" by millmen) which are formed on copper plates when grease and other impurities are present in the battery water, do not appear when Muntz metal is used, and such discolorations as occur on these plates can be better removed by dilute sulphuric acid than by potassium cyanide. At the Saxon Mill, New Zealand, the copper plates formerly required 7 lbs. of cyanide, costing 23s., per month to keep them in order, while the Muntz metal plates, by which they were replaced, could be kept clean by 5 lbs. of sulphuric acid per month, the cost being 3s. 4d. It is stated, however, that, in the treatment of highly acid ores, which have been weathered for some time so that they contain large quantities of soluble sulphates, or in cases where the battery water contains acids, copper plates are less affected than Muntz metal, over which a scum is rapidly formed. This is not the experience in the Thames Valley, N.Z., where Muntz metal is preferred in spite of the extremely acid nature of the water and ore.

Generally, it may be stated that Muntz metal plates are cheaper, wear better, and require less attention than copper. In dressing new Muntz metal plates the following method is adopted in New Zealand:—The surface of the plate is scoured with fine, clean sand; then it is rinsed with water, and washed with a dilute (1 to 6) solution of sulphuric acid. Mercury is then applied and rubbed in with a flannel mop until it wets the surface of the plate (*i.e.*, amalgamates with it) in one or more places, after which the mop is given a circular movement, passing through these spots, until the amalgamation of the surface spreads from them over the whole plate.

The discoloration of the Muntz metal plates is prevented by the weak electric current produced, as has been already stated. The same effect can, according to Aaron, be obtained when ordinary copper plates are in use, by placing them in contact with iron or some other metal which is positive to copper. Strips of iron bolted to the top and sides of the plate are said to be sufficient for the purpose, the copper being in that case unaffected by the acidity of the water, which causes oxidation and dissolution of the iron only. Janin's experience does not support these views.

Shaking Copper Plates.—A shaking copper plate is recommended by some of the best authorities to be used either below or in place of the ordinary amalgamating tables, especially in cases where these do not appear to give good results. An ordinary fixed copper plate requires an inclination of from $1\frac{1}{2}$ to 2 inches per foot, in order to keep it clear of sand, when the plate is of the same width as the battery. If, however, the plate is subjected to a short rapid shake, the sand is kept from packing, and amalgamation is well performed with a grade of only $\frac{1}{4}$ to $\frac{1}{2}$ inch per foot, or the amount of water needed with the pulp may be greatly reduced and better contact thus obtained. For these plates, silver-plated copper is the material employed. They are affixed to a light wooden frame which is moved by a crank-shaft, revolving 180 to 200 times per minute, placed on one side, with a throw of 1 inch at right angles to the direction of the flow of the pulp. In some mills, a longitudinal shake is given to the plate instead of this side shake. The frame may be hung on rods from above, but is more conveniently supported on four short iron springs, forming rocking legs. The width of the tables should be made as great as possible, while the length is of less importance, as, the thinner the current of pulp flowing over them, the better the chance of the gold particles coming in contact with the plates and being retained. These shaking plates were first used in Montana in 1878, and have since been employed at several Californian mills, giving satisfactory results. It is advantageous to add to them an amalgam- and mercury-saver. A simple device for this purpose is to nail a strip of wood, half an inch thick, across the copper plate near the top, thus forming a shallow riffle, the angle of which is soon filled with sulphides and coarse sand, which are kept in agitation by the movement of the table. This is stated by W. M'Dermott and P. W. Duffield[*] to be the most effective contrivance yet devised for catching quicksilver and hard amalgam. If the inside copper

---

[*] *Gold Amalgamation*, London and New York, 1890, p. 16.

plate should become hard by accident or neglect, chips of amalgam escaping through the screens are retained in this riffle, and, becoming spherical by rolling up and down under the effect of the shaking motion, increase in size just as a snowball does when rolled in snow.

Corrugated Plates.—Another device to increase the catching power of copper plates is to make them in a corrugated form so that they will hold little pools of mercury. Where much float gold is believed to be passing off, this is supplemented by the addition of a second plate placed above and facing the other with its amalgamated surface downwards. The pulp is allowed just sufficient room to pass between them, in contact with both, and there is little doubt that in this way some fine gold is caught that would otherwise escape.

Mercury Wells.—Another method of saving mercury and amalgam, which would otherwise be lost in the tailings, consists in the application of mercury wells or riffles. A mercury well or riffle consists of a shallow gutter filled with mercury, over the surface of which the pulp flows or through which it is forced to pass by suitable machinery. Attwood's amalgamator, formerly much used in California, was a machine of the latter class. Such wells are even now often used in modern mills. It is believed by some of the best authorities that such devices catch no more mercury than could be secured by either well-arranged stationary or shaking plates, while the practice of placing a well between the screens and the amalgamating tables is especially to be condemned, as it prevents proper supervision being kept over the feeding of mercury into the battery, over-feeding being difficult to detect under these circumstances.

Galvanic Action in Amalgamation.—In amalgamation in the mortar, on plates, or in pans, not only are free metals absorbed, but the dissolved salts, and, to a less extent, the insoluble compounds of the heavy metals, are reduced and amalgamated, chiefly by galvanic action. The copper of the plates, or the iron of the mortar or pan, constitutes the positive element, and all metals less oxidisable than this reacting metal are reduced by it, and are then amalgamated by the mercury. In this way iron reduces both lead and copper, although, if these are present in the form of undecomposed sulphides, this action will be very slight. Now, if lead is introduced into the amalgam, the latter becomes pasty, and is subjected to considerable losses, and copper has an equally harmful effect. It is for this reason that the arrastra is found to be better than the stamp battery or even the pan for certain plumbiferous ores. This action of iron is of course enormously increased if the ores are subjected to a chloridising roast before being amalgamated, as in the Reese River process for the treatment of auriferous silver ores.

In a number of mills, this galvanic action is increased by the passage of a weak electric current through the charge by means of a dynamo. The amalgamated plates or the walls of the pan are connected with the negative pole, while the positive pole is formed of a plate of lead or iron dipping into the pulp. Under these conditions the mercury is still further protected from attack, and remains bright and lively, but the deposition of base metals in it is favoured, and the stronger the current the more this action is induced. Consequently, such methods are attended with the best results when dealing with ores containing little or no copper, lead, &c., since in these cases the strength of current can be increased, and the mercury kept clean, without any ill effects. The principle is made use of in Bazin's centrifugal amalgamator, Molloy's hydrogen amalgamator, and other similar machines.

Designolle Process of Amalgamation.*—In this process a solution of mercuric chloride is used. It was tried at the Haile Mine, South Carolina, the method being as follows:—Charges of 600 lbs. of roasted ore were placed in cast-iron barrels with 1,000 lbs. of cast-iron balls, or in pans. The barrel was partly filled with water, and 1 gallon was added of a solution containing 1·7 per cent. of mercuric chloride, the same amount of hydrochloric acid, and twice as much salt, if the ore contained less than 15 dwts. of gold per ton. This would be equivalent to about 10 parts of mercury to 1 part of gold. The barrel was rotated for twenty minutes and then discharged into a settler, and the suspended amalgam caught on copper plates. The mercury was supposed to be reduced by the iron thus—

$$HgCl_2 + Fe = FeCl_2 + Hg,$$

and the metallic mercury thus freed amalgamated with the gold. If no common salt was present, some mercurous chloride was formed according to the equation—

$$2HgCl_2 + Fe = Hg_2Cl_2 + FeCl_2,$$

and the subsequent reduction of the insoluble calomel by iron was not complete. Hydrochloric acid was supposed to hasten the amalgamation by setting up some electrolytic action.

The total cost of the process at the Haile Mine was said to be only 35 cents, or 1s. 7½d., per ton, but it was abandoned when the percentage of iron in the material under treatment increased, owing to improved methods of concentration. Large quantities of oxide of iron were then amalgamated, and rendered the resulting mass harder to treat than the ore itself. By repeated washing, settling, and regrinding with fresh mercury, it could be partially purified, but not without a loss of gold. It is stated that 87 per cent. of the gold in the ore was extracted.

* This account is abridged from that given by Louis Janin, Junr., in *Mineral Industry* in 1894, p. 346.

The Clean-up.—The amalgam, both on the inside and outside plates, does not accumulate evenly, but in ridges and knots which serve as nuclei for the collection of more. It is not advisable to allow the coating of amalgam to become very thick, since, although the plates catch better as the amalgam accumulates, losses may be experienced by scouring. The wiping of the outside plates usually takes from 10 to 15 minutes for each battery. The amalgam so obtained is ground with more mercury in a clean-up pan in order to soften it, the skimmings from the mercury wells, &c., being added to the charge. The inside plates are not wiped until the amalgam stands up in ridges on it; the operation may be necessary as often as twice a week, but it usually takes place twice a month, when a general clean-up is made. All amalgam, however obtained, unless it is already hard and dry, is usually at once separated from the excess of mercury contained in it by being squeezed in filter-bags, and the pasty residue alone passed to the clean-up pan or the retort.

In cleaning-up, the stamps are hung up, two batteries at a time, the screens, inside plates, and dies are all taken out, and the "headings," or contents of the mortar, consisting of the pulp, mercury, sulphides, and pieces of iron and steel, amounting in all to a quantity sufficient to fill two or three buckets, are carefully scraped out and fed into the mortar of one of the other batteries, which has not yet been cleaned up. In California, the headings from the last batteries are panned, the iron removed by a magnet, and the remainder ground with mercury in the clean-up pan. In Gilpin County, however, it is not panned, but merely returned to the battery on restarting. Amalgam is found adhering to the inside of the mortar and to the dies, and is carefully detached and added to the clean-up pan, and all the plates are well scraped with a sharp-edged piece of hard rubber, care being taken not to scratch them. The plates are then redressed and put back, the batteries restarted, and the next ones stopped and cleaned up. Three men can clean up forty stamps in from five to seven hours, ten stamps being thus idle for the whole of this time.

The amalgam obtained from the batteries, outside plates, mercury wells, or sluices, is rarely clean enough for immediate retorting; it is usually found to contain mixed with it grains of sand, pyrites, magnetite and other minerals, together with fragments of iron and other foreign substances. The skimmings from mercury wells are still more impure. These materials must be purified by grinding with fresh mercury and washing before they can be passed to the retort. The scraps of iron consist of fragments of shoes, dies, shovels, picks, hammers, and drills: they are knocked about in the mortar until a quantity of gold and amalgam has been driven into their interstices. At the Jefferson Mill, Yuba County, California, about $\frac{1}{2}$ ton of such scrap, picked out by hand or by a magnet, had accumulated in 1885. It was attacked by warm dilute sulphuric acid until the surface had all been dissolved off, and the residue was then well washed, and gold to the value of $3,000 thus recovered. The shoes, dies, &c., which were too large for this treatment, were boiled in water for half an hour, and then struck by a hammer, when the gold dropped out.* In Colorado, and in small mills generally, the dirty amalgam is ground in a mortar by hand with fresh mercury and hot water, until it is reduced to an even thin consistency, when the dirty water is poured off, and the mercury poured backwards and forwards from one clean porcelain basin to another until the pyrites, dirt, &c., have risen to the surface, when they are skimmed off. The clean mercury is then squeezed through canvas or wash-leather, when the greater part of the gold and silver contained in it, together with about one and a-half times its weight in mercury, remains in the bag, the rest of the mercury, with a small quantity of the precious metals dissolved in it, passing through. The skimmings obtained are put back into the mortar, and re-ground by themselves with fresh mercury.

In large mills a clean-up revolving barrel is often employed to mix the amalgam. This is made of iron, and is similar in construction to chlorinating barrels, but without the lead lining. At the Plumas Eureka Mill, the barrel is 3 feet in diameter and 4 feet long, and revolves twenty times a minute; the charge is 700 lbs. of amalgam and 20 lbs. of mercury, or more if the amalgam is very rich. A dozen or more pieces of iron, such as worn out battery shoe shanks, are put into the barrel, which is filled nearly up to the top with water, and then revolved for from six to twelve hours. The use of the iron is to help to mix the amalgam and mercury, but it causes some loss by flouring, and is omitted with advantage in Australian mills. The barrel is then opened and washed out with water, the tailings being run over amalgamated plates and through a mercury well or some other form of amalgam-saver, after which the amalgam is scooped out of the barrel and squeezed again in wash leather.

The Clean-up Pan is even more extensively used than the barrel. One of the oldest in use in the United States, the Knox Pan, is still a great favourite, especially for treating battery sands, skimmings, &c. It consists (Fig. 30) of an iron pan, 5 feet in diameter and 14 inches deep. Wooden or iron shoes are attached to the arms, g, which make from twelve to fourteen revolutions per minute. Iron shoes are considered better for brightening or polishing the particles of gold contained in the pyrites, and so rendering them fit for amalgamation. The charge for this pan is about 300 lbs. of concentrates, skimmings, &c. Concentrates are treated in this way only when the quantities of them are small, while at the same time they

* *Sixth Report Cal. State Min.*, 1886.

contain mercury and amalgam. The charge is made into a pulp with water and ground for three or four hours, after which 50 lbs. of mercury are added, and mixing is carried on for a few hours longer, before the pulp is diluted, settled, and discharged. The tailings suspended in the water are often caught in settling pits, and either sold or subjected to further treatment on the mill, as they are frequently of high value. When the Knox pan is required for mixing dirty or impure amalgam with additional mercury, the wooden shoes are used, as shown in the figure. The waste matter is then washed off with a stream of water as before, and the amalgam retorted.

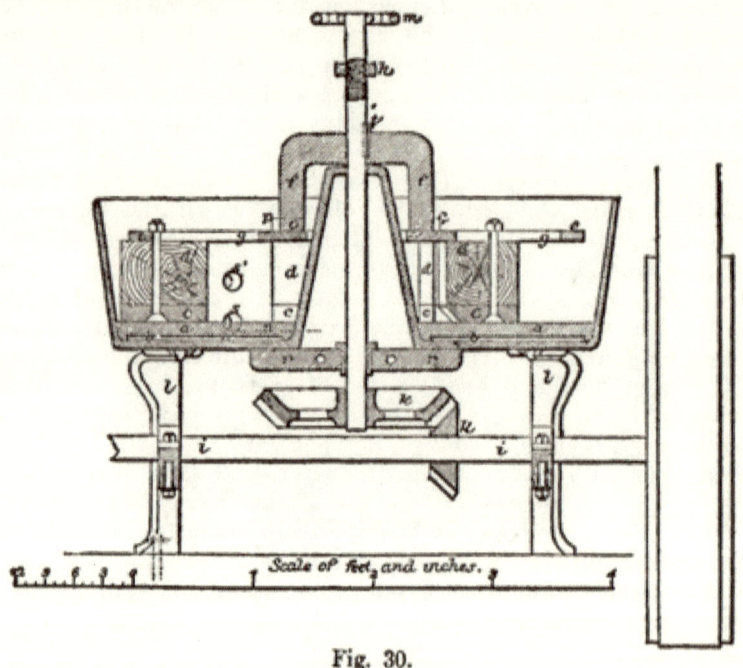

Fig. 30.

The position from which the greater part of the gold is obtained in a clean-up varies according to the ore and the method of treatment. In California from 50 to 80 per cent. of the gold saved is caught on the single plate inside the battery, the remainder being caught on the outside plates and the sluice plates, or being contained in the concentrates. In the Grass Valley, at the Original Empire and the North Star Mills, from 70 to 85 per cent. is caught inside the battery. The amalgam from the battery plates is usually richer than that from the outside plates, especially if the gold is coarse. The reason for this is that coarse gold, being easily amalgamable, is almost all caught on the inside plates, while fine gold, even if amalgamated in the battery, forms a more fluid amalgam which passes through the screens and is caught outside. Coarse gold forms a richer and stiffer amalgam than fine gold, because it is more easily taken up by amalgam which is already almost saturated with gold. At the two mills last named the value of the plate amalgam is $4.50 per oz., and that of the battery amalgam $8.50 per oz.

About every three or six months the plates are scraped with a sharp iron chisel or palette knife, or scaled by chipping with a hammer and chisel. At the North Star Mill, California, the plates are immersed in boiling water so as to soften the amalgam before they are scraped. They are sometimes "sweated" in California by heating them over a wood fire before scraping them. At the Empire Mill, Grass Valley, California, the sweating of the outside and apron plates of four batteries produced bullion to the value of $19,000.* These plates had been down for eighteen months, and the ore which had been run over them averaged 18 dwts. of free gold per ton. After scaling and sweating, the plates may require replating. In course of time they are worn out, the copper becoming brittle and worn into holes, but they usually contain gold enough when discarded to pay for a new set.

Several methods are in use for recovering the gold from old plates. For example, they may be dissolved in nitric acid, when the gold is left nearly pure. A more economical method of detaching the gold, much used in Australia, is described by Mr. M'Cutcheon as follows:—The plates are placed on the hearth of a reverberatory furnace, or on a fire made with logs in the open air, and the mercury expelled at a gentle heat. If the temperature is too high, the gold sinks into the copper at once, and the copper must then be dissolved. After the mercury has been driven off, the plate appears to be more or less coated with gold on one side. This surface is treated with hydrochloric acid for eight or ten hours, and the plate is then replaced on the hearth and exposed to a dull red heat until well blackened. On plunging it into cold water, the gold now scales off, and is collected and freed from copper by boiling in nitric acid.

Retorting.—The solid amalgam, which is retained in the canvas or wash-leather filters, usually contains from 30 to 45 per cent. of gold and silver, according to the state of division of the gold present in the ore, and also to the degree of care exercised in squeezing out the excess of mercury. For separating the gold from the mercury there are two kinds of retorts in general use—the *pot-shaped* retort, which is sometimes cast with trunnions to swing on supports, in small mills; and the cylindrical retort, shown in Fig. 31, in larger mills. The pasty amalgam is rolled up into balls or kneaded into cakes, and squeezed into the pot-shaped retort, and often rammed down with a

* *Eighth Report Cal. State Min.*, 1888, p. 714.

bolt-head, although this course is deprecated by some managers, who prefer to leave the amalgam as spongy and open in texture as possible, believing that a uniform product is thus obtained more rapidly at a lower temperature and without so much loss. In the horizontal retort the amalgam is placed in iron trays divided into compartments by partitions. In either case, the retorted metal is prevented from adhering to the iron, either by laying it on three or four thicknesses of paper, the ashes of which remain beneath the amalgam, or by covering the iron trays with a coating of whitewash. The former plan is preferred. The mercury is condensed in cooling tubes passed through water; the loss through volatilisation is usually very

Fig. 31.

small, and may be taken as being about one grain of gold per pound of mercury.

The charge is heated slowly until the boiling point of mercury is reached, when the fire is checked, and the retort kept at an even temperature for one or two hours, or until the bulk of the mercury has been driven off. The retort is then raised gradually to a bright red heat to expel the remainder, and after cooling it is opened, the trays are withdrawn, and the retorted metal is loosened by a chisel, if necessary, and turned out on a table.

In retorting amalgam containing considerable quantities of base materials, there is a danger of the vent being choked up by condensation of solid material. The retort should be so arranged that a rod can be passed through the condensing pipe so as to clear it of obstructions, if necessary. The front of the retort is luted on carefully with chalk or wood-ashes and salt, and firmly clamped so as to be quite tight, otherwise a loss of mercury is incurred. In all retorts the lid is turned and ground so as to fit on perfectly. The condensing pipe should not have an open end dipping freely into water, as in that case a sudden cooling of the retort would cause the water to be sucked in, and an explosion would occur. The open end of the pipe may be enclosed in a rubber bag immersed in water, in which case the amount of distention of the bag will indicate the progress of the operation.

The pot-shaped retort requires no brick fittings, and can be heated over an assay furnace or forge fire, or in a fire built on the ground, when it is placed on a tripod stand. In the latter case the fire is lit at the top and burns slowly downwards. The pot-shaped retort is not filled to more than two-thirds its capacity, and must be heated very gradually at first.

The retorted metal is porous and from 500 to 950 fine in gold, the remainder being in general chiefly silver, with base metals and sulphides in smaller quantities. The gold is melted in crucibles with carbonate of soda and borax, and suffers a further loss in weight, due to the volatilisation of a small quantity of mercury, which is obstinately retained until the melting takes place. This amount is usually not more than from 0·5 to 1 part per 1,000 of the retorted metal, if the retorting has been carefully performed.

Loss of Mercury.—The loss of mercury is due to two causes, "flouring," or minute mechanical sub-division, due to excessive stamping or grinding, and "sickening," or extreme sub-division caused by chemical means. In the latter case, a coating of some impurity is formed over the minute globules of mercury, which are thereby prevented from coalescing, from taking up gold and silver, or from being caught by the plates and wells, as the coating prevents all contact between the mercury and other bodies. The impurity may be an oxide, sulphate, sulphide, or arsenide of some base metal, either originally present in the mercury, or taken up from the ore by it; occasionally the mercury itself may be partly converted into a sulphate or other salt, although this latter condition is not common. The employment of mercury, which contains no base metals dissolved in it, will reduce the loss due to sickening, but such pure mercury is not always obtainable (except by very careful distillation, in which the first and last portions condensed are rejected), and it soon takes up fresh impurities when used with sulphuretted ores. The base metals usually present in mercury are rapidly oxidised in the air, especially in contact with water; the oxidation is made much more rapid by the presence of any acid in the water, and this acidity (due to the presence of acid sulphates from decomposing pyrites) is rarely quite absent from battery and mine waters. The metallic oxides thus formed are not soluble in mercury, and they float on its surface in the form of little black scales, which soon form a coating. A surface coating may also be formed of grease. The use of sodium amalgam in preventing the sickening of mercury is noticed below (p. 137). One of the impurities in mercury most to be feared is lead, as the amalgam of this metal tends to separate out of the bath of mercury in which it is dissolved. According to Prof. J. Cosmo Newbery, it rises to the surface by degrees, taking with it any gold amalgam that may have been formed, and floats as a frothy scum, coating the mercury and preventing any further action by it, whilst it is readily powdered and carried away in suspension by a current of water flowing over it, so that the gold contained in it is lost.

Sickening of the mercury is also promoted by base minerals present in the ore. Most gangues, except heavy spar, hydrous silicates, &c., have no action on the quicksilver; even clean cubical iron pyrites, and other iron and copper pyrites, if they are undecomposed, are harmless, although the materials last named cause sickening when partly decomposed. The other sulphides are all more or less harmful, their action being, however, much less energetic than the compounds of arsenic and antimony. J. Cosmo Newbery conducted a number of experiments in Australia many years ago, to determine the action of some of the base metals on mercury, and found that compounds of arsenic and antimony are particularly harmful, and that if gold containing metallic arsenic is amalgamated, the resulting amalgam is black and powdery, and floats on the mercury, being coated by black metallic arsenic, which separates out and refuses to unite with mercury. Arsenical pyrites seems to act in the same way as metallic arsenic, a large amount of black sickened mercury being produced by it, the action being especially energetic if the pyrites is partly decomposed. The black coating is, in this case, a mixture of pyrites, arsenic, and mercury, in a very finely divided state. Sodium amalgam acts beneficially when arsenic is causing loss of mercury.

Sulphide of antimony breaks the mercury into black powder even more quickly than arsenic, some sulphide of mercury being formed if there is any trituration, whilst the antimony forms an amalgam. The action of sodium amalgam on this mixture is of no avail, as sodic sulphide is formed, more antimony amalgam produced, and sulphuretted hydrogen set free, the results on the amalgamation of the gold being very disastrous. Bismuth sulphide acts similarly, but with less rapidity.

Floured mercury is perfectly white in appearance, like flour, sickened mercury, as already stated, being blackish. If this floured mercury is examined with a lens, it is seen to consist of a number of minute particles—many of them microscopic—each of which is perfectly bright and pure, shining like a mirror. They are prevented from coming into contact and coalescing by being surrounded by films of air, concerning which Prof. Huntington has remarked—" If you scratch the film of air, the particles run together and form one globule." Floured mercury is readily carried away and lost in the tailings, but if passed through and agitated with a large body of clean mercury much of it is at once absorbed in the mass. The loss through flouring is experienced in the milling of refractory and free-milling ores alike.

In California the total loss of mercury varies from $\frac{1}{5}$ to 1 oz. of mercury per ton of ore crushed, the mean being about $\frac{1}{2}$ oz. per ton. Most of the mercury is lost as such and not in the form of amalgam, as is proved by the fact that where the largest proportion of mercury is fed into the battery the greatest loss takes place but the highest percentage of gold is recovered. Thus, at the North Star and Empire Mills the greatest loss in the State occurs, 1 oz. of mercury being lost per ton, but over 90 per cent. of the gold is extracted. In the Blackhawk Mills, Colorado, where base ores are crushed, containing from 12 to 20 per cent. of pyrites, the loss of mercury is from $\frac{1}{5}$ to $\frac{1}{2}$ oz. of mercury per ton. The mills in which the greatest loss of mercury occurs have the deepest discharge, the ore and mercury being in these cases pounded together for a greater length of time before being ejected from the mortar, so that more flouring takes place.

Many suggestions have been made at various times for the reduction of the loss of mercury. The use of certain chemicals in keeping it clean and lively and in neutralising the bad effect of base minerals has already been noticed. Sodium amalgam was suggested by Prof. Wm. Crookes, F.R.S., many years ago as likely to effect this purpose. It is prepared by heating a basin or iron flask of mercury to about 300° F., and dropping in little pieces of sodium not larger than a pea, one by one. Each addition causes a slight explosion and a bright flash of flame. The sodium may be added with less loss and less danger to the operator if the mercury is kept at a somewhat lower temperature and the sodium stirred into the mercury with an iron pestle or pressed below its surface with a spatula. When about 3 per cent. of sodium has been added to the mercury, the reaction becomes less active and the amalgam is then poured out upon a slab or shallow dish, allowed to cool and solidify, and then broken up and kept in stoppered bottles under naphtha. When it is necessary to revivify a quantity of mercury, a few small pieces of the amalgam are added to it and stirred in, or are previously dissolved in clean mercury before being added to the impure stuff. The strong affinity of sodium for oxygen enables it to reduce the oxides of the base metals which are coating the mercury globules, and as sodic oxide is very soluble in water it is at once removed in solution, neutralising part of the acidity of the water at the same time. The base metals are redissolved by the mercury which is then in good condition to take up the precious metals or to be caught on amalgamated surfaces or in riffles. Sodium amalgam is not much used except in amalgamating-pans or in mercury wells or riffles—*i.e.*, wherever large bodies of mercury can be directly acted on by it. It is of comparatively small value when added to the mortar of a stamp battery, although this use of it is not unknown.

Loss of Gold.—The losses of gold in amalgamation may be ranged under the following heads:—

1. Loss of free gold or amalgam due to a want of proper care in amalgamation.
2. Loss of gold or sulphurets imbedded in particles of rock.
3. Loss of gold which "floats" in water and is carried away with the slimes.
4. Loss of gold which is not in a condition to be directly amalgamated.

The latter heading may be subdivided into two, viz.:—

(*a*) Loss of gold contained in sulphurets.

(*b*) Loss of free gold, which is prevented from being amalgamated by being coated with a film of some mineral ("rusty" gold), or with grease, or by being in an unsuitable physical condition (hammered gold).

In the preparation of gold ores for amalgamation every care must be taken that the course best suited to each particular case is being pursued. In some instances, which are not common, the whole of the gold may be present in a form in which it can be directly amalgamated. In general, however, the gold is present in two or more forms, one capable and the others not capable of amalgamation. In such cases there is no reason to be dissatisfied with the action of an amalgamating machine if it extracts a high percentage of the free gold, even though the total extraction obtained by it is comparatively low. It has frequently been declared that the greater part of the loss of free gold, which really takes place, is not due so much to the imperfection of the various amalgamating machines which are now at the disposal of the metallurgist, as to the lack of care and facility of resource on the part of the men in charge of them. It is probable that no two ores can be treated to the best advantage under exactly the same conditions. A millman experienced in the treatment of the ores of one district may be quite at fault when attempting to amalgamate an ore unlike those to which he has been accustomed. A silver mill, in particular, has been pronounced to be the worst possible school for a gold amalgamator, whose work must be closer in proportion as his amalgam is richer than that obtained from silver ores. At the risk of recapitulating much that has been already said, it may be worth while to direct the student's attention to a few general rules which can be applied in many cases. In the first place the stamps and screens must be such as are calculated to produce the largest possible output, without rendering the pulp unsuitable for the processes of amalgamation or of concentration, or both, which are to follow. The ideal crushing would be to "crack the nut and leave the kernel entire," or in other words, to liberate the particles of gold without breaking them. It was formerly believed that a light stamp working rapidly was best adapted for this, and even quite recently light fast stamps, with narrow heads, only 4 inches across, have been re-introduced experimentally. On the whole, however, the tendency is in favour of heavy stamps working fast with a low drop, as producing the maximum output with the minimum amount of slimes. In connection with this, the small production of slimes by the steam stamp (described on p. 149) should be noted.

The subject of delivery is closely connected with that of crushing and must be considered at the same time. The screens are not usually placed quite close to the level of the pulp in the mortar on account of the rapid wear caused by the violent projection of pulp against them when in that position. Their height above the dies is varied according to the ore, the delivery being slower in proportion to this depth of discharge. The best size for the mesh of the screens must be determined by direct experiment. It has often been contended that, as the crushing must be fine enough to liberate the particles of free gold from their matrix, therefore the size of the screens depends on the state of division of the precious metal in the ore. Nevertheless it does not follow that an ore must be reduced so that the gangue is as fine as the gold particles contained in it, although this is sometimes erroneously assumed. On the contrary, even when the gold is of extreme fineness, coarse crushing through 20 or 30-mesh screens, may be the best practical method to adopt, since otherwise the sliming of the ore may cause the loss of valuable mineral which should be obtainable by concentration, besides reducing the yield on the plates. In the course of the crushing, much of the ore will have been reduced to a comparatively fine state of division, and probably this portion will be found after amalgamation to contain but little gold; from this, the coarser material, in which the gold is still locked up, may be separated by sizing in "pointed boxes" (see p. 170) and reground in an amalgamating mill or pan. If, on the other hand, the slimes are found to be as rich as the coarse sand, it is obvious that no finer crushing is required, as the output would be thereby diminished without any corresponding increase in the yield per ton. If the slimes, after separation of the sulphides, are found to contain more free gold than the coarse sand does, this fact points to the conclusion that a coarser screen might be used without detriment, and experiments in this direction should be made, and the limit of economy thus found by trial.

The evils of overstamping, due to slowness of discharge, have been often dwelt on, and probably are frequently exaggerated. It is true that the excessive production of slime thus caused is an undoubted evil, but, setting this aside, it has been frequently asserted that particles of gold are reduced in size by overstamping, so that they will float off in suspension in water, whilst, even if not reduced in size, they are hammered and flattened so as to be rendered incapable of amalgamation. Other authorities, however, consider it doubtful whether much loss is due to either of these causes. Small particles of gold, it is contended, are not likely to suffer further comminution when mixed with larger particles of gangue, which would usually prevent them from coming into contact with the shoes and dies; these small particles, too, would probably only receive a single blow, which would throw them off the die, and a large piece of gold, which might remain on the die while several blows were struck, would not be easily lost even if partially subdivided. As regards hammered gold, it does not seem to be beyond doubt that flattening and hardening alone will prevent gold from being amalgamated. Prof. T. Egleston has described a number of experiments[*] which tend to show that amalgamation is retarded by this treatment of gold, but, on attempting to repeat these experiments at the Royal Mint, the author could not obtain similar results to those of Professor Egleston. Pieces of pure gold when subjected to repeated blows with a clean 7-lb. hammer on a clean anvil occasionally showed a disinclination to amalgamate, but, if these pieces were washed with dilute ammonia, so as to remove any grease that might be adhering to them, they were instantly wetted by mercury and were dissolved by it at about the same rate as clean annealed gold. It thus appeared that, in these cases at least, grease from the fingers of the operator formed a more potent preventive of amalgamation than the hardness of the gold. Moreover, gold-leaf which has been subjected to an extended course of hammering is readily amalgamable. The matter may perhaps be left as an open question for the present.

There is still much difference of opinion as to the desirability of attempting to amalgamate the gold inside the battery. It is considered by some authorities that the addition of mercury to the mortar is a mistake, and that no copper plates should be put inside. In spite of numerous practical illustrations of the soundness of the contrary view, they hold that no machine can be successful at once as a crusher and an amalgamator. The practice of adding mercury to the mortar when no inside plates

---

[*] *Metallurgy of Gold, Silver, and Mercury in the United States*, 1887, vol. ii., p. 586.

are used is certainly not now much favoured, although it is still adhered to in Ballarat and some other districts in Australia. In treating rich ores, however, when the gold is coarse and of high standard, there does not seem to be any valid objection to be raised against catching the gold on inside plates in a concentrated form, instead of letting it all go to the outside. In this case mercury must be added to the mortar, and this is probably not so important a cause of loss of mercury by flouring and sickening as is often assumed. If a decomposing ore is mixed with lime beforehand, and the acidity of the water used is corrected, the conditions do not appear to be favourable to the production of salts injurious to the mercury, and the latter when charged in is probably almost instantly washed through the screens or else dashed against and retained by the plates, the condition of which is thereby improved by the softening of the amalgam, so that gold is more readily caught on it. The mercury does not remain on the die, subjected to repeated blows, which would no doubt cause much flouring. Finely divided gold, however, particularly if it is of low standard, containing much silver, requires more mercury for its amalgamation than coarse gold, and in this case it is difficult to keep the plates in good order, so that it is usually advantageous to save the extra trouble and labour, caused by looking after and cleaning-up inside plates, by putting all the plates outside. This has been the experience in a number of mills, including that of the Montana Company, where the inside plates have been entirely discarded.

Amalgamation outside the battery has also been the subject of much discussion and some careful investigation. Numerous amalgamating machines have been patented, the inventors in every case praising their own contrivances and decrying the copper plate, but the latter has not as yet been superseded, and its principle is applied to almost all its more promising rivals.

To secure successful amalgamation it is necessary that the particles of gold should be brought into absolute contact with the mercury. This contact is obtained in one of three ways, viz.:—

1. The mercury and ore are ground together in pans, arrastras and similar machines, contact being secured by pressing the gold and mercury together.

2. The ore is allowed to flow over or even through a bath of liquid mercury, or the endeavour is made to ensure contact by letting the pulp fall from a height upon the mercury.

3. The ore is allowed to flow over either stationary or shaking amalgamated copper plates, drops being introduced between the plates to break up the pulp and to assist in catching the amalgam.

The first method is undoubtedly the best for ensuring contact, but the operation is tedious and, in most cases, unnecessary. Opinions are divided as to the relative merits of the two latter methods, but the great majority of metallurgists are advocates of the superiority of the plates. They point out that in a mercury bath, in spite of the first impression to the contrary felt by every one on approaching the subject, contact with the ore is very difficult to obtain. If wet pulp is introduced at the bottom of a bath of mercury, it rises to the surface in lumps, surrounded by films of water, and dry pulp is still more effectually protected from contact with the mercury by films of air. If a thin stream of pulp is run over the surface of a bath of mercury of sufficient size, the chances of the particles of gold coming in contact with the quicksilver (by settling through the stream) are greatly improved, but in this case the more convenient copper plate could be substituted for the bath. Moreover, it is well known that rich gold or silver amalgam catches gold more readily than pure mercury, and, whilst the surface of the plates can readily be covered with such pasty amalgam, it would involve a large and unnecessary sinking of capital to keep any considerable percentage of precious metals in the baths. For these reasons, and for the practical reason that plates are found to work better than baths, the use of the latter has been gradually more and more restricted, until they are now only to be found in narrow wells and riffles for the purpose of catching hard amalgam and floured mercury, and as purely supplementary aids to the plates, whilst in the most modern mills they are often dispensed with altogether. When a proper disposition of the plates is made, it is rare to find amalgamable gold escaping into the tailings. Even if the latter contain several pennyweights of gold to the ton, this does not show that the amalgamation effected on the plates is unsatisfactory, and, where tailings have to be reground or roasted, or treated in any way that may alter the condition of the gold, before a further extraction by mercury is obtained, no proof is afforded that the plates are not doing their work.

The loss of finely divided or "float" gold, particularly when it cannot be checked by the use of swinging plates, is often another name for the loss of slimed sulphides. Many examples have been adduced by the vendors of patent amalgamating or chlorinating machinery of the large percentage of the gold in the ores crushed in particular mills, which has been carried away suspended in water in a form not easily recoverable by settling. In the majority of these cases, however, no attempt seems to have been made to distinguish between the values contained in slimed sulphides and those existing as particles of free gold. Where this is not done there are no grounds for the assumption that any free gold is escaping at all. Thus G. M'Dougal, of Grass Valley, California, found[*] that a gallon of water, in the stream, ¾ mile below two mills, contained on

[*] *Mines West of the Rocky Mountains,* by R. W. Raymond.

an average 1·18 cents worth of gold. He called this "float" gold, but did not try to find out its physical condition, and it was very likely contained in sulphides. Again at the Spring Gully Mine, in Queensland, the tailings from the battery, if settled in the ordinary way by running off the water, were found to contain 7 dwts. of gold per ton, but, if carefully filtered, assayed 15 dwts. All such examples prove only that the slimes are rich, not that "float" gold is being lost, and although it is of course likely that some finely divided gold is carried away in suspension in water during the treatment of many ores, nevertheless, if sufficient care were taken in ascertaining this loss, it would probably prove to be less than is generally believed.

The following scheme of examining tailings with a view to determine the causes and amount of loss is given by M'Dermott & Duffield * :—

Small samples are taken at intervals from the waste outflow of the mill, until a bucketful is collected; this is allowed to settle for several hours, the clear water is decanted, preferably through a filter, and the remainder evaporated to dryness. Care must be taken to avoid spilling anything out of the vessels containing the samples. The sample having been well mixed, portions are treated as follows—

A. One part is panned and examined for free gold, amalgam, and quicksilver. If these are present, it is probably the fault of the millman, and nothing further need be done until this state of things is remedied.

B. The tailings are sized on brass screens, and the coarse, medium, and fine materials (the latter consisting, say, of that portion which passes a 100-mesh screen) are weighed and assayed separately, the coarser portions being reground and panned to find whether their values are in free gold or in sulphides.

C. The sulphides are separated from the tailings on a vanning shovel or batea, and are weighed and assayed. It may thus be determined whether they are worth saving, and the size of the mesh used for the screens will depend largely on this, and on the nature of the sulphides, which will in many cases be badly slimed and difficult to catch if the ore is finely crushed.

D. The loss due to fine or "float" gold may be determined by assaying the slimes after the sulphides have been carefully removed by concentration. This requires much skill and patience, but can in almost all cases be successfully accomplished by the vanning shovel. The concentrates may be examined under the microscope for fine specks of gold, but these and the fine sulphides can be recovered by concentration on suitable machinery. The assay value of the tailings from the vanning shovel will give some idea of the amount of float gold which is being lost. It will usually be found to be smaller than may be expected. If it

* *Gold Amalgamation*, London and New York, 1890, p. 7.

is large, the use of some system of amalgamation more perfect than that by copper plates (such as pan-amalgamation) or of a method of smelting, or of a wet method, may be considered, if the advantages appear sufficient to pay for the presumably increased cost.

Non-Amalgamable Gold.—The appearance in the tailings of free gold, which is not especially finely divided, but, nevertheless, is not in a condition to be amalgamated, may be regarded as a rare occurrence, but deserves some consideration. Amalgamation is in these cases prevented by the existence of a thin film of some neutral substance over the surface of the gold. The film may be so thin as to be transparent, but it is enough to prevent contact between the gold and the mercury. The disastrous effect of a film of grease covering gold particles has already been remarked upon. It is said to have been a fruitful source of loss in the treatment of certain ores in the Transvaal that they were impregnated with mineral oil. The effects of grease may be combated by the use of chemicals (caustic alkalies, potassic cyanide, &c.), but it is, of course, better to use every precaution to avoid the introduction into the pulp of oil from bearings, guides, &c., or contained in steam from the boiler. Losses in amalgamation are also caused by the greasy substances contained in some ores, such as the powdered hydrated silicates of magnesia and of alumina, which cause frothing, and coat the gold with a slime which prevents the action of the mercury.

Other films are formed of oxide of iron, compounds of sulphur, arsenic, &c., or of silica. Some years ago J. Hankey, of San Francisco, had a collection of particles of native gold which appeared as bright and lustrous as usual, but were coated by thin translucent films of red oxide of iron. These particles of "rusty" gold could not be wetted by mercury, but, if a piece were snipped off one end, the mercury seized on the fractured surface at once. Such gold seems to be rare in nature.

In 1867, William Skey, of the Geological Survey of New Zealand, after a series of experiments on the ores and tailings of the Thames Valley, came to the conclusion that the bright gold particles which refused to amalgamate were always coated by some compound of sulphur. He found that gold takes up sulphur from sulphides of ammonium or sodium, or from sulphuretted hydrogen, when brought in contact with their solutions, and that after this the gold refuses to amalgamate. He supposed that these compounds of sulphur were often formed by the action of acidulated water on the minerals in ores, and that consequently "a large area of the natural surfaces of native gold is covered with a thin film of auriferous sulphide, and that the greater part of the gold which escapes amalgamation at the battery consists of this sulphurised gold."

Gold in Pyrites.—In the preceding pages no account has been taken of the loss of gold which is contained in pyrites, as it has been assumed that these are saved by concentration if they are valuable, and this subject is dealt with in Chapter ix. Nevertheless, as this gold comes under the head of non-amalgamable gold, its physical state and the causes of its disinclination to unite with mercury may conveniently be considered here. In general, pyrites yields an extremely low percentage of its gold contents if it is run over the amalgamated plates, and if it is ground very fine in a pan with mercury the percentage extraction is better. However, in general, pyrites, even if it is reduced to the very finest slimes, and its prolonged contact with mercury ensured by continuous grinding, cannot be made to yield more than about 40 per cent. of its gold, and this is at the cost of much mercury, which is floured or sickened during the process. Isolated instances of better results are on record, but must be regarded as exceptional cases. Among the old processes used for the amalgamation of the gold in pyrites may be mentioned the treatment in revolving wooden barrels with mercury, as practised at the St. John del Rey Mine, and the practice of leaving the pyrites to be decomposed by weathering before grinding it with mercury. This method of oxidation seems to be decidedly inferior to the alternative plan of roasting the sulphides, by which the oxidation is rendered more complete and the particles of gold agglomerated to some extent. However, the amalgamation of pyrites, even when roasted, is far from perfect, part of the gold still remaining in a condition unfit for extraction in this way. The ores, which have been met with in various parts of the world, consisting mainly of limonite or hydrated oxide of iron, and in most cases believed to be the result of decomposition of pyritic ores by atmospheric agencies, are also extremely refractory, causing the mercury to sicken rapidly, and yielding only about the same percentage of gold as can be obtained from unoxidised pyrites.

The most celebrated case of this kind is that of the Mount Morgan ore, in Queensland, which is an ironstone gossan consisting of silicious brown iron ore, derived according to one view from the decomposition of pyrites. Although the gold appears to be free, it cannot be amalgamated, yielding only about 30 per cent. when crushed in batteries and subjected to prolonged grinding in pans with mercury. When the ore is dehydrated by roasting in reverberatory furnaces the extremely fine particles of gold are agglomerated, and between 80 and 90 per cent. can then be extracted by amalgamation, the remainder being presumably coated by oxides of iron. The richness of the ore, however, makes even this result unsatisfactory, and a process of chlorination has been adopted in practice. A similar case was noticed by Mr. Mactear [*] in South America, where a limonite ore which

---

[*] *Mining Journal*, Jan. 24, 1893, p. 70.

only yielded from 35 to 40 per cent. of its gold when treated in Huntington pans, was made to yield between 85 and 90 per cent. by merely subjecting it to a dehydrating calcination before amalgamating it. Louis Janin, Jr., mentions another case * in the ores of the Southern Cross Mine, Deer Lodge County, Montana, which consist of limonite derived from the alteration of pyrites. In panning large samples, only one or two specks of gold could be seen, although the ore contained from 1 to 2 ounces per ton. This ore yielded only about 40 per cent. on being amalgamated, but over 90 per cent. was dissolved out by leaching the raw ore with cyanide of potassium, and similar results were obtained by chlorination. Here the ore was thoroughly decomposed, but yet the gold would not amalgamate to a much greater extent than if it were still contained in the original pyrites, whilst the chemicals at once dissolved it.

In seeking to explain the behaviour of gold in pyrites, various theories have been propounded. According to one of these, the gold is supposed to exist in the pyrites in the form of sulphide combined with sulphides of iron, silver, copper, &c., and the refusal of the gold to amalgamate is explained in this way, auric sulphide not being acted on by mercury. Some observers have endeavoured to dissolve gold out of pyrites by the action of alkaline sulphides, and when, after many attempts, this was at length successfully accomplished, it was put forward as additional evidence that the gold must have been in the state of sulphide, although metallic gold is known to be soluble in these menstrua.

The balance of evidence, however, seems to be in favour of the theory that gold, at any rate in great part, exists in pyrites in the metallic state. Although the metal is generally invisible in undecomposed crystals of pyrites, it becomes visible when such crystals are oxidised either by air and water in nature, or by means of nitric acid, or by being roasted or subjected to deflagration with nitre. As a result of such decomposition, particles of bright lustrous gold, angular and ragged in shape, but of considerable size, often become apparent. These particles may be separated from the oxides of iron by washing, and the use of nitric acid, followed by panning, is frequently resorted to in order to detect gold in pyrites. Moreover, although usually invisible, gold can sometimes be seen in unroasted pyrites. As long ago as the year 1874, Richard Daintree and Latta found specimens of cubical pyrites,† in which gold could be seen under a microscope gilding the cleavage planes of the crystals. Again, G. Melville-Attwood, on examining crystals of auriferous pyrites from California in 1881,‡ found that the faces of the crystals were gilded in some places, and that here and there

* *Mineral Industry* for 1892, p. 249.
† *Proc. Royal Soc. of New South Wales*, 1874; paper on "Iron Pyrites."
‡ *Precious Metals in the United States*, 1881, p. 604.

little specks or drops of gold occurred, partially imbedded in the pyrites. These films were too thin to be detected by an ordinary lens, so that it did not seem surprising that such impalpable material could not be taken up by mercury. Louis Janin, Jr., more recently * found crystals of pyrites in a porphyritic gangue from the Republic of Colombia, which had gold in small globules on their surfaces. Lastly, it has long been known that crystals of pyrites are often found adhering to an amalgamated plate, the particles of gold on their surfaces having been amalgamated. It seems likely, in view of all these facts, that some of the gold at any rate is in the metallic state, and its refusal to amalgamate is not very surprising, when it is remembered how completely a thin coating of certain sulphurised compounds prevents amalgamation, and how readily sulphuretted hydrogen would be evolved from decomposing pyrites. Some authorities have contended that the metallic gold is disseminated mechanically through the mass of pyrites, but the action of potassic cyanide, in dissolving the whole of the gold out of comparatively coarsely crushed pyrites, seems to point to the correctness of the view that the interior of the crystal is not auriferous, the deposition of the gold being superficial, so that the enrichment of the pyrites is confined to its crystalline faces, and possibly, but not probably, to its cleavage planes.

The following details † of a microscopical examination by Prof. Morton, of the condition in which pyrites is left after being leached with cyanide, confirms this view to some extent:—

"Upon the ordinary auriferous sulphide of iron, or arsenical pyrites, the solution of potassium cyanide acts readily, not by dissolving the sulphuret, but by attacking the gold upon its exposed edges, and eating its way into the cubes by a slow advance, dissolving out the gold as it goes. An examination with the microscope of the pyrites after the gold has been removed, suggests the method of the operation. A sample of very rich pyrites, from a mine north of Redding, was treated with a weak solution, containing less than two-tenths of 1 per cent. of cyanide, for 168 hours; the assay showed a complete extraction of the gold; as the sulphurets showed no change in their appearance to the naked eye, some of them were placed under the microscope.

"There is no change visible in the form of the crystals as a whole; along the fractured faces the mispickel looks clean and unaltered, showing the silvery-white colour and intense refraction of the arseno-pyrite. Upon the faces of the crystals appear dark lines, short, and parallel to each other. In places they are crowded close together; in other parts they are at considerable distances, but always in parallel lines. The lines vary in length, being from four or five to over a hundred times their width; the

* *Mineral Industry*, 1892, p. 249. † *Loc. cit.*

lines are very irregular and often broken. These lines are fissures in the pyrites, and extend so deep into it that the microscope does not reveal their depth. By using the higher powers the walls of one of the fissures were seen to be completely honeycombed, looking somewhat like two empty honeycombs set opposite each other; evidently the mineral removed was crystallised along its contact walls at least. As the raw or untreated pyrites does not show any such fissuring, but, upon the contrary, shows a surface marked only by striation lines common to pyrites, I assume that the fissuring in the treated sample is caused by the solution acting upon some soluble mineral, probably gold, arranged in plates, occurring in groups, but which, by its colour and isomorphism and the extreme tenuity of its lines, is undistinguishable from the mass of pyrites enclosing it."

CHAPTER VIII.

OTHER FORMS OF CRUSHING AND AMALGAMATING MACHINERY.

Special Forms of Stamps.—Since within certain limits and under certain conditions the capacity of a stamp battery depends on the number of blows given per minute and on the momentum of the fall, various contrivances have been suggested with a view to increase both of these. Among these special stamps Husband's pneumatic stamp, the Ball Steam stamp, and the Elephant stamp may be noticed.

Husband's Pneumatic Stamp consists usually of two stamps, the stems of which are attached to pistons of small diameter working in pneumatic cylinders, D (Fig. 32). These cylinders have a reciprocating up and down motion given to them by the revolution of a crank shaft, C, above them, with which they are connected by iron rods, E, affixed to trunnions, d^1, on the cylinders. When a cylinder is raised by the crank shaft, the air in it, below the piston, is compressed and the stamp stem thus lifted. Similarly, the downward motion of the cylinder causes a compression of the air above the piston, which urges the stamp downwards with greater velocity than it would have by virtue of its weight alone. There is also a contrivance for rotating the stamps so as to give even wearing of shoes and dies. These stamps have not been much used except on tin-ores in Cornwall. Their output is from 20 to 30 tons per head per day through a 36-mesh screen, the power required being about 20 to 25 H.P. per head. One of the chief difficulties encountered in attempting fine crushing with these stamps of enormous capacity is the effect on the screens. Wire screens do not stand the excessive impact in a satisfactory manner, while Russia-iron screens, if punched for

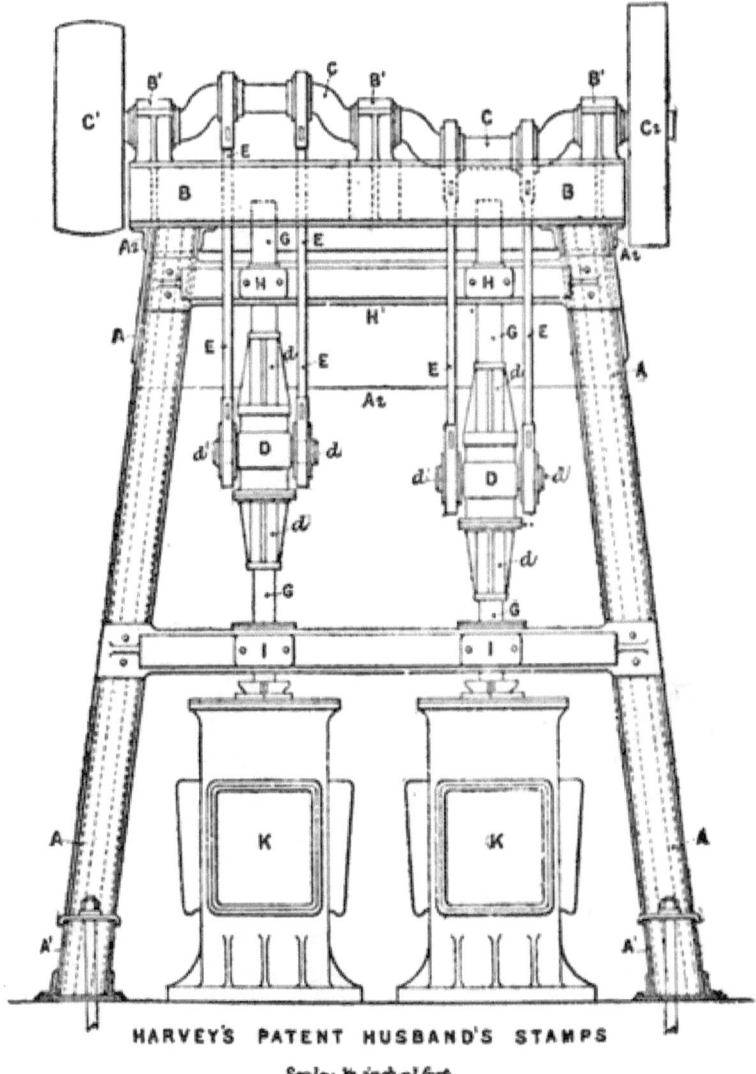

HARVEY'S PATENT HUSBAND'S STAMPS

Scale: ½ inch = 1 foot.

Fig. 32.

fine crushing, are more effective when thin, and are then of little durability.

Steam Stamps.—The ordinary form of steam stamp consists of a direct-acting vertical engine, having a steam cylinder and slide valve at the top, the piston-rod being rigidly connected with the stamp. Each stamp head works in a separate rectangular cast-iron mortar, with screens both at the front and the back, and sometimes all round. The screens are of Russia sheet-iron with punched holes of about $\frac{1}{5}$ to $\frac{1}{10}$ inch in diameter, the steam stamp being best adapted for coarse crushing. The speed

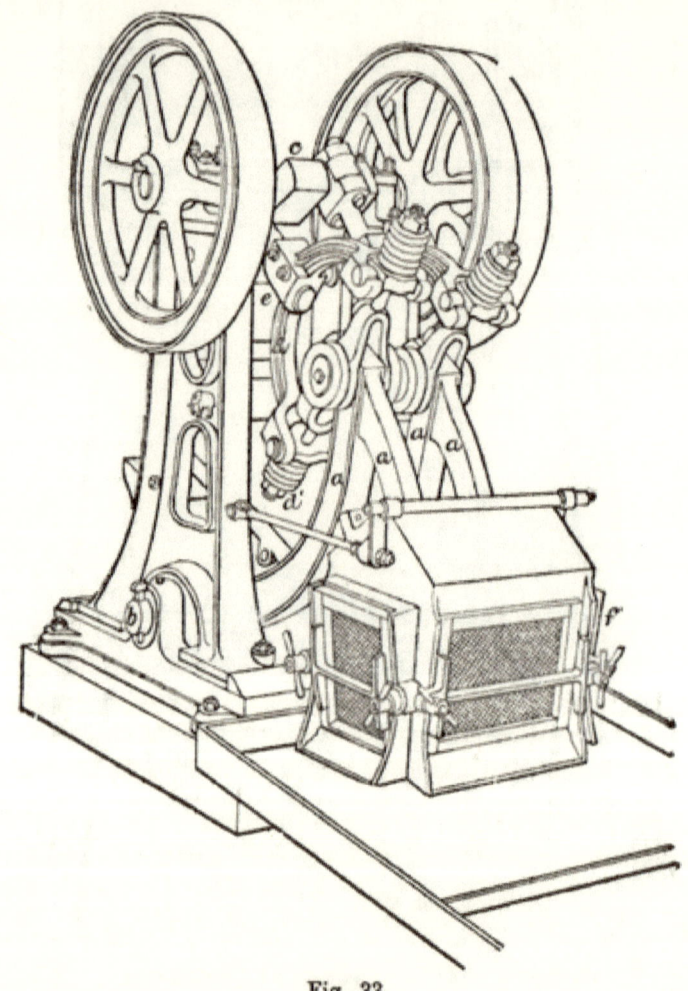

Fig. 33.

of working is from 90 to 100 blows per minute, and the out-turn is from 100 tons to as much as 225 tons of ore per head in twenty-four hours. It is obvious that gold could not be economically saved on plates inside the mortar of one of these stamps, and, as a matter of fact, until recently they were only employed in coarsely crushing the copper ores of the Lake Superior region.

Nevertheless, as the curious fact seems to be well established that these stamps, with their heavy blow, do not make so much slimes as the ordinary gravitation stamp, it is to be expected that they will be largely used in the future for crushing before concentration. They are already in use to crush silver ores in Montana, and gold ores at the Homestake mine, in the Black Hills, through a 30-mesh screen. As their capacity is so great their use is limited to cases in which large quantities of ore are available, one Ball steam stamp, such as is used in the Lake Superior district, being equal to at least fifty head of gravitation stamps.

The chief advantages of the steam stamp are economy of space and labour. With gravitation stamps more work is thrown on the rock-breaker, but it is doubtful whether there is any waste of power, as the whole of the work can be done by means of a single steam engine. The advantage of subdividing the work among a number of batteries is that stoppages for repairs and breakages affect only a small part of the crushing capacity at one time. Also, it has been pointed out that water-power can be directly applied to gravitation stamps with little loss of efficiency, whilst it is more difficult and wasteful to apply water-power to compress air for use in place of steam in the Ball stamp.

The Elephant Stamp consists of a bent or compound lever of hammered-iron or steel (a, a, Fig. 33), one end of which is pivoted on a fulcrum-pin, b, and the other end is fitted with stamp shoes. The figure represents a battery of two stamps. The levers are connected with the crank-shaft, c, by strong semi-circular springs, d, secured to short connecting rods, e. The stamp heads work in a mortar, f, fitted with dies in the usual way, and having discharge screens on three sides, and coarsely crushed quartz is fed in through an inclined shoot at the back. The action of the bow-springs, d, introduced between the crank shaft and the lever, is to receive and store up the force of the recoil from the blows of the stamps, and by giving it out at the next descent of the crank to effect a certain saving of power; in addition to this the springs act as cushions, taking up the jarring effect of the blows, and so diminishing the wear and tear of the machine. A mill, consisting of two heads of Elephant stamps, is said to be capable of stamping from 12 to 15 tons of hard gold-quartz in twenty-four hours through a 30-mesh screen. The advantages over gravitation stamps consist in the lightness, cheapness, ease of transportation and erection of Elephant stamps, and the driving power required, and they are consequently especially suitable for prospecting purposes, and in preliminary work during the development of new mines. It was found, however, in India that the lateral wear and tear at the top of the mortar-box by the levers which carried the crushing heads was very great, and frequent renewals were necessary.

The constant position of the heads, moreover, makes the wear very uneven, and the machines require much oil, whilst their capacity rapidly falls off as they are being used.

The Huntington Mill.—The Huntington Roller Mill, having now been in steady use for several years in the United States and elsewhere, has conclusively proved its claim to rank with stamps as a practical machine for fine crushing. It consists of an iron pan, at the top of which a ring, B (Fig. 34), is set, and attached to this are three stems, D, each of which has a steel shoe, E, fastened to it. The stems are suspended from the ring and are free to swing in a radial direction, as well as to rotate round their own axes, whilst the whole ring, B, with the stems

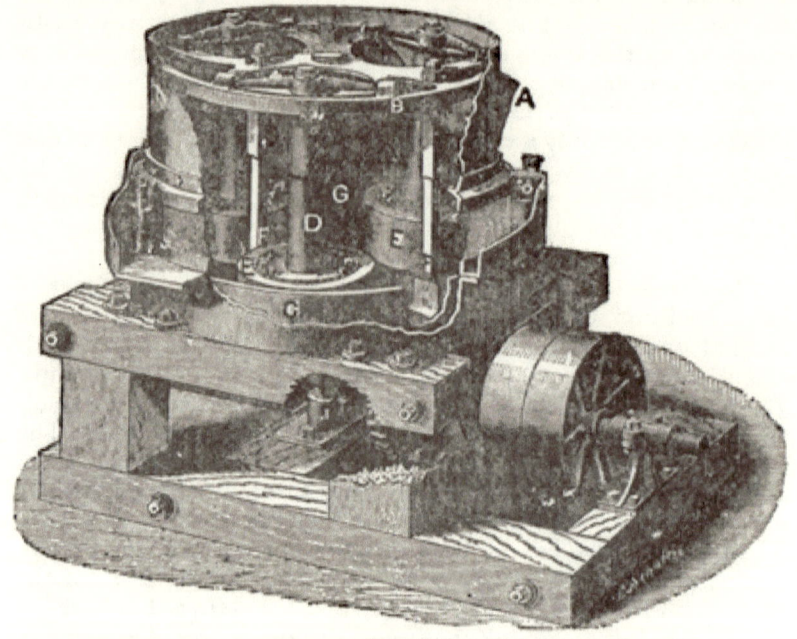

Fig. 34.

and shoes, revolves round the central shaft, G. The shoes or rollers, as they are called, are thus driven outwards by centrifugal force and press against the replaceable ring die, C. The most modern yoke for the suspension of the rollers is shown in Fig. 35.* In front of each roller is a scraper, F, which keeps the ore from packing. The rollers are suspended with their bases at the distance of 1 inch from the bottom of the pan, which can also be replaced when worn. The lowest part of the screen is situated a little above the top of the rollers, and outside it there is a deep gutter into which the ore is discharged,

* This figure is taken from Mr. A. Harper Curtis's paper on "Gold Quartz Reduction." *Proc. Civil Engineers*, vol. cviii., part ii., 1892.

and from which a passage leads to the copper-plate tables. The ore and water being fed into the mill through the hopper, A, generally by an automatic feeder, the rotating rollers and the scrapers throw the ore against the sides where it is crushed to any required degree of fineness by the centrifugal force of the rollers acting against the ring die. From 17 to 25 lbs. of mercury are placed in the bottom of the pan, the clearance below the rollers permitting them to pass freely over the mercury without coming in contact with it, so that it is not stirred up and

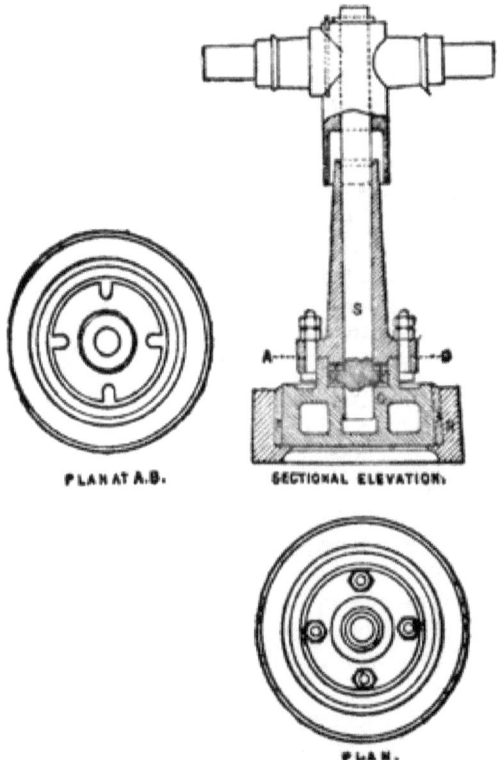

Fig. 35.
Scale, ¼ inch = 1 foot.

"floured," but the motion is such as to bring the pulp in contact with the quicksilver. The speed of the mill is from 45 to 75 revolutions per minute. The ore should be broken in rockbreakers to a maximum size equal to that of a walnut, or, better still, of a cobnut, before being fed in. The action of the rollers is one of impact rather than of grinding, the ore being granulated without the production of much slimes. The free gold, as soon as it is liberated from its matrix, is in great part amalgamated and retained by the mercury at the bottom of the pan, the remainder being caught on the plates outside the mill. Coarse gold is caught inside and fine gold outside the mill, but the yield inside is comparatively small when ores with high percentages of sulphides are in course of treatment.

The mill is particularly adapted for the treatment of ores containing brittle sulphides, which, if pulverised by stamps, are liable to become "slimed," and so to be in an unsuitable condition for concentration. It is also suitable for argillaceous quartzes, which yield their gold more readily under the "puddling" action of the rollers than when pounded by stamps. Moreover, the Huntington mill does much more satisfactory work than stamps on soft ores or in regrinding coarse tailings. The reason for this lies, of course, in the relatively large amount of screen area in the mill and its consequent high efficiency of discharge, a point in which stamps are decidedly inferior to it. In the experiments at the Metacom Mill, California, already quoted, it was proved that stamps will take as long to pass a ton of tailings through a screen as a ton of the original rock, but with the Huntington mill the finer and softer the material, the more rapidly it is passed through. As the splash is heavy against the sides the wear of the screens is somewhat rapid, but they can be very quickly replaced.

The mill is made in three sizes—viz., $3\frac{1}{2}$, 5, and 6 feet in diameter respectively—and the capacity of a 5 feet mill, the one which is most commonly in use, is from 10 to 20 tons of rock per day through a 30-mesh screen, the power required being from 10 to 12 H.P. The weight of the ring die in the mill is 611 lbs., and of each of the roller shells, which are replaceable, about 140 lbs. The wear and tear on these replaceable parts is very great, amounting to about 14 oz. of steel per ton of rock crushed when soft ores, previously broken small, are being treated. If large pieces of hard quartz are fed into the mill or if the mill is overfed, the mercury is splashed against the screens and passes through with the pulp, and when by accident pieces of iron or steel are introduced, the ring die is occasionally broken. Another source of disaster in Huntington mills lies in the use of acidulated water, such as that derived from mines or encountered when decomposing pyritic ores are treated; the mill is rapidly corroded and rendered unfit for work by such water.

The chief advantages supposed to be gained by the use of Huntington mills instead of stamps may be thus epitomised:—

1. *Reduced First Cost.*—The cost for the same capacity is not more than two-thirds that of stamps, even at the manufacturers' shops, while the difference in favour of the mill is even more in outlying districts from its light weight, and corresponding low freight, and from the cheapness of its erection.

2. *Saving of Power.*—The mill is said to run with about one-half the power per ton of ore crushed.

3. *The wear and cost of renewals is less* for the mill than for

stamps, the cost being from twopence to threepence per ton of ore for the former, against about fivepence or sixpence for the latter.

4. *There is less loss by flouring* of amalgam and quicksilver, while the good discharge and absence of grinding leaves the pulp in a better condition for concentration.

The first three advantages appear to refer only to such soft and brittle ores as are especially suited to the Huntington mill. The mill requires to be set to work in an intelligent manner by experienced and skilful hands, and watched carefully. The dangers of over-feeding have been already alluded to. One difficulty in automatic feeding is that self-feeding, such as is carried on by stamps, is impossible. The automatic feeder must work separately and be set to feed a certain weight of ore per hour, this weight having been determined by trial. If, after this, there is any change in the hardness of the rock, no automatic change in the rate of feeding takes place, and the machine may be choked up or run at below its maximum capacity unless watched and the feeder regulated. Another difficulty is in the quantity of water to be added. An excess of water, making thin pulp, does not favour internal amalgamation, and it may be stated that, in general, the pulp should be kept as thick as possible, consistent with its prompt discharge through the screens when sufficiently fine. If the pulp is too thick to run easily over the copper plates outside the mill, water may be added there by means of a perforated pipe. The rate of running should be as high as possible, since, if the other conditions are the same, the crushing power varies as the cube of the number of revolutions per minute.

A few examples are appended of the results obtained in actual practice by this excellent machine. At the Spanish Mine, Nevada County, California, there are four Huntington mills, three of 5 feet diameter and one of 4 feet diameter. The ore is free-milling and is passed through a Blake stone-breaker and thence to the mills. The four mills run at 58 revolutions per minute, and pulverise 35 tons of ore each in twenty-four hours, to pass through a slot screen equal to 20-mesh. The pulp is passed over the usual amalgamated plates after leaving the mills, and $\frac{1}{4}$ oz. of mercury is added with each ton of ore. Forty-five per cent. of the gold recovered comes from the inside of the mill, where the amalgam obtained is much richer than that from the plates. The loss of quicksilver is from $\frac{1}{15}$ to $\frac{1}{36}$ oz. per ton of ore, and the total cost of milling is about one shilling per ton, while for the month of November, 1887, it was only tenpence per ton. The ore is a soft talcose slate, containing streaks and veins of ferruginous quartz, carrying gold. The chief trouble in working lies in the frequent re-adjustment of feed which is found necessary. In a special test run of one month 42·4 per cent. of the gold contents was extracted, and the remainder lost in the tailings. This poor result was probably due to over-feeding, but profits were made, nevertheless, although the ore yielded only a little over 1 dwt. of gold per ton in 1887 and 1888. Twenty-two horse-power were used by the Huntington mills in crushing from 120 to 140 tons per day.

At the Shaw Mine, El Dorado County, a 5-foot mill, making 50 revolutions per minute, pulverised 10 to 12 tons per day, so as to pass through a 25- to 30-mesh screen. At the Mathines-Creek Mine, in the same county, a 5-foot mill pulverised 9 to 10 tons per twenty-four hours, so as to pass a screen equal to a 40-mesh. At the Monte Christo Mine, Mono County, two 5-foot mills, running at from 65 to 75 revolutions per minute, pulverised 2 tons of ore per hour to pass a screen equal to a 40-mesh; 25 lbs. of quicksilver were charged into each mill at the commencement of the run, and about $\frac{1}{2}$ oz. more added each half hour.

Although cases have been adduced in which 90 per cent. of the gold contents of an ore crushed in the Huntington mill was retained inside the machine, this is decidedly exceptional, and the mill is probably inferior to the stamp mill as an amalgamator on many ores, although its product is better adapted for treatment on copper plates and by concentration.

Crawford Ball Mill.—This mill (see Fig. 36 *) consists of an annular trough, divided concentrically by a vertical slit, which passes through its deepest part all round; the outer part of the trough is fixed to the framework, whilst the inner portion, having a circular spindle passing up through it, is capable of being revolved by bevel gearing as in the Huntington mill. In the annular trough, which is 4 feet in diameter, are placed eight chilled cast-iron or hard-steel balls, each about 8 inches in diameter; two of these are shown in position in the figure. Below the vertical slit in the trough there is another vessel, in which the mercury for amalgamation is placed; this vessel communicates with the trough by the narrow continuous slit passing all round the machine. The ore, previously partly crushed, is fed into the mill through the hopper, and the central spindle, carrying with it the inner half of the annular trough, is revolved about 180 times per minute. This causes the 8-inch balls to revolve, each on its own axis, and all of them around the annular trough, and the ore is ground by them. A current of water is introduced into the lower cavity of the machine, and passing over the surface of the mercury moves up through the vertical slit in the bottom of the trough, and through the ore which is in process of being crushed, and overflows at the top. The direction of movement of the pulp is shown by the arrows.

The idea is that, as the ore is crushed sufficiently fine, it will rise and pass away with the current of water, and that the particles of gold will fall down through the slit at the bottom

* From *Proc. Inst. Civil Eng.*, vol cviii., part ii., 1892.

of the trough, and coming in contact with the mercury, will be amalgamated there. It is claimed that the current of water may be regulated so that gold only can fall through it, while not only quartz but sulphides also are carried up by it, and not permitted to come in contact with the mercury, which is thereby kept quite clean and preserved from agitation, so that the loss due to flouring and sickening is entirely obviated.

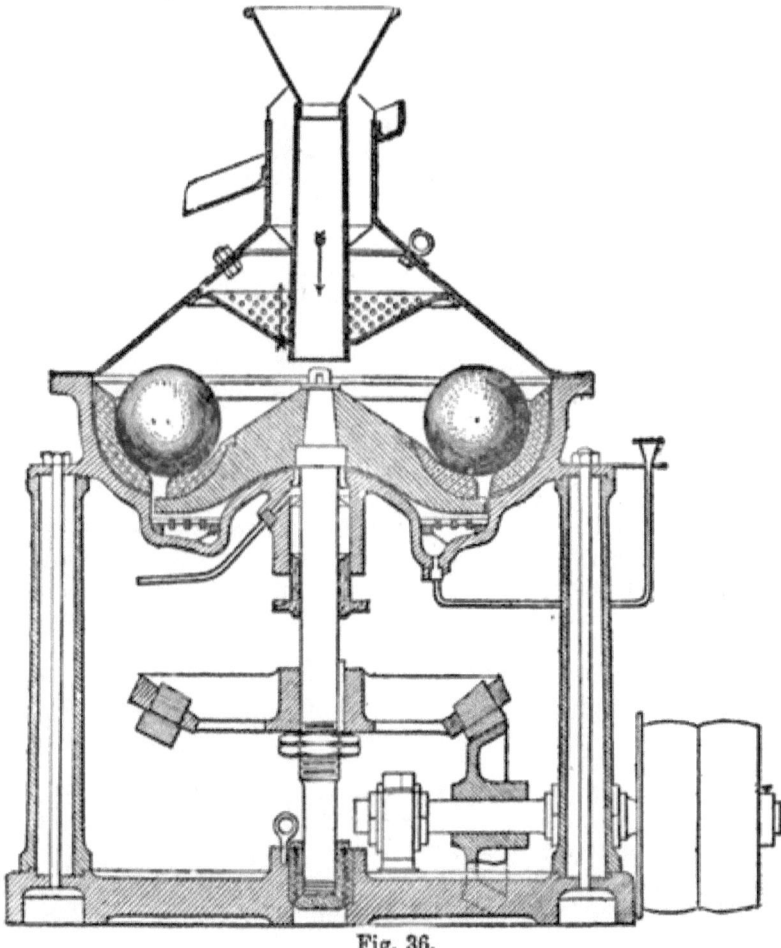

Fig. 36.

It is obvious that rather too much is expected from this machine. Since the effect of the current of water is exactly the same as that in the ascending-current classifiers (see p. 170), it is clear that "equal-falling" particles will not be separated from each other by its action. Consequently, only the very coarsest particles of gold, which could not escape amalgamation in any machine yet devised, can be separated from sulphides in this way. Moreover, it seems likely that the agitation produced by the movement of the balls would occasionally permit the escape of large particles of ore before they were finally comminuted, and, what is of much greater importance, the rapid swirl of the water and pulp, travelling in places at little less than 24 feet per second, which is the speed of the periphery of the moving part of the trough, would appear to effectually prevent that perfect separation by gravity of heavy and light particles, which is essential to the successful working of the machine. The machine is of little practical value, and has only been described as a type of ball mills.

Other Ball Mills.—Among other mills in which the crushing is done by iron balls may be mentioned the Cyclops mill, in which a single large ball, 12 inches in diameter, is carried round in a vertical plane by frictional contact with a pair of flexible discs fitted on a horizontal revolving shaft. The grinding is done between the ball and a grooved circular path in a vertical plane, against which it is pressed by centrifugal force. The ball makes from 250 to 600 revolutions per minute, and the largest sized mill is said to crush over 40 tons per day. It is fitted for both wet and dry crushing, but records do not seem to exist of successful work done on a large scale during any considerable length of time.

Speaking generally, most of the ball machines which have been devised have already been proved to be very uneconomical owing chiefly to the great wear and tear of the grinding surfaces, and the enormous driving power required. It is difficult to obtain a ball which can be relied on to wear evenly, and in many cases the effectiveness of the machine is materially reduced after it has been in operation for a very short space of time.

Amalgamation Pans.—An amalgamation pan consists of a circular cast-iron pan, provided on the inside with a renewable false bottom of cast iron—constituting the lower grinding surface—and a "muller," or upper grinding surface (d, Fig. 37), attached to a vertical revolving spindle, g, which is set in motion by bevel wheels, t, placed below the pan. The muller grinds to impalpable pulp ore which has been already reduced to a coarse powder by stamps, and also mixes the ore with mercury, introduced into the bottom of the pan, and so amalgamates the gold and silver. The origin of the pan is clearly to be traced to the Mexican arrastra, and some of the varieties of the pan are merely slightly modified arrastras. One variety consists of a sectional wrought-iron pan, fitted with a granite grinding bottom and with granite mullers, which are attached to a vertical spindle rotated by hand or by animal power. The Berdan pan also forms a transition between the arrastra and the modern pan, see p. 205.

In work on gold ores the use of amalgamating pans is mainly limited to regrinding skimmings, blanket sands, and concentrates obtained in working a stamp mill. Silver ores in many cases do not readily yield their precious metal if merely treated in a stamp battery and run over copper plates. These ores are then crushed in a battery, roasted with salt if necessary, and then amalgamated in pans. Silver ores containing considerable quantities of gold are often similarly treated, but with purely gold ores it is seldom necessary to resort to this process, and a detailed description will, therefore, be more in place in the volume devoted to the metallurgy of silver, only a brief account being given below.

Gold ores which do not yield a fair percentage of their values by copper-plate amalgamation are occasionally treated in pans. In such cases the ore may be roasted or treated raw. As already stated, p. 145, it is seldom advantageous to roast a gold ore before amalgamation, since, although in a roasted pyritic ore specks of free gold may often be detected where none were visible in the raw ore, a part of the precious metal usually appears after roasting to be difficult to bring in contact with mercury. The cause of this is not always easy to discover, but it may sometimes be due to the coating of gold by thin films of iron oxide or other mineral. Moreover, the addition of salt to a gold ore in the roasting furnace, as is pointed out in the chapter on chlorination, is often attended by appreciable losses by volatilisation. These two causes are sufficient to account for the low percentage of gold usually extracted when an auriferous silver ore is treated by roasting with salt and pan-amalgamation. When no base metals are present a gold ore may sometimes prove to be satisfactorily handled by roasting and amalgamation, but such cases are exceptional.

Pan-amalgamation, whether the ores treated are raw or roasted, may be conducted in one of two ways. The older system is to crush wet in the stamp mill, and collect the ore in large shallow settling pits or pointed boxes (see p. 168). A sufficiently dry pulp having been obtained by draining, it is dug out by hand and charged into the pans. The modern system, which is already extensively in use, is called the Boss Continuous System, after the name of its inventor, M. P. Boss. It is briefly described on p. 162.

Old System. — The amalgamating pans in use are very numerous, and vary greatly in form. The shape of the bottom was formerly much in dispute, flat, cone-shaped, and hemispherical bottoms each having its advocates, but it is now generally believed that flat-bottomed pans are the best, wearing more evenly and doing more work. The pans are often heated, so as to increase the rate of amalgamation by means of steam led through a chamber below a false bottom in the pan, but the more economical device of introducing steam into the pulp itself has also at all times been in use. The objections to the latter course are that the pulp may be so much diluted that amalgamation is checked, and that oil is liable to be introduced with the steam with equally disastrous results. When the ore is roasted before being treated in the pan, it is in some mills charged in hot, hot water being added also, and as the pan is covered up and is still warm from the previous charge, it remains at a sufficiently high temperature throughout the operation without further treatment. The grinding of the ore by the muller is an additional source of heat.

One of the common forms of amalgamating pans is shown in Fig. 37. This pan is 5 feet in diameter, with cast-iron bottom, a,

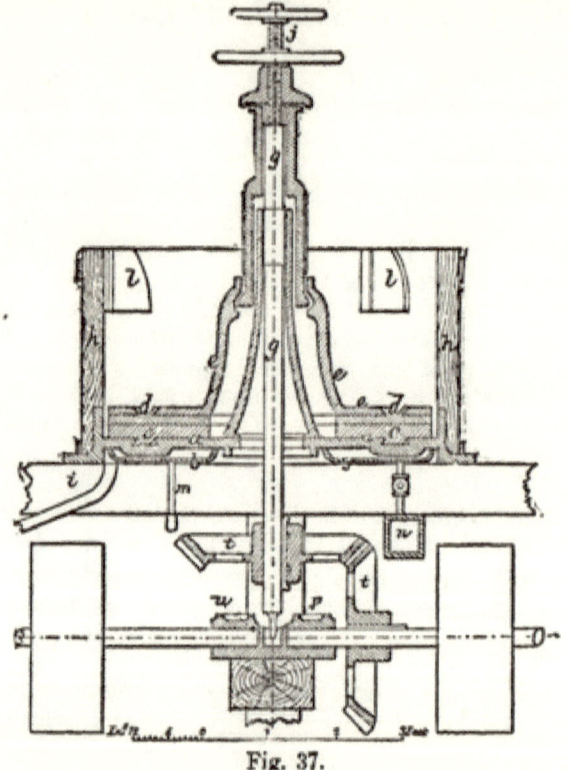

Fig. 37.

and wooden sides, h. The mullers are shown resting on the cast-iron dies, c, which protect the bottom from wear, whilst replaceable shoes attached to the lower surface of the mullers are also shown. The shoes and dies can be kept in contact while the spindle, g, is rotated, so that the ore can be ground, or the muller can be raised by rotating the hand-wheel and centre screw, j, on the top of the spindle, so that only circulation and mixing of the charge take place. In some pans copper plates, l, are introduced, being attached to the side walls and projecting into the interior. These plates are intended both to mix the pulp and to catch the amalgam, much of which is retained on them. The more usual system is to employ separate vessels called settlers for the collection of the quicksilver and amalgam, after the pans are discharged. The speed of the muller is usually from 65 to 75 revolutions per minute; below the muller the pulp is continually worked from the centre of the pan to the circumference, being returned towards the centre above the muller and passing down through the latter by inclined slots which terminate near the centre. In Fig. 37, which represents the form known as the Patton pan, n is the main through which steam is passed into the chamber, b, to heat the pulp, and m is the outlet pipe.

The method of operation is as follows:—The charge of ore is introduced with the mullers raised slightly and kept revolving, water being added at the same time in quantities sufficient to make the pulp of a pasty consistency, so that globules of mercury remain suspended in it without subsiding. The mullers are then lowered and the ore ground for from two to four hours, after which the mullers are raised and the mercury added gradually, and thoroughly mixed with the pulp for 6 to 8 hours longer. The object in raising the mullers is to prevent the sulphides from being ground up with mercury, which would cause considerable losses by flouring and sickening. Nevertheless this raising of the mullers is not an invariable practice. When the amalgamation is thought to be complete, water is introduced to dilute the pulp, and the whole is discharged into a settler; or else the diluted pulp is stirred by the raised muller at a reduced rate of speed until the globules of mercury have re-united and sunk to the bottom, when the pulp is gradually run off, beginning at the top, usually by pulling out in succession plugs set in the side of the pan at different levels. The discharge takes place into a bucket or tub, where some of the mercury accidentally carried over is caught. The bulk of the mercury in some mills is drawn off from the bottom of the pan before the pulp is discharged.

In order to facilitate amalgamation various chemicals have been recommended as desirable additions to the charge. At the present day it is recognised that most of these are either useless or absolutely harmful, and only salt, sulphate of copper, nitre, cyanide of potassium, lime, and sodium amalgam are now used. In treating gold ores, cyanide of potassium and sodium amalgam are added to keep the mercury clean and lively, but the latter chemical is now comparatively rarely resorted to. Salt and sulphate of copper are chiefly added to silver ores, their use having been suggested by the Patio process. They are believed to decompose certain base minerals, and so to prevent the sickening of mercury, which would otherwise be caused by their presence, and also to liberate silver from some of its compounds and thus render it capable of amalgamation. The use of lime is of course to neutralise any acid sulphates of iron, &c., which may be formed by the partial decomposition of the ore, and so to prevent the sickening of the mercury. If added when the pulp is diluted, lime is said to be efficacious in assisting the mercury to collect together and settle.*

The Boss Continuous System.—In this system of pan-amalgamation the pulp is continuously run direct from the stamp battery through a series of pans arranged so that each overflows into the next one, which is placed at a slightly lower level. The first

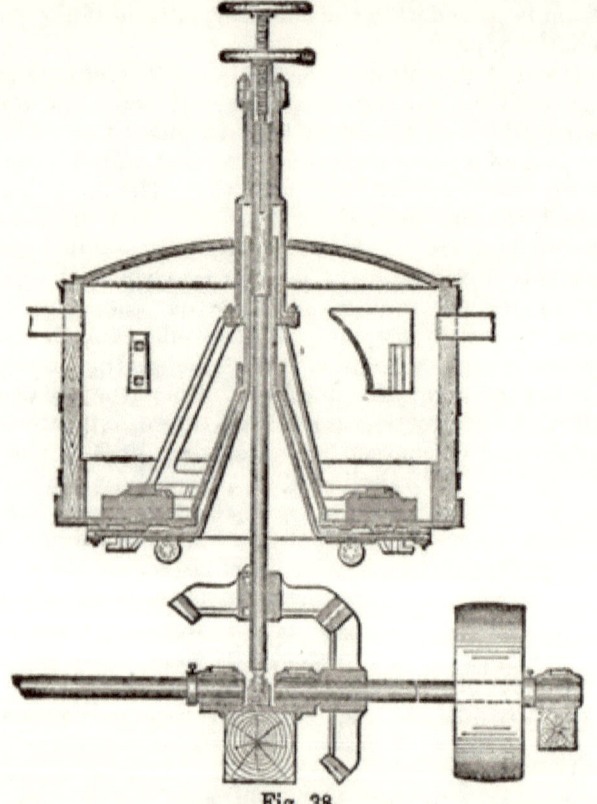

Fig. 38.
Scale, 1 inch = 3 feet.

two or three pans are arranged as grinders, the battery pulp not being fine enough for complete amalgamation, and the pulp is then passed through a series of amalgamating-pans supplied with mercury, after which the mercury and amalgam are separated from the ore in settlers, which are larger pans in which the pulp is diluted and stirred less vigorously. The tailings overflow

* For a full account of the chemical reactions involved in pan-amalgamation, the student is referred to the *Report of the United States Survey of the Fortieth Parallel*, vol. iii., chap. v.

from the settlers, and are run to waste or led over concentrators. The number of pans arranged in series through which the pulp must pass, in order to yield a fair percentage of its precious metals, is determined by experiment for each particular ore. It is obvious that the consistency of the pulp must be thinner than has usually been considered desirable for successful amalgamation, and, as a matter of fact, its volume is usually doubled by the introduction of the continuous process, but in spite of this the percentage of extraction is not lower than by the old method. By the Boss system there is a large saving in labour, fuel, and in wear and tear; the settling pits or pointed boxes are dispensed with, and no movement of the pulp by hand is needed. The mercury is collected in wells and pumped up into tanks, whence it is fed automatically into the amalgamating pans. One of the pans in use in this process is the Boss Standard Pan, shown in Fig. 38. It will be noticed that there is a steam chamber below the false bottom of the pan, extending up into the conical space in the centre, for warming the pulp.

Concentration before Pan-Amalgamation.—Some reference must be made to this subject and also to that of concentration after pan-amalgamation, as much attention will probably be directed to them in the near future. In most cases it is desirable first to remove the sulphides by concentration and then to treat the tailings in pans. In this way the objectionable, but often valuable, minerals, which complicate the reactions in the pan and cause losses in mercury and amalgam by sickening, are prevented from doing mischief, whilst they are saved by concentration after the preliminary crushing in the battery, more effectually than if they are also subjected to the grinding and sliming action of the mullers. After concentration, the tailings of course contain too much water for immediate pan-amalgamation, and the excess of water must be removed by settling in pointed boxes. The chief disadvantage in this course is that, where the ore contains chlorides, sulphides and other compounds of silver, the slimes which remain in suspension in the water are the richest part of the pulp. It is, therefore, necessary to collect them as perfectly as possible by settling, but, in these cases, a certain amount of loss is unavoidable. The settled tailings are now in good condition for amalgamation, and can be treated expeditiously and effectively. If, on the other hand, the ore is very rich in silver, so that the loss caused by immediate concentration is heavy, it is first treated in pans, and the product is then passed over Frue vanners or other slime tables.

Treatment of Concentrates in the Pan.—Some details of the treatment of concentrates by pan-amalgamation are given in Chap. x. in the accounts of work in special localities. It is, however, a survival of old methods, and does not represent the most modern practice, in which concentrates are either smelted or treated by wet methods. Whether they have been previously roasted or not, the treatment of concentrates in pans is seldom attended by the successful extraction of a high percentage of the gold. A recourse to this method is justifiable only when the concentrates are not in sufficient quantity to warrant the erection of a chlorination plant, and when there is no smelting works near where they can be sold. Under these circumstances a stone arrastra usually gives better results in treating roasted concentrates than an iron pan. In Australia the method is often adopted of employing a large excess of mercury and little water, and of keeping the roasted material from contact with iron, and in some experiments conducted in Mexico, C. A. Stetefeldt found that by the use of gold amalgam instead of mercury, and by grinding in stone vessels, a high percentage of gold was extracted from low grade ores. The chief advantage in roasting lies in the reduction of the loss of mercury which is effected by the elimination of the compounds of sulphur, the percentage of extraction of gold not being, as a rule, much affected. This subject is discussed under the heading Gold in Pyrites, on p. 145.

Molloy's Hydrogen Amalgamator.—A description may be given of this machine as it has gained the approval of some well-known metallurgists, although it has never been highly esteemed by practical men, and, in spite of the fact that it has been before the public for several years, has not yet had any success in prolonged operations on a large scale. It consists of a circular pan of cast iron, about 40 inches in diameter and 4 inches in depth, which is nearly filled with a bath of mercury. In the centre is a bottomless ebonite box, the sides of which dip into the mercury for about $\frac{1}{4}$ inch. This is called the anode box, and is filled with an aqueous solution, which in the earlier form contained sulphate, but in the later form carbonate of soda. A leaden plate dips into this solution, and is connected with the positive pole of a battery, the negative pole of which is connected with the mercury bath. A wooden disc, nearly as large as the pan, touching the mercury near the periphery of the basin, but riding free above it towards the centre, is rotated, when the machine is at work, at a speed of not more than sixteen revolutions per minute, and crushed pulp and water is fed in near the centre of the machine, but outside the anode box. This pulp travels outwards in widening circles under the influence of the rotating disc, which causes it to come in contact with the mercury, and overflows at the edge, leaving its precious metals amalgamated with the bath of quicksilver. An electric current passed through the machine is supposed to decompose the carbonate of soda, liberating free sodium and hydrogen over the whole surface of the mercury, which is thereby kept clean and lively. It is doubtful, however, if any free sodium is ever present in the bath. If formed it would be at once oxidised by the water, liberating hydrogen; but this element, when nascent, is doubtless of great efficacy in reducing base oxides which may be contaminating the mercury, and so in keeping the latter from sickening. The absence of all grinding action prevents any considerable loss from occurring by flouring, consequently, the loss of mercury in operating the machine would probably be small.

The value of this machine depends on the success of the efforts to induce perfect contact between the mercury and the gold particles contained in the ore. It does not seem likely that this contact is obtained. The machine is said to be able to treat from 10 to 20 tons per day, and, therefore, to deal with the product of from five to ten head of stamps. The area of the amalgamating surface, however, over which the pulp passes is only about 8 square feet, whilst the area of the copper plates used in the stamp battery in the ordinary way would be from, say, 32 to over 100 square feet. The layer of pulp between the disc and the mercury, if this quantity is treated, must therefore be much thicker than the corresponding layer flowing over copper plates, so that there is less chance of every part of the pulp being brought in contact with the amalgamating surface in the former case. As already intimated, the machine has not yet been proved a success in prolonged operations on a large scale.

Jordan's Amalgamator.—This machine is intended for the treatment of auriferous tailings. It consists of a number of shallow amalgamated copper pans fixed one below the other on a vertical axis, which is revolved slowly by bevelled gearing placed at the top. Between each two pans is a circular shelf, projecting from the wall of the surrounding casing. The tailings, in the form of fine pulp, are fed in to the first pan, and, by the centrifugal force exerted by the revolution of this pan, the pulp is caused to pursue a spiral course, until, after several revolutions, it falls over the edge on to the shelf below. The pulp then flows back on the shelf to the centre of the machine and into the second pan. In this way it traverses the surfaces of all the pans, one after the other, and finally falls into a conical separator, which, it is claimed, saves any heavy material and also any mercury that has escaped. The machine has not yet passed into use on any extended scale, although it has been tried at a few mills for short periods of time. It seems well adapted to catch fine gold if the amount treated is not too large to admit of perfect and prolonged contact with the plates.

CHAPTER IX.

CONCENTRATION IN STAMP MILLS.

Concentration.—The object of concentration is the separation of the heavy valuable mineral from the light worthless gangue. In Europe, where careful attention has been paid to the mechanical treatment of ores for some hundreds of years, complications are introduced by the fact that various base minerals must be separated from one another, an ore being subdivided into several products. Almost all gold ores, however, only require separation into two parts—the "concentrates," in which the precious metal is contained, and the "tailings," which are thrown away. Consequently, the German machinery has not in general proved applicable to gold ores without modification, and most of the best appliances now in use have had their origin in the United States. The German system of coarse crushing, sizing by means of screens, and concentrating on jigs, is not, as a rule, applicable to gold ores proper, although much used on auriferous lead, zinc, and copper ores. This system will not be described here, where only the methods in use for the treatment of battery sands, after passing over the amalgamated plates, will be considered, and some reference made to the machines in use for the mechanical treatment of pulp, either before or after pan-amalgamation.

All the concentrating machines depend for their action on the effect of a difference of densities on the fall of bodies in a fluid. The fluid employed in almost every instance is water, although several machines have been devised in which air is used as the concentrating medium. The use of any other fluid than these may be safely condemned as chimerical. It has often been proposed to use some fluid which shall have a lower density than the valuable mineral to be saved, but a higher density than the worthless gangue; the mineral would then sink, while the gangue would remain floating. The high cost of any such fluid is sufficient to put this method out of the question, without discussing any further disadvantages.

The fall in water of solid materials takes place according to two laws, one applicable to very shallow strata, through which the particles fall with increasing velocity, while the other is true when the depth of the water is considerable, so that the particles for the greater part of their course proceed at their maximum velocity. In shallow water the fall is almost entirely according to density,* so that those machines which utilise only the first instants of the fall will have great efficacy in concentrating. It is this fact which has necessitated the use of shallow currents in concentrating tables, sluices, &c. In almost all these machines the fine sand and slime is brought into suspension in water, and the liquid is then run over an inclined surface. The deposit of sand, which is thus formed on the table, tends to become enriched in heavy minerals, because the stream moves faster at the surface of the water, where the lighter particles remain, than it does next the bed, where the heavy particles have settled. The deposit is continually worked up and brought again into suspension by a rake or broom, or by a series of shakes or blows imparted to the apparatus, so that the effect mentioned above is repeated frequently. If the stirring up is violently performed all the slime and very fine particles are kept in suspension in the water and carried away and lost, a slow stream of water and very slight agitation being favourable to their retention in the deposit which is formed. When the stream of water is rapid and voluminous, fine material, whether heavy or light, is swept away and lost in the tailings, whilst if too small a stream of water is used, much worthless sand is deposited with the concentrates. It follows that the amount of water used must be regulated according to the work which it is proposed to do. Frequently, it happens that clear water must be added to the pulp to dilute it sufficiently. On the other hand, it occasionally happens that the pulp is too thin, especially after hydraulic sizing, and water is then removed by means of *settling boxes*, described below.

Another operation, which it is often of the utmost importance to perform before concentration, is that of classification by size. The necessity of this is obvious, when it is remembered that a shallow stream of water swift enough to carry down fine sulphides, might be powerless to move a pebble of quartz. The usual method of classification is based on the varying rates of fall of particles through a deep column of water. In this way *equal-falling* particles are obtained together, and since a sphere of galena is equal-falling with a sphere of quartz of four times the diameter, it follows that the sizing cannot be perfectly performed by this method. Nevertheless, such sizing as is possible in this

* At the very beginning of the fall, before the velocity has had time enough to become great, the motion varies nearly as $g\left(1 - \frac{\delta}{D}\right)$, where δ is the density of the medium (in this case, water), and D the density of the solid particle. When the depth is sufficient for the velocity to attain its maximum, this maximum velocity will be the same for all particles for which $a(D - \delta)$ has the same value, where a^2 is the area of the section of the particle at right angles to the line of fall. Such particles are called *equal-falling*. For full investigations of these formulæ the student is referred to Callon's *Lectures on Mining*, English edition, vol. iii., p. 47.

way, when efficiently performed, is of great assistance as a preparation for treatment by shallow-stream concentrators. Some sizing machines are described below, p. 169.

Settling Boxes.—These are for the purpose of allowing the sand and mineral in suspension in a flowing current to settle, so that the part of the water not needed in the subsequent treatment of the ore may be run off; if it is desirable, this excess of water is available for use over again. These boxes will deliver a small stream of thick concentrated pulp at the bottom and give an overflow of almost clear water at the top. The boxes manufactured by Messrs. Fraser & Chalmers are shown in plan and

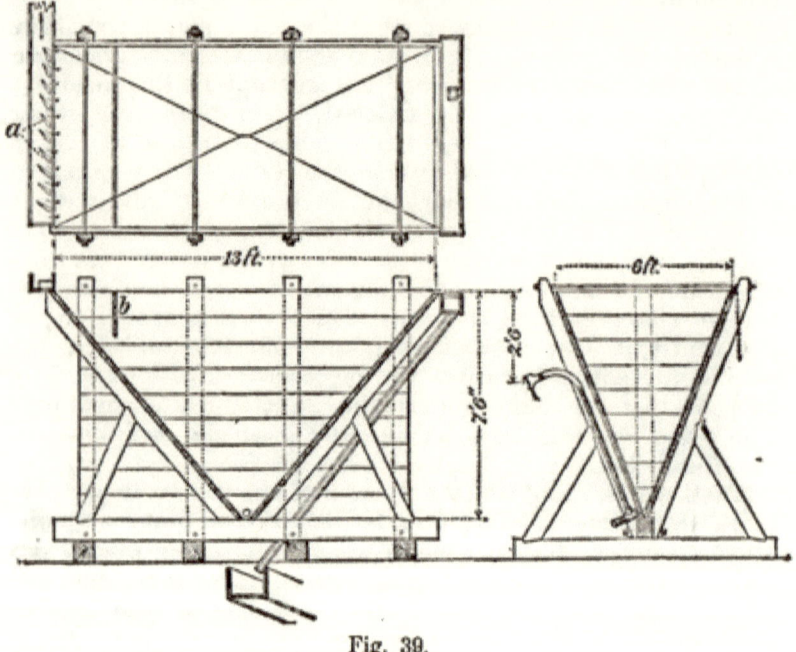

Fig. 39.

elevation in Fig. 39. The main points observed in the construction of these boxes are as follows:—

The sides are at an angle of at least 50° to the horizontal, to insure the uninterrupted descent of the slimes. The pulp is delivered evenly across the end of the box by means of some such arrangement as the movable tongues (a, Fig. 39). Surface currents are prevented and the incoming pulp thoroughly mixed with the mass of the water in the box by the partition, b, which extends across the box near the inflow end. The discharge is made by cutting the other end of the box from 2 to 3 inches lower than the sides; this overflow is made perfectly level so that the water flows out in an even sheet. The discharge pipe for the pulp is at the bottom of the box. The siphon discharge shown is the most convenient, as the aperture need not be so much contracted as if direct discharge is used; and, besides this, less fall in the mill site is required. A contracted orifice is liable to be choked up, and a smaller diameter than $1\frac{1}{2}$ to 2 inches is to be deprecated.

These boxes have the advantage over the ordinary settling-pits used to retain tailings and to catch pulp for pan-amalgamation, that they do not require to be dug out, while the settling, owing to the elimination of surface currents, is more perfect. One large box of 12 feet by 6 feet at the top and 7 feet 6 inches deep is sufficient to settle the product from five or ten stamps, according to the degree of separation of water and ore required.

If the pulp is required for treatment in amalgamating pans, such settling boxes do not give material of sufficiently thick consistency for the purpose. To adapt them to this use M'Dermott suggests that the pulp should be allowed to flow through two small square tanks placed above the pointed boxes, so as to catch the heavy sand. As soon as one tank is full, the pulp is turned to pass through the other, while the first is dug out after a short draining. The slimes are caught in the pointed boxes, and after mixture with the drier heavy sand the pulp is not too thin for amalgamation in pans. Concentration before amalgamation is, however, as yet seldom resorted to.

Classification according to Size.—Classification according to size, although seldom attempted by mill-managers in the treatment of battery pulp, is desirable as a preliminary to concentration by almost every machine; these work more efficiently if engaged in the separation of equal sized particles of different specific gravities than if pieces of all sizes are mixed up together.

The sizing machines most commonly employed are of two kinds, viz. :—

1. Revolving or flat shaking screens, which are suitable for coarsely ground ores.
2. Hydraulic classifiers—*i.e.*, boxes containing ascending currents of water; these are suitable for finely pulverised ores.

1. Sizing by **screens** is almost always conducted on wet material, spraying jets of water being used to carry the small ore through. Flat shaking screens offer advantages in cheapness and simplicity, but revolving cylindrical screens, inclined at a slight angle to facilitate discharge, are in more general use. It is customary to employ screens of the minimum fineness of 8 to 16 mesh. Where the sizing of finer particles than could be treated by this mesh is required, hydraulic classifiers are used, since their cost in repairs and renewals is far less than that of screens. The latter are seldom used in the sizing of battery sands, but are recommended for the purpose by Rosales.[*]

[*] *Loss of Gold in the Reduction of Auriferous Veinstone in Victoria.* By H. Rosales, Melbourne, 1895. This exhaustive and closely reasoned treatise will well repay a careful study.

2. **Hydraulic sizers** were introduced by Prof. P. von Rittinger more than forty years ago, for use in the Hartz, and from their shape were known as *Spitzkasten* or *pointed boxes*. These boxes have the shape of inverted pyramids, the stream of unsized pulp entering at one side and flowing out at the other, whilst there is also a small discharge at the apex. The current slackening on entering the box, the heavier and larger particles in suspension at once begin to settle, and, escaping the influence of the current, fall quietly to the bottom of the box where they are discharged. It is evident that in this classification the particles which reach the bottom will not all be of the same size. A piece of galena of specific gravity 7·5 will fall through water at a faster rate than a piece of quartz of the same size, but only one-third of the weight. The rate of fall is dependent both on the size of the particles and on their densities, according to a well-known law.* By applying this law it is easily shown that in water a sphere of iron pyrites (specific gravity 4·9) of $1\frac{2}{3}$ mm. in diameter will fall at the same rate as a sphere of quartz (specific gravity 2·6) of 4 mm. in diameter, and a sphere of galena of 1 mm. in diameter. In each successive box, then, a number of approximately equal-falling particles are removed, the closeness with which the subdivision is made varying with the size of the box, and the corresponding extent to which the current carrying the pulp is checked on entering it. The larger the box the more material is collected by it, and, therefore, the more heterogeneous the particles caught. In a small box the material collected consists more nearly of truly equal-falling particles.

These equal-falling particles however are, in any box, whatever its size may be, carried away through the aperture in its apex by muddy water containing material of all sizes down to the finest slime, and it was to eliminate this material that the *ascending current* was introduced. This is a current of clear water, which enters at the apex of the box in greater quantity than can be discharged by the outflow near the same spot, so that there is an upward current of water into the box. The result is that no muddy water is discharged below, but only the particles of ore which have weight enough to drop through the ascending current, so that by regulating the strength of this, any desired class of ore can be obtained. In Fig. 40, a form much used in America is shown. The sliding partition assists the settling, by causing the pulp to pass downwards, rapid surface currents across the box to the overflow being thus prevented. The discharge of the heavy particles is effected through A, the clear water pipe itself, by the arrangement shown. The launder above the box supplies the clear water current, and shows the head of water used, which must be kept constant to ensure uniformity of results.

* See footnote on p. 167.

The chief defect of the early forms of pyramidal boxes was that, as the area of the vat became larger and larger towards the top, the velocity of the ascending water naturally became less and less, so that many particles were able to settle down below the level of the overflow, but were stopped by the increasing force of the current, so that an accumulation of ore took place half way up the box, and ultimately became so great as to interfere with the classification. Several remedies have been devised for this defect, of which one of the simplest and most effectual is to make the box pyramidal below, but with vertical sides in the

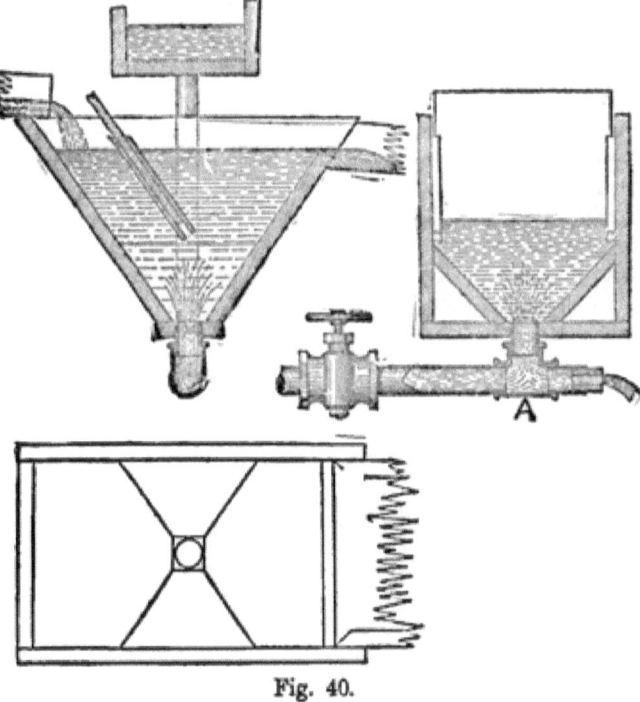

Fig. 40.

upper part. This construction is partly carried out in the box described above, which, however, though given as a type, is not theoretically the most perfect of its kind. A slime-pit is added to catch the stuff which is too light to settle in the boxes, in all cases in which these slimes are of sufficient value to pay for treatment. The number of boxes used depends on the number of classes of material which it is deemed advisable to make. Usually two or three classes are sufficient, and, as already stated, in the majority of cases no sizing is attempted.

Early Concentrating Machinery.—One of the oldest and most primitive machines employed in concentration of fine sand by means of a shallow stream of water was the German *buddle*, which has a distinct but imperfect resemblance to the Long-Tom described on p. 46.* Canvas tables were used below these buddles in Germany, and probably suggested the use of *blanket-strakes* or *tables*, which were adopted in the early days of the gold fields of the United States and Australia, and are still retained in some places. The rough surface of the blanketing seems to be particularly efficacious in catching and holding thin plates and spangles of free gold or of sulphides, which are readily washed off smooth surfaces by a current of water, and these rough appliances, although almost useless for catching slimes, still find favour where considerations of economy prevent the purchase of modern high-priced concentrators. The blanketing is usually in strips of 16 or 18 inches wide, and several feet long, and is nailed or stretched on wooden frames, which have an inclination of about $\frac{1}{2}$ inch per foot.

At intervals of about half an hour, when a quantity of mineral has already been collected on the rough surface, the blankets are taken off and washed in tubs of water, where a deposit collects which is afterwards dug out. At the St. John del Rey Mine, the framework supporting the blankets was hung on pivots above a shallow tank. When it was necessary to clean them, the framework was turned so that the upper surface of the blanket was inclined downwards, and the mineral washed off its surface by a hose, much time and labour being thus saved.

At this mine the trays supporting the blankets were 18 inches wide and 30 feet long, with a fall of 1 inch per foot. The upper 16 feet were covered with bullock's skins, tanned with the hair on them, and in lengths of 26 inches; below these were a series of blankets or baize cloths of the same length, made of coarse wool with a long nap. The fall from the battery box upon the tray was 4 inches, a screen being placed across the end to break the fall of the water, and cause it to strike the tray nearly at right angles. About 90 per cent. of the gold contained in the ore was caught on these blankets. The blanket sands contained 95 per cent. of sulphides, and were so fine that 90 per cent. of them passed through a 100-mesh sieve. They were amalgamated in revolving wooden barrels, yielding 96 per cent. of their assay value, but this was due to the fact that very little gold was contained in the pyrites, most of it being present in the form of free particles.

Blanket sluices have been declared unsuitable for catching fine sulphides, and their concentrates are usually contaminated by admixture with much sand. If set at a proper inclination they will save fine amalgam and free gold, but even in this respect they are less satisfactory than shaking copper plates and riffles.

* For a description of the similar buddle formerly used in Colorado for the treatment of battery sands, see Raymond's *Mines, Mills, and Furnaces of the Pacific States.* New York, 1871, p. 357.

Riffled sluices were employed at the same time as blankets for effecting rough concentration. The riffles were formed of half-inch strips of wood nailed across the sluice box, the grade of which was about three-quarters of an inch to the foot. As soon as the concentrates had accumulated until they reached the top of the riffle, another strip was nailed on, and the process was repeated until the bed of concentrates was several inches thick, when they were scraped out and a fresh start made. Similar to this device was the *raising-gate concentrator*, which was practically a riffled sluice in which the riffle was raised continuously by machinery, instead of being adjusted at intervals by hand.

The Round Buddle was invented in Cornwall, where it is still used in dressing the tin ores to the exclusion of almost every other concentrator. There are two varieties.

1. The convex round buddle, in which the ore and water are added at the centre of the machine, and flow down over the surface to the periphery.

2. The concave buddle, in which the pulp is added at the periphery of the machine and flows down to the centre.

In both cases revolving arms, carrying brushes, pass over the surface and stir the deposit as it is being formed, and the spouts distributing the ore also rotate so as to deliver the pulp evenly. The buddles are from 12 to 18 feet in diameter.

These machines are not continuous in action, and after the deposit has accumulated to a depth of a few inches the operation is suspended and the deposit dug out. The "headings" or material on the upper 12 or 18 inches of the inclined surface are kept separate, and the stuff near the bottom of the slope is called the "tailings." Round buddles are not adapted to obtain a finished product in one operation. The headings and tailings must, as a rule, be subjected to further treatment, and between them is a large quantity of material which differs little from the original ore. Thus handling and re-handling of the stuff is necessitated, and it is on this account that these machines are not now much used on gold ores. The principle on which they depend is favourable to the collection of slimes, and the modern improved buddles are perhaps better adapted for fine ores than any other machines, except those employing a travelling belt. Modifications have been proposed to adapt these buddles to the treatment of gold ores by adding riffles containing mercury, and by other devices, but have not found much favour. One of the chief changes, which was proposed in the United States, is to keep the brushes and ore-spouts stationary, and to rotate the inclined bed. During the earlier part of the revolution an attempt is made to separate the pulp into headings, middlings, and tailings, and each of these is washed into a separate launder by strong jets of water before the completion of one turn, the table being left clean for further work. A saving of labour in digging out the material is thus effected. The inclination of a buddle is usually from 1 to $1\frac{1}{2}$ inches per foot. A device formerly much in vogue in Western America consisted in the use of a cylinder of iron in the centre of a concave buddle, through which the discharge took place. This cylinder was made to rise slowly as the concentrates accumulated on the buddle, the effect being similar to that obtained in the sluices with self-raising gates.

For working large quantities daily of low grade ore containing a high percentage of pyrites, the revolving buddle is still used. When the percentage of pyrites is low, or if the ore is rich, so that closer work is desirable, or if the slimes are worth catching, *revolving-belt tables* are substituted for the older machines. The advantages of the revolving buddles consist in low first cost, large capacity, and ability to stand large quantities of water, which are always present in the pulp after the operations of screening and hydraulic sizing.

Centrifugal Concentrators. — Among concentrators with circular revolving beds, which were apparently suggested by the revolving buddle, may be mentioned Hendy's and Duncan's machines. *Hendy's concentrator* was one of the earliest forms employed in California, having been patented in 1868, and is still retained in some mills. It consists of a shallow cast-iron pan, 5 feet or 6 feet in diameter, supported in the centre by a vertical shaft. A rapid horizontal oscillating motion is given to the pan by the revolution of a crank shaft, which is joined to the periphery of the pan by a connecting-rod. The bottom of the pan slopes gently downwards from the centre to the periphery, and there is a basin-shaped depression in the centre, and an annular gutter running round the outside. The pulp is fed into the pan near the periphery, and a depth of about 3 inches of material is maintained. In consequence of the rapid oscillating motion the heavy particles move outwards by centrifugal force and accumulate in the peripheral gutter, while the lighter particles are discharged into the central circular basin and are removed by opening discharge gates at intervals.

The *Duncan concentrator* resembles the Hendy pan in consisting of a circular iron pan, in which the heavy sulphides are moved to the periphery by centrifugal force. The pan, however, revolves, making $8\frac{1}{2}$ turns per minute, while the pulp is supposed to make about 3 revolutions in the pan, thus allowing time for the pyrites and gangue to separate before the former settles at the bottom towards the sides, where it remains until discharged. The lighter particles remain more or less in suspension in the water and are carried to the centre, where their removal from the pan takes place. This machine is used to some extent in California, and, like the Hendy pan, is effective in the separation of coarse sulphides. Neither of these concentrators do good work in saving slimes, for the reason that the movement of the concentrates towards the periphery, depending on centrifugal force, takes place in consequence of the superior mass of the heavy minerals. With slimes, the mass of the particles of pyrites is so small that it is not enough to enable them to overcome the effects of the stream of water which carries them towards the centre, and they are, consequently, swept away with the tailings.

Percussion Tables.—In these machines the work of keeping the pulp in a state of agitation, done by the rakes or brushes in the German and Cornish buddles described above, is effected by longitudinal shakes imparted to the table. The table is made of wood or sheet iron, the surface being as smooth as possible, and the sides being flanged. It is hung by chains or in some similar manner, so as to be capable of limited movement, and receives a number of blows delivered on its upper end. These blows are given by cams acting through rods, or else the table is pushed forward against the action of strong springs by cams on a revolving shaft, and then being suddenly released is thrown back violently by the springs against a fixed horizontal beam. The movement of the pulp depends on the inertia of the particles, which are thrown backward by the blow given to the table, the amount of movement varying with their mass, and depending, therefore, both on their size and density. The vibrations produced by the percussion also perform the work of the rakes in destroying the cohesion between the particles, and a stream of water washes them down. The result is that the larger and heavier particles may be made to travel up the table in the direction in which they are thrown by the blow, by regulating the quantity of water, while the smaller and lighter particles are carried down. It is obvious, however, that the inertia of the fine particles of heavy minerals will not be sufficient to move them against the current, and, in consequence, all the slimes, both rich and poor, will be carried down and lost in the tailings.

Gilpin County "Gilt Edge Concentrator."—This variety of shaking table was devised in Colorado, and has displaced the blanket sluices at almost all the mills at Blackhawk. It consists (Fig. 41) essentially of a cast-iron or copper table, 7 feet long and 3 feet wide, divided into two equal sections by a 4-inch square bumping-beam. The table has raised edges, and its inclination is about 4 inches in $5\frac{1}{2}$ feet at its lower end, the remaining $1\frac{1}{2}$ feet at the head having a somewhat steeper grade. The table is hung by iron rods to an iron frame, the length of the rods being altered by screw threads, so as to regulate the inclination to the required amount. A shaft with double cams, A, making 65 revolutions per minute, enables 130 blows per minute to be given to the table in the following manner; on being released by the cam, the table is forced forward by the strong spring, B, so that its head strikes against the solid beam, C, which is firmly united to the rest of the frame. The pulp coming from the copper plates is fed on to the table near its upper end by a distributing box, D, and is spread out and kept in agitation by the rapid blows. The sulphides settle to the bottom of the pulp, and are thrown forward by the shock, and eventually discharged over the head of the table at the left hand of the figure, while the gangue is carried down by the water and discharged at the other end. One machine is enough to concentrate the pulp from five stamps. If the table consists of amalgamated copper plates, it is of some use for catching free gold also, treating about 8 cwts. per hour. This machine, like other

Fig. 41.

percussion tables, is very effective for separating coarse pyrites, but almost all the slimes are lost in the tailings. This is of less importance, as the sulphides with which it has to deal are not of high value, containing usually only from 10 to 12 dwts. of gold per ton. The high price of the Frue vanner and other slime tables prevents their introduction on this field, as they cost more than three times as much as the Gilt Edge concentrator for equal capacity.

The Frue Vanner.—This machine is described in detail as being typical of the shaking travelling-belt concentrators. Brief accounts of some other forms are given subsequently. Machines of this class are especially adapted for treating finely-crushed battery sands which do not contain a large percentage of "mineral" (that is, sulphides and other heavy materials). They are frequently set to concentrate unsized pulp coming straight from the amalgamating tables, and although they often do good work under these circumstances, when the screens used are equal to 40-mesh sieves or finer, nevertheless if the pulp contains coarse stuff the results are not so satisfactory. Rosales recommends * that the tailings from the plates should in general be passed over percussion tables, and should then be separated into four classes, viz. :—Coarse sand, medium sand (which is retained on a 40-mesh sieve), fine sand (which will all pass through a 40-mesh sieve), and slimes. The coarse sand may often be thrown away as worthless, although, if necessary, it is reground, and the other classes may be advantageously treated separately on some form of travelling-belt table or round buddle.

The Frue vanner (Fig. 42) consists essentially of an endless rubber belt, mounted on a frame, with its upper surface slightly inclined to the horizon, and subjected to two movements, a slow constant longitudinal movement and a slight and rapid side shake. The belt forms the bed or plane on which the dressing of the ore is effected, being an inclined plane, 12 feet long, and bounded down the two sides by projecting rubber flanges, which prevent the water and sand from dropping over the sides. An arrangement of rollers permits of the belt being slowly revolved in the direction of its length and *up* the incline; thus, though the dimensions of the working plane remain always the same, its surface is constantly travelling. The crushed rock in a stream of water is delivered near the upper end of the belt by means of the sand distributor, No. 1, Fig. 42, and flows down the belt towards its lower end. Now, as the inclination at which the belt is set is only from 3 to 6 inches on the 12 feet, and as the stream of water is not large and spreads over the whole width of 4 feet, it is obvious that, if it were not for the movements of the belt, much of the crushed rock contained in the water would settle on the belt, while the water and the finer and lighter particles of sand would alone reach the foot of the table and drop over into a waste launder.

In order to separate the heavy metallic minerals from the accompanying gangue or rock, a second stream of water is applied, whilst a gentle side shake is given to the belt, to keep the sand in a state of agitation, prevent it from "packing," and facilitate the sorting process. The water distributor, 2, is placed about 1 foot above the pulp distributor, and delivers small jets of water, 3 inches apart, over the entire width of the belt. The side shake thoroughly mixes this water with the pulp, spreads the whole uniformly on the belt, and enables the heavy particles of mineral to settle through the sand and cling to the belt, when they are carried up by it past the small jets of water and deposited in the collecting tank, while the lighter gangue is carried down by the stream and delivered into the tailings launder.

The Frue vanner is shown in plan in Fig. 42*a*, in side elevation in Fig. 42*b*, and in end elevation in Fig. 42*c*. The following description is abridged from that given by the manufacturers:—

* *Loc. cit.*

178 THE METALLURGY OF GOLD.

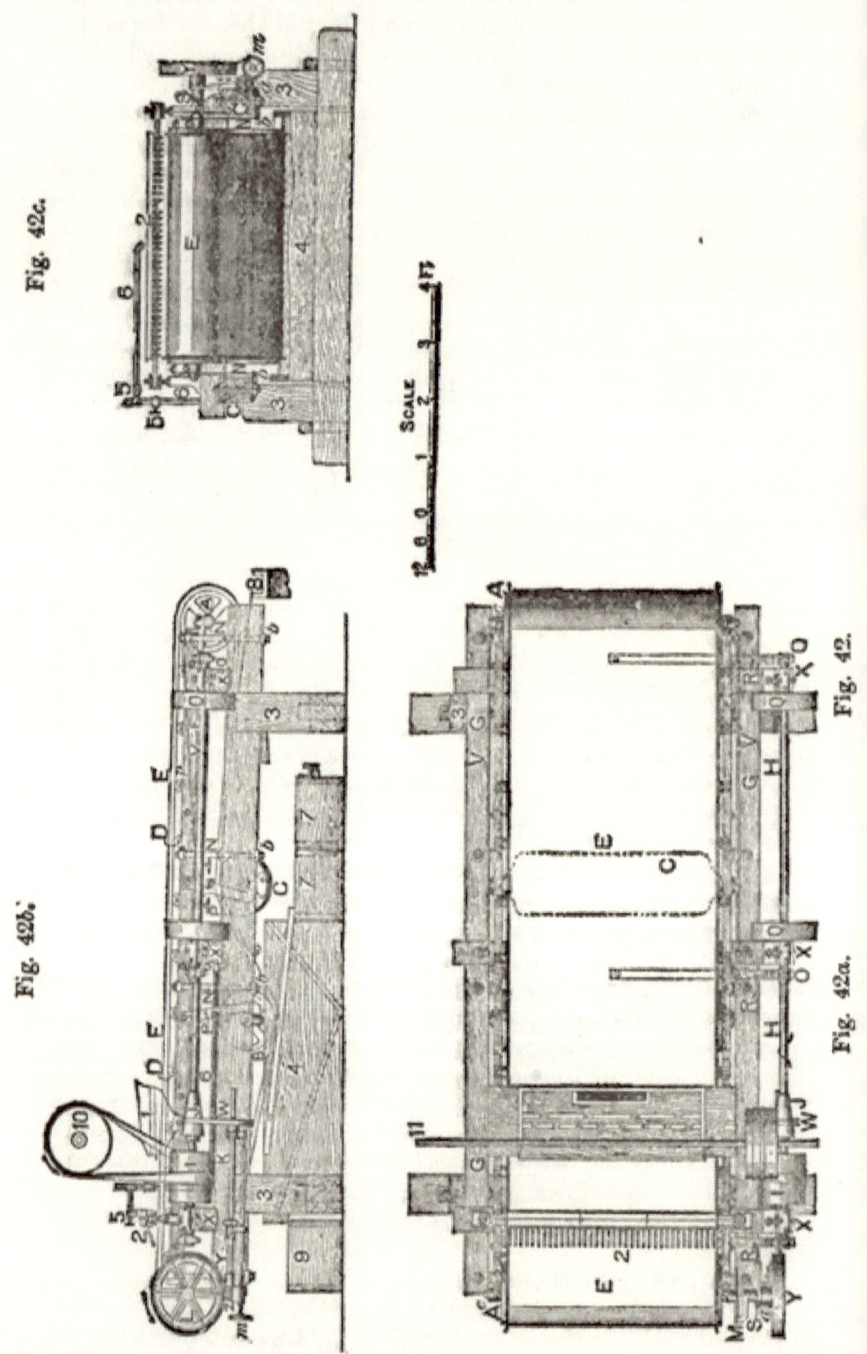

A A are the main rollers that carry the belt and form the ends of the table. Each roller is 50 inches long and 13 inches in diameter, and is made of galvanised sheet iron. B and C are of the same diameter, and are made in the same way as A A. The roller part of C is shorter than that of A A and B, and also has rounded edges, the upper surface of the belt with its flanges passing over it. The belt E passes through water underneath B, depositing its concentrations in the box, No. 4; and then, passing out of the water, the belt, E, passes over C, the tightening roller. By means of the hand screws, B and C can be adjusted on either side, thus tightening and also controlling the belt.

The boxes holding A A in place have slots and adjusting screws, so that, by moving them out or in, A A can be made to exercise an influence on the travel of the belt, E; when, as sometimes happens, it travels too much towards one side, this tendency can be stopped most quickly by altering the screws on one end or the other of A A; the change of position of B or C also controls the belt.

D D, &c., are small galvanised iron rollers, which support the belt, E, and cause it to form the surface of an evenly inclined plane table. This moving and shaking table has a frame, F, of ash, bolted together, with A and A as its extremities. The frame is braced by five cross-pieces.

The belt, E, is 4 feet wide, and $27\frac{1}{2}$ feet in entire length; it is an endless belt of rubber with high flanges at the sides.

G G is the stationary frame. This is bound together by three cross-timbers, which are extended on one side to support the crank shaft, H.

F is supported on G G by uprights, N, &c., four on each side. The bearings of A, the upper or head roller, are higher than those of A, the foot roller, so that the former is a trifle higher than the regular plane of the table, and the first small roller, D, should be raised by a corresponding amount.

The shape of the lower or bottom bearings of the uprights, N, &c., can be understood by examining b, as shown in the end elevation and also partly in Fig. 42b. This lower bearing, b, extends across G, underneath, and is supported by a bolt passing through G. A lug on the upper side and on the outside end of b rests on G; and b hangs on the head of the bolt, and is kept stationary by the weight of N and its load. By striking with a hammer the face of b shown in the elevation, b is moved, changing the position of the lower bearing, and thus making N more or less vertical. By thus moving the lower supports of N, &c., the sand corners on the belt hereafter explained are regulated.

The cranks attached to the crank shaft, H, are $\frac{1}{2}$ inch out of centre, thus giving a throw of 1 inch, which is the amount of the lateral throw. I is the driving pulley that forms with its belt the entire connection with the power. J is a cone pulley on the crank shaft, H. By shifting the small leather belt connecting J and W, the uphill travel of the main belt, E, is increased or diminished at will; the pulley, W, is moved by the hand-screw, *m*. R, R, R are three flat steel spring connections bolted underneath the cross-pieces of the frame, F, and attached to the cranks of the shaft, H. These springs give the quick lateral motion—about 200 per minute.

No. 2 is the clear water distributor, and is a wooden trough which is supplied with water by a pipe, and the water discharges on the belt in drops through grooves 3 inches apart.

No. 1 is the ore-spreader, which moves with F, and delivers the ore and water evenly on the belt.

There is also a copper well that fits in (and shakes with) the ore spreader as shown in the drawing. This is used in concentrating gold ores, for saving amalgam and quicksilver escaping from the silvered plates above, and can be taken out and emptied at any time. Into this well falls all the pulp from the battery. Its ends are lower than the wooden blocks of the spreader, so that the pulp passes over the ends of the well and is evenly distributed.

For some gold ores it is desirable to use on the ore-spreader a silvered copper plate the size of the spreader, and, when this is used, the wooden blocks of the spreader are fastened to a movable frame on top, so that they can be removed when the plate is cleaned-up, once or twice a month. No. 4 is the concentration box, in which the water is kept at the right height to wash the surface of the belt as it passes through. No. 8 is a section of the launder to carry off the tailings.

Method of Working.—The ore is fed with water on the belt, E, by means of the spreader, No. 1, which distributes it uniformly. A small amount of clear water is added by No. 2. The depth of the sand and water is kept constant at from $\frac{3}{8}$ to $\frac{1}{2}$ inch.

The main shaft, H, is given the proper speed for each kind of ore; this usually varies from 180 to 200 revolutions of the crank shaft per minute, the former speed being for light, fine "slimes," the latter for somewhat coarser materials. The effect of increasing the speed of the side shake is to increase the percentage of the material which is being discharged as tailings, and so to tend to loss of pyrites. The diminution of the speed, on the other hand, tends to the production of concentrates containing much sand.

The speed of the uphill travel of the belt varies from 2 to 12 feet per minute, and the grade or inclination from 3 to 6 inches in 12 feet, according to the ore. If the ore treated be poor in pyrites, the upward motion of the belt should not exceed 20 inches per minute; if richer, the speed is increased accordingly, and in agreement with the inclination of the belt, being greater as this inclination increases, but usually not exceeding $3\frac{1}{2}$ feet per minute. The inclination can be changed at will by wedges at the foot of the machine, these wedges being under the lower end of G, G, and resting on shoulders of uprights from the main timber of the mill. The motion, the water used, the grade, and the uphill travel is regulated for every ore individually, and must unfortunately be adjusted with every change in the pulp, if good work is to be maintained.

In treating ore coming direct from the stamps, too much water may possibly be present in the sand for proper treatment by the machine. In such a case there should be a box between the stamps and the concentrator, from which the sand with the proper amount of water can be drawn from the bottom, whilst the superfluous water will pass away from the top of the box; but as sulphides will also pass away with this water, settling tanks should be provided, and the settlings can be worked from time to time as they accumulate.

The main body of the belt suffers hardly any wear at all, since it merely moves its own weight slowly around the freely revolving rollers; the life of the belt is lengthened by taking the precaution of keeping it clean from sand at every point except the working surface, so that sand cannot come between the belt and the various rollers.

The concentration box, No. 4, which is kept full of water, and through which E passes, may be of any size or depth desired. Though not indispensable, it is best to have a few jets of water playing above and underneath on the belt as it emerges from the water in No. 4, so as to wash back any fine material adhering to the belt, and as such a method will cause an overflow in No. 4, the waste water, being full of finely divided mineral, should be settled carefully in the boxes, Nos. 7, 7, 7. Every few hours the concentrations may be scraped out with a hoe into the box, No. 9, and if this box be on wheels, it can be readily run on a track to the place where the concentrations are stored.

The amount of water used on the machine is from 1 to $1\frac{1}{2}$ gallons per minute of clear water at the head, and from $1\frac{1}{2}$ to 3 gallons per minute with the pulp.

The capacity per day of twenty-four hours is usually put down as about 6 tons of material, fine enough to pass a 40-mesh screen, but the machine does better work when set to treat a smaller amount. In California, in general, two Frue vanners treat the product of each battery of five stamps. Where the gangue is light, only one Frue vanner is sometimes used for five stamps. Sizing of the material is frequently omitted, the pulp passing direct from the stamps to the copper plates, and thence to the vanners. Rosales, however, points out[*] that this is a mistake; the coarse sand should be removed by sizing, as otherwise it interferes with the successful conduct of the work, caus-

---

[*] *Reduction of Auriferous Veinstone in Victoria*, 1895.

ing serious losses of sulphides. If there are large quantities of slimes in the pulp, the capacity of the machines is proportionately reduced. The machine requires careful watching by a skilled workman, as any change in the condition of the pulp necessitates readjustment of the belt.

It is very important to use the proper quantity of water with the pulp from the stamps, and this should be carefully regulated. There should be formed on each side of the belt a slight corner of sand—*i.e.*, there should be, on each side, sand with less water in it than there is in the remainder of the pulp on the belt. If there is not a slight sand corner, the corner will be sloppy, and there will be a loss. Sloppy corners are caused by using too much water with the pulp from the stamps passing on to No. 1.

Frequently, on the other hand, there may not be enough water with the pulp from the stamps, and as a result the sand corners formed will be too wide. The remedy for this is to use more water in the pulp coming on to No. 1.

As regards the proper amount of water to be used in the water spreader, No. 2, use just enough (no more) to keep the field between No. 1 and No. 2 covered, so that no points (or fingers) of sand shall show on the surface. The whole width of the belt between the water spreader and ore spreader should be kept quite wet. If dry streaks or points occur, and water, as a consequence, runs in streaks at the junction of the wet and dry channels, sulphides will be picked up and "floated" away on the surface of the water; this "floating" of pyrites is caused by its dryness, not by its lightness; it has been coated with a film of air. When the proper amount of water has been fixed, the carrying over of the clean concentrates past the jets of No. 2 should be accomplished and regulated by the uphill travel only.

Frequently, the sand and water on the belt will be distributed unevenly, the sand working to one side of the belt, and making a broad heavy corner, while the other is sloppy. This is caused by the belt not being horizontal from side to side; it is levelled by raising or lowering one side, by altering the position of the supports, N.

W. M'Dermott makes the following observations on the appearance of the ore on the belt:—

"Should the discharge of concentrates exceed the quantity of sulphides falling on the belt, sand or rock will be found close up to the jets of water, and by and by passing them. If the uphill travel be too slow, the sulphides collect below No. 2, forming a great "head," extending towards No. 1, and even below, in which latter case an increased loss of sulphides will assuredly take place in the waste. When working properly, a small head is always kept below the jets of clear water, and the sulphides come over clean and regularly."

Riffle Surfaced Belts for the Frue Vanner.—The smooth belt vanner, just described, is essentially a slime machine adapted to the very finest material. The special value of the vanner in recovering fine sulphides without previous sizing has been illustrated as follows:—*

"If a flat heavy object—say a sixpence—is put on a Frue vanner while in operation, it will not pass over with the fine material steadily delivered past the water jets. The shaking motion, by cutting away the supporting adhesion of the belt surface, owing to the inertia of the coin, allows the light current of water on the edge of the coin to wash it backwards with the sand. The same action applies to coarse particles of sand mixed with fine particles of mineral. Thus the effect of mere mass is such that the Frue vanner with smooth belt is not adapted to saving coarse mineral; an excess of wash water at the head drives down the coarser mineral, but the finest clings to the belts and safely passes the water jets."

In order to overcome this defect a roughened belt was introduced, having a number of depressions or riffles on its surface. In moving down this belt the coarse sulphides and gangue pass into these depressions, and then are compelled to proceed over a *rising* surface. At these points the separation of the coarse sulphides and coarse sand is perfectly performed, the heavy particles remaining in the depressions, while the lighter are washed away. The belt then in moving upwards carries the pyrites with it. In other respects the machine closely resembles the older smooth belt form, to which it is inferior in its power of catching slimes, although better adapted for coarse material. The riffle-belted Frue vanner is worked at a greater inclination, with a faster upward motion, a slightly faster shake, and with more water than the ordinary vanner, and consequently it treats more pulp, one machine usually taking the product of five stamps. It offers great advantages over the smooth belt whenever the percentage of sulphides is very high, as the riffles collect and carry up much more material.

The Embrey Concentrator.—This machine, like the Frue vanner, consists of an endless belt with flanges along its edges; it differs mainly in having an end-shake instead of a side-shake, the motion being parallel to the length and travel of the belt, instead of at right angles to it. It is shown in Fig. 43, where the light wooden-framed form is shown. This is cheaper but less compact than the iron-framed variety, and is recommended for use wherever floor space is not of the first importance. It is unnecessary to describe it in detail, as it differs little from the Frue vanner. It is necessary to shake this machine at a faster rate than the side-shaking machines, the usual speed being from 230 to 240 revolutions per minute, although 200 per minute is

* *Gold Amalgamation*, p. 71. M'Dermott and Duffield, New York, 1890.

occasionally enough. If an attempt is made to treat an ore at too low a rate of speed, complete separation cannot be effected. More water is used than on the Frue vanner, while the inclination and rate of upward movement of the belt are both greater. In consequence of this, more pulp is treated, the amount varying from 6 to 10 tons per day of twenty-four hours, and, therefore, either one or two machines are used to each five stamps, according to the character of the ore and the closeness of concentration required. Fewer Embrey concentrators than Frue vanners are in use, owing to the fact that the sale of the latter is pushed more by the company which owns both the patents.

Among other concentrators with endless inclined rubber belts may be mentioned the Triumph, which, like the Embrey, has a longitudinal shaking motion applied to the belt, the crank shaft revolving at the rate of 235 to 240 times per minute. The

Fig. 43.

forward motion of the belt is regulated by a friction roller, instead of a cone pulley, which is used in the Frue vanner and the Embrey concentrator. It has also an amalgam saver, consisting of an iron trough containing quicksilver, which is stirred by iron teeth attached to a slowly revolving horizontal shaft. It is in wide use in California (where it seems to be preferred to the Embrey machine) and in Australia.

The Lührig Vanner.—This machine was invented about five years ago, and has already met with great success in many instances in the concentration of auriferous tailings from stamp batteries. It bears a considerable resemblance to the Frue vanner and is thought by many experts to be destined to supersede it. In the Lührig vanner the endless travelling india-rubber band is not flanged at the sides, and has a slight side inclination, which is, therefore, at right angles to the direction of travel; the latter is horizontal, not inclined; by an arrangement of cams and springs, end-blows are given to the framework carrying the belt. The pulp from the batteries is collected in settling boxes placed overhead and delivered on to the belt through a small distributing box situated above its right-hand upper corner, the belt being driven from right to left. Clear water is supplied through a perforated pipe fixed diagonally across the belt. The pulp moves across the belt from the higher side to the lower, this motion being assisted by the clear water; the light particles of gangue are washed down in this direction at a faster rate than the heavy particles of pyrites. On the other hand, the travel of the belt and the end-blows move the ore in the direction of the length of the belt. The results of the combined motions are as follows:—

1. Tailings pass off the table nearly opposite the distributing box at the right-hand end.
2. A middle product, containing both sulphides and gangue, is delivered near the middle of the side of the belt.
3. Clean concentrates are delivered near the left-hand end of the belt, having travelled the greatest distance before being washed off at the side.

Each of these three products is delivered into a separate hopper, and by a simple arrangement of sliding plates the exact points on the belt at which the delivery of the middle product is divided from those of the headings and tailings can be altered so that the percentage of each product can be regulated.

Details of the dimensions, capacity, &c., of these machines are as follows:—

The india-rubber belt is about 4 feet wide and 19 feet in total length; the belt travels from 18 to 20 feet per minute; the total quantity of clean water used is from $4\frac{1}{2}$ to 5 gallons per minute; the supporting framework is 5 feet 6 inches high, 6 feet broad, and 16 feet long; the number of percussions is from 150 to 210 per minute; the extent of motion to and fro is from $\frac{1}{8}$ inch to $1\frac{1}{2}$ inches, according to the nature of the ore. The amount of ore capable of being treated in twenty-four hours is stated to be 4 to 5 tons, if the concentrates are iron pyrites, and 3 to 4 tons if they are galena and blende. At the May Consolidated Mine, 7 to 8 tons per day were being treated on each machine in 1893. At a test run in Glasgow, 2,898 lbs. of a pyritic gold ore were treated in four hours five minutes, and the re-treatment of the "middlings" occupied one and a-half hours more; this is equal to a rate of $6\frac{1}{4}$ tons per day. In the compound system the middlings from several machines are delivered at once to a vanner placed at a lower level, and time thus saved. The power required is given by the Lührig Company as $\frac{1}{16}$ H.P. per table, and by Mr. H. D. Griffiths, A.R.S.M., of Johannesburg, as $\frac{1}{8}$ H.P. in the case of the machines at the May Consolidated Mine.* Three single tables are required for each battery of five

---
* *South African Mng. Journ.*, August 19, 1893.

stamps on the Witwatersrand, this being the same as in the case of the Frue vanners on the same field.

In comparing the Lührig vanners with other travelling belt machines, the chief points to be noted are that in the former case:—

1. The direction of travel is horizontal.
2. Three or more different products are drawn off continuously instead of only two.
3. The water and pulp flow across the belt instead of in the direction of its length.

The machine resembles the Frue vanner in that alterations are required in the inclination of the belt, the rate of travel, the number of blows per minute, the force of each blow, and the amount of clean water and pulp, with every change in the constitution of the ore, so that the machines require constant supervision. The regulation of these is, however, very simple in the Lührig vanner.

The results obtained in various countries have been excellent. The following is the result of a trial run on an American gold ore from the Eureka and Excelsior Mine, Oregon:—

	Weight Treated.	Value.			Percentage.
	Lbs.				
Raw Ore,	760·00	@ $10·50 ⅌ ton	=	$3·99	
I. Concentrates—Heads,	38·33	,, 178·57 ,,	=	3·42	85·71 } 90·47
II. ,, ,,	14·51	,, 26·93 ,,	=	·19	4·76
Tailings,	523·08	,, ·20 ,,	=	·052	1·31
Slimes,	12·00	,, 2·70 ,,	=	·016	·40
Unaccounted for,	172·08				7·82
	760·00				100·00

It will be observed that 90·47 per cent. of the gold and silver in the ore is saved in the concentrates; the Frue vanner is stated to have saved only 60 per cent. in this case, and, as a result of the run, the Lührig vanner was adopted at the mine. At the May Consolidated Mine, Johannesburg, the best result obtained on unsized pulp was the saving of 77 per cent. of the gold in the concentrates, but the tailings seldom contained much more than 2 dwts. of gold, and sometimes only 1 dwt. This result would have been better if the material had been previously classified.

In 1892, after exhaustive investigation, Prof. J. Cosmo Newbery reported to the Victorian Government that many of the heaps of tailings in Australia, which contained from 1½ to 3 dwts. of gold per ton, could be concentrated by this vanner, and the concentrates roasted and chlorinated at a profit. Many of the vanners are now in use in Victoria.

Hartz Jigs.—These machines differ in principle from all those previously described, inasmuch as the particles are separated by their fall through a somewhat deeper column of water than is the case on inclined tables, while a series of blows from below, causing waves moving upwards, continually brings the particles into suspension, and allows them to drop again. The initial period of the fall in water, during which the motion depends chiefly on density, is thus continually reproduced, and the result is a perfect separation of heavy from light particles of ore when working on any materials except the finest pulp. Jigs consist of sieves supporting beds of ore, which are completely immersed in water; the ore is raised and allowed to fall by a quick succession of currents of water caused by the sudden action of a piston below, which is so worked that the upward movement or pulsation resembles that produced by a blow, while the downward movement is gradual. Under these conditions the heavy particles work downwards and pass through the sieve, while the lighter gangue is carried away horizontally by a stream of water introduced either from below or from above. Such machines are especially suitable for coarse ores.

In the Hartz jig a layer of coarse heavy particles are spread on the sieve to prevent too much of the ore from passing through. The stuff is fed in regularly at the head of the jig, and the strokes of the piston raise both the bed of heavy particles and the ore. The heaviest grains of ore find their way during the downstroke into the interstices of the bed, gradually pass through it, and coming to the screen, fall through into the tank below. The lighter particles cannot descend, and are gradually washed over the end partition by the continuous supply of water. Two products are, therefore, given, neither requiring further treatment if the conditions are favourable, and the machine properly adjusted. The wire meshes of the screen are always much larger than the ore treated, and the bed is composed of material of as nearly as possible the same density as the concentrates to be obtained, and is usually from $\frac{1}{2}$ to 1 inch in depth. The number of strokes of the piston per minute is from 60 to 80 with coarse sand of $\frac{1}{12}$ inch in diameter, and 200, 300, or even 400 with very fine sand, approaching slimes. The length of stroke varies under the same conditions from $\frac{2}{5}$ to $\frac{1}{5}$ inch, and in the case of very fine, almost impalpable sand, the stroke may be diminished till it becomes a mere tremor. Some of the highest authorities on concentration have stated their belief that for enriching even very fine sand, the Hartz jig is the simplest and most economical machine yet invented, and requires the least amount of labour.* It is obvious that it is not adapted to save rich slimes, and up to the present it has not been widely used for treating gold ores.

* Callon's *Lectures on Mining.* Eng. edition, vol. iii., p. 103.

Pneumatic Jig.—In this machine, which was devised by S. R. Krom, air is used instead of water for the separating medium. The ore is fed into a tank, the discharge from the bottom of which is regulated by a revolving wheel. The puffs of air are delivered through a number of vertical wire gauze tubes, which pass through the ore bed nearly to its top. No less than 450 to 500 puffs of air per minute are given, and the result is that the heavy particles alone descend through the interspaces between the wire gauze tubes, while the lighter material is discharged horizontally. It is necessary for the success of this machine that the ore should be quite dry, so that the particles may be completely independent of each other; as a rule the ore would have to be dried in special furnaces. The air jig has not passed into any extensive use, although it might succeed in hot climates, where water is scarce, and where only a pneumatic machine could be used. Careful sizing is necessary as a preliminary to successful working of this machine.

Clarkson & Stanfield's Concentrator.—Another dry concentrator, although one constructed on a totally different principle, is the centrifugal machine, which was devised by T. Clarkson & R. Stanfield. The action of this machine is based upon the joint operation of the three powers — centrifugal force, atmospheric resistance, and gravitation—and the machine somewhat resembles a "Catherine-wheel" working horizontally. Pulverised ore is fed on to the surface of a rapidly rotating disc about 20 inches in diameter, provided at the periphery with a raised rim, which is perforated by a multitude of small radial holes, through which the ore is thrown by centrifugal force. The particles of rich and heavy ore, by reason of their superior inertia, are thrown to a greater distance, while the worthless particles, being lighter, are more quickly overpowered by the forces of atmospheric resistance and gravitation, and thus fall short. The ejected ore is then collected in annular compartments arranged concentrically round the central disc, the process going on continuously while the ore is being shot out. Means are also provided for regulating either the centrifugal force or the atmospheric resistance, according to the nature of the ore.

It is stated by the inventors that in one of these machines, 5 feet in diameter, 50 tons of ore can be concentrated in twenty-four hours, only 3 H.P. being required. In this machine, the centrifugal force is proportional to the mass of the particles, whilst the atmospheric resistance is proportional to their cross section. It may be shown, therefore, that all particles for which the product $a D$ (diameter × density) is approximately the same will fall into the same compartment. But the value of $a D$ is nearly the same for all particles that are *equal-falling* in air; for in this medium, δ is so small in comparison with D (see footnote on p. 167), that it may be neglected, and, therefore, the formula

$a(D - \delta)$ becomes $a\,D$. Now, the difference in size between equal-falling bodies of different specific gravity is less for air than for water; for example, the diameters of spheres of galena and of quartz, which are equal, falling in water and in air, are in the ratio of $1 : 4{\cdot}1$ and of $1 : 2{\cdot}9$ respectively. Hence it would follow that, according to theory, this machine would be less effective as a concentrator than the wet sizing machine described on p. 170. However, the inventors have obtained good results in trial runs on both a large and a small scale.

A complete mill on the Clarkson-Stanfield system has been erected at the Castell Carndochan Mine, near Bala, North Wales, for the treatment of a low-grade gold ore, containing about 3 dwts. gold per ton. The mine owners have reported that, in the treatment of a parcel of 700 tons of ore, about 12 tons of rich concentrates of two grades were obtained, containing about 72 per cent. of the gold in the ore. The total cost of drying, milling, classifying by sieves, concentrating, cartage, &c., was certified as being 5s. 6·83d. per ton. The fuel used in drying the ore was 40 lbs. of coal per ton, and the motive power is water, and therefore costs little. The concentrates are sold to smelters.

Samples of ore from Coolgardie have also been successfully treated, very rich concentrates and tailings suitable for cyaniding being obtained.

CHAPTER X.

STAMP BATTERY PRACTICE IN PARTICULAR LOCALITIES.

Battery Practice in California.—The ores treated are quartzose, containing from $3\frac{1}{2}$ dwts. to $1\frac{1}{2}$ oz. of free gold per ton, the average being from 10 to 12 dwts. Some iron pyrites is also present, sometimes containing traces of other sulphides. The amount of pyrites varies from 1 to 5 per cent.; it is usually massive, and contains from 4 to $4\frac{1}{2}$ ozs. of gold per ton. The pyrites is in somewhat rare instances crystalline, and is then of little value. The gold is seldom visible, and is in such a finely divided state that it cannot be saved by mere gravitation methods, the use of mercury being essential.

The obsolete practice formerly pursued at Grass Valley, and in other districts of California, has been fully described by G. F. Deetken.* Up to a recent date it had not been entirely superseded,† and may be briefly described as follows:—The ore is fed into the batteries by hand, and no rock-breakers are used. The stamps weigh 800 lbs., and have a fall of 10 inches, the depth of discharge being only 3 inches. No mercury is added to the battery. After crushing, the pulp is run over a set of blanket-strakes, where the ore is subjected to a rough concentration. The tailings are passed through the so-called Eureka rubber, where the "rusty" gold, ground between sliding plates of iron, is brightened and rendered amalgamable. Subsequently the sand is run through narrow sluices, lined with amalgamated copper plates, where an effort is made to catch the fine free gold, although the depth and speed of the current renders the operation very incomplete. The tailings are then concentrated in round buddles, tossing tubs, &c., and the concentrates subjected to chlorination.

The blanket concentrates are treated in an Atwood's amalgamator, in which they are made to pass through two mercury wells or baths, being forced under the surface by revolving paddle wheels, and then they are treated with the ordinary tailings in the Eureka rubber. The skimmings of the mercury wells are treated in a small slowly revolving pan, the Knox pan, in which

---

* *Mineral Resources West of the Rocky Mountains*, 1873, pp. 319-345.
† According to the latest reports, this method has now been abandoned everywhere.

they are ground with mercury between iron mullers and dies for several hours, and the tailings from it are roasted and chlorinated.

The whole process bears a strong resemblance to the present Australian practice on free-milling ores containing coarse gold. It was uneconomical, costing about $2 or 8s. per ton, without reckoning the cost of chlorinating the concentrates, and only from 70 to 75 per cent. of the gold was extracted.

The present method of treatment* consists briefly in rapid crushing in narrow mortars with heavy stamps working fast with a low drop, the depth of discharge being moderate. Mercury is fed into the mortar box, and the gold is saved on copper plates situated both on the inside and the outside of the mortar, whilst the tailings are concentrated on shaking tables. Mercury wells and the various accessory amalgamating machines formerly attached to most mills are now falling into disuse, and do not form part of the most approved modern machinery.

The stamps weigh from 750 lbs. to as much as 1,100 lbs., the later practice being to make them of a weight of not less than 950 lbs., whilst the height of the drop has been gradually diminished, until the average is now only about 6 inches. According to the most modern view, from 4 to $4\frac{1}{2}$ inches is quite enough for ordinary ores; this height was first used at the Pacific Mill of the Plymouth Consolidated Mining Company, some three years ago, but other companies have been quick to follow their example. The heavy low-drop stamp, delivering over 100 blows per minute, has a duty much higher than the more old-fashioned stamps, which still form a great majority of the 3,500 head now in use in the State. The softer the ore the lower the drop should be, the limit being passed only when a good splash of the pulp can no longer be obtained, or when the momentum becomes insufficient to break down the larger lumps of quartz rapidly.

The mortars are narrower and less roomy than those in use in Colorado, and the depth of discharge is kept constant at about 6 inches. Below the screen is a single amalgamated copper plate about $4\frac{1}{2}$ inches wide, inclined outwards at an angle of about 45°. Mercury is fed into the mortar at intervals of one hour. The Blake rock-breaker and some form of ore feeder are in almost universal use; among the varieties of the latter, the Champion and Tulloch's machines are most favoured, although Stanford's and the roller feeders are still to be seen. The screens are inclined outwards at an angle of about 10°, and consist either of Russia sheet iron or of brass wire cloth. The horizontal burr-slot screen is perhaps that most in use, but others are equally suitable for certain classes of ore. The size

---

* The following is a brief digest of the full account given by J. H. Hammond, M.E., entitled "The Milling of Gold Ores in California," published in the *Eighth Report of the California State Mineralogist*, 1888.

of slot most in use is No. 6, which is ·027 inch in width, with about 180 holes to the square inch. The stamping is finer with low grade ores, because the gold in these is in a finer state of division, and the object of the stamping is to release the particles of gold from the rock in which they are enclosed, without reducing the ore to an unnecessarily finely divided condition. The crushing is kept coarse if the sulphides contain an appreciable percentage of the gold, since the finer the stamping the more difficult close concentration becomes, and, as the sulphides constitute the most brittle portion of the ore, they are always reduced to a finer state of division than the gangue in which they are enclosed. In general, it is better to crush too coarsely than too finely, for if the latter error is fallen into, not only is the output diminished and a larger proportion of the sulphides reduced to slimes, with the result that more is lost in the tailings, but the particles of gold are supposed to be subjected to repeated hammering, and converted partly into non-amalgamable and partly into "float" gold, and so lost in great measure. This question of over-stamping has already been discussed, p. 140, and an opinion there stated that the evils caused by it have been greatly overrated. Most of the ores are crushed much finer than might be supposed from the size of the meshes of the screen. If the ores are made to pass through a 30-mesh screen, it is found on trial that more than 80 per cent. of the material will usually pass through a 60-mesh screen also, and 50 per cent. will pass through a 120-mesh screen, much of it being impalpable slime. The low drop, already referred to, acts in the direction of minimising the quantity of slime, and has a most beneficial effect in facilitating the concentration of the sulphides.

The high speed of the stamps, besides increasing the crushing capacity, is advantageous on account of the good splash thus created, which is beneficial to the battery amalgamation, besides being essential to rapid discharge. The duty of the stamps is on an average $2\frac{1}{4}$ tons per head per day, but is considerably more in the most modern mills. It is lower on clayey or on very hard ores, and cannot conveniently be increased beyond about 4 tons, since the battery is an amalgamating as well as a crushing machine, and the greatest output may not be compatible with the highest percentage of extraction. The amount of water used is from 1,000 to 2,400 gallons per ton, most being required with clayey and heavily sulphuretted ores. In addition to this from 1 to 2 gallons per minute are supplied to each concentrator for feed water, and from $\frac{1}{2}$ to 1 gallon per minute for wash water.

The amalgamating plates outside the battery are in three sets, the apron plates, the tables, and the sluice plates; the united length of these varies from 10 to 30 feet, according to the requirements of the ore. They are almost invariably made of copper which is electro-plated with silver. The details concerning the preparation and care of these plates have already been given, pp. 122-125. The coarse gold is chiefly saved inside the battery; the finely divided particles, forming a fluid amalgam, are splashed through the screen and caught on the outside plates. From 50 to 80 per cent. of the gold is caught inside the battery, and the amount of free gold lost in the tailings is not more than 5 to 8 per cent. in the best mills. The tendency is in the direction of using more feed water with a lower inclination to the plates, as by this means the ore is brought into more intimate contact with them, being rolled over and over instead of swimming down them, and the danger of scouring is diminished. Other details concerning Californian stamp mills have been already given in the general description in Chapters vi. and vii.

Sizing of the tailings before concentration is not attempted, and this point deserves more attention than it has hitherto received at the hands of Californian millmen. Moreover, means should be adopted to get rid of the free gold and floured amalgam contained in the sands before passing them over the shaking tables, which are ill adapted for catching these materials. The two concentrators in most general use in the State are the Frue vanner and the Triumph concentrator, both of which have been already described. Next to these the Golden Gate and the Duncan machines are favoured, while Hendy's pan is still to be found doing good work, although it was one of the earliest contrivances used in the Pacific States.

The concentrates are treated by roasting and chlorination, generally by the vat process. Barrel chlorination has as yet made little headway in California, owing in great part to the fact that only small quantities of sulphides are produced at any one mill, whilst the distances between the establishments are very great. Chlorination costs about $10 per ton, and an average of 92 per cent. of the gold is thus extracted from the concentrates, which are usually worth from $80 to $100 per ton. After concentration, the tailings are in many cases run through a length of 100 to 200 feet of blanket sluices, where an effort is made to catch the free gold, amalgam, and sulphides still remaining in the tailings. These sluices are cheap and require little attention, but only account for a small percentage of the total gold recovered. The blanket sands are ground with mercury in a Chilian mill or arrastra, driven by water or steam power.

An excellent addition to a mill is afforded by the automatic tailings sampler erected at the original Empire Mill. This machine, by an arrangement of cogs, dips a bucket into the falling stream of tailings at stated intervals and takes a sample; all these samples are kept separate. They are usually taken at intervals of an hour. Most mills sample their tailings on very rare occasions. The average percentage of extraction does not fall below 85 per cent. in any but the oldest and worst appointed mills in the State, or in those running on unusually refractory ore.

The usual cost of milling in a 40-stamp mill, treating 80 tons per day, is as follows:—

		Cost per Ton.
I. Labour—		
(a) In the mill,		20¼ cents.
(b) Assaying, retorting, melting, and superintendence,		2¼ ,,
II. Supplies—		
Castings,		7 to 10 cents.
Mercury,		1½ to 4 ,,
Lubricants, screens, illuminants, and miscellaneous,		4 to 8 ,,
Total,		35¼ to 45 ,,
	or,	1s. 6d. to 1s. 10½d.

The cost of water power, if it is purchased, must be added to this, and, if steam is used, about 11 cents per ton must be added for additional labour, oil, &c., besides the cost of fuel. Interest on capital, and the cost of chlorination are not included in this estimate.

The power required in such a mill is distributed as follows:—

Rock-breakers,	12 H.P.
40 stamps,	66 ,,
16 concentrators,	8 ,,
8 shaking tables,	2½ ,,
1 clean-up pan,	1¼ ,,
1 clean-up amalgamating barrel and batea,	2 ,,
Total,	92 ,,

The poorest ore which can be mined and milled at a profit must contain from 2 to 5 dwts. of gold per ton.

In 1889 there were about 3,000 stamp heads existing in California, of which not more than about 2,000 were in active operation, crushing 4,000 tons of ore per day. Of these stamps about 60 per cent. were run by water power, and the remainder by steam. The cost of milling varied from 39 cents to $2 per ton, the mean cost being about $1 per ton.[*] In 1892, about 4,000 head of stamps were in position in California, 3,500 of which were at work. The lowest cost of milling was that at the Dalmatia Mine, El Dorado County, where the total cost of mining and milling was 43 cents per ton. In 1851, when quartz mining was first begun at Grass Valley, no ore was considered worth treating unless it yielded $40 per ton by the wasteful process then in vogue.[†] In 1873, according to Deetken, the cost had fallen to a minimum of $2 per ton.

[*] *Ninth Report Cal. State Min.*, 1889, p. 25.
[†] *Eleventh Report Cal. State Min.*, 1892, p. 19.

Method of Working employed in the Blackhawk Mills, Gilpin Co., Colorado.—The ore obtained in this district and supplied to the mills is often called refractory, although the greater part of the gold contained in it is extracted by amalgamation. It is refractory in the sense that it yields its gold less readily than the typical Californian free-milling ores. It contains, in general, from 10 to 20 per cent. of metallic sulphides, and from 15 to 20 per cent. of quartz, the remainder

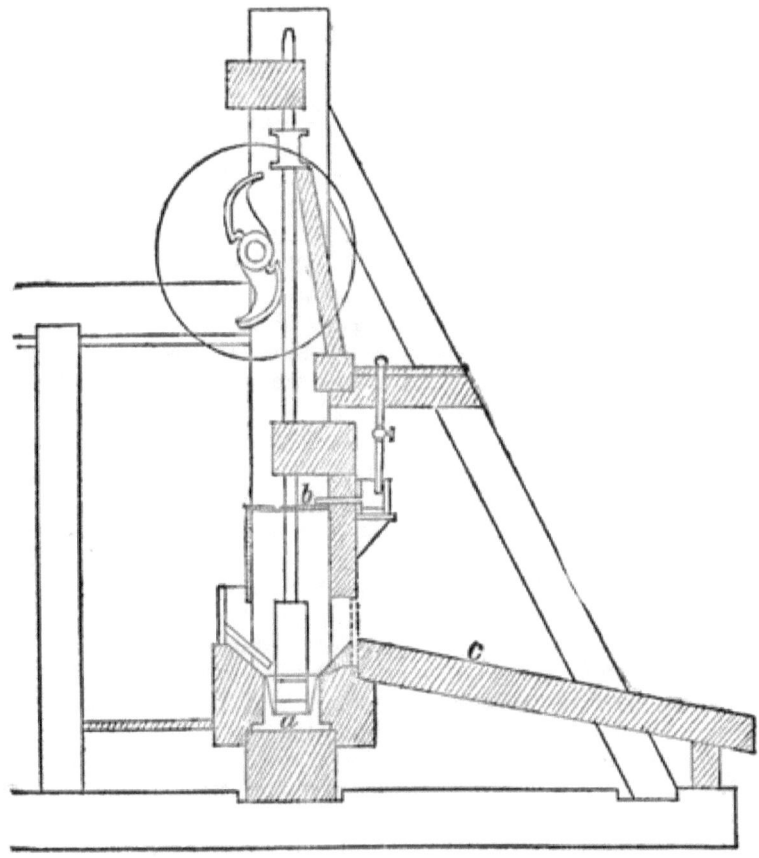

Fig. 44.

of the gangue being chiefly felspathic matter derived from the country rock. The metallic sulphides are chiefly copper and iron pyrites, but antimonial grey copper, arsenical pyrites, and galena are also present, and blende and carbonate of iron occur sometimes. The greater part of the gold is free, being enclosed in the quartz, but about 25 per cent. of the gold and 33 per cent. of the silver are locked up in the sulphides, which are saved by concentration, and shipped to the neighbouring smelters in Denver. The gold is in a very finely divided state, the ore often giving no "colours" when subjected to ordinary panning, although in the mill it will yield a good return. The average milling ore contains about $\frac{1}{2}$ oz. of gold and 2 or 3 ozs. of silver per ton.

The method used consists briefly in slow crushing and amalgamation in the battery, where as much gold is saved as possible, followed by the passage of the pulp successively over a set of amalgamated plates, then over blankets, and finally over concentrating machines, the tailings being allowed to run into the stream. A sectional view of the battery used is shown in Fig. 44. It will be observed that only a shallow light iron mortar, a, is used, the sides consisting of a housing of wood. The water is introduced by the pipe, b, and allowed to run down the stem of the stamp. The ore after passing the screens flows over the amalgamated plates fixed on the tables, c. The stamps are light, weighing from 500 to 600 lbs. each, and fall at the extremely slow rate of from 26 to 32 drops per minute, the height of the drop being from 16 to 18 inches. This is greater than that adopted in any other part of the world, being four times as great as that adopted in the latest Californian practice. The depth of discharge is 13 inches when new dies have just been put in, and increases to 15 or 16 inches as they wear down. According to the most modern view, however, this depth should be diminished, and at the mill latest erected it is only 11 inches. The shoes and dies are small (8 inches in diameter) owing to the lightness of the stamps, and the screens are fine, a burr slot being used which corresponds to a 50-mesh wire screen. The screen surface is normal, the dimensions being $4\frac{1}{2}$ feet by 8 inches for each battery of five stamps. Under these conditions the output is of course low, being on an average about 1 ton per stamp head in twenty-four hours. The feeding is done by hand, one man feeding twenty-five stamps, and there are no rock-breakers, their absence being accounted for by the age of the mills and the lack of fall in the ground in the rear of the batteries, which interferes with their introduction at the present day. Nevertheless, rock-breakers and automatic feeding would reduce the cost of crushing considerably, as the cost of hand-feeding alone is $450 per battery in a year, while the first cost of automatic feeding machines would only be from $500 to $1,000, and the cost for maintenance, power, and repairs nominal.

The amount of water added to the battery is from $1\frac{1}{4}$ to 2 gallons per stamp per minute. The amount of mercury used varies of course with the richness of the ore. When ore containing $\frac{1}{2}$ oz. of gold per ton is being crushed, about $2\frac{1}{2}$ ozs. of mercury are added to each battery per day in small quantities at intervals of an hour, while about an equal amount ($2\frac{1}{2}$ ozs.) is required to dress the plates, both inside and outside. Of this quantity one-third is put on the back inside plate, one-fourth on the front inside plate, and the remainder on the outside plates. The loss of mercury from all sources is about 20 or 25 per cent. of that added each day, and, with $\frac{1}{8}$ oz. ore, is about $\frac{1}{4}$ oz. of mercury per ton crushed.

Two sets of amalgamated plates are used inside the battery, each $4\frac{1}{2}$ feet long, the front one being 6 inches wide, and the back one 12 inches. The front plate is nearly vertical, but the back one is set at an angle of 40°. There is only a single set of plates outside, which are 4 feet wide by 12 feet long, and have the high grade of $2\frac{1}{8}$ inches per foot. The plates are dressed every twelve hours with a weak solution of cyanide of potassium (2 ozs. in 3 gallons of water), this being the only chemical used on the mill. The consumption of cyanide is about 2 ozs. per battery of five stamps in three days. Its use is necessitated by the acid nature of the water, which already contains sulphates and carbonates when it comes from the mines, and is further contaminated by ferrous sulphate, formed by the oxidation of the pyrites in the ore and then dissolved. This water soon forms a film on the surface of the copper plates, and also rapidly corrodes the screens, which otherwise, from their unusual height above the dies, would last a long time. Nevertheless no effort is made to counteract the bad influence of this water by adding alkalies, limestone, or lime water to it or to the ore.

After leaving the plates the ore passes over the "blanket-strips," which are 3 feet long and 18 inches wide. They are washed every four hours, or every two hours if the ore is rich. The blankets serve to catch any escaping mercury, amalgam, "rusty" gold, and the heaviest pyrites, together with pieces of rock to which gold is attached. Probably this work would be done equally well by the concentrators, by which an operation would be saved. The blanket-sands are sold to the smelters with the other concentrates. The concentrators consist of shaking tables, constructed in the locality; they have been already described at p. 175. The concentrates usually contain about 15 dwts. of gold and 5 ozs. of silver per ton.

The amalgam obtained is divided fairly equally between the front inside plate, the back inside plate, and the outside plates. In retorting it the interior of the iron retort is chalked or lined with paper to prevent the gold sticking to it, the latter method being considered the better. The balls of dry squeezed amalgam are put into the retort, broken with an iron rod and then pressed down until hard and uniform. A large bolt with a nut at the end is often used for the purpose. The cover is then put on and luted down with clay. The variety of retort used is similar to the bell-shaped one mentioned on p. 133. One hundred parts of amalgam yield from 33 to 50 parts of retorted metal, the average being about 40; the amount depends on the richness of the ore and the coarseness of the gold. The bullion contains from 750 to 850 parts of fine gold, and almost all the remainder is silver, about 10 parts per 1,000 being base metal. According to T. A. Rickard,* the late manager of the New California Mine and Mill, from whose descriptions many of the details already given have been taken, the results obtained by his mill on an ore containing 10 dwts. of gold and 2¼ ozs. of silver per ton, were as follows :—

	Gold.	Silver.
Saved on the plates,	70·4 per cent.	42·6 per cent.
Saved in the concentrates,	23·4 ,,	31·4 ,,
Lost in the tailings,	6·2 ,,	26 ,,

The cost of milling was 84 cents per ton in 1890, and 78 cents per ton in 1891. This did not include the cost of shipping and smelting the concentrates. Power is supplied for four months in the year by water, for four months (in the dead of winter) by steam, and for the remaining four months by the two combined. Labour in 1890 cost $3 per day per man; in 1893, about $2.50.

In this locality, the slow drop and deep discharge are more accentuated than in any other part of the world, although, in Calaveras County, California, an almost identical course was formerly pursued with the refractory ores found there.

The pulverisation is even finer than is indicated by the size of the screen slots, owing to the retention of the pulp in the mortar after it has been crushed, which is due to the small area of discharge as compared with that of wire screens, and to the depth of discharge and the slowness of the drop. A large percentage of the pulp is fine enough to pass through a 100-mesh sieve, the pyrites being especially finely crushed. This fact is of importance from two points of view. On the one hand, as the fine free gold is chiefly in intimate association with the pyrites, the fine crushing of the latter enables the gold to be separated from it by the mercury, and this result is advantageous; but, on the other hand, the excessive fineness of the pyrites makes it exceedingly difficult to separate from the ore by concentration, and, although the shaking tables do excellent work and save a fair proportion, a somewhat serious percentage is perforce allowed to run to waste. So difficult to catch is it, that it is carried many miles by the stream, and can always be panned out of the sand which is deposited where Clear Creek debouches on the plains, at a distance of over 30 miles from the mills.

* *Eng. and Mng. Journ.*, Sept. 10, 1892.

The use of the slow drop is intended to enable the gold and pyrites to settle a little by gravity through the lighter particles of gangue between each blow, and so to keep them longer in the battery, and thus increase the chances of amalgamating the gold. The wide roomy mortar prevents a violent splash against the screens and plates, so that little scouring of the latter takes place. There can be no doubt that, on the whole, the method of working in Gilpin County, which was slowly and painfully elaborated about twenty years ago, is the best that could be applied to the particular ores which occur in the district.

Battery Practice on Free-Milling Ores in Australia and New Zealand.—Australian practice as a whole, if some of the new mills in Queensland are excepted, owes little to the experience gained in the United States, and pursues widely different methods from those in use there. Modern methods in both countries are based almost entirely on the processes in use in Central Europe in the first half of this century, and have been modified to some extent in different directions in various localities, the changes being in most cases beneficial, and especially suited to the class of ore in course of treatment. The process employed on the free-milling coarse-gold ores in Victoria, New South Wales, and New Zealand has perhaps undergone less variation from the original Transylvanian type than that employed in any other part of the world, chiefly owing to the extreme simplicity of these ores, and the ease with which a high percentage of the gold can be extracted.

The ores consist of white quartz, containing large grains and plates of remarkably pure gold, which is thus often visible, and sometimes occurs in big pieces weighing several pennyweights. Inclusions of country rock (slate, &c.) are common, and in these cases the sulphides often slightly increase in quantity, but rarely form more than from $\frac{1}{2}$ to $\frac{3}{4}$ per cent. of the rock, and consist chiefly of iron or arsenical iron pyrites. The free gold is not intimately associated with these pyrites, and the ore may even fall off in value when the quantity of sulphides increases. The poorest ores treated contain 4 or 5 dwts. of free gold per ton, an amount which includes from 85 to over 99 per cent. of the total gold contents, and the concentrates contain from 10 dwts. to 4 ozs. or more of gold per ton. These concentrates are saved in some districts and treated by roasting and chlorination, while in other places they are allowed to run to waste, especially if they contain less than 1 oz. of gold.

The method of treatment consists in crushing the ore somewhat coarsely in rectangular mortar boxes, with stamps of medium weight, running at a medium rate of speed, and then in passing the pulp through mercury wells and over blanket strakes. The mortar sand, which is collected periodically, is amalgamated in revolving barrels, together with the blanket concentrates, but no mercury is added to the mortars, and amalgamated plates are conspicuous by their absence. The following details concerning the machinery, &c., employed at certain mills in Victoria and New Zealand are chiefly derived from T. A. Rickard's writings* on the subject, to which the reader is referred for a more complete account.

Rock-breakers and automatic feeding machines are seldom used, and the hand-feeding is often done very carelessly. The mortar boxes are of rectangular shape, have no amalgamated plates inside them, and are considered by American experts to be, in many cases, unsuited for internal amalgamation, as the mercury and amalgam do not collect out of reach of the falling stamps, which would, therefore, cause flouring. It is in this way that the fact may be explained that a smaller saving of gold is effected when mercury is fed into the mortar boxes, than when the customary practice is adhered to, since the mercury and amalgam are subjected in the former case to conditions under which excessive flouring takes place, and the mercury is lost in the tailings, together with such gold as it may have taken up. On the other hand, Australian millmen point out that the mortar boxes are spacious enough for the retention of the coarse gold which collects by gravity in them, without being pounded to fragments by the stamps, and is removed at intervals by a scoop, and that the collection of this gold is retarded and not aided by the addition of mercury. The mortar boxes are from 14 to 16 inches wide, and spaces are usually left between the dies, and between these and the ends and sides of the boxes, which would certainly be far too roomy if the battery were intended only for a crushing machine and not as an apparatus for collecting coarse gold also. Australian managers prefer blankets to copper plates, their reason for this preference, as well as for the retention of their mortar boxes, being based, no doubt, on the coarseness of the gold to be saved.

The weight of the stamps is usually about 800 or 900 lbs., 8 cwts. being a favourite weight; the diameter of the shoe is 10 inches, the height of the drop is $7\frac{1}{2}$ or 8 inches, and the speed about eighty blows per minute. The depth of discharge is low, being usually about 4 inches, and this is kept constant by packing up the die with sand or by putting in thin iron plates below it, as it is worn down. The screens are round-punched Russia iron or sheet copper, the latter lasting much longer; they are pierced with from 81 to 180 holes per square inch. The screens are of greater area than those used in America, being about 12 inches in vertical height instead of 8 inches. Double discharge is often used and would always be advantageous, the duty being from $2\frac{1}{2}$ to $3\frac{1}{2}$ tons per stamp per day at the mills where double dis-

* "Variations in Gold Milling," *Eng. and Mng. Journ.*, Jan., Feb., and March, 1893.

charge is used, and little more than half this amount in the single discharge batteries. The amount of water used is very large—*e.g.*, 8 gallons per minute per stamp head at Clunes, where there is a double discharge and where a low inclination is given to the blankets. This quantity of water is four or five times as much as that supplied in California.

A double discharge battery in use at the South Clunes United Mill is shown in section in Fig. 45; A and B are the screens, and the launder, C, conveys the discharge from the back to mix with that coming from the front screen. The perforated iron plate, D, serves to distribute the ore evenly over the apron, E, over which the pulp flows; the latter subsequently passes through the mercury wells, F and G, being compelled to pass through the bath of quicksilver by the vertical boards, H and K. The wells are made of iron, which keeps the mercury lively, and are 3 inches in width and 4 inches deep; the ore has a fall of 10 inches upon the surface of the quicksilver. Each well contains about 50 lbs. of mercury, and is usually cleaned up once a week by passing its contents through canvas and so separating the amalgam; the wells are also skimmed at short intervals. They are omitted altogether in some mills, gold being saved only by gravity in the mortar and by concentration on blanket strakes.

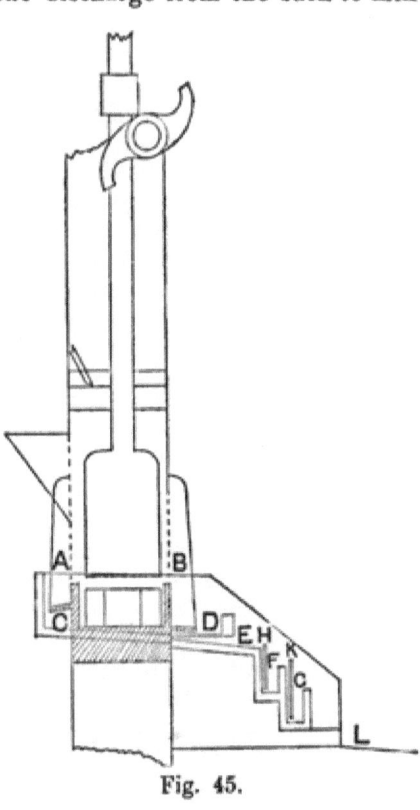

Fig. 45.

The blanket tables, L, should have a united width equal to that of the screens; they are usually divided into widths of 18 inches, and consist of three or four lengths of 6 feet each, over which the ore passes in succession; the grade at which they are placed is from 1 to 1½ inches per foot. Green baize is the material most commonly used, the colour enabling specks of gold to be readily seen. The blankets in the top row are washed in tubs every half hour or hour, according to the richness of the ore, and those in the lower rows at longer intervals. The blankets are sometimes succeeded by concentrators, which usually consist of some modification of the Cornish buddle. At some mills the experiment has been made, with considerable success, of replacing the lowest row of blankets by amalgamated plates. The blankets are well adapted to catch coarse gold, but the fine gold escapes them, and is caught on the plates, which are efficient for this purpose, though perhaps of no greater utility than blankets in arresting coarse particles.

In cleaning-up, the screens are taken out, the contents of the mortars are passed through a No. 4 sieve, and the material that remains on this is returned to the battery and used to pack round the dies. The fine material—one or two buckets full from each mortar box at each fortnightly clean-up—contains a quantity of free gold, and is treated with mercury in the amalgamating barrel, together with the blanket concentrates. The skimmings from the wells, and occasionally the blanket concentrates also, are treated in Berdan pans, and the concentrates from the buddles are either roasted and amalgamated in Chilian mills, or, more frequently, roasted and chlorinated in revolving barrels.

The amalgamating barrel, mentioned above, is hung on trunnions, and revolved at the rate of about 16 revolutions per minute. From 5 to 20 cwts. of concentrates, battery sand, &c., according to the size of the barrel, are charged in together with 75 lbs. of mercury and some cold water, the chemicals used being sulphate of copper and wood ashes. After running for about ten hours, the barrel is discharged into a tank and its contents allowed to settle, the process being sometimes aided by passing the pulp through a series of wells, as, for example, at the South Clunes United Mill where there are three drops, the heights of which are 12, 9, and 6 inches respectively, the material falling each time into a bath of mercury. The battery sand is usually kept separate from the blanket concentrates, as it is much richer than the latter. The tailings from the barrel and the Berdan pans are sometimes run over a few feet of copper plates to catch any finely divided amalgam that may be contained in them, and are sometimes concentrated on Rittinger shaking tables.

The percentage of gold recovered varies greatly according to the nature of the ore and to the method of treatment adopted. At many of the mills no steps are taken to ascertain the extent of the loss, and, in the best mills, only from 80 to 85 per cent. of the gold is saved. More than half of the gold which is recovered comes from the mortar boxes, where it is merely collected by gravity, and the remainder is derived mainly from treating the blanket concentrates. At Clunes, where mercury wells are used, the relative amounts of gold obtained from the various appliances are approximately as follows:—From the mortar box, 45 per cent.; from the wells, 30 per cent.; from the blankets, 15 per cent.;

and from the skimmings of wells, 10 per cent. At Otago, where mercury wells are not used, the distribution is as follows:— Mortar box, 60 per cent.; blankets, 33 per cent.; copper plates, 7 per cent.

The loss of mercury would be very small, if it were not for the flouring that takes place in the treatment of the skimmings, &c., in Berdan pans. Loss by flouring also takes place if the amalgamating barrel is run too rapidly. Owing chiefly to this latter cause, the loss per ton of ore crushed is 8 dwts. of mercury at Otago, while at Clunes it is about 3 dwts. At the South Clunes United Mill the loss for the eight years (1884-92) averaged only $5\frac{1}{2}$ grains of mercury per ton, this being probably the smallest loss on record in a stamp mill. This fact proves the often-repeated assertion that a bath of mercury can be manipulated with very little loss. If metallic copper be present in the ore, as is sometimes the case, the copper amalgam formed floats on the surface of the mercury, and is readily carried away with the tailings, the loss being thus increased to 2 or 3 ozs. of mercury per ton of rock crushed. The cost of treatment is, in some well appointed mills, only from 2s. to 2s. 6d. per ton.

Method of Working in the Thames Valley, N.Z.[*]—The ore in this district may be regarded as on the border line between "refractory" and "free-milling" ores. It varies greatly in hardness and composition, and consists of "stringers" of quartz running through a more or less decomposed and brecciated hornblende-andesite. The gold is often in visible specks and threads in the quartz, but is also largely associated with copper and iron pyrites, blende, galena, stibnite, &c., the quantity of sulphides varying from $\frac{1}{2}$ to 10 per cent. of the ore, and averaging from 2 to 3 per cent. Silver occurs as argentite, &c., while tellurides of both gold and silver are sometimes found. The average amounts extracted from the ordinary ore are—gold, 5 to 10 dwts. per ton, silver, 2 to 5 dwts.; but a considerable percentage of the gold recovered comes from "specimen" ore, which is often worth some thousands of pounds per ton.

No rock-breakers are used, grizzlies or sizing screens being also unknown, and the feeding is done by hand. The small stuff is shovelled into the mortar, and big pieces of rocks are rolled into the feed opening, and rammed in by a few blows of the sledge hammer, if they stick there. The results of this primitive practice are seen in the excessive rate of wear of the shoes and dies —i.e., 22 ozs. of iron per ton of ore crushed, 14·5 ozs. for the shoe, and 7·5 ozs. for the die. The feeding is very high, the mortar boxes being choked up with ore, so that the output is still further reduced from this cause. The mortar box is spacious, approaching in design the Gilpin County pattern, but, as no attempt is made

---

[*] The following account is partly based on the description given by T. A. Rickard in the *Engineering and Mining Journal*, Dec. 3 and 10, 1892.

to amalgamate the ore before it leaves the battery, a narrower shape might be used with advantage. The stamps weigh from 620 to 840 lbs., the faces of the shoes and dies being $9\frac{1}{2}$ inches in diameter, while the dies are 4 inches deep. Both are made of white cast iron, only the shoes being chilled. The height of the drop is 8 or 9 inches, and the number of drops per minute is 75. The depth of discharge, though varying as the dies wear, is on an average only from 2 to 3 inches, the bottom of the screen being placed on a level with the surface of the pulp in the mortar, or even below it, the bottom being on a level with the top of the dies in one mill. The screens are placed in a vertical position, and are made of round-punched Russia sheet iron, having from 148 to 180 holes per square inch.

The wearing of the screens is very rapid, owing partly to the small depth of discharge, the ore being flung almost horizontally from the surface of the dies and striking the screens with great force, but still more owing to the acidity of the battery water. A grating usually lasts about six days, or while about 50 tons of ore are crushed. Copper gratings might be used with advantage. The mine waters contain an unusual quantity of the protosulphates of iron, copper, manganese, and aluminium, and, consequently, when the millstuff is very wet, the screens are corroded with great rapidity, while the soft surface ore, in which the sulphides have been largely decomposed, wears out the screens faster than the hard deep-level ore. It is unfortunate that lime water and limestone are not readily obtainable in the district, as an addition of either of these would undoubtedly lengthen the life of the screens. The output is from $1\frac{1}{2}$ to $1\frac{3}{4}$ tons per stamp per day, which is low, considering that no battery amalgamation is attempted. A fast-running heavy stamp, with double discharge, would probably double the capacity of the mills, and leave the gold in a better condition for amalgamation.

The pulp, after being ejected from the mortar, is run over amalgamated tables, 7 feet long and $4\frac{1}{2}$ feet wide. On the upper part of these tables, next the battery, is a plate $2\frac{1}{2}$ feet long, succeeded by a well $2\frac{1}{2}$ inches wide and 2 inches deep, filled with mercury. Below this is another length of 18 inches of amalgamated plate, and then a succession of three more "ripples" or wells, not filled with mercury. The total length of the plates is thus only 4 feet, and the only other means adopted to catch the gold consist in the use of the ripples and some blanket strakes, by which some of the free gold and pyrites are caught. After passing over the blankets, the tailings are allowed to run into the sea. The "blind ripples"—*i.e.*, those not containing mercury—are cleaned with a scoop every half hour, the heavy sand and pyrites so obtained going to the pans together with the material caught on the blankets, which are washed every hour or even oftener. The blankets cost 6 shillings per square yard, and last for three months. The plates consist sometimes of copper, but more usually of Muntz metal; they are cleaned, wiped, and dressed every four hours, the wells being skimmed with a cloth at the same time. The mercury in the wells is squeezed through wash leather once a week.

The use of the mercury wells serves but little purpose as, in accordance with general experience elsewhere, the sulphides cause a scum to form over the surface of the bath after a few minutes, and no further amalgamation takes place until it is skimmed off.

The pans used for the treatment of the blanket concentrates are chiefly Berdans, with a few Watson and Denny's. The Berdan pans are furnished with a drag, fixed to one side of the sloping bottom, instead of a ball, as is usual. This is a useful modification, as grinding and amalgamation are kept separate, while in the ordinary Berdan pan, the ball remains at the bottom of the cone where the mercury collects, and is the cause of a considerable loss by flouring. The drag consists of two parts, the lower part being the slipper or shoe, weighing 196 lbs., to which the boss or top, weighing 230 lbs., is rigidly keyed. The shoe wears out in four months. The Berdan pans are $4\frac{1}{2}$ feet in diameter, with a depth of 9 inches, the bottom having an inclination towards the centre of 1 in $2\frac{2}{3}$; they work at 23 revolutions per minute, and the amalgam is removed and strained every twenty-four hours.

The greater part of the amalgam is obtained from the plates, the percentages being—from the plates about 75 per cent., from the wells 5 to 10 per cent., from the pans about 15 per cent. One of the features of the district is the "specimen" stamp attached to each mill, by which the valuable specimen ore is crushed, the gold being caught chiefly in the mortar box, while all the tailings are saved and treated in the pans.

The amalgam yields about 40 to 50 per cent. of retorted metal, which on melting yields bullion about 600 to 675 fine in gold, the remainder being almost entirely silver. The loss of mercury is high, owing to excessive flouring in the pans; it varies from 15 dwts. to 1 oz. per ton of ore crushed. The cost of working is about four shillings per ton, water power being used and paid for.

The absence of concentrators to save the sulphides renders the process somewhat wasteful. The sulphides from these tailings often contain from 15 to 25 dwts. of gold per ton, and a much greater amount of silver. Even the free gold is stated to be by no means completely extracted, nearly 50 per cent. passing away in the tailings. There is certainly an accumulation of tailings on the foreshore estimated to amount to over 1,000,000 tons, which are believed to average about $\frac{1}{2}$ oz. of bullion per ton. A few years ago a large number of samples were collected from various parts of the foreshore and sent to England, where they were assayed by the author, and were found to contain on an average about 2 dwts. of gold and 6 dwts. of silver per ton. A large proportion of the sulphides was in the state of slimes and difficult to save by concentration, but, on reducing the samples in the ratio of 10 to 1 by panning, the residue was found to contain $\frac{1}{2}$ oz. of gold per ton, while if reduced in the ratio of 28 to 1 the residue of pure sulphides contained 1 oz. 6 dwts. of gold and 14 dwts. of silver per ton. These results seem to point to the fact that the extraction of the free gold is not so incomplete as is usually believed, although the total loss is considerable, owing to the value of the sulphides.

Battery Practice in Dakota.—The method adopted in the Black Hills consists in battery and outside plate amalgamation, followed by treatment by a few mercury traps and rifles, and by blanket sluices, no systematic concentration being attempted. The reasons for this are that the ore is free milling and the greater part of the gold is sufficiently coarse to be caught inside the battery, whilst the sulphides from most of the lodes are not rich enough to pay for treatment, and so are allowed to run to waste, although they could be saved at a trifling cost. The sulphides from the Homestake Mine contain gold and silver to the value of only $5 or $6 per ton, whilst the poorest ores which can be treated at a profit, either by the pyritic smelters or the barrel chlorination works in the neighbourhood, yield from $8 to $12 in gold per ton. At the Caledonia Mill, however, the ore treated contains about 4 per cent. of pyrites, which assay about $90 to the ton, and concentration is absolutely necessary for successful working. There are 740 stamps running continuously in the Black Hills, most of which are under the management of the Homestake Company; 3,000 tons of ore are crushed per day. The yield in bullion is from 3 to 4 dwts. of gold and nearly 1 dwt. of silver per ton. When surface ore was being treated the tailings often contained less than $\frac{1}{2}$ dwt. of gold, and even as late as the year 1889, from 70 to 80 per cent. of the total contents were extracted, but the increase in the amount of pyrites in the ore, due to the greater depth of the workings, has of late years resulted in an increased loss in the tailings and a diminished yield on the plates, so that the extraction probably does not average more than from 50 to 70 per cent., and the tailings often contain as much as 2 to $2\frac{1}{2}$ dwts. of gold per ton.

The Homestake Mill is arranged in such a way that the ore need not be touched by hand after it is let fall into the gigantic ore-bins; consequently eight men can do all the work in connection with the battery of 200 head of stamps. The ore is passed over grizzlies, thence through Blake's or Gates crushers, each Blake (of which the dimensions of the mouth are 9 × 15 inches) being set to supply 20 stamps with ore passing through a 1½ inch ring. From the crushers the ore passes through Challenge ore feeders into the batteries, which are arranged in two rows set back to back with the ore feeders between them, the distance between the faces being 44 feet. The stamps weigh 850 lbs., their fall is 8 to 9 inches, and they make 90 drops per minute. The screen used is No. 7, and the output is about 4 to 4½ tons per stamp per day. There is one copper plate inside the mortar below the screens, and about 1 oz. of mercury per day per stamp head is added with the ore. The shoes and dies are of cast iron and last about two months. It has been shown by experiments in the Father de Smet Mill* that for crushing low grade ores a square-bottomed die is the best, and that there is no economy in using the dies after their surfaces have become perceptibly irregular by wearing. It is accordingly desirable to cast them as thin as practicable and to replace them often—at least once a month.

The amalgamated plates placed outside the mortar are about 10 feet long and 4½ feet wide; they have an inclination of 2 inches to the foot, the grade being kept as high as is practicable consistent with good amalgamation, owing to the scarcity and high cost of the water. At the lower end of these plates the ore is passed through a mercury trap or well, and the discharge is then narrowed and the pulp is passed through sluices from 60 to 100 feet long and 2 feet wide, lined with copper plates, and broken at intervals by mercury traps. Hungarian riffles are also used, and at the Caledonia Mill, where the sulphides are valuable, an imperfect concentration is effected by means of blankets. The battery plates are dressed every day, and any excess of amalgam then found is removed; the sluice plates are dressed every four days. The battery is cleaned up fortnightly, and the sluices, mercury wells, riffles, &c., monthly. The cost of milling is 83 cents (3s. 8d.) per ton, the water alone costing from 50 to 57 cents per stamp per day, or 12 to 13 cents per ton of ore. The pulp is run through settling pits and the water used over again during the winter months, when the scarcity of water is felt most severely. It is then warmed before use. The fineness of the bullion is, in general, 820 in gold and 170 in silver.

Stamp Battery Practice in the Transvaal.†—On the Witwatersrand the ores consist almost entirely of the now familiar "banket" formation, a conglomerate consisting of water-worn pebbles of translucent quartz set in a matrix composed mainly of oxides of iron and silica. In the ore near the surface the gold is almost entirely free, existing in the cement, the pebbles being generally barren. At a moderate

---

* *Trans. Am. Inst. Min. Eng.*, vol. x., p. 97.
† This account is mainly based on information kindly supplied by Mr. S. H. Farrar, M.I.C.E., F.G.S., M.I.M.E., of Johannesburg.

depth, however, the oxides of iron gradually give place to sulphides, but even then the greater part of the gold can be extracted by amalgamation. The ores vary greatly in richness, some containing less than 1 dwt. of gold per ton, and others as much as 2 or 3 ozs., the average being about 15 dwts.

The method of treatment consists in crushing through somewhat coarse screens with a small depth of discharge, thus obtaining a very large output; the gold is caught partly on plates inside the mortar boxes, partly on copper tables placed outside. The pulp is concentrated in various ways, blanket strakes, Frue vanners, and *spitzluten* being in common use. The Lührig vanner is being introduced, and has already met with a considerable amount of success. The concentrates, if they contain little or no sulphides, are ground in pans with mercury; if pyritic, they are either roasted and chlorinated or treated with cyanide. The tailings from the batteries or the concentrators are treated by the cyanide process.

There are two types of stamp mills in existence on the Rand, both of which have been very successful, although differing somewhat in their construction. These are the American mills, constructed chiefly by Messrs. Fraser and Chalmers, and the English mills sent out by the Sandycroft manufactory, and by other well-known English firms. A brief account of some of the details of these mills is given below.

Rock-breakers are in use at almost all the mills on the Rand, both the Blake-Marsden and the Gates crushers being in wide use. In general, one Blake-Marsden machine with a feed-opening of 15 by 9 inches is required for every twenty stamps. The rock-breakers are generally contained in separate rock houses, placed at the mine. The advantages are that the rock-breakers can be placed nearer the ground, and are less expensive to erect, while the stuff in the battery ore-bins is more completely mixed. Self-feeders to the batteries are universal, the Challenge, Tulloch, and Sandycroft machines being chiefly used; they are arranged so as to keep the ore at a depth of 1 to 2 inches on the dies.

The mortar blocks are usually of 14-inch square timber, from 9 to 15 feet long, and are laid on concrete. It has been found that the usual practice of filling the space between the blocks under one mortar, and those under the next, with well-tamped earth is less satisfactory in giving stability than that of filling the space with solid masonry. The latter course has accordingly been adopted at the best mills. The mortar blocks are covered with blanketing or sheets of india-rubber or lead before the mortars are bolted to them. The English mortars are cast about 4 feet high and 4 feet 8 inches long; the width inside at the level of the dies is $10\frac{3}{4}$ inches, that of the American mortar being 11 inches. Both mortars widen considerably towards the upper part, the walls not being vertical at any point. The thickness of the mortar at the bottom is 6 to 9 inches, and the inside is usually lined with wrought-iron plates to protect it from wear. The American mortars are furnished with amalgamated copper plates inside, both on the feed and discharge sides, but the earlier English mortars on the discharge side only. The inclination of these plates to the horizon is usually from 70° to 80°. Steel wire mesh screens are almost invariably used, punched Russia iron plates being rare. There are from 100 to 1,200 holes per square inch of the screen, the usual number being 600 to 900 (*i.e.*, 25 to 30 holes to the linear inch). The depth of discharge is about 3 to 5 inches, and the scouring action on the front inside plates is consequently so violent that but little amalgam accumulates there, and, in the opinion of many experts, these plates might advantageously be dispensed with.

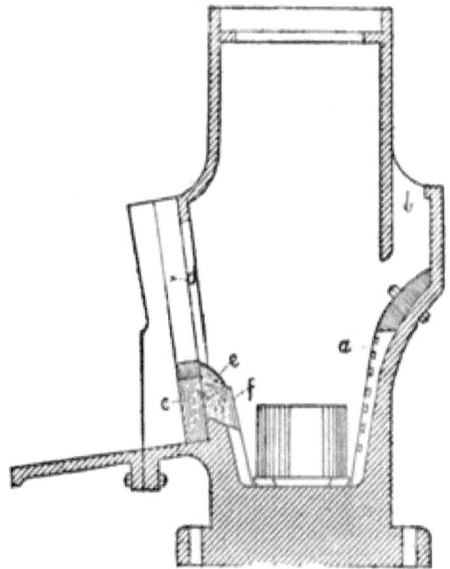

Fig. 45*a*.

A section of the mortar used at the Croesus mine is shown in Fig. 45*a*. *a* is a cast-steel lining plate with slots or recesses in it for collecting the amalgam; *b* is the feed opening; *c* the wooden blocks for carrying *e*, the copper plate; *d* is the screen opening; and *f* a steel plate.

The weight of the stamp in the earlier mills was usually about 850 lbs. in the American, and from 700 to 750 lbs. in the English mills. In the later mills, which represent in every respect the very best modern practice, the stamps are of 1,000 lbs. weight or upwards, while in the Modderfontein 40-stamp battery the weight is 1,250 lbs., the heaviest gravitation stamp yet in action. The actual weights of the parts of two English stamps in use were found to be as follows:—

Nominal weight of Stamp,			700 lbs.	750 lbs.
Actual ,,	Stem,		238 ,,	287 ,,
,, ,,	Head,		197 ,,	197 ,,
,, ,,	Tappet,		115 ,,	115 ,,
,, ,,	Shoe,		180 ,,	180 ,,
Total actual weight of Stamp,			730 ,,	779 ,,

The usual height of drop varies from 6 to 9 inches, and the speed is usually from 90 to 95 drops per minute. The shoes and dies are 9 inches in diameter, and the shank of the shoe in English mills is 5 inches long and 3 inches wide. The shoe is often kept in work until the butt is worn to about 1 inch thick. The dies are now sometimes made in two pieces, a new boss being fitted into the footplate as soon as the old one has been worn out. Steel is almost exclusively used for shoes and dies, manganese steel being preferred. The order of drop is usually 1.3.5.2.4 in the English mills, but by some authorities the orders 1.4.2.5.3 and 1.5.3.2.4 are considered better.

The cam-shaft is from $4\frac{3}{4}$ to 6 inches in diameter, according to the weight of the stamps. A separate cam-shaft is used for each battery of five stamps. The cam-pulleys in the American mills resemble that described on p. 108; in the earlier English mills they were often of wrought iron, 6 or 7 feet in diameter, with round arms. Many of these have proved defective, becoming crystalline and breaking after running for a short time only. Wooden pulleys are now generally used.

The water for ten stamps is supplied through a 3-inch gas pipe, from which two $1\frac{1}{4}$-inch pipes branch to each mortar. The pipes are sometimes suspended from the roof instead of being attached to the battery posts; the objection to the latter course is that the pipes work loose owing to vibration. The amount of water supplied varies in the De Kaap gold field from 100 to 200 gallons per stamp per hour, or from 1,600 to 3,200 gallons per ton of ore crushed. On the Witwatersrand the amount of water per ton of ore varies from 1,600 to 3,500 gallons, and, according to Hatch and Chalmers, averages about 2,000 gallons per ton crushed. Mercury is added to the mortar boxes in small quantities every few minutes, the total addition amounting to a spoonful in two hours.

The arrangement of the amalgamated copper plates outside the mortars differs somewhat in different mills. In some, a single plate of best Lake Superior copper is used, $4\frac{1}{2}$ to 5 feet wide, and from 8 to 10 feet long, with a mercury well at the lower end. In others, the lip of the mortar box is covered with a sloping copper plate from 12 to 14 inches wide, running the whole length of the mortar. The pulp flows over this "lip-plate" on to a second copper strip, the "splash-plate," 14 inches wide, and as long as the lip-plate, but inclined in the reverse direction so as to throw the pulp back towards the mortar. It then falls into an iron trough or distributor of semi-circular cross-section, perforated with small holes for the purpose of distributing the pulp over the plates. The latter are about $4\frac{1}{2}$ feet wide and $\frac{1}{8}$ inch thick; the pulp flows over 4 or 5 plates in succession, each $2\frac{1}{2}$ feet long. They consist of copper, which is seldom coated with electro-deposited silver. There is a shallow riffle at the lower end of the first plate, and a drop of 1 inch between each pair of the rest; sometimes the plates simply overlap, so that the drops are of only $\frac{1}{8}$ inch. At the lower end of the series of plates is placed a mercury trough. The tables for the support of the plates are made of heavy timber so as to be steady; these timbers are not fastened to any of the battery posts or sills, as the vibration caused by the stamps would interfere with amalgamation. The grade of the plates varies from 1 to 2 inches per foot, and can be altered by means of wooden wedges or of adjusting screws.

The plates at the Sheba mill are treated as follows, according to an account furnished by Mr. H. Nichols, A.R.S.M. :—The outside plates are usually dressed every hour if the rock is of ordinary richness (i.e., containing from 10 to 15 dwts. of gold per ton). In dressing, the amalgamator brushes all the quicksilver and thin amalgam up to the top of the plate and distributes it there. If the rock crushed has been rich (containing say 1 oz. per ton), quicksilver must be sprinkled on the plates and worked into the hard amalgam with a brush, but if it has been only 10-dwt. ore, the plates only require fresh mercury once in two or three hours. The outside plates may require scraping once every day or every other day if the ore is rich; the inside plates are scraped only on the occasion of the monthly clean-up. The yield from each of the two sets of plates is about the same. A retort of 2,500 ozs. of squeezed amalgam yields about 750 ozs. of gold. Cyanide of potassium is added in the mortar boxes in small quantities when the plates are discoloured by salts of copper; and soda is used to remove grease. When a screen is badly broken, the five stamps connected with it are at once hung up, as much loss may be incurred by allowing coarsely-ground rock to pass over the plates.

On the Rand, the amount of gold caught on the plates is usually between 60 and 70 per cent. of the gold in the ore, and the average is about 65 per cent. A higher percentage can be easily caught, but only if the output is decreased, and this sacrifice is less worth making, for the reason that most of the gold which is allowed to escape from the tables is extracted subsequently by cyanide. The percentage extracted from pyritic ore is but little less than that from the oxidised banket, the more refractory minerals being in general absent from the ore. The amount of gold caught inside the battery is from 10 to

40 per cent. of the total saved by amalgamation if inside plates are used, and 4 or 5 per cent. if there are none, and the greater part of the remainder is retained on the first two feet of tables. This part is scraped every day, but the lower parts are left undisturbed for longer periods, varying up to a week. At the Robinson mine the plates are dressed every four hours, but amalgam is removed only every other morning (Hatch and Chalmers).

The method of concentration at different mills varies; the following details refer to the Simmer & Jack mill of 100 stamps, when the ore was still free-milling, and a rough method of concentration sufficed. Here blanket tables, 16 feet long and of the same width as the copper plates, are placed immediately below the latter. The blanket tables are divided longitudinally into three parts by wooden "fillets." The blankets are washed every two hours and the concentrates collected in wooden boxes running on wheels. One division of each table is washed while the pulp is deflected so as to run over the other two divisions. The blanket-sands amount to about 6 per cent. of the tailings; they are treated by grinding with mercury in four pans, which together treat 24 tons per day of twenty-four hours. The charge for each pan is 2,600 lbs., five charges being treated in twenty-four hours. The pans are of cast iron, 5 feet in diameter, and have double bottoms into which steam can be passed for heating purposes; the mullers make from 50 to 60 revolutions per minute. The discharge takes place through a pipe at the bottom of the side of the pan, and the pulp is run into settlers, of which there are two, each 7 feet in diameter. The pulp is here diluted with water and stirred by four arms revolving at the rate of fifteen turns per minute. There are a number of holes at different levels in the side of the settler and the diluted pulp is run off through these as soon as the mercury has settled sufficiently.

Where the ore is pyritic, this method does not suffice, and vanners or hydraulic classifiers (*Spitzluten*, see p. 170) are employed. It is usual to set three Frue vanners to treat the pulp coming from five stamps. There has been much discussion on the respective merits of these two classes of machines in treating battery pulp on the Rand. Where vanners are used, the concentrates have usually been roasted and chlorinated, and the tailings treated by cyanide. The growing tendency to treat even concentrates by cyanide, however, throws doubt on the advantage of using vanners at all, for if both classes of materials into which the pulp is separated are ultimately treated by the same process, there is no *prima facie* reason for effecting the separation at all, and, on the other hand, *spitzluten* afford a product in an excellent condition for cyaniding. The pulp coming from the plates has about 40 per cent. of its gold locked up in the pyrites, and 60 per cent. in the form of finely-divided free gold which cannot be removed by the vanning machines, but is readily extracted by cyanide. It follows that concentration and chlorination are not sufficient of themselves, although the cyanide process may be, and for this reason a Committee of the Johannesburg Chamber of Mines reported in May, 1895, that close concentration is usually quite unnecessary. Doubtless the best percentage extraction would be obtained in almost every case by classifying the pulp, concentrating the various classes separately, chlorinating the concentrates, and cyaniding the tailings. It would appear, however, that, under present conditions, the greatest profit is gained by rough classification in *spitzluten*, followed merely by cyanide treatment.

The treatment of amalgam presents no unusual features. At the Simmer & Jack mill there is a cast-iron clean-up pan, $3\frac{1}{2}$ feet in diameter, with mullers making 60 revolutions per minute. The hard amalgam, skimmings, and headings from the battery are ground in this pan, with fresh mercury, for three or four hours; after this the mercury is strained through chamois leather, and hard balls of amalgam, which contain from 33 to 35 per cent. of gold, are thus obtained ready for retorting.

The life of the various parts of the mills is approximately as follows:—The shoes last for two to four months, the dies from three to four months, and the screens about fourteen days; when banket is being crushed the wear is less than if quartz is in course of treatment. Tappets have been in use for three years without showing any signs of wear. The stems of the stamps become crystalline and the ends usually break off after about twelve or eighteen months' use, but this can be avoided by annealing at regular intervals. The mortar boxes last at least four years, the cam-shafts about three years, and the guides twelve months.

The amount of ore crushed is usually over 4 tons of banket per stamp per diem. In 1891 the average output was from $2\frac{1}{2}$ to $3\frac{1}{2}$ tons per day; but is now greater (viz., about $4\frac{1}{4}$ tons) owing to the greater average weight of the stamps. At the May Consolidated Mine the 1,150-lb. stamps crush $5\frac{1}{2}$ tons per day, and Hatch and Chalmers consider that, in a few years time, from 5 to 6 tons will be a common result. The large output is due partly to the nature of the rock, banket being softer than ordinary quartz, and partly to the small depth of discharge and the coarseness of the screens.

In present practice there is really no difference between English and American mills as regards design. It is generally acknowledged that stamps of less weight than 950 lbs. should not be used, and the tendency is to increase this weight. The fittings, such as shoes and dies, are almost universally made of English steel. It may safely be said that English mills made by the leading makers compare favourably with the best American mills.

Taking an average of the whole of the operations on the Rand, the yield of gold on the amalgamated plates is gradually diminishing, having been 13·5 dwts. per ton in 1890, 11·3 dwts. in 1891, 9·77 dwts. in 1892, 9·59 dwts. in 1893, 9·52 dwts. in 1894, and less than 8 dwts. at the end of 1895. In the last-named year the thirty-seven leading companies on the Rand crushed 2,485,311 tons of ore, and obtained 1,182,794 ozs. of gold on the plates (or 64 per cent. of the total gold contents), 37,138 ozs. by concentration and 486,082 ozs. from the tailings by cyanide treatment.

The cost at the best mills is from 3s. to 5s. per ton for crushing and amalgamating, and about 4s. to 6s. per ton for treatment of tailings by the cyanide process. Concentration and chlorination probably cost about 3s. to 3s. 6d. per ton of ore. The following details of the work done in two mills in twelve months in 1894-95 are given as examples of what is possible with good management:—

	Tons of Ore.	Gold Recovered.		Cost per Ton.
		Total.	Per Ton.	
ROBINSON COY.'S MILL. Treated in Stamp Battery,	107,935	110,962 ozs.	1 oz. 0¼ dwt.	3s. 11d.
,, by cyanide, .	70,526	20,213 ,,	5·7 dwts.	3s. 10½d.
CROWN MILL. Treated in Stamp Battery,	200,785	73,339 ,,	7·3 ,,	3s. 0d.
,, by cyanide, .	138,800	42,749 ,,	6·16 ,,	4s. 3d.

At the Robinson Mill, 4·60 tons of ore were crushed per stamp per day, and at the Crown Mill 5 tons.

CHAPTER XI.

CHLORINATION: THE PREPARATION OF ORE FOR TREATMENT.

The Plattner Process.—The value of chlorine, as an agent for the extraction of gold from certain ores and from almost all concentrates, has now been recognised for many years, and its use is gradually extending, although it is doubtful whether its application will ever be as general as appeared probable before the introduction of the cyanide process. Its use was first suggested by Dr. John Percy, F.R.S., at the Swansea meeting of the British Association, held in August, 1848, in a paper * embodying the results of experiments carried out in the year 1846. Simultaneously, in 1848, Prof. C. F. Plattner, Assay Master at the Royal Freiberg Smelting Works, applied chlorine gas to the assay of the Reichenstein residues, and proposed that a similar method of treatment should be adopted on a large scale. These residues were the result of treating the Reichenstein ore with the object of extracting the arsenic. They consisted chiefly of oxides of iron and oxidised arsenical compounds, and had been roasted in the course of the process for the extraction of the arsenic. The residues had been accumulating for more than a century, and contained from 15 dwts. to 1 oz. of gold per ton; they were considered too poor to smelt, while they could not be made to yield the gold contained in them by amalgamation.

Prof. Plattner's suggestion was followed up by investigations made by Dr. Duflos in 1848,† and by Lange in 1849.‡ Dr. Duflos compared the results obtained by treating the residues with chlorine water by percolation in a stationary vat, and by agitation in a revolving barrel; and as these results were the same, he recommended the stationary vat as being more economical. He also obtained identical results with chlorine-water and with dilute solutions of chloride of lime and hydrochloric acid mixed together. On the other hand, Lange found that gaseous chlorine, applied to the ore in the same manner as had been used by Plattner in assaying, was a more efficient agent than a solution of chlorine in water, and it seems to have been in accordance with his advice that the first chlorination works, that at Reichenstein, was established in 1849. The chlorinating vessels

---

* *Phil. Mag.*, 1853, vol. xxxvi., pp. 1-8.
† *Erdm. Journ. Prak. Chem.*, vol. xlviii., 1849, pp. 65-70.
‡ *Karsten's Archiv.*, vol. xxiv., 1851, pp. 396-429.

were small earthenware pots, and the precipitant employed was sulphuretted hydrogen. Plattner subsequently recommended wooden vats coated with pitch, and ferrous sulphate as a precipitant, and although these were not at first used at Reichenstein, they were adopted by Mr. G. F. Deetken in 1857, when he introduced the system into California. The system ascribed to Plattner consists of the following operations :—

1. The concentrates or residues are subjected to a "dead" roast in a reverberatory furnace.

2. The roasted ore is slightly damped with water and charged into wooden vats, holding from 1 ton upwards. The vats have false bottoms consisting of filter beds of gravel or of cloth. Chlorine gas, generated in another vessel, is introduced at the bottom of the vat, and rises through the ore, permeating every part of it. The vat is then closed up and left undisturbed for twenty-four hours or more, by which time all the gold is converted into soluble chloride of gold.

3. The soluble salts are then washed out by water, which is allowed to flow on to the surface of the ore, and, passing through it, drains through the filter bed. When all the gold has thus been removed in solution, the tailings are thrown away.

4. The solution of gold chloride is acted on by ferrous sulphate, or some other suitable reagent, by which the gold is precipitated; the particles of the precious metal are allowed to settle, and then are collected and melted down.

Plattner Process at Reichenstein.—The following description of the original process employed at Reichenstein is an abstract of the account given in Kerl's *Hüttenkunde*, vol. iv., p. 372, 1865. The material treated consisted of the residues obtained by roasting arsenical iron pyrites for the production of white arsenic, which was volatilised and condensed in brick chambers. There were forty-eight earthen chlorination pots, each holding 150 lbs. of ore. These pots were strengthened with iron hoops, and suspended on two journals, so that they could be discharged by inverting them.

The lower part of the pots was of a conical shape, and this part was filled with pebbles and sand covered by a perforated earthen plate, the function of which was to prevent the ore from mixing with the filter bed. The ore filled the cylindrical part of the pot above the earthen plate. The chlorine was generated by the action of hydrochloric and sulphuric acids on manganese dioxide in earthenware vessels, and was conveyed thence to the ore-pots through leaden pipes. The gas was introduced below the filter bed, and passed upwards through the ore for an hour; a wooden cover was then fitted on, but not luted down until chlorine had been passed for from six to seven hours longer, after which all joints were luted down with dough, and the vat left until the next day. The cover was then removed and water, at a temperature of from 64° to 77° F., poured on, and allowed to percolate through the ore and filter bed by gravity. The liquid coming from twenty-four pots was conveyed to four vats, the first one being filled with solution before the second was used, and so on; the contents of the fourth vat, being too poor for precipitation, was used over again for leaching. The leaching was stopped when 90 cubic feet of water had passed through the total charge of 3,600 lbs., this being at the rate of 312 gallons per ton of 2,000 lbs. The liquid from the first three vats was drawn off into twenty glass globes, which were heated on a sand bath so as to raise the temperature to 77° F. Sulphuretted hydrogen, obtained from fused sulphide of lead and sulphuric acid, was passed through until the saturation point was reached, when the liquid was left to settle until the next day; after this the clear supernatant liquid was passed through sawdust filters to catch all the sulphide of gold, which might still have been in suspension. The sulphides were refined by dissolving them in acids, precipitating the metallic gold by ferrous sulphate, and melting it in clay crucibles with nitre and borax. The amount of arsenides treated daily was 3,600 lbs., containing about $\frac{5}{8}$ oz. of gold per ton, so that only about 250 ozs. of gold were extracted yearly, and it is difficult to believe that the enterprise could have been a commercial success.

MODERN PRACTICE IN CHLORINATION.

The original vat process as described above has been subjected to great changes, many of which originated in the United States. The most notable alterations are an increase in the size of the vats and the substitution of wood for earthenware in their construction, and the use of ferrous sulphate instead of sulphuretted hydrogen to precipitate the filtered liquors. This process, as altered, is still in very extended use for dealing with roasted concentrates, but in a few special cases, where large quantities of material are available for treatment, the barrel process has been adopted. The barrel process of chlorination, which was first used on the large scale in the United States, differs from the vat or Plattner process chiefly in the fact that the chlorination is effected in revolving barrels, instead of in stationary vats, while the operations of chlorination and lixiviation are usually, though not invariably, performed in different vessels. The two processes will be described under the following headings:—

1. Crushing.
2. Roasting.
3. Chlorination and Lixiviation.
4. Precipitation of the gold from its solution and production of bullion.

Vat and barrel chlorination differ essentially only in the methods described under the third heading, although the differences in the conditions under which they are applied involve variations in the treatment under the other headings. The sections devoted to crushing and roasting include descriptions of the methods used to prepare ore for chlorination both in the vat and the barrel. The remaining operations are described separately for each of the two processes, and a general account of the precipitation of gold is added.

CRUSHING.

In those cases in which gold ores are treated by crushing and amalgamation, and the whole of the tailings, or only the concentrates obtained from them, are subsequently chlorinated, the method of crushing will be determined by the considerations already discussed in the chapters on amalgamation and concentration, and will depend partly on the state of aggregation of the free gold. If, on the other hand, ores are to be treated in the first instance by chlorination, special regard may be paid to the method of crushing as affecting the suitability of the crushed product for leaching. There are two somewhat opposite conditions to be fulfilled, viz.:—

1. The crushed product should be fine enough to admit of the whole of the gold being laid open to the attack of the chlorine.

2. It must be coarse enough to allow all the soluble chloride of gold thus formed to be washed out of the ore rapidly and easily.

It is quite impossible to fulfil completely both of these conditions on a large scale, although in the laboratory some ores, when treated with the greatest care and patience, may be made to yield the whole of their gold. In practice it is better to aim at an extraction of from 85 to 95 per cent. (98 to 99 per cent. being attained in some rare cases), and to do this at as low a cost as possible. With this object in view, the ore is kept as coarse as possible, and is usually passed through screens with only from 8 to 30 holes to the linear inch. The chief point to be attended to is the attainment of uniformity in the product, as any considerable proportion of slimes enormously increases the difficulties of leaching. Moreover, since the crushed ore must almost invariably be roasted, it is a great advantage to adopt some method of dry crushing. Proposals have been made to subject the ore to wet-crushing by stamps or other machines, and to collect the pulp in settling-pits, and dry it, either in the roasting furnace or in a separate furnace. It appears that this method has never been tried, and there are several practical objections to it.

Dry crushing by means of stamps has been found to answer well at Park City, Utah, and at other places in preparing silver ores for treatment by the hyposulphite leaching process, but many ores would form too much impalpable powder for this system to be applicable. In order to obtain a perfectly uniform product in any machine, it would be necessary for every particle that had been reduced to the required degree of fineness to be instantly separated from those which were still too large. If the screening is not instantaneous and thorough, the small particles are ground still finer while the larger pieces are in process of reduction to the required fineness. Of all the machines for dry crushing which have been designed with a view to effect this object, rolls have come into the most general use, and give the most satisfaction. They are used, among other places, at Mount Morgan, the largest chlorination mill in the world, and at the Dakota Mills. Rock-breakers are now always employed to reduce the ore to the size of a hen's egg, or smaller, before feeding it to the rolls.

The following description of Krom's rolls, the best known modern make of high-speed fine-crushing rolls, is chiefly derived from Mr. A. H. Curtis' paper on "Gold Quartz Reduction."* Crushing rolls made by several other manufacturers in various countries might be described almost in the same words, and give nearly if not quite equally good results in practice, while they are much lower in price.

The rolls are placed with their faces a short distance apart, this distance varying with the degree of fineness to which the ore must be reduced. The main driving power is applied to the shaft of only one roll, by means of belt pulleys, the other roll being driven only with sufficient force to ensure that the rollers will always take hold of the ore, and also to keep them in motion when no ore is passing between them. If all the power is applied to turn one roll only, the feeding must be continuous, or the free-running roll would come to a standstill. On recommencing the feeding, a blow would then be given to the stationary roll, in the attempt made by the other roll to set it instantaneously in motion at full speed, and this jar would have a disastrous effect on the machinery.

In the older forms, tooth-gearing was used instead of belt pulleys for the application of the power, and the two rolls were compelled to revolve at an equal rate of speed by gear-wheels, connecting together, placed on the axles. The advantages of belted rolls are, that a higher speed can be easily attained, and also that if the rolls were to become jammed from any cause, the belts would slip or be thrown off, while the tooth-gearing would be broken. Geared rolls, however, are still largely used for coarse crushing. The rolls have crushing tires made of forged steel, and are firmly and simply secured to the shafts by means of cone-shaped heads. Chilled-iron rolls are still often used,

* *Proc. Inst. Civil Eng.*, session 1891-92, vol. cviii., part ii.

being much cheaper; but the wearing of the faces is more rapid and less uniform than in the case of steel rolls. Emery wheels for levelling the unevenly worn faces of chilled-iron rolls are recommended by some makers. These crushing tires can be taken off and replaced when they are worn out. In the older forms of Krom's rolls the crushing strain is taken up by powerful springs, which press the rolls towards one another; when particularly hard fragments are passing through the rolls, they are forced apart against the action of the springs. These springs are now dispensed with. It is desirable, in order to keep the wear of the faces even, that the rolls should always be kept parallel, and special appliances, such as Krom's swinging pillow blocks, have been introduced by some makers to ensure this. The hopper is specially designed to spread the ore evenly across the crushing face, and the rolls, screens, elevators, &c., are all securely boxed in with a wooden housing. This last precaution is necessary in order to prevent loss by floating dust, which otherwise may be large, the richest part of the ore thus passing off, and not only making the atmosphere of the mill insupportable, but having a disastrous effect on the bearings of the machinery. Rolls are usually from 12 to 16 inches across the face, and from 22 to 36 inches in diameter.

Corrugated rolls are used to some extent in Australia, but though the crushing surface is increased in this way, they are considered in the United States to be of little value, wearing unevenly and soon getting out of order.

The method of crushing usually adopted in mills where rolls are employed may be described in general terms as follows. It may be supposed that the ore is to be passed through a 20-mesh screen:—

The ore is put through a rock-breaker, and passed at once to classifying screens, by which it is divided into four classes of material. One of these consists of ore that will pass a 20-mesh screen, and this amount (usually small) is separated as finished product; another small portion is returned to the rock-breaker by some form of elevator, being too coarse for the rolls, and the greater part is passed to one or other of two pairs of rolls. Each of these pairs of rolls is occupied in crushing material of a particular degree of fineness, reducing it to a further degree of fineness. The material fed to them is classified accordingly. After each passage through the coarse rolls, a certain amount of finished product is obtained, and the remainder is classified, part being returned to the same rolls, and part being sent on to the fine rolls, the product from which is also classified. Revolving screens are in general use for classification. The use of two rock-breakers, one for coarse crushing, and the other to take its product and reduce it to the size of broad-beans, is often recommended. The number of pairs of rolls to be used in succession depends on circumstances: two is the usual number, a third pair being occasionally added. One rock-breaker and two pairs of rolls will suffice for the reduction of most ores to a size sufficiently fine to pass through a 30-mesh screen, but the crushing can be done more economically in the long run by still more gradual reduction.

Mr. John E. Rothwell, late manager of the Golden Reward Chlorination Mill, Dakota, gives it as his opinion[*] that "two sets of rolls are sufficient, but three will do better. The ore should come to the coarse rolls not coarser than $\frac{3}{4}$-inch mesh, and these rolls should be set about $\frac{3}{8}$ inch apart. The middle rolls are set about $\frac{3}{16}$ inch or less apart, and the fine rolls about as far apart as the size to which the ore has to be crushed. If only two sets are used, the coarse are set a little closer than with three, and the fine remain the same.

"The springs should be set up so tight that they will not give to the hardest pieces of ore, but will allow a piece of steel or iron to pass through without throwing the belts. The periphery speed of the rolls should be about the same as, or a little faster than, the falling speed of the ore, and the ore should be fed in an even sheet across the surface of the roll; little trouble will then be experienced in keeping the surfaces true and in producing a granular pulp, carrying but a small percentage of dust. If rolls were made of larger diameter and narrower, the result would be a still more gradual reduction, and possibly a greater capacity. I have used those of $39\frac{1}{2}$ inches (1 meter) in diameter and 12 and 15 inches face."

"It is stated by users of the Krom rolls (which claim to be essentially fine pulverisers) that the faces of the tires wear evenly and do not become grooved, and that they have a long life. At the Bertrand Mining Coy.'s Lixiviation Mill, Nevada, where Krom's rolls are used for pulverising silver ores with a quartz gangue, 15,000 tons of ore were crushed before it was found necessary to put new tires on the finishing rolls; while after a further crushing of 5,000 tons of ore, the tires of the roughing rolls were still expected to be good for two or three months' work. In neither case were the tires found to be at all grooved, the reason for their renewal being that they had become worn too thin for further work. The rolls in the Bertrand Mill are stated to crush 50 tons of hard quartzose ore in twelve hours to pass through a 16-mesh screen; while in the Mount Cory Mill, Nevada, 50 tons are reduced to 30-mesh in the same time."[†] In these mills silver ores are crushed for treatment by roasting and lixiviation with a hyposulphite solution. It is possible to crush quartz to pass through screens of from 30- to 70-mesh by rolls, the degree of fineness to which the

---

[*] *Eng. and Mng. Journ.*, Feb. 7, 1891.
[†] Curtis on "Gold Quartz Reduction." *Proc. Inst. Civil Eng.*, 1892.

rock can be pulverised depending on the number of times the product, after screening, is returned to the hopper. The very fine dust, which is inevitably produced in greater or less quantity, according to the character of the rock, may be separated by exhaustion with a fan, driven at carefully regulated speed, and in this way the sliming is reduced and the resulting loss of gold in the subsequent operations minimised. The average speed of fine-crushing rolls is from 80 to 100 revolutions per minute, the two rolls being driven at equal rates of speed.

Crushing by rolls at Rapid City, Dakota.—The following details concerning the crushing plant at the Rapid City Chlorination Works, S. Dakota, which was fitted up in 1891, have been supplied by Mr. D. Dennes, formerly the foreman of the mill. "Gates crushers were used for some months, after which they were replaced by a large Blake crusher, which is found to work more economically and with less cost for repairs. The crushed material from the Blake is passed through a rotating cylindrical drying furnace, and then screened, part being passed to the ore-bin set aside for finished product, a little going back to the rock-breaker, and the remainder going to the first pair of rolls. These have steel tires and are set at $\frac{3}{16}$ inch apart; they are driven at a speed of 90 revolutions per minute by means of toothed gearing, both rolls being driven independently. The objections to the use of belts for the coarse rolls are stated to be that a specially hard piece of ore may throw the belt off and so interrupt the work. The peripheral speed of the rolls is 14 feet per second. The product of the coarse rolls is classified by screening, and the bulk of it sent at once to the fine rolls, which are steel-faced belted rolls driven at 155 revolutions per minute. The two faces just touch, but each roll is driven by a separate belt, and one is revolved at the rate of two revolutions per minute faster than the other. This arrangement has a remarkably beneficial effect in keeping the wearing surfaces even. The tires are replaced alternately, so that there is always one new and one old tire working together. Almost all the product of these rolls passes at once through a 16-mesh wire screen, which is the finest screen used.

"The revolving screens of Messrs. Fraser & Chalmers were used for six months, and then discarded. Flat rectangular screens are now used, measuring 6 feet by 12 feet, set in a wooden frame which is subjected to a reciprocating motion, striking against a rubber pad placed close to one of its long sides. The screen is slightly inclined, so that the ore, which is fed evenly across the screen, travels down it. This apparently retrograde step in screening was justified by a considerable improvement in efficiency and a reduction of expense. The revolving screens were found to wear out very rapidly, and the repairs were costly. The weight of ore causes the metal filter-cloth to sag down between the bars constituting the framework; when this part of the cylinder reaches the top of its course, the cloth sags in the opposite direction from its own weight. The result of this double motion is that the sieve breaks along a line close by and parallel to the framework bars. When this breakage occurs, repairs are tedious and expensive, and the mill must be stopped while they are being done, while flat screens can be lifted out, and a new one put in, by two men in a few minutes."

From 100 to 120 tons of ore are stated to have been crushed in this mill per day, but as the mill was in continuous operation for a short time only, and is now shut down, it is not possible to draw definite conclusions from the results obtained. The speed of the rolls seems to have been excessive, and the difficulties with the revolving screens do not appear to have been felt in other mills. Probably they were due to the fact that wire-cloth was used for the screens, instead of the more serviceable punched sheet-iron.

Comparison between Rolls and Stamps.—As the subsequent treatment of an ore determines its method of crushing, no accurate general comparison of stamps and rolls can be made. A comparison is only possible in the special cases where both methods of crushing are applicable. Wet crushing by rolls need not be considered, as it is not practised; even where the advocates of rolls wish to replace stamps by rolls for wet crushing and amalgamation, they propose that the ore should be crushed dry and then wetted down. Dry crushing by stamps is usually about one-third more expensive than wet crushing, as the capacity falls off to that extent, in spite of double discharge, assisted occasionally by currents of air given by blowers, which are designed to carry the crushed ore against the screens. The amount of slimes made is also large. Advantages in the use of rolls for dry crushing have been stated to be "the fewness of the wearing parts, and consequent small cost of repairs; the great efficiency of the process, in that the ore escapes from the rolls immediately it is crushed, so that over-crushing is unlikely to occur, and the great capacity of rolls, effective work being constantly done, and the amount of crushing surface brought into contact with the ore per minute being very large. The prime cost of the rolls is considerably less than that of stamps of the same capacity." *

The capacity of rolls has perhaps been frequently overestimated owing to the assumption having been made, that the product of equal crushing surfaces must be the same. Thus, Mr. Curtis observes in the paper already cited:—"As an index of the capacity of Krom's rolls, it may be stated that two sets of 20-inch (diameter) rolls, with faces 15 inches

* Curtis, *loc. cit.*

long, give rather more effective crushing surface than fifty gravitation stamps, each 8 inches in diameter, falling at the rate of ninety drops per minute. In making this calculation, the average diameter of the rolls is taken as only 24 inches, so as to allow for their gradual wearing, while their speed is taken at 100 revolutions per minute."

Although the calculated crushing surface is as stated, it does not follow that the capacity of the two sets of machinery compared with one another is the same, since it will depend on the pressure as well as the crushing surface. No doubt the pressure at the moment of impact of a 900-lb. stamp is enormously greater than that exercised by the faces of the rolls.

On this point Mr. J. Richards, M.E., writes:*—"The coefficient of (crushing) effect in such machines (as revolving rollers and reciprocating machines) is as the area of the acting surfaces, and the speed with which they approach each other. The mistake in respect of the crushing power of rollers comes from confounding their circumferential velocity with the working one— that is, with the parallel velocity at which the surfaces approach and leave each other," that is, the velocity at right angles to the crushing surface. Mr. Richards elsewhere shows that a Blake machine in this way actually outruns a revolving roller, although its crushing face travels more slowly than the periphery of the roll. He proceeds:—"The same principle holds good in respect to stamps, the crushing surfaces having a parallel approach, while with rotary machinery the approach is not parallel, except on an imaginary line at the centre." As a matter of fact, the product of two pairs of rolls amounts to only about 40 to 50 per cent. of that which would be given by a fifty-stamp mill on the same ore. In the absence of exact comparative data regarding the same ore, no precise figures regarding the relative capacities can be given.

A point to be noticed is that, in machines which act by pressure applied on the principle of the lever, such as reciprocating-jaw crushers or rolls, the whole force necessary to crush the lumps of quartz is transmitted to the fulcrum, this fulcrum being represented in the case of rolls by the bearing surface at the axle. The consequence is that the frame must be made very strong and heavy, and the axle bearings attended to with great care, if rapid wear is to be avoided.

Drying the Ore.—Although rolls are essentially dry crushers, nevertheless, if the ore is nearly pure quartz, a small percentage of moisture in it is not a serious disadvantage. If, however, the moist ore is argillaceous, the product from the rolls will not readily pass through the screen, and the latter soon becomes clogged, frequent stoppages for cleaning being thus necessitated. It has been stated that 5 or 6 per cent. of moisture present in an

* *Prod. of Gold and Silver in the United States*, 1880, p. 369.

ore does not seriously interfere with crushing, but this does not accord with the experience in some mills. At any rate, in every mill where dry crushing is used, some means of artificially drying the ore is adopted.

The oldest method was to spread the ore, after the large lumps had been removed by a grizzly and crushed to $1\frac{1}{2}$ inch size, on large flat areas heated from below by flues from the roasting furnace. The floor was usually covered with iron plates. After being dried, the ore was shovelled up and passed to the crushing mill. This plan involved much additional handling of the ore, and was a source of ill-health among labourers, besides requiring great floor space. It has been superseded by the adoption of inclined continuous-discharge, revolving iron cylinders, similar to the Howell-White furnace, but not lined with bricks. The ore is passed through these, and is dried by the products of combustion of a fire, which are also passed through it. One such cylinder, of 3 feet in diameter and 18 feet long, will dry from 30 to 40 tons of ore per day, at a small cost for fuel and power.

An alternative furnace—viz., Stetefeldt's shelf drying kiln—was described in a paper read by the inventor at the meeting of the American Institute of Mining Engineers, held at Roanoke, Virginia, in June, 1883. In principle it resembles the Hasenclever furnace, a number of shelves being arranged zig-zag above each other in a vertical shaft, down which the ore slides, falling from shelf to shelf, while the products of combustion from a furnace rise through it. It is 21 feet high, and dries from 30 to 50 tons of ore per day. It is in wide use in the United States, and although its first cost is considerable, its working expenses are said to be lower than those of the rotary furnaces.

ROASTING.

The operation of roasting, as a preliminary to chlorination, has for its object the expulsion of the sulphur, arsenic, antimony and other volatile substances existing in the ore, and the oxidation of the metals left behind, so as to leave nothing (except metallic gold) which can combine with chlorine when the ore is subsequently treated with it in aqueous solution. For this purpose the ore is heated in a furnace, through which a current of air is passed, salt being added if oxide of copper, lime, magnesia, &c., are present. Ores containing much pyrites might be freed from most of their sulphur by pile roasting, and then subjected to fine crushing and a dead roast in a reverberatory furnace, but the extra cost of handling would probably exceed the saving due to the smaller consumption of fuel. This system has not been tried in chlorination mills on an extensive scale. The ordinary reverberatory furnace, worked by hand labour, is in wider use than any other, especially where only a few tons, or less, of concentrates are to be treated per day. Various mechanical furnaces, capable of handling large quantities of ore, have been devised to supersede the old-fashioned contrivance, and some of these will be described in the sequel.

Reverberatory Furnaces.—The construction of the ordinary reverberatory furnace is too well known to need detailed description here.* It consists of a vaulted chamber, containing the ore; through this chamber, the flames and products of combustion from a furnace and a current of air are made to pass in a horizontal direction above the ore, which is thus heated. The ore is also stirred by hand with iron rakes, which are passed through small working doors. The hearth of the vault (also called the "laboratory" of the furnace) is formed of bricks placed on edge (not flatwise, except where economy is studied rather than durability), as close together as possible. No mortar is used, but a little clay is plastered between the bricks. The height of the furnace hearth is about $3\frac{1}{2}$ feet above the floor of the building, on which the labourers stand, and the space underneath the hearth is either occupied by vaults or filled with well tamped rubble. The arch is usually one course of bricks (8 inches) thick; the height between it and the hearth is, in long furnaces, about 24 inches near the bridge, and gradually diminishes towards the other end. This height is less in short furnaces. The best fire-bricks are used for the fire-box and bridge, and for the hearth and arch of the first few feet of the "laboratory." The remainder is made of common brick. It is necessary to have a damper in the flue to regulate the draught; the aperture of the flue should not be on a level with the hearth, as in that case the loss by dusting is increased. The brickwork of the furnace is supported by longitudinal and transverse iron braces. The working doors have cast-iron frames, and are about 15 inches wide and 9 inches high. The fuel used must of course be a long-flame coal, or wood; short-flame coal and coke are inadmissible.

For the particular purpose of roasting pyritic ores before chlorination, the temperature on the working floor of the furnace must be low when the ore is first charged in, and high in the later stages. If a single small floor is used, the fire must be alternately checked and urged to secure these conditions. Moreover, when the roasting is nearly complete, the high temperature required renders the gases passing into the flue very hot, and so a corresponding waste of fuel results. To prevent this waste, it is customary for roasting furnaces to be built with a very long hearth, or to have several successive hearths, so as to utilise the

* See the *Introduction to the Study of Metallurgy*, by Prof. Roberts-Austen, 1894, p. 207. Also Küstel's *Roasting of Gold and Silver Ores*, 1880, pp. 80-89; and Grüner's *Traité de Metallurgie*, p. 265.

waste heat from the portion of the working floor next the fire. Furnaces with three floors at slightly different levels are much favoured; in these a charge remains for a few hours on each of the floors in succession. It is first placed on the floor farthest removed from the fire, and, after a time, is raked down on to the middle hearth, and thence to that nearest the fire, fresh charges being put on the spaces just cleared, so that there are always three charges in the furnace in various stages of oxidation.

The most usual form in the Western States of America is the "4-hearth," in which the length of the hearth is four times its width, so that the dimensions are, say, width 15 feet, length 60 feet. In this case there should be eight working doors on each side. Instead of three floors at different levels, a single continuous floor, gently sloping from the flue towards the fire, is in use at many works in Australia, Mexico and the United States. At Suter Creek a continuous-hearth furnace is 12 feet wide and 80 feet long, and the mineral is worked in three distinct parts, as though there were three floors. The angle of slope is made large in some Australian furnaces, so that in the course of the "rabbling" or stirring, the ore continually travels towards the fire-box. Furnaces with two or three superposed floors are also used to a limited extent; the lowest floor is next the fire-box, and communicates by a vertical flue with the floor above, and so on. The ore is charged-in on the top floor, and after a time is raked down through the vertical flue on to the next floor. In this case the floors are heated by the gases passing below them as well as above them, and fuel is economised, but the furnaces are costly to build and to keep in repair.

It was proposed many years ago to insert a drop of 10 feet between the finishing floor and the floor next to it. The charge, when already red hot, would thus fall vertically downwards in a thin shower against a current of hot air. Some furnaces are said to have been built in this way, but it seems that none are now in existence. The principle is excellent, and is utilised in the Stetefeldt furnace.

When the pyrites to be roasted are rich, it may be an advantage to build dust chambers to ordinary reverberatory furnaces. The amount of gold contained in the dust thus recovered is usually only 1 or 2 per cent. of that contained in the ore, so that in some cases it may be a long while before the dust chamber pays for itself, even if that point is ever reached.

The operation of roasting pyrites in an ordinary furnace with three floors may be described as follows:—The furnace being hot, and the flame from the fire-box reaching completely across the first floor, the ore is charged-in on the third floor and spread out by the rabbling tool. The weight of the charge may be taken as from 12 to 18 pounds per square foot of floor space, varying according to the nature of the ore, a high percentage of sulphur necessitating small charges. The layer of ore is 2 or 3 inches deep. It is not spread quite evenly, but made to form a series of parallel ridges by means of the rabbling tool, so as to increase the surface exposed to the air. The working doors may be closed at first to heat the ore quickly. Moisture is at once given off in great quantities, and the sulphur soon begins to burn with a blue flame. When this is seen to take place, all the working doors are opened, and the charge is energetically rabbled, with little intermission, until the sulphur flame disappears. If this is not done, clots are formed which are afterwards difficult to break up. The air for the combustion of the sulphur is supplied by holes in the fire-bridge and from the working doors. The flames and heated products of combustion from the fire tend to rise above the colder air, and move along next to the arch of the furnace, while the air forms a sheet between these gases and the ore. In practice, although the air is introduced below them, nevertheless the reducing gases from the fire partially mix with the air, and greatly reduce its oxidising power. Moreover, the combustion of the sulphur in the ore on the first two floors further reduces the amount of free oxygen present in the current of air, and roasting on the third floor is, therefore, largely dependent on air derived from the working doors.

When the sulphur flames have abated, and the charge has been heated almost to redness, it is transferred to the middle floor, where it is raised to a dull red heat and most of the oxidation is performed; during this stage the ore swells considerably, so as to occupy much more than its original bulk. All the lumps previously formed should be broken up on this floor. Rabbling is continued until the ore is uniformly dull throughout, so that, on turning it over, the fresh surfaces appear but little brighter than that which has been exposed for some time. The charge is then transferred to the floor next the fire. There is now little risk of the formation of lumps, and the charge may be allowed to reach a bright red heat. Rabbling is of less importance than before, as little oxidation takes place, the chief reaction which occurs being the decomposition of the sulphates already formed. As long as this is still going on, the ore emits the odour of sulphur dioxide. When no further odour can be detected, and the ore can be piled up so as to maintain a vertical face, shows no bright specks on its glowing surface, emits no sparks if some of it is tossed up by the working tool, and is inclined to become black very readily from cooling, the charge is said to be "dead" or "sweet," and is ready to be withdrawn. It should be observed that, when the ore contains much sulphur, its particles at a low red heat appear less coherent than when cold, and flow almost like water so that the charge cannot be made to form a heap with steep sides. Care must therefore be taken when the ore is on the middle floor to prevent any part of the charge from flowing out of the working doors, which it is very liable to do when being rabbled.

Küstel states* that the best means of rapidly ascertaining whether a charge is completely roasted is to throw a little of the ore into some water, and then to plunge a bright iron rod into the liquid. If the rod remains bright the ore is ready for withdrawal, but if sulphates still remain undecomposed, the surface of the iron will be darkened. This is not a safe test with all classes of ore, as the presence of sulphate of iron in the water would not be detected in this way. A more trustworthy and equally simple test is to add a few drops of chloride of barium to the water. A white cloud, which consists of $BaSO_4$, indicates the presence of soluble sulphates. In most cases the water need not be filtered before it is tested, but even when filtration is necessary the whole operation can be performed in two or three minutes. The charge is withdrawn by a scraper, and falls by gravity through a hole in the floor of the furnace near the working door into a pit below. This hole is covered by a plate while roasting is being performed.

The time of roasting depends chiefly on the ore, but may be shortened by more continuous rabbling than most workmen can perform, the work being somewhat exhausting. In a three-floor furnace, concentrates with 15 per cent. of sulphur usually remain eight hours on each of the three floors. The fuel used is either wood or flaming-coal. If the flame from the coal is not long enough to reach across the first floor, this will not be heated uniformly; in that case the part of the charge next the fire-bridge is finished first and must be moved away, while that from the other end of the floor is brought up nearer to the fire. This causes a great increase in the labour, besides occasionally leading to the withdrawal of a charge of which a part is not quite "dead." The draught is regulated by the damper in the flue leading to the dust chamber, and by opening and closing the working doors.

Chemistry of Oxidising Roasting. — Professor Roberts-Austen discusses as follows † the roasting of a "mixture consisting of sulphides mainly of iron and copper, with some sulphide of lead, small quantities of arsenic and antimony as arsenides, antimonides, and sulpho-salts, usually with copper as a base. The temperature of the furnace in which the operation is to be performed is gradually raised, the atmosphere being an oxidising one. The first effect of the elevation of the temperature is to distil off sulphur, reducing the sulphides to a lower stage of sulphurisation. This sulphur burns in the furnace

---

* *Roasting of Gold and Silver Ores.* San Francisco, 1880.
† Presidential address to the Chemical Section, British Association, Cardiff Meeting, 1891.

atmosphere to sulphurous anhydride (SO_2), and coming in contact with the material undergoing oxidation is converted into sulphuric anhydride (SO_3). It should be noted that the material of the brickwork does not intervene in the reactions, except by its presence as a hot porous mass, but its influence is, nevertheless, considerable. The roasting of these sulphides presents a good case for the study of chemical equilibrium. As soon as the sulphurous anhydride reaches a certain tension, the oxidation of the sulphide is arrested, even though an excess of oxygen be present, and the oxidation is not resumed until the actions of the draught change the conditions of the atmosphere of the furnace, when the lower sulphides remaining are slowly oxidised, the copper sulphide being converted into copper sulphate, mainly by the intervention of the sulphuric anhydride, formed as indicated. Probably by far the greater part of the iron sulphide only becomes sulphate for a very brief period, being decomposed into the oxides of iron, mainly ferric oxide, the sulphur passing off. Any silver sulphide that is present would have been converted into metallic silver at the outset were it not for the simultaneous presence of other sulphides, notably those of copper and of iron, which enables the silver sulphide to become converted into sulphate. The lead sulphide is also converted into sulphate at this low temperature (viz., about 500°). The heat is now raised still further with a view to split up the sulphate of copper, the decomposition of which leaves oxide of copper. If, as in this case, the bases are weak, the sulphuric anhydride escapes mainly as such; but when the sulphates of stronger bases are decomposed the sulphuric anhydride is to a great extent decomposed into a mixture of sulphurous anhydride and oxygen. The sulphuric anhydride, resulting from the decomposition of this copper sulphate, converts the silver into sulphate, and maintains it as such, just as, in turn, at a lower temperature, the copper itself had been maintained in the form of sulphate by the sulphuric anhydride eliminated from the iron sulphide. When only a little of the copper sulphate remains undecomposed, the silver sulphate begins to split up (viz., at about 700°) . . . partly by the direct action of heat alone, and partly by reactions such as those shown in the following equations :—

$$Ag_2SO_4 + 4Fe_3O_4 = 2Ag + 6Fe_2O_3 + SO_2$$
$$Ag_2SO_4 + Cu_2O = 2Ag + CuSO_4 + CuO$$

The charge still contains lead sulphate, which cannot be completely decomposed at any temperature attainable in the roasting furnace except in the presence of silica. . . . The elimination of arsenic and antimony gives rise to problems of much interest, and again confronts the smelter with a case of chemical equilibrium. For the sake of brevity it will be well for the present to limit the consideration to the removal of antimony, which may be supposed to be present as sulphide. Some sulphide of antimony is distilled off, but this is not its only mode of escape. An attempt to remove antimony by rapid oxidation would be attended with the danger of converting it into insoluble antimoniates of the metals present in the charge. In the early stages of the roasting it is, therefore, necessary to employ a very low temperature, and the presence of steam is found to be useful as a source of hydrogen, which removes sulphur as hydrogen sulphide, the gas being freely evolved. The reaction

$$Sb_2S_3 + 3H_2 = 3H_2S + 2Sb$$

between hydrogen and sulphide of antimony is, however, endothermic, and could not, therefore, take place without the aid which is afforded by external heat. The facts appear to be as follows:—Sulphide of antimony, when heated, dissociates, and the tension of the sulphur vapour would produce a state of equilibrium if the sulphur thus liberated were not seized by the hydrogen, and removed from the system. The equilibrium is thus destroyed, and fresh sulphide is dissociated. The general result being that the equilibrium is continually restored and destroyed until the sulphide is decomposed. The antimony combines with oxygen and escapes as volatile oxide, as does also the arsenic, a portion of which is volatilised as sulphide.

"The main object of the process which has been considered is the formation of soluble sulphate of silver." The reactions, however, are precisely similar in an ordinary oxidising roast.

The following remarks on the decomposition of the various minerals present in complex ores may be of use in assisting the student to understand the reactions which proceed in the roasting furnace:—

1. *Iron Pyrites*, FeS_2.—On heating this compound, sulphur is volatilised, the reactions being probably expressed thus:—

$$3FeS_2 = Fe_3S_4 + S_2$$
$$7FeS_2 = Fe_7S_8 + 3S_2$$

The sulphur burns to SO_2, which is partly converted by the heated quartz, &c.,* into SO_3, uniting with the free oxygen present. The ferrous sulphate formed by this sulphuric acid is split up by the heat and the ferrous oxide (FeO) converted into ferric oxide (Fe_2O_3) which gives the ore a red colour when cold. If the temperature of the part of the charge next the fire-bridge has been too high, or if the charge is kept too long in the furnace, some magnetic oxide is formed, thus:—

$$3Fe_2O_3 = 2Fe_3O_4 + O$$

This is an undesirable change, as the magnetic oxide is acted on by chlorine far more readily than the sesquioxide.

* Plattner's *Metallurgische Röstprozesse*, Freiburg, 1856.

2. *Copper Pyrites.*—The decomposition of the copper sulphate formed in the furnace leaves a mixture of cuprous and cupric oxides, both soluble in chlorine.

3. *Galena*, PbS.—The presence of this mineral in any but small quantities is very detrimental, as both lead sulphate and lead silicate (formed by its decomposition in the presence of silica) are very fusible, and, at the temperature required to split up copper sulphate, cause the ore to become pasty and form lumps. Roasting must be performed very slowly and cautiously to avoid this effect.

4. *Arsenical Pyrites*, FeAsS.—Arseniates of iron, copper, lead, &c., when formed are not easily decomposed, as they resist a high temperature, and are only slowly converted into sulphates by sulphuric acid at a red heat. It is, therefore, desirable to avoid their formation, and with this end in view the precautions which have been already mentioned in the extract from Prof. Roberts-Austen's address are taken.

5. *Antimonial Sulphides* are still more difficult to deal with, the antimoniates formed being less easily decomposed than arseniates. Their formation is avoided in the manner already described.

6. *Blende*, ZnS, forms oxide and sulphate of zinc, of which the latter can only be split up by a very high temperature. At a bright red heat a basic sulphate is formed which is converted to oxide at a white heat. If blende is roasted at a high temperature and with a plentiful supply of air, sulphate of zinc is not formed to a large extent.

Elimination of Arsenic and Antimony.—Mr. H. M. Howe, in explaining how this is effected, distinguishes three horizontal zones in the ore :* (1) the upper surface, where oxidation is only slightly hindered by sulphurous and sulphuric acids and by the products of combustion of the fuel; (2) the middle layers, where oxidation proceeds to a very limited extent; (3) the lowest layers, where "a pellet of ore is simply exposed to the action of the other pellets with which it is in contact, of volatilised sulphur, and of sulphurous and sulphuric anhydrides generated by the action of sulphur on previously formed metallic oxides." He proceeds—"The expulsion of arsenic and antimony as sulphides is favoured in the middle and lower zones by the presence of volatilised sulphur, mixed with sulphurous acid and at most a very limited supply of free oxygen and sulphuric acid. In the upper part of the middle layer, to which a small amount of free oxygen penetrates, we have the gently oxidising conditions favourable to the formation of arsenious acid and trioxide of antimony. In the upper zone the stronger oxidising conditions rather favour the formation of fixed arseniates and antimoniates, though, even here, part of the arsenic and antimony may

* Copper Smelting, *U.S.A. Geol. Survey*, Washington, 1885.

volatilise and escape while passing through their intermediate volatile condition of arsenious acid and trioxide of antimony." On stirring the mass, these arseniates and antimoniates, being exposed to the reducing action of volatilised sulphur and undecomposed sulphides in the lower zones, may again be converted into volatile oxides. Protoxide of iron, suboxide of copper, and sulphurous acid are also efficacious in reducing arsenic acid, higher oxides of iron and copper, and sulphuric acid being formed. "Thus, every individual atom of arsenic may travel forth and back many times through the volatile condition, being oxidised at the surface and reduced below the surface, . . . and every time it arrives at this volatile condition an opportunity is offered it to volatilise and escape." If a small quantity of coal or coke dust is mixed with the ore, after it has been completely oxidised, and the air excluded, the arseniates and antimoniates are again reduced to the lower oxides, and, if they are "carried past the volatile state," *i.e.*, reduced to metals, they may be again passed through it by an oxidising atmosphere. "Of course the expulsion of arsenic and antimony is favoured by the presence of a large proportion of pyrites, both because the sulphur distilled from the pyrites tends to drag them off as sulphides, and because the presence of the pyrites prolongs the roasting, and thus increases the number of times which the arsenic and antimony pass back and forth past their volatile conditions; hence, it is sometimes desirable to mix pyrites with impure ores to further the expulsion of their impurities."

The Use of Salt in Roasting.—Certain ores require the addition of salt in roasting in order to chloridise material which would otherwise absorb chlorine when the ore came to be "gassed," and so cause additional expense as well as inconvenience. If silver as well as gold is to be extracted from the ore, the addition of salt is necessary in order to form chloride of silver in the furnace, since metallic silver is not attacked by chlorine at the highest temperature ever employed in the leaching vat. The silver chloride is then dissolved out by hyposulphite of soda or some other solvent either before or after the extraction of the gold.

Even if no silver is present, an ore must be roasted with salt if it contains much copper (as sulphide, or as an oxidised salt), lime, magnesia, or other substance which, after being subjected to an oxidising roasting, is rapidly attacked by chlorine at ordinary temperatures. The salt is usually added towards the end of the operation, when no sulphides and only a small percentage of sulphates are left undecomposed; sometimes, however, the ore and salt are mixed before charging in. To some sulphides only 5 pounds of salt per ton of ore are added, but others require as much as 90 pounds per ton. The weight of salt added must be at least six to eight times that of the silver present in the ore.

If a large amount of salt is used, it is desirable to leach the roasted ore with water, before treating it with chlorine gas, in order to remove the coating of soluble sulphates and chlorides remaining on the surface of the granules of ore.

The chemical action of the salt is due to a double decomposition between it and the sulphates of the heavy metals, by which sulphate of soda and the chlorides of the heavy metals are produced. The following general equation approximately represents the reaction:—

$$2NaCl + RSO_4 = RCl_2 + Na_2SO_4$$

Chlorine is also set free by the action of sulphuric anhydride on salt, and the presence of water vapour induces the formation of much hydrochloric acid. These gases act directly on the several constituents in the ore, forming chlorides and oxychlorides. The metallic chlorides and oxychlorides formed are in many cases volatile (e.g., the compounds of copper, iron, lead, arsenic, antimony, &c.), and, in passing off, the volatile compounds carry away with them varying proportions of gold and silver, which, as a rule, are not recoverable in the dust chambers. The chloride of copper is especially active in causing these losses.

Other reactions which probably take place are as follows:—

1. Ferrous sulphate, acted on by salt at a red heat in presence of air, yields hydrochloric acid and chlorine, which act on the gold and silver, while ferric sesquioxide and sodic sulphate are produced.

2. Ferric chloride, Fe_2Cl_6, is also produced at the same time. This is volatile, and chloridises silver with great energy at a red heat, sesquioxide of iron being produced.

3. Cupric chloride, $CuCl_2$, is easily decomposed into cuprous chloride, Cu_2Cl_2, and free chlorine, or into the oxychloride, $Cu_2O.Cl_2$, and free chlorine. The vapours of $CuCl_2$ thus give rise to further supplies of nascent chlorine available for the chlorination of the silver.

4. Arsenic and antimony form volatile chlorides which are decomposed by means of oxygen and water vapour, yielding arsenious and antimonious acids and nascent chlorine or hydrochloric acid.

It is thus obvious that the presence of base minerals is advantageous in that they may cause nascent chlorine to be set free in the presence of silver in all parts of the furnace. On the other hand, the loss of gold is increased by any increase in the quantities either of silver or of the base metals, since in the former case the time of the roasting is prolonged. The best chloridising effect is obtained in a highly oxidising atmosphere, so that very little sulphur is required in the ore, and, if much is present, the practice of eliminating the greater part before adding the salt is not likely to be attended with any diminution in the percentage of silver chloride formed. Moreover, the water of crystallisation in the salt promotes the formation of hydrochloric acid. When salt is used in roasting, the ore should be allowed to cool slowly in heaps after being withdrawn from the furnace. If treated in this way, a higher percentage of the silver, &c., is found to be chloridised than if the ore is wetted down at once, or even spread out to cool in a thin layer. Chlorine continues to be evolved for a long time after the withdrawal of the charge has taken place, the heaps smelling strongly of the gas.

Losses of Gold in Roasting.—Plattner proved in 1856* that in the oxidising roasting of ordinary auriferous pyrites, a loss of gold can take place only when the operation is carried on so rapidly that fine particles are carried off mechanically by the draught. This conclusion, as far as sulphides and arsenides are concerned, has been confirmed by Küstel,[†] and by Prof. S. B. Christy,[‡] but the latter adds that it is extremely difficult to prevent all mechanical loss by dusting, which is caused by even a moderate draught. Küstel records the loss of 20 per cent. of the gold present during the oxidising roasting of certain tellurides of gold and silver, and states that this is not a mechanical loss, but is due to volatilisation. The effect of tellurium on the volatilisation of metallic gold is shown on p. 7.

The losses of gold which are sustained when salt is added to the furnace charge have been fully investigated by Prof. S. B. Christy,§ and the following account is mainly derived from his paper on the subject. Küstel had previously found that a telluride ore, on being roasted with 4 per cent. of salt, lost 8 per cent. of its gold before the ore was red hot. Aaron ‖ found that certain ores, consisting of simple pyrites, suffered great loss of gold in roasting with salt which had been added at the commencement of the operation; only a small part of this gold was condensed in the flue, in which was found a yellowish fluffy precipitate, consisting largely of chlorides of copper and iron, and containing nearly 30 ozs. of gold to the ton. He found that the loss was greatly reduced by diminishing the quantity of salt, and by reserving it until the dead roasting was nearly complete.

In the chloridising roasting of a Mexican ore, consisting mainly of magnetite and pyrites with 3·5 to 7 per cent. of chalcopyrite, Mr. C. A. Stetefeldt found the losses of gold to be from 42·8 to 93 per cent. of the total gold contained. He states¶ that "there is no doubt that the volatilisation of the gold takes place with

---

* *Metallurgische Röstprozesse*, Freiburg, p. 128.
† *Roasting of Gold and Silver Ores*, 1880, p. 56.
‡ *Trans. Am. Inst. Mng. Eng.*, 1888.
§ *Loc. cit.*
‖ *Leaching of Gold and Silver Ores*, 1881, p. 121.
¶ *Trans. Am. Inst. Mng. Eng.*, vol. xiv., p. 339.

that of the copper chlorides. The loss increased with the quantity of these chlorides formed and volatilised." He further shows, however, that the presence of copper chloride is not the only possible cause of loss, since an ore consisting of hard white quartz, intimately mixed with about 7 per cent. of calcite and a little pyrites, lost 70 to 80 per cent. of its silver, and 68 to 85 per cent. of its gold, when roasted with 5 per cent. of salt. When subjected to an oxidising roast, no loss of gold took place. The reason for the extraordinary behaviour of this ore was not discovered.

Prof. Christy found that, in the ores on which he experimented on a small scale in a muffle furnace, a greater loss was sustained by adding the salt near the end of the roasting operation, than by mixing the same weight of salt with the ore at the start. He explained that this is due to the fact that the amount of gold volatilised varies with the amount of chlorine which comes in contact with it. When the salt is added at the start, the chlorine is at first removed by the sulphur as fast as it is formed, escaping as chloride of sulphur, and thus the gold is protected from attack. When the salt is added after a long oxidising roast, the chlorine is rapidly generated (the ore being red hot and containing large quantities of sulphates), and the gold is no longer protected from attack by the sulphur. The loss of gold is also in all cases increased by working at a higher temperature, owing to the larger amount of chlorine generated, and to the increase in the volatility of the gold. It is apparent from the results given on p. 21 that the temperature used in chloridising roasting must be very carefully regulated, the loss of gold being increased far more by high temperature than by a lengthening of the time in the furnace. Moreover, the salt must be reduced to the least possible quantity.

The advantage found to be gained in practice by adding salt near the end of the operation is due to the fact that, in the continuous roasting of ores in long-bedded furnaces, the gases given off from the finishing floor pass over a great length of comparatively cold, unsalted, and unoxidised ore before reaching the flue. The quantity of gold chloride mixed with the chlorine which is evolved from the red-hot ore as soon as the salt is added is no doubt large, but the SO_2 from the colder ore, and the steam from the fuel, "offer excellent means for the reduction of the chloride of gold right within the furnace, while the most efficient means probably is the pyrites themselves," which have been proved to be readily capable of condensing gold on their surface. If all the salt is added at the start, there is a continued volatilisation of chloride of gold throughout the furnace, and a less favourable opportunity for it to condense. The difference between the results in the muffle and in the reverberatory furnace is thus explained.

At Nevada City, at the Merrifield Mine, and in other works in the neighbourhood, the old-fashioned long furnace, with a single step separating the finishing hearth from the rest of the furnace, was still used in 1888.* These furnaces are from 55 to 65 feet long, holding from 6 to 9 tons, and producing about 3 tons of roasted ore per day, so that the ore remains in the furnace from two to three days. The custom there was to give the ore a long oxidising roast at a low red heat, ending at a low cherry-red heat, and then, when the ore reached the finishing floor, the temperature was slightly lowered, and the salt added. The salt was stirred thoroughly into the ore, and as soon as it was "dissolved" by the roasted ore—*i.e.*, in about half an hour—the charge was drawn into the cooling pit. This lowering of the temperature is evidently of great importance in reducing the loss, while the duration of the roasting is regarded as less material, so long as no salt is present. These mills are on custom work, charging $15 to $17 per ton of ore for treatment, and guaranteeing a yield of 90 per cent. of the gold and 60 per cent. of the silver. Their method of roasting seems to be considered in California as that best suited to concentrates containing a high percentage of sulphur, but their loss in roasting has not been ascertained. The best method of roasting any particular ore, however, cannot be determined by any general rule, and exhaustive experiments must be made in every case before a definite course of procedure is finally adopted.

At one of the Californian chlorination mills it was found by experiment in 1882 that nearly 50 per cent. of the gold and 28 per cent. of the silver was being lost by volatilisation. In this case the pyrites was roasted on two hearths for thirty-six hours, 1 per cent. of salt being added four hours before the charge was drawn. The reason for the great loss was thought by Professor Christy to be the high temperature of roasting, particularly on the charging-in floor.

The variation of the loss in different ores which are treated precisely alike is doubtless due partly to the presence or absence of metals forming volatile chlorides which carry off the gold, and partly to the physical condition of the latter, the volatilisation being greater if it is in a state of minute subdivision.

MECHANICAL FURNACES.

The furnaces which have been designed with the object of saving the labour necessary to work the reverberatory furnaces may be divided into four classes, viz. :—

* *Trans. Am. Inst. Mng. Eng.*, vol. xiv., p. 340.

1. Stationary hearth furnaces, supplied with iron hoes moved by machinery by which the ore is rabbled. The O'Hara, Spence, and Pearce Turret furnaces are examples of this class.

2. Rotating-bed furnaces, in which the hoes or stirrers are stationary, while the bed supporting the ore revolves, so that the latter is stirred by the hoes. An example of this class used to roast gold ores as a preliminary to chlorination is afforded by the furnace used at the Bunker Hill Mine, California.

3. Rotating cylindrical furnaces, which consist of brick-lined iron cylinders capable of being rotated, so that the ore is tumbled over and over by their motion while it is being roasted. Examples are the Brückner, the White-Howell, and the Hofmann furnaces.

4. Shaft furnaces, in which the powdered ore falls by gravity, in a shower, through an ascending column of hot air, the oxidation being effected in the course of the fall. The Stetefeldt furnace, which is the only one based on this principle, is not used for dead roasting, as it is not adapted for the purpose. It is used for the chloridising roasting of silver ores, and will not be described in this volume.

In mechanical furnaces the consumption of fuel is often much greater than that in the long-bedded reverberatory furnace, where it is usually from 10 to 20 per cent. of the weight of the ore if flaming coal is used.

1. Furnaces with Mechanical Stirrers.—*O'Hara Furnace.*—This is the oldest mechanical furnace, and it bears a great resemblance to the old-fashioned reverberatory furnace. It has two superposed hearths, in each of which the arch is very low, so as to confine the heat close to the ore. An endless chain, set in motion by suitable machinery, passes through the furnace, resting on the upper hearth, and returns along the lower hearth. Attached to the chain at proper intervals are iron frames of a triangular shape; on these frames are a number of ploughs or hoes set at an angle, so that one set of hoes turns the ore to the centre, and the next set turns it in an opposite direction towards the walls. The ploughs thus stir the ore thoroughly, and at the same time move it gradually towards the fire. The ore falls from the upper to the lower hearth by gravity, and similarly falls from the lower hearth into a pit when it arrives at the hottest place in the furnace. The ore is from five to ten hours in the furnace, according to the amount of sulphur contained in it. In the modern form the hearths are each 8 feet wide and 90 feet long, and the capacity is about 35 tons per day, at an expenditure of about $2\frac{1}{2}$ H.P. The ore is not roasted dead, however, in this case, about 6 per cent. of sulphur remaining in it. No less than twenty-three of these furnaces are now in operation in the United States, although none are used in chlorination mills.*

* *Eng. and Mng. Journ.*, May 20, 1893, p. 463.

Spence Furnace.—This is perhaps the best shelf furnace yet devised, and has been used successfully in the roasting of copper sulphides in the United States. It consists of a series of four or five superposed hearths, communicating with one another by means of vertical passages at alternate ends of the furnace. The rabbling is done by rakes having a reciprocating motion longitudinally in the furnace and armed with teeth of triangular section having the apices pointing in the opposite direction to that in which the ore travels. When the rakes move in the direction in which these apices point, the ore is only stirred, but when the return movement is made, the flat sides of the teeth push some of the ore along the floor of the furnace, and a part falls through the vertical passages on to the lower hearths. The ore is charged in through a hopper and discharged roasted into a pit. The number of floors may be varied according to the character of the ore to be treated. The motion of the rakes should not be continuous, as in that case wearing of the teeth becomes very rapid. It is better to set them in motion (by racks and pinions) for a few minutes, and then to withdraw them for a while in order to allow them to cool. It is claimed for this furnace that if once made hot, the combustion of the sulphur in pyritic ores supplies the place of other fuel, so that the roasting is perfectly performed with no further cost than that of power for the rakes. It is of course obvious that without an extra fire, the lowest hearth, which is not heated from below and which is the place where the air is admitted, would tend to become the coldest, and would then be in no way adapted for the decomposition of the sulphates formed on the upper hearths, a process which involves endothermic reactions, and, therefore, requires external heat. The Spence furnace without a fire may, however, be valuable for the production of sulphates of iron or copper.

At the Treadwell Mine, Alaska, Brückner cylinders were formerly used,[*] but discarded on account of the amount of dust made and carried into the flue, and the large amount of fuel consumed, and Spence furnaces introduced. Each furnace had four hearths, the lowest being strongly heated to decompose the sulphates. The ore remained in the furnace for sixteen hours, and 3 per cent. of salt was added on the hearth next above the lowest one. The rakes of the upper hearths lasted a long time, but those of the finishing hearth were worn out in three months, and were often broken sooner. These furnaces were not found to be of any use until converted from muffle into reverberatory furnaces so that the products of combustion of the fire passed directly over the ore to be roasted. Six of the double Spence furnaces were built at a great expense, and the cost of roasting in them was found to be less than half that in the Brückner, but

[*] *Eng. and Mng. Journ.*, April 11, 1891.

their capacity was small and the amount of fuel required was found to be very great. An ordinary reverberatory furnace was, therefore, built, and the results obtained were so satisfactory that three others were added, and the Spence furnaces thrown out of work. They are apparently not now used at any chlorination mill, but might possibly be adapted to some ores.

One of the chief causes of difficulty and expense in working furnaces with mechanical stirring apparatus is that the iron hoes gradually become heated, and they are then rapidly corroded by the sulphur in the ore. In hand stirring, the rabbling tool is withdrawn as soon as it is hot, and allowed to cool, while another is substituted for it meanwhile, thus prolonging its life. To imitate this action, the Spence hoes are sometimes arranged to work for a while and then to be withdrawn completely to cool. In the O'Hara furnace, also, it is better to have only one hearth, the hoes passing completely outside the furnace on the return journey. In spite of such arrangements, however, the trouble and expense caused by the wearing of the hoes are very great.

Pearce Turret Furnace.—This consists of an ordinary reverberatory hearth built in an annular form. In the centre of the circular space surrounded by the hearth is a vertical iron column carrying four hollow horizontal arms projecting through a slot into the reverberatory hearth which they cross transversely. The column revolves and the arms carry rabble blades which traverse the hearth, stirring the ore and moving it round the circle by degrees. "Air is forced through the hollow arms and is discharged against the rabble blades, performing the double duty of cooling the iron work and of furnishing heated air for the oxidation of the ore." The ore is discharged automatically after passing once round the furnace. Two or more fireplaces are used. These furnaces are very economical, and are now preferred to any others in Western America for roasting.

2. Rotating Bed Furnaces.—*Rotary Pan Furnace.*—This is used at the Bunker Hill Mill, California, to desulphurise concentrates containing much sulphur and small quantities of arsenic, antimony, and lead. The stationary hearth is 7 feet wide and 18 feet long, and has two working doors. At the end of the stationary hearth is a drop of 6 inches on to a horizontal revolving hearth, made of iron lined with fire-brick, 12 feet in diameter, with a discharge hole in the centre. This is next the fire-place. The hearth revolves by means of gear-wheels placed beneath it, at the rate of one turn per minute. The charge remains on it for eight hours, and is then discharged through the central aperture. The capacity of the furnace is 2 tons per day, the fuel required being $1\frac{1}{4}$ cords of wood. In Fig. 46, a similar but larger furnace is shown.

MECHANICAL FURNACES.

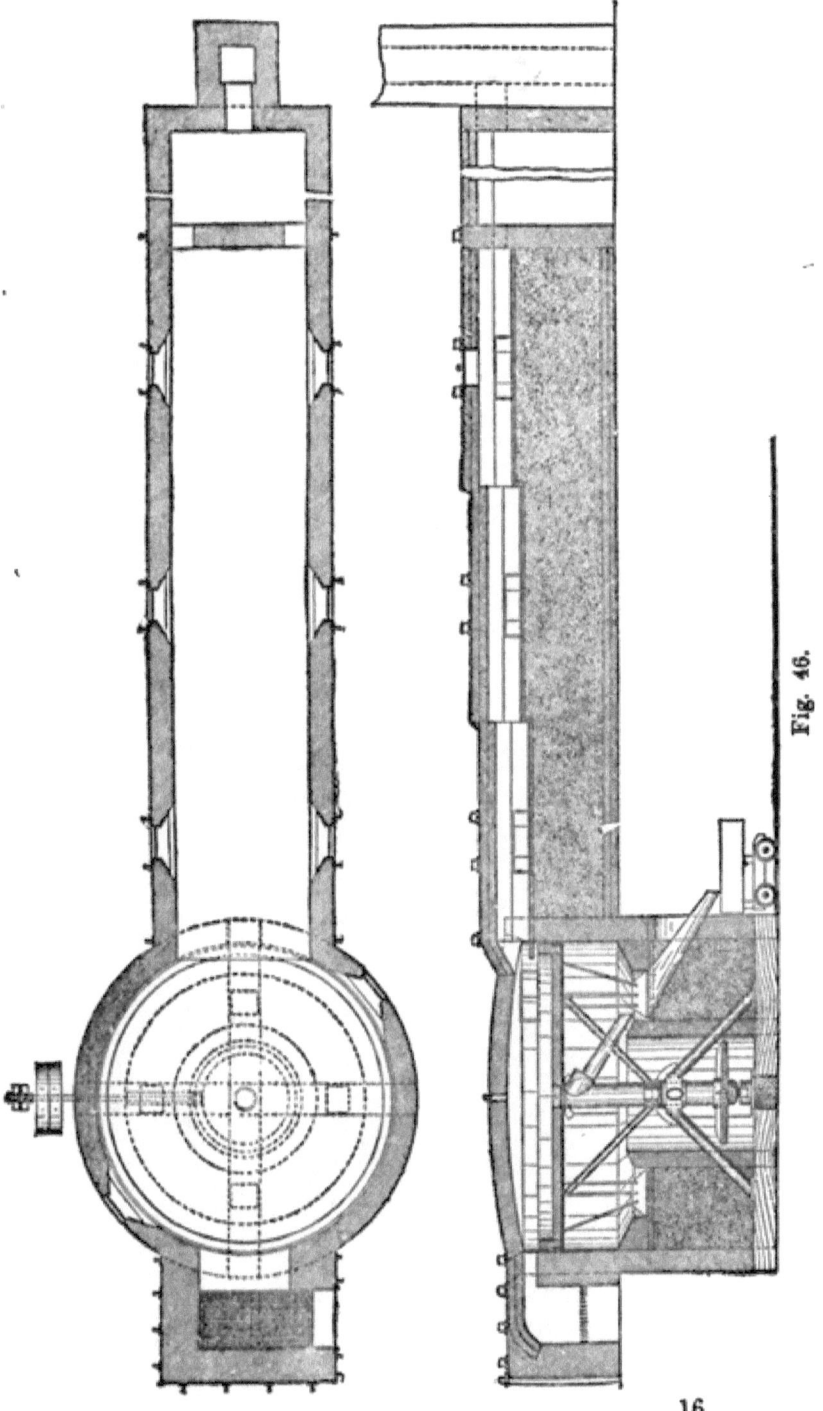

Fig. 46.

242 THE METALLURGY OF GOLD.

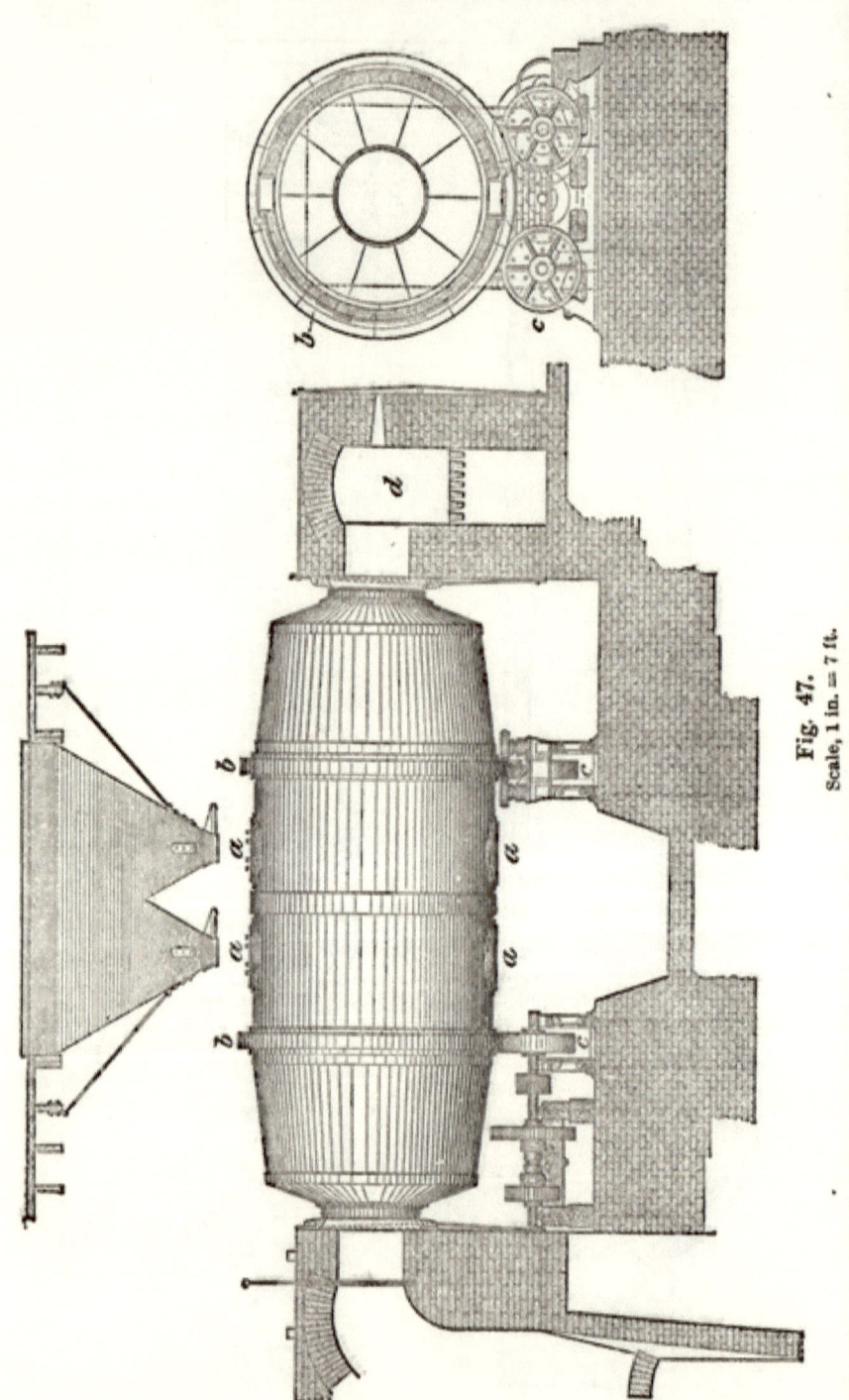

Fig. 47.
Scale, 1 in. = 7 ft.

3. **Revolving Cylindrical Furnaces.**—These furnaces have come into more general use than other mechanical furnaces, but are employed at very few gold chlorination mills, since their capacity is too large for the modest requirements of most of these establishments, many of which treat only a few tons per week. Revolving cylinders treat from 10 to 50 tons of ore per day.

The Brückner Cylinder.—This furnace was first introduced in Colorado in 1867, and is now, after alterations and improvements, extensively used in the United States, chiefly in the chloridising roasting of silver ores, although it is also suitable for dead roasting. In its latest form it consists (Fig. 47) of a cylinder of boiler plate iron, lined with fire bricks. The ends are partly closed, leaving central apertures 2 feet in diameter. There are also four receiving and discharging openings (a, a), closed by hinged doors. The discharge is effected into a pit, or into hot-ore cars placed underneath. The cylinder revolves on two chilled-iron friction rings (b, b), resting upon four chilled-iron carrying rollers (c, c). The rotation is caused by friction between the rings and the rollers, which are driven by a belt and pulley. The older method of turning the cylinder by means of a gear-wheel, connecting with teeth set on the cylinder, has been abandoned.

The lining is one brick ($2\frac{1}{2}$ inch) thick in the middle of the cylinder, but additional layers are added near the ends until the circle is contracted down to the size of the openings in the ends, which are also lined; each layer falls short of the preceding one by about 2 inches, so that the lined ends have a conical form. The mortar used consists of one-third fire-clay and two-thirds crushed fire-brick. In the old form six iron pipes passed through the cylinder in a plane at an angle of 15° to the axis, and perforated plates uniting them formed a diaphragm which assisted in stirring the ore. The circular flange surrounding the opening at one end of the cylinder connects loosely with the fire-box, d; the other end connects with an opening leading to dust chambers and the stack. The dust chambers (not shown in the figure) are large, and must be carefully attended to, as the amount carried over is usually 10 per cent. of the ore, and may rise to 25 per cent. when the mineral is light. The heavier particles collected in the first chamber are charged into the furnace again; the lighter dust is unsuitable for this purpose, and must be subjected to special treatment. In the improved Brückner the diaphragm is discarded, as it is rapidly corroded by the sulphur, and increases the quantity of material carried into the dust chambers. The furnaces are now usually made 7 feet in diameter and 18 feet long, a full charge being from 6 to 8 tons. This large size has given greater satisfaction than that formerly employed—viz., 6 by 12 feet—the full charge of which was from 3 to 4 tons of ore. As a charge can be kept in the furnace as long as is necessary, the Brückner is particularly useful when very base ores are to be roasted, or when the ores vary greatly in composition, so that the duration of roasting is not constant. The mixing of the ore is rendered more perfect by the conical shape of the ends, which causes the ore to be thrown backwards and forwards, changing its position frequently and exposing new surfaces to the action of the fire.

Brückner's cylinder has been found suitable for the dead oxidising roasting of pyritic ores, a little charcoal being sometimes added towards the end of the operation to reduce the sulphate of copper formed. It has the disadvantage of being too hot at first, and not hot enough at the finish if the fire is kept uniform, and consequently pyritic ore tends to form into balls. These may contain sulphates, which are not then easy to decompose except with great waste of time and fuel. The consumption of fuel is also rendered greater by the shortness of the furnace, as a large amount of waste heat passes out into the dust chamber, and is not utilised, a state of things exactly similar to that which occurs in the reverberatory furnace with only one floor.

An important improvement in the Brückner furnaces in use at the Portland mill, Deadwood, Dakota, has lately been made by the Manager, Mr. Hickock. Between the fire-box, which is mounted on wheels, and the revolving cylinder is placed a short iron cylinder, 18 inches long, rigidly connected with the firebridge and closely fitting the throat of the revolving cylinder. In the lower side of the small cylinder is an aperture, from 8 inches to 10 inches square, closed by a sliding door. This is useful in regulating the draught, without cooling down the fire. By its use a sheet of cold air can be made to pass immediately above the ore and below the products of combustion of the fire. It is stated that a considerable saving of fuel has been effected since the introduction of this simple device, and that a more perfect roast can be obtained in one-third less time than was formerly possible.

The Hofmann Furnace.—This furnace was designed by H. O. Hofmann as an improvement on the Brückner cylinder, which has the disadvantage of being much hotter at one end than at the other. The result of this is that the ore near the fire is exposed to a higher temperature than that near the flue, and, consequently, it is finished much sooner, so that fuel is wasted while finishing the ore in the colder part of the furnace, and, in the case of antimony ores, if the temperature is high enough to roast the ore next the flue, the portion close to the fire becomes caked and suffers a considerable loss of silver by volatilisation. The Hofmann furnace (Fig. 48) has a fireplace and flue at each end. The flues are between the fireplaces and the cylinder, and open downwards into dust chambers, C, built beneath, which are connected with the main stack. The arrangements at each end are

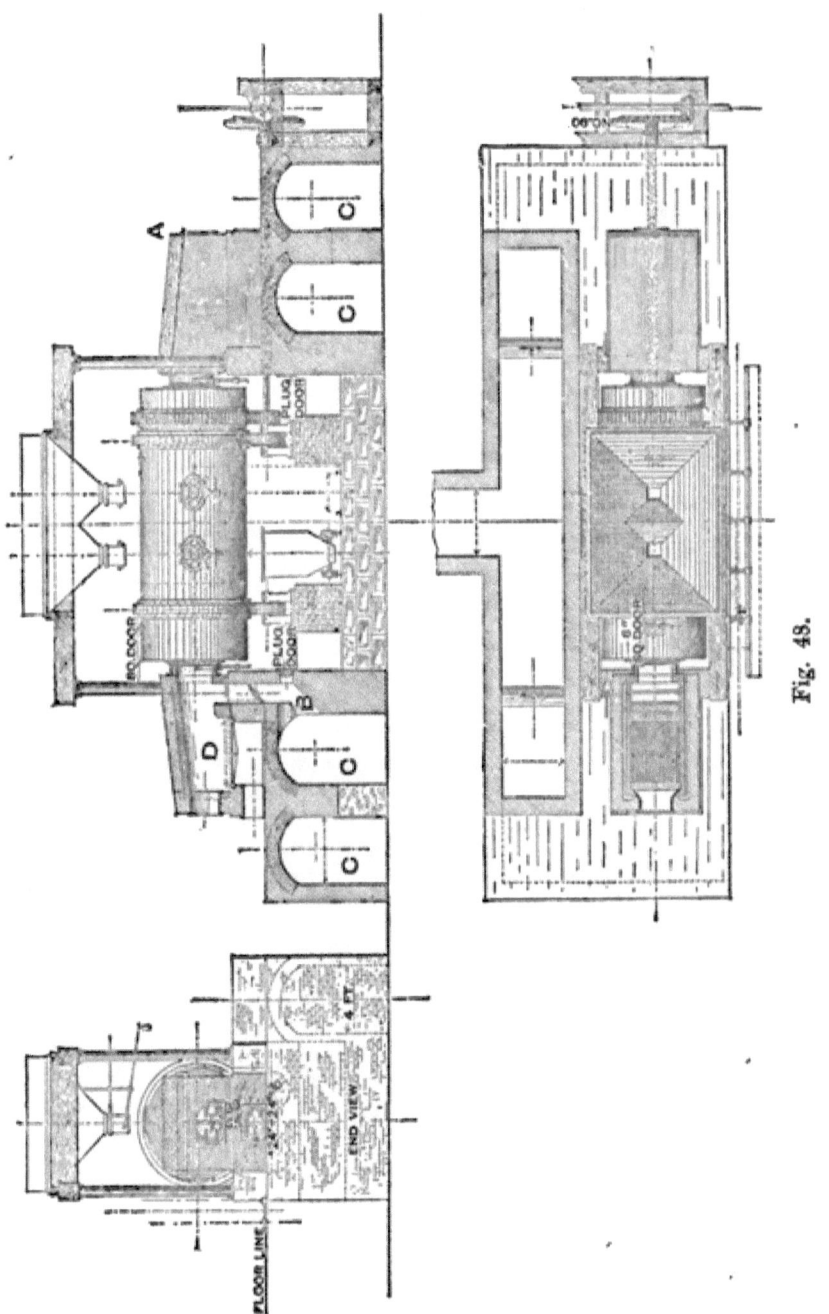

Fig. 48.

the same, and are such that, by means of dampers, the current of the air and gases can be made to pass through the furnace in either direction. The fire is lighted first at one end, A, and the dampers arranged so that the draught passes through the revolving cylinder and down the flue, B, at the other end. After a few hours the fire is lighted at the other end, in the fireplace, D, and the position of the dampers is reversed. The alternate heating from the two ends is regularly performed until the charge is completely roasted. In this way more uniform heating is obtained, both halves of the charge being raised to the required temperature without any portion being overheated. It is stated that the formation of balls is diminished by this system, and, in particular, the furnace is found to answer well when treating ores which require either a very low roasting temperature, or a very high one. Thus, for example, antimonious ores, which cake readily, are said to be successfully treated by the use of moderate fires; and ores containing very little sulphur, and so requiring a higher temperature, can readily be made as hot as is necessary by the alternate use of large fires, without using much fuel. When one fire is in operation the other is allowed to go down. By closing one of the large dampers near the main flue, and opening the damper of the corresponding descending flue, and the "plug door" in connection with it, a current of fresh air can be introduced into the furnace beneath the flame from the fire, thus greatly increasing the rate of oxidation of the ore. This arrangement is especially useful when ores carrying a high percentage of sulphur are being treated, their tendency to form balls being in this way greatly diminished.

It is clear that the use of much larger cylinders than those of the Brückner furnace is rendered possible by the presence of a fire at each end. Nevertheless, the evil of heating the ore unequally is only palliated; the ore in the middle of the furnace is never heated so strongly as that at the two ends, whilst in the case of continuous-discharge inclined cylinders, every particle of ore is heated similarly. Moreover, the use of two fires, which are alternately checked and urged, must cause a considerable waste of fuel. The principle is obviously inferior to that of the White-Howell furnace.

One incidental advantage of this furnace over all other revolving cylinders lies in the absence of loss from the carrying away of dust at the moment of charging-in. In other cylinders the ore falls in a shower into the furnace through a strong draught, by which much of the dust is carried off into the flue. When charging the Hofmann furnace, however, the dampers in both the descending flues are left open, so that the draught passes through both fireplaces direct to the dust chambers, without entering the furnace at all.

The White and White-Howell Roasting Furnaces.—The *White*

furnace consists of a long cast-iron revolving cylinder, lined with fire-brick, and inclined towards the fire end. The cylinder is bound by four cast-iron rings which rest on friction wheels, serving as supports, and it is revolved by other friction wheels driven by a shaft and pulleys. Crushed ore is fed into the furnace continuously at the upper end, passes through it by gravity, and is continuously and automatically discharged by falling into a pit through an opening in the floor close to the fire. The time occupied by the ore in passing through the furnace depends on the angle of inclination of the cylinder which can be changed, so that the time of roasting can be shortened or lengthened according to the nature of the ore. Ores containing a high percentage of sulphur require to be subjected to heat for a longer time than those with little sulphur, and the angle of inclination is reduced in such cases. The average inclination is about one in twenty. The cylinder is lined with fire-brick throughout, and projecting bricks raise portions of the ore and drop it through the flames, thus assisting the oxidation.

The advantages claimed for this furnace are that it is continuous in its operation, discharging its product regularly into the pit at the lower end, and this roasted ore can be allowed to accumulate and be withdrawn as required. The ore is submitted to a gradually increasing temperature, the most favourable conditions for dead roasting being thus obtained. The usual size of this furnace is about $4\frac{1}{2}$ feet in diameter and 27 feet long, the capacity being usually stated at from 20 to 30 tons per day. When employed for the dead roasting of pyritic ores its capacity is less, and in some cases the passage through a 27-foot cylinder is insufficient to eliminate the whole of the sulphur. In such cases the ore is made to traverse successively two similar cylinders, the combined length of which is sometimes over 60 feet. Great care should be exercised in keeping the ore supplied to the furnaces as uniform as possible, so that when once the rate of feed and the proper angle of inclination for the ore have been determined, no further alterations are needed in order to continue to give a perfectly roasted product.

In the *White-Howell furnace* (Fig. 49) only the enlarged part next the fire-box is lined with fire-brick, the remainder being left unlined. Cast-iron spirally arranged shelves assist in raising and showering the pulp through the flames. To both White and White-Howell furnaces an auxiliary fire is often added for roasting the dust which escapes from the main furnace. The dust, when it has been completely roasted, is shovelled out and mixed with the main bulk of the ore by hand. This arrangement is decidedly inferior to that suggested by John E. Rothwell,[*] who uses a hopper-shaped dust chamber with its bottom con-

[*] *Mineral Industry for* 1892, p. 234.

sisting of an inclined cast-iron plate projecting about 8 inches into the upper end of the cylinder. The dust carried out of the cylinder settles in this chamber, and, as it accumulates, slides down the sides and mixes with the fresh ore. The ore is thus kept more uniform than if re-mixed by hand, and some saving in labour is also effected. Rothwell uses a cylinder 36 feet long and 5 feet in diameter, with an inclination of 14 inches only, rotating once per minute. The lining is of fire-brick, six inches thick.

Use of Producer Gas in Roasting.—The use of producer gas in roasting may be mentioned, as it bids fair to completely replace solid fuel for the purpose at some future time, great saving of expense being thus effected. It was introduced at the Holden Mill, Aspen, Colorado, in 1891, and has now completely displaced other fuel there for both drying and roasting.* The gas plant consists of two "Taylor" revolving-bottom gas producers, one 6 feet and the other 7 feet in diameter. The

* W. S. Morse in *Trans. Am. Inst. Mng. Eng.*, Montreal meeting, 1893.

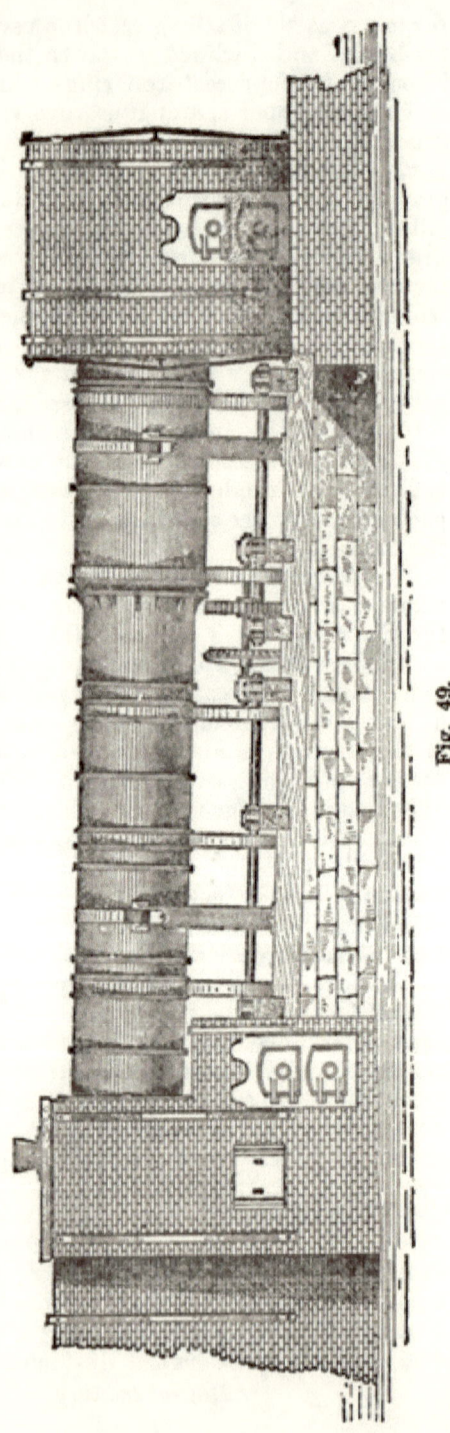

Fig. 49.

coal used is a mixture of one-third "Sunshine" nuts and two-thirds "Newcastle" nuts and pea-coal, the analyses of which are as follows:—

	Sunshine.	Newcastle.
Water,	2·8	1·7
Volatile matter,	36·3	37·95
Fixed carbon,	37·1	48·6
Ash,	23·8	11·6
	100·0	99·85

The drying plant consists of four Stetefeldt double shelf driers, and the roasting furnace is a Stetefeldt shaft furnace. The ore contains 8·10 per cent. of sulphur before roasting and 0·2 per cent. as sulphide afterwards. The coal consumed is 72·22 lbs. per ton of ore dried and 117·44 lbs. per ton roasted, the cost for the two being 29·37 cents per ton of ore. All the above figures are averages for a run of over a year—viz., from November, 1891, to January, 1893. The heat obtained in the Stetefeldt furnace could be made sufficient to fuse the ore; and there is, therefore, no doubt that similar producers could give heat enough to desulphurise ores in reverberatory furnaces.

CHAPTER XII.

CHLORINATION: THE VAT PROCESS.

The Vats.—The vats used for impregnating the ore with chlorine are usually, in California, about 7 feet in diameter, and are made of staves 3 feet long and 2 inches thick, which consist of the best split sugar-pine. They are coated with a mixture of pitch and tar to protect them from the corrosive action of the chlorine. Before being used for the first time, the vats are thoroughly soaked with water to diminish their absorptive action on the chloride of gold solution, but all wood brought in contact with this solution is nevertheless invariably impregnated with a certain amount of gold. This may be recovered after the vats, &c., are worn out, by burning them and fusing the ashes with suitable fluxes. The false bottom generally consists of quartz pebbles, the lowest layer being of the size of hazel nuts, and each successive layer consisting of finer material until, at the top, a thin layer of fine sand (passing a 20-mesh sieve, but retained on a 60-mesh sieve) is spread evenly over the surface. The thickness of the filter bed (which is not shown in the figure) is usually from 6 to 12 inches. It is supported on boards (A, Fig. 50), 1 inch thick, in which numerous ½-inch auger holes are drilled; these boards rest on wooden strips (not shown in the figure), 3 inches wide and 1 inch thick, which do not reach the edge of the vat, and so keep a clear space 1 inch deep, just above the true bottom of the vat, in which the solution can accumulate. The solution is drawn off by a leaden pipe fitted with a stopcock, preferably of stoneware; the pipe should be level with the bottom of the vat, which may with advantage be made with a slight fall towards the outlet to prevent any liquid being left in it. Deetken states * that fine sea-shells (consisting

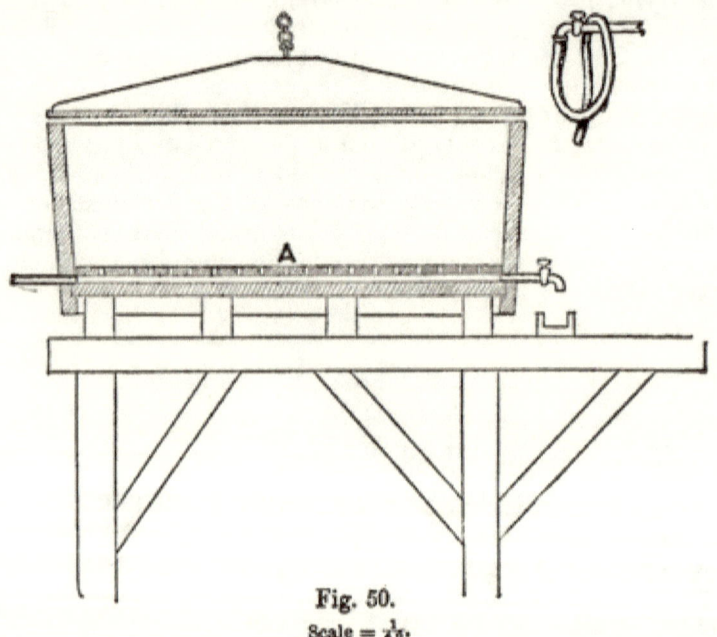

Fig. 50.
Scale = $\frac{1}{15}$.

of carbonate of lime) have been used instead of quartz pebbles for the filter bed without any prejudicial result. Talcose rocks, and particularly silicates of alumina, must not be used on account of their power of absorbing the chlorine. For the same reason sulphides, magnetic iron oxide, metallic iron, fragments of wood or other organic matter, or, briefly, any substances capable of being acted on by chlorine or of reducing the chloride of gold must be carefully excluded from the filter bed.

The surface of the filter bed may be covered with boards, not fitted closely together, but made into a framework by cross-pieces and pierced with many auger holes. This cover is useful when

* *Mineral Resources of the Rocky Mountains*, 1873, p. 342.

the tailings are being cleared out, otherwise, in shovelling away the ore, the surface of the filter bed is partly removed also. Messrs. MacArthur & Forrest's devices to avoid this are given on p. 310. Filter cloths of canvas, burlap, or cocoa-nut fibre matting are also frequently used above the filter bed, stretched tightly over a framework of wood which accurately fits the inside of the vat. The space between the canvas and the wall of the vat is packed with hemp or other material closely tamped down. Filter-cloths of every material, except asbestos, are soon rotted and demoralised by the action of the chlorine, and their use is frequently dispensed with. Wool lasts longer than cotton.

Charging-in the Ore.—When the vats are ready to be charged, a layer of dry ore is spread over the false bottom, and time given for the water from the filter bed to be drawn up into this layer by capillary attraction. If attention is not paid to this point, the lowest layer of ore becomes too wet from the combined effect of the water added to it before charging-in and that absorbed from the false bottom. The result is that the passage of the chlorine through the mass is resisted, and there is a great increase in the consumption of the gas. Deetken states that the whole of the usual charge of gas may be thus consumed, not rising more than a few inches above the bottom. The greater part of the charge is damped by sprinkling with water and thorough mixing. The amount of the water added varies with the nature of the ore, but the usual amount is from 6 to 12 per cent. for roasted ores. If it is made too wet, dry ore to the required amount is mixed with it. A good rough method of ascertaining when it is of the proper degree of dampness is to compress some in the hand; balls of ore should be readily formed in this way, but should be just dry enough to crumble up again. The reason for the addition of water is that perfectly dry chlorine has scarcely any action on metallic gold at ordinary temperatures, and up to a certain point an increase in the amount of water present raises the rate of solubility of the gold. The limit of the amount of water that can be added is, however, determined by physical conditions, as the mass must be of loose porous texture in order to permit the gas to readily permeate through every portion of it. In order to promote this porous texture and uniform dampness, the ore is shovelled upon, and made to pass through, a sieve of four holes to the linear inch. This sieve may be conveniently made to slide on rollers on iron rails placed above the vat, and the ore, shaken through it, falls into the vat in a light shower. Although left undisturbed as far as possible, the charge must be levelled off with a rake occasionally.

When the vat is filled to within 6 inches of the top, the surface of the ore is made concave or saucer-shaped, higher at the sides than in the centre. The cover, usually of wood, is then lowered on to the vat by means of a chain and pulley, and the rim luted with a mixture of wet clay and sand, or, more usually in former times, with dough. These joints are kept moist during the "gassing" process by wet rags. The gas is introduced through a lead pipe, which is shown on the left-hand side of Fig. 50, passing into the vat below the false bottom. A small hole is left in the cover through which the displaced air may escape, and the issuing gases are tested from time to time by means of a rag tied to a stick and moistened with dilute ammonia. As soon as chlorine is found to be coming off freely, the hole is plugged, but the current of gas is not stopped until after the lapse of one or two hours more, when the charge is supposed to be saturated with the gas, the total time required for the impregnation being usually from five to eight hours.

Generation of the Chlorine.—The chlorine is generated in air-tight vessels of lead fitted with a stirring apparatus passing through the lid and worked from the outside. The gas necessary for a 3-ton charge of roasted concentrates may be generated in a leaden vessel of 20 inches in diameter and 12 inches deep. The joints of this generator and of all pipes traversed by the gas or by liquids carrying it in solution must be made by melting the lead, no solder being used except pure lead. The "burning together" of the joints is usually effected by an air-hydrogen or coal gas blowpipe jet. The cover of the chlorine generator is made gas-tight by a water joint, 2 inches deep, as are also the apertures in the lid for the passage of the revolving stirrer (which is usually made of hard wood) and for the delivery tube. Heat is applied by placing the generator on a sand bath standing on a perforated arch over a fire-place. The sand bath is often replaced with advantage by a water bath, as the heat required is not more than $90°$ F., and sudden heating causes inconveniently tumultuous generation of gas. The charge for 3 tons of ore consists of 20 to 24 pounds of rock salt, 15 to 20 pounds of manganese dioxide, containing 70 per cent. of available material, and 35 pounds of oil of vitriol of $66°$ B., diluted with half its weight of water. The cover is usually removed to introduce the solids and the water, but the acid is added, half a gallon at a time, through a siphon. At Heywood's Works, California, the acid is contained in a lead vessel furnished with a glass stopcock, from which a small continuous stream is allowed to fall into the siphon.

The outlet tube is of lead, but connections in pipes are often made by short pieces of indiarubber-tubing, well greased on the inside. These resist the action of chlorine fairly well. The gas is passed through the wash-bottle shown in Fig. 51. It is usually a large glass bottle or carboy with its bottom removed, supported in a lead-lined box filled with water. The gas is made to pass through about half an inch of water. The use of the wash bottle is partly to free the chlorine from hydrochloric acid or other impurities with which it is contaminated, but mainly to give an indication of the rate of flow of the gas; it is desirable that this should be as uniform as possible, as otherwise leakage is more difficult to prevent. As soon as the current of gas falls off, fresh acid is added and the vessel stirred. The wash-bottle,

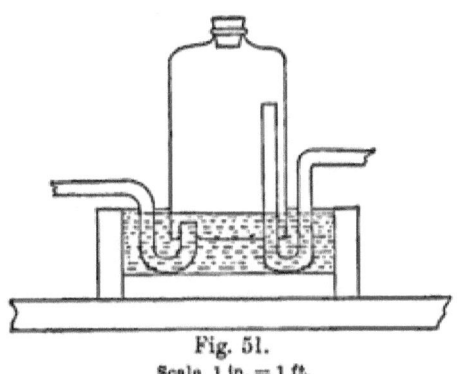

Fig. 51.
Scale, 1 in. = 1 ft.

as usually constructed, would be quite inadequate to free the gas completely from hydrochloric acid, whilst, even if it were approximately eliminated, some more would be speedily formed by the decomposition of water by the chlorine. It is, therefore, fortunate that this elimination is not absolutely necessary, in spite of the customary declaration to the contrary which generally appears in the descriptions of the process. Thus it is frequently stated that hydrochloric acid will act on any sulphides left undecomposed in the ore, generating sulphuretted hydrogen which would precipitate the gold already dissolved in the impregnation tank. This statement would not need any criticism, if it were not for the fact that it has often been made, but has apparently never been contradicted. No doubt matters might be arranged for the above reactions to take place if sufficient care were taken, but in practice they are not to be feared for the following reasons:—

(1) If undecomposed sulphides were present in the roasted ore, they would be attacked with much greater violence by chlorine than by hydrochloric acid, and before any gold could be dissolved, the whole of the sulphides present would be converted into sulphates, according to reactions, the final effect of which is shown in the following equation:—

$$R_2S + 4Cl_2 + 4H_2O = R_2SO_4 + 8HCl.$$

In some cases, no doubt, chlorides would be formed. (2) If the sulphides were not oxidised by the chlorine, they would be almost equally efficacious with sulphuretted hydrogen in precipitating the gold, so that even in this case hydrochloric acid would do no harm. (3) If an excess of chlorine is present, sulphuretted hydrogen could scarcely be said to be formed at all under any circumstances, as, if it were formed, it would be instantly decomposed by the chlorine. (4) If gold were precipitated by sulphuretted hydrogen in the impregnation vat, it would be in such a finely divided state that it would be re-dissolved in chlorine in a very short time, provided the gas were in excess. The only disadvantage due to the presence of much hydrochloric acid in the gas lies in the fact that certain metallic oxides (oxides of iron, copper, &c.) are much more readily soluble in the acid than in chlorine, chlorides and water being formed, and the resulting solution will be contaminated with these chlorides, so that special precautions are necessitated to prevent the bullion from becoming base.

Impregnation of the Ore.—The ore is allowed to remain impregnated with the gas for from twenty-four to forty-eight hours, the continued presence of a strong excess of gas being ascertained at intervals by removing the plug from the cover and applying the ammonia test. When, as is usually the case, there is a large excess of gas when the impregnation is at an end, it may be disposed of in one of several ways. It may be dissolved by adding water before raising the cover (the usual method of procedure), or it may be withdrawn by aspiration and discharged outside the building, or stored in a gasometer for use in a subsequent charge. If the cover is raised before getting rid of part of the excess of the gas, the atmosphere of the mill is rendered unbearable for several minutes.

The time of impregnation varies according to the size of the particles of gold, the fineness of the metal, and the temperature employed. Chlorine has a very slow action on pure gold, the rate increasing gradually with the temperature up to 100°. In order to obtain some data for the rate of solution of gold by chlorine at different temperatures below 100°, the following experiments were conducted by the author in the laboratory of the Royal Mint. Comparisons were made at the same time between chlorine, bromine, and cyanide of potassium, in order to determine their relative efficiency. The gold used consisted of "cornets" weighing about a half gramme each. These cornets offered a large surface to attack, being porous in texture, and consisted of gold 999·3 parts, and silver 0·7 part. They had all been prepared together by cupellation, parting, and annealing, so that their physical state must have been very similar. The results are given in the table on the next page.

The results obtained simultaneously by the use of potassic cyanide are given at p. 336. The amount of liquid used was in all cases 30 c.c. These results tend to show that both chlorine and bromine dissolve gold more rapidly at 50° to 60° C. than at ordinary temperatures, that bromine is more rapid in its action than chlorine, and that both are considerably more rapid than potassium cyanide, particularly at the higher temperatures employed. In all cases an excess of the reagent was present at the end of the operation, but the strength of the solutions obviously fell off during the experiments. The results of these experiments point to the desirability of further investigations on the subject.

Solvent.	Strength.	Time of Action.	Temperature.	Amount of Gold Dissolved per 1,000 Treated.
Chlorine,	Saturated solution in water,	1 hour.	15° C.	2·3
Bromine,	Pure,	,,	,,	169·0
,,	Mixed with 20 times its weight of water,	,,	,,	83·0
,,	,, ,,	,,	50° C.	157·1
Chlorine,	Saturated solution in water,	5 hours.	15° C.	57·6
Bromine,	Aqueous solution containing 0·2 per cent. bromine,	,,	,,	58·1
,,	,, ,,	,,	50° C.	101·0
Chlorine,	Saturated solution in water,	1¼ hours.	60° C.	44·9
Bromine,	1 per cent. aqueous solution,	,,	,,	64·6

The fact that the length of time occupied depends on the state of division of the gold needs no demonstration. The influence of the fineness and chemical composition of the particles of gold alloy present in the ore was pointed out by Deetken and Küstel. Fine gold is acted on more slowly than that of low standard. The alloys containing base metals (copper, &c.) are dissolved very rapidly, and small quantities even of silver appear to increase the rate of solution, but if the percentage of silver is increased beyond a certain amount, an insoluble coating of chloride of silver is formed over the granule, and further action is checked or completely stopped.

Reactions in the Impregnation Vat.—The amount of chlorine to be used depends mainly on the substances present, other than gold, by which chlorine is absorbed. A small amount is invariably converted into hydrochloric acid by the decomposition of the water present, the extent to which this reaction proceeds being increased by light and heat. If any sulphides are present they are oxidised by the chlorine in presence of water, sulphates and hydrochloric acid being formed, as follows:—

$$4Cl_2 + 4H_2O = 8HCl + 2O_2$$
$$CuS + 2O_2 = CuSO_4$$

These equations do not represent the whole of the reactions that take place. Some oxygen appears to be liberated, but the subject needs investigation. If considerable quantities of hydrochloric acid are thus formed, certain sulphates are converted into chlorides and sulphuric acid is set free, the reactions being influenced by the mass of the reagents present. Protosulphates or any other protosalts present are converted almost instantaneously to persalts by the chlorine, as follows :—

$$6FeSO_4 + 3Cl_2 = 2Fe_2(SO_4)_3 + Fe_2Cl_6$$

It is obvious from these reactions that great waste of chlorine in the impregnation vat is caused by imperfect roasting, 1 per cent. of unoxidised sulphur present in pyrites converting 8·9 per cent. of chlorine (or about 200 lbs. per ton of ore) into hydrochloric acid. This simple calculation is sufficient to show the impolicy of neglecting to roast the ore dead and then trying to retrieve the error by increasing the allowance of chlorine. Moreover, it demonstrates the uselessness of eliminating the hydrochloric acid from the chlorine before mixing it with the ore, and expecting in that way to prevent the ill effects produced by sulphides. The fact that many sulphides are almost instantly oxidised by even very dilute solutions of chlorine has been proved by a series of laboratory experiments by the author. These experiments would have been quite unnecessary if it were not that some chemists engaged in chlorination still appear to doubt the rapidity of the reaction. The oxidation of protosalts is almost as rapid, although the percentage waste of chlorine is considerably less; thus 1 per cent. of sulphur present in the ore as ferrous sulphate will convert 1·1 per cent. of chlorine (or 24·6 lbs. per ton of ore) into hydrochloric acid.

Sulphate of copper ($CuSO_4$) does not appear to be acted on by chlorine, but, nevertheless, whenever it is present in a roasted ore, chlorination seems to be rendered impracticable. This is possibly due to the fact that some sulphate of iron accompanies it. Whenever sulphates of these metals are left in the roasted ore by accident or design it is necessary to remove them by a preliminary leaching with water before the chlorine is introduced. Of course if the ore is chlorinated in tubs by gas, it must be partially dried and sieved back into the tub before impregnation can be attempted.

Organic matter is also oxidised by chlorine, although much more slowly. At ordinary temperatures, pitch and tar are almost unaffected, and the fibres of matting, canvas, &c., are acted on very gradually. Pieces of decaying wood or dried leaves must not be introduced with the water into the leaching vat, and if surface water is used it should always be carefully strained before being run in. A rough analysis of the water employed will often be serviceable, as it is frequently strongly alkaline in dry countries, and may be softened with advantage.

The absorption of chlorine by metallic oxides is the most frequent cause of waste, and, in the vat process, there are usually no efforts made to prevent this. Well roasted sesquioxide of iron (Fe_2O_3) is scarcely attacked by chlorine, especially if the temperature attained in the furnace has been high. If any magnetic oxide (Fe_3O_4), however, has been formed from over heating, or has been originally present in the ore, the absorption of chlorine is considerably greater, ferric chloride being formed and dissolved. Protoxide of iron is instantly converted into a mixture of chloride and sesquioxide of iron. Hydrochloric acid acts more rapidly than chlorine on all these oxides, but is nevertheless very slow in dissolving the well roasted sesquioxide. Metallic iron, which is sometimes accidentally introduced, is dissolved at once by both chlorine and HCl. The oxides of copper and zinc are quickly dissolved by chlorine, and still more readily by HCl. Lime and magnesia also readily absorb chlorine, forming hypochlorites, chlorates and chlorides, but hypochlorites are decomposed by any acid which may be present.

If any appreciable quantity of oxides capable of absorbing chlorine are present, it is cheaper to dissolve them by adding dilute sulphuric acid to the ore, and then, if possible, to leach out the soluble sulphates formed, before subjecting the ore to the action of the gas.

Amount of Chlorine required.—The amount of chlorine required varies greatly, both with the nature of the ore and the manner in which it is roasted. In order to roast pyrites dead, a long time in the furnace terminating at a high temperature is necessary, and the addition of salt may be desirable in order to chloridise oxides which would otherwise absorb the more expensive chlorine in the impregnation vat. These conditions in the furnace, however, may cause enormous losses by volatilisation, the endeavour to save a few pounds of chlorine in the vat causing the loss of 30 or 40 per cent. of the gold in the furnace. In ores where the percentage of copper, &c., is not large, and where, in consequence, salt need not be used in the furnace, the roasting may be finished at a high temperature without any disadvantage, and the consumption of chlorine may be thus reduced to a very low point. Thus certain ores from Dakota, containing only 1 or 2 per cent. of sulphur, and consisting chiefly of silica, were chlorinated by the author in revolving barrels, using only $3\frac{1}{2}$ pounds of chlorine per 2,000 pounds of ore. Even in this case, there was a strong excess of chlorine in the ore after the solution was complete, and the amount used could probably have been still further reduced without lowering the percentage extraction of gold. This ore contained 10 dwts. of gold to the ton, and over 80 per cent. was extracted. This was an extreme case, and it is seldom that so little chlorine is sufficient. Mr. Butters states[*] that at his mill at Kennel, California, where all descriptions of concentrates and

[*] *Eng. and Mng. Journ.*, Dec. 20, 1890.

pyrites were treated by the vat process, the average consumption of chlorine was 12 pounds per 2,000 pounds of ore. At Deloro, Canada, the amount used in the barrel process was from 12 to 18 pounds per 2,000 pounds of ore, but at the Haile Mine, South Carolina, only about 3 or 4 pounds per ton. In giving the amounts of chlorine which are used both in the vat and barrel processes, side by side, the intention is to show that the quantity absorbed depends on the nature of the material treated and not on the process used and the amount of water present, which are immaterial within certain limits as far as this point is concerned.

Leaching the Charge.—When it is judged that the impregnation has lasted long enough for all the gold to be dissolved, the excess of chlorine gas is removed, the lid is taken off, and water is added to the charge to wash out the soluble chloride of gold. The water may be added from below, and is then either allowed to overflow at the top, or is subsequently drawn off again at the bottom, the inflow being suspended. It is far more usual, however, to pour on water at the top, and let it flow out at the bottom.

The water must be added carefully, as otherwise the ore may pack unevenly, and channels may be formed through the mass, and the leaching thus rendered imperfect. Water is usually run from a tap on to a layer of gunny-sacking placed over the ore, by which it is distributed in a fairly even manner. It has been proposed to attach a coil of lead pipes, pierced with small holes, underneath the cover, and so sprinkle the water all over the ore in fine jets. In any case, water is added until it forms a layer 2 or 3 inches deep above the surface of the ore, and it is then allowed to stand until gas bubbles have ceased to rise through it, which happens in about half an hour. The stopcock below the false bottom is then opened, and the yellow- or blue-coloured solution (coloured by salts of iron, gold and copper), which should have a strong odour of chlorine, is run slowly through a filter, consisting of a canvas bag, into a small barrel about 18 inches in diameter and 2 feet deep, the overflow of which passes, by means of a launder or by rubber hose, to the precipitating tanks. Some of the ore and sand, escaping with the solution, is deposited in the canvas bag and barrel, but, if much slimes are present in the charge, either the canvas bag becomes clogged or the solution still remains turbid when it enters the precipitating vats. Water is supplied on the top of the ore as fast as it is drained away below, care being taken not to let the surface of the ore emerge from the liquid. The leaching is continued as long as any trace of gold can be detected in the issuing liquid by protosulphate of iron. As has been explained elsewhere (p. 25), a reaction is visible in clear solutions so long as the liquid contains more than two-thirds of a pennyweight of gold per ton of water, or one part of gold in one million of water. The strongly-coloured turbid solutions usually encountered in mills, however, are not capable of yielding distinct reactions, unless they contain much larger amounts of gold than this. In particular, when large quantities of copper salts are present in the solution, their strong bluish tints mask the slight discoloration due to a precipitate of a small quantity of metallic gold, and, moreover, they appear to interfere with the precipitation itself, in some cases at least preventing it from taking place.* It is always advisable to filter the solution by asbestos or filter paper before testing the liquid. Filter paper may be used, since it is very slow in precipitating gold from dilute solutions, even if they are neutral, and this action is completely stopped if free chlorine is present. It is better to test the clear liquid with stannous chloride under the conditions given at p. 26, since the presence of salts of copper does not seem to interfere with the reaction in this case, and, moreover, the amount of gold present can be determined in very dilute solutions with much greater accuracy than if ferrous sulphate is used.

Since a few minutes longer time occupied in leaching is of small moment, while the extraction of a few more grains of soluble gold from a ton of ore may be of the utmost importance in the long run, it is advisable to continue to leach at any rate until the water contains less than 1 part of gold in 5,000,000 (about 3 grains per ton), a point which can easily be determined by means of stannous chloride properly manipulated. The last charges of wash-water should not be mixed with the strong solution, but stored in other vats and used again for the first washings of other charges. In this way the amount of wash-water does not become excessive, although the tailings are cleaned more effectually than is usually the case. Re-precipitation of dissolved gold in storage vats or impregnation vats is not to be feared, so long as there is an excess of chlorine present in the liquid, and this can easily be ensured by adding a small quantity to any solutions not smelling strongly of the gas.

The amount of water used varies according to the richness of the ore and the method of leaching adopted. It is usually about 2 tons of water to 1 ton of ore, but in most cases part of the water is used again in the next charge.

Precipitation of the Gold.—The precipitating vat is of the same materials as the leaching tubs, and may be from 5 to 7 feet in diameter and 3 feet deep. There is no false bottom, and the vat is often made wider at the bottom than at the top to prevent any adherence of the gold to the sides. The wood is protected by a coating of pitch or paraffin-paint, or is left without paint of any kind. The vat receives a smooth finish inside to facilitate perfect cleaning, and is set perfectly level to avoid loss of gold

* Letter from Mr. Butters, *Eng. and Mng. Journ.*, Dec. 20, 1890.

while the waste liquor is being drawn off. The precipitating solution of protosulphate of iron (the reagent which has been chiefly used in practice in the vat process) is usually introduced into the precipitating vat at the beginning of the filtering operation. Care must be taken not to introduce a wasteful quantity, as it is better to make up for any deficiency when the gold-solution has all been run in. This operation is conducted in such a way as to impart a circular motion to the contents of the vat, so that the solutions are mixed without hand-stirring, but the latter is often resorted to in addition, in order to make the precipitate settle better; flat wooden staves with round handles are used for the purpose. The contents of the vat may be tested with solutions of gold chloride and of ferrous sulphate to determine whether the precipitation is complete, and the precipitant present in excess.

If lead or lime is present, dissolved in the solution, it will be precipitated as an insoluble sulphate on the addition of the ferrous sulphate, and thus render the gold-precipitate impure and less easy to treat. The amount of lead in the solution is usually small, unless hot water has been used for leaching, and most of the lead chloride is, in any case, separated by the canvas filter. The usual method of removing the lime is to add sulphuric acid to the gold solution, and to let it stand for a few hours, when calcic sulphate crystallises out, forming a crust on the sides and bottom of the vat. The liquid is then drawn off and transferred to another vat for the precipitation of the gold. Instead of this method, Nelson A. Ferry, E.M., recommends* the addition of molasses to the leach, before the addition of the sulphate of iron. He dissolves 1 gallon of molasses in 30 or 40 gallons of water, and keeps it for use, determining the quantity to be added by a laboratory experiment. He states that in this way the precipitation of the lime is prevented, but a large excess of the ferrous sulphate should be avoided, and the liquid kept slightly acid. The gold then sometimes comes down flocculent at first, but soon changes to its normal condition.

The ferrous sulphate is usually prepared on the mill by dissolving iron in sulphuric acid. When precipitation is complete the liquid is allowed to remain at rest for some time, in order to allow the gold to settle to the bottom. The old practice was to leave it "overnight," but the length of time allowed has of late years been greatly extended. Thus Mr. Chas. Butters † states that forty-eight hours is usually sufficient, but that sixty hours is better, and the determination of the extent to which the settling has progressed may be made by tapping the solution at various heights and filtering the liquid thus obtained. When a quart of liquid, drawn from a point 2 inches above the bottom

* *Eng. and Mng. Journ.*, Nov. 28, 1885.
† *Eng. and Mng. Journ.*, Dec. 20, 1890.

of the vat, gives only a slight dark stain to a No. 7 Swedish filter paper, on being passed through it, the settling may be regarded as complete. Mr. C. H. Aáron quotes instances* where, after forty-eight hours settling, as much gold remained in suspension in the liquid which was drawn off as was equivalent to 50 cents per ton of the ore treated. The conclusion may be drawn from these statements that a certain amount of gold is inevitably lost by being carried away in suspension, but with care and patience the loss may be reduced to a low percentage, and even at the present day, in spite of the introduction of many other precipitants, ferrous sulphate is probably as widely used as it has been at any time in the past.

When the waste liquid has been drawn off by a floating siphon, more ferrous sulphate and fresh solutions from the leaching vat are poured into the vat, and the process repeated until enough gold has accumulated at the bottom to warrant a clean-up. This may take place at intervals of from a fortnight to three months. The clear liquid is drawn off as closely as possible, and the slime scooped out and filtered through paper, or, by means of a press, through canvas. Finally, the vat is thoroughly cleaned out by rinsing it with water which is run off through a plug-hole, level with the bottom, into a wash-tub. The gold precipitate is then dried carefully and fused in graphite pots, with salt, sand, nitre, borax, &c., as fluxes, according to the requirements of the case. If the precipitate contains any considerable amount of impurities (such as oxides and basic salts of iron), which is usually the case, it may be treated with hydrochloric acid before fusion. The bullion produced varies from 920 to 990 fine, the alloying metals consisting chiefly of iron and lead.

Cost of Working.—The cost of treating concentrates or ores by the Plattner process depends chiefly on the cost of roasting. In 1867, the total cost in California was stated by Küstel to be $14.55 per ton, but in 1872 it had been reduced to $11, the expense of roasting being in each case about two-thirds of the whole. At the works of the Plymouth Consolidated Mining Company, California, in 1886,† the cost of treating 100 tons per month was $9.40 per ton (the roasting accounting for $4.60 per ton, or nearly one half), and at the Providence Mine, in the same State, it was about $6.30, without including the expenses of general supervision, interest on first cost, and depreciation of plant.

It was estimated in 1888 ‡ that, generally, throughout California, a plant, capable of treating 6 tons of concentrates daily, cost from $6,000 to $7,000 for its erection, while the cost of extraction was about $10 per ton, and the proportion of the gold

---

* *Cal. State Mineralogist*, 1888, p. 836.
† *Trans. Am. Inst. Ming. Eng.*, 1886.
‡ *Eighth Report, Cal. State Min.*, 1888.

extracted was from 90 to 92 per cent. It has, however, been more recently stated by G. F. Deetken that the cost of vat chlorination under favourable conditions need not exceed from $3 to $4 per ton, but he gives no details or facts in support of his statement, and it seems probable that the conditions required are too favourable to be expected.

CHAPTER XIII.

CHLORINATION: THE BARREL PROCESS.

THE use of revolving barrels for chlorinating ores was perhaps suggested by the old Freiberg method of barrel amalgamation. It has already been mentioned, p. 215, that Dr. Duflos used a revolving barrel in some of his experiments at Breslau, in 1848, and obtained results almost identical with those given by the vat percolation method. He, therefore, preferred the latter as being cheaper. The next mention of revolving barrels seems to have been contained in a patent taken out by Mr. De Lacy in Victoria, in 1864. The process thus patented appears to have been tried there, probably on a small scale, but certainly never passed into general use and was soon forgotten. It need not be described here.* In 1877, Dr. Howell Mears, of Philadelphia, patented a process which had some points of resemblance to that described by De Lacy. This process was gradually improved in practice, after having been adopted by several mines in the United States, and in particular the improvements introduced by Mr. A. Thies, of South Carolina, which were everywhere adopted, caused the name of the Thies process to be applied to the amended method of procedure. In 1887 the barrel process was re-introduced with some modifications into Australia by Prof. J. Cosmo Newbery and Mr. C. J. T. Vautin, who applied it to the ore of the Mount Morgan Mine, where it was worked for some years with conspicuous success. Messrs. Newbery and Vautin obtained patents in many countries for their improvements, and the method of treatment of ores by barrel chlorination is now known in Australia as the Newbery-Vautin process. Considerable improvements have been effected in the practice during the last few years, these improvements having been largely due to the attention which was called to the subject by the efforts of Mr. Claude Vautin.

* For description *vide* O'Driscoll on *Gold Ores*, where is to be found a curious history of the patent literature of chlorination. This work must be read with caution, as it appears to attach too much importance to one of the later patent processes.

The Mears Process.—In this process the roasted ore is charged into cylindrical barrels, together with enough water to make an easily flowing pulp; chlorine is then forced in under pressure, and the barrel, which must be air-tight, is revolved until the gold has been dissolved. The barrel is then opened and discharged by gravity into a leaching vat below, where the soluble gold is washed out and precipitated by any of the known methods. Dr. Mears became aware by an accident of the increase caused by the use of compressed chlorine in the rapidity with which gold is dissolved by the gas. He was experimenting on the action of chlorine on roasted pyrites, when the discharge pipe of his apparatus becoming stopped up, the gas accumulated in the vessel until it was burst by the pressure. He then found that the ore was perfectly chlorinated, although it had been subjected to the action of the gas for a few minutes only.

The Mears process as practised on a large scale may be described as follows:—Cylindrical iron barrels lined with sheet lead and mounted on hollow trunnions are used, most of the general arrangements being similar to those shown in Fig. 52, p. 265, in which the Thies barrel is depicted. An aperture of from 6 inches to 1 foot in diameter, situated in the middle of the length of the barrel, serves for charging-in and discharging the ore; the aperture is covered by an iron plate which is kept tightly screwed down while the chlorination is in progress. The ore and water are introduced through this aperture, and the plate then screwed on. The chlorine is generated in vessels similar to those used in the vat process, and stored in a large gas-holder made of lead. From this gas-holder the chlorine is pumped direct into the barrel through a leaden pipe passing through the hollow trunnion, and continued into the barrel as the "gooseneck," which is a pipe passing vertically upwards, close to the end of the barrel, starting from the centre, and terminating in a hook-shaped curve near the top. The gooseneck remains stationary and vertical when the barrel is revolved; it is made of iron so as to be strong enough to withstand the weight of ore which presses against it on revolving the barrel, and is lined inside and outside with lead to protect the iron from the action of the chlorine. The object of the gooseneck is to prevent the ore and water from flowing into and clogging the pipe intended for the introduction of the chlorine.

When it has been charged, the barrel is exhausted of air by a steam-jet exhaust pump, and chlorine is then pumped in until a pressure of 40 or 50 pounds to the square inch is attained. Another method of introducing the chlorine is to fill a large receiver with the gas at high pressure, and then to connect it with the pipe leading to the gooseneck, when the gas rushes into the barrel. The effect of the high pressure in the barrel, however it has been obtained, is to make a strong solution of chlorine in contact with the ore. The previous exhaustion of the air is of little importance, as the quantity present in the barrel is small, at least three-quarters of the space inside the barrel being usually occupied by the charge of ore and water, while it is doubtful if such air as is present does any harm. Messrs. Newbery and Vautin, indeed, subsequently claimed that the presence of a large quantity of air is actually beneficial.

The barrel is then revolved at the rate of six to ten revolutions per minute by a belt and pulley (the latter being fixed on a continuation of the trunnion), or by means of cog-wheels, or best of all by a friction clutch. The ore is kept constantly stirred and tumbled about by the revolution of the barrel, the advantages gained from this course being as follows:—

1. Every particle of ore is exposed equally to the action of the chlorine.

2. Native gold is usually alloyed with more or less silver. If this metal is present in large excess, the dissolution of the gold is stopped after a certain point has been reached, by the formation over it of an insoluble coating, usually supposed to consist of chloride of silver. Such alloys cannot be dissolved by aqua regia unless this silver chloride is removed at intervals by friction or by solution in ammonia. It is doubtful whether chloride of silver is formed to any extent in the cold by free chlorine, although no doubt the nascent chlorine produced in aqua regia is more potent, but, whether the coating formed consists of silver or of chloride of silver, it must in either case be in a powdery condition, and so is readily removed by the mechanical attrition of the particles of ore against one another, caused by the rotation of the barrel. A clean surface of gold is thus continually offered to the action of the chlorine. Coarse particles of gold may also be covered and protected by a coating of undissolved chloride of gold in the vat process, where so little water is present, but, in the barrel, the larger amount of water present instantly dissolves all such soluble salts.

The pressure of gas inside the barrel is tested from time to time by opening a valve to which a pressure-gauge is connected. A pop-valve may also be used for the same purpose. When chlorination is believed to be complete, the excess of gas is drawn off by the exhaust apparatus, and stored up or discharged outside the building. The barrel is filled with water, and the contents then discharged into a filtering vat, where the solution is separated from the ore and precipitated as usual.

This description applies in great part to the other barrel chlorination processes which differ little. The process was formerly in use at the Phœnix and Haile Mines, in Carolina, and at Bunker Hill Mine, California. The chief disadvantages in the process were the rapidity with which the gooseneck wore out, and the great strength and corresponding cost of the barrels rendered necessary by the high pressure used. It was also found to be difficult to keep the stuffing-boxes in the hollow trunnions from leaking, and the cost of repairs was excessive. Modifications were made by Mr. Adolph Thies, at Bunker Hill, with a view to remove these objections to the process, and his modifications have been everywhere adopted. The amended method of procedure may be called by the name of its inventor.

The Adolph Thies Process.—Mr. Adolph Thies was in

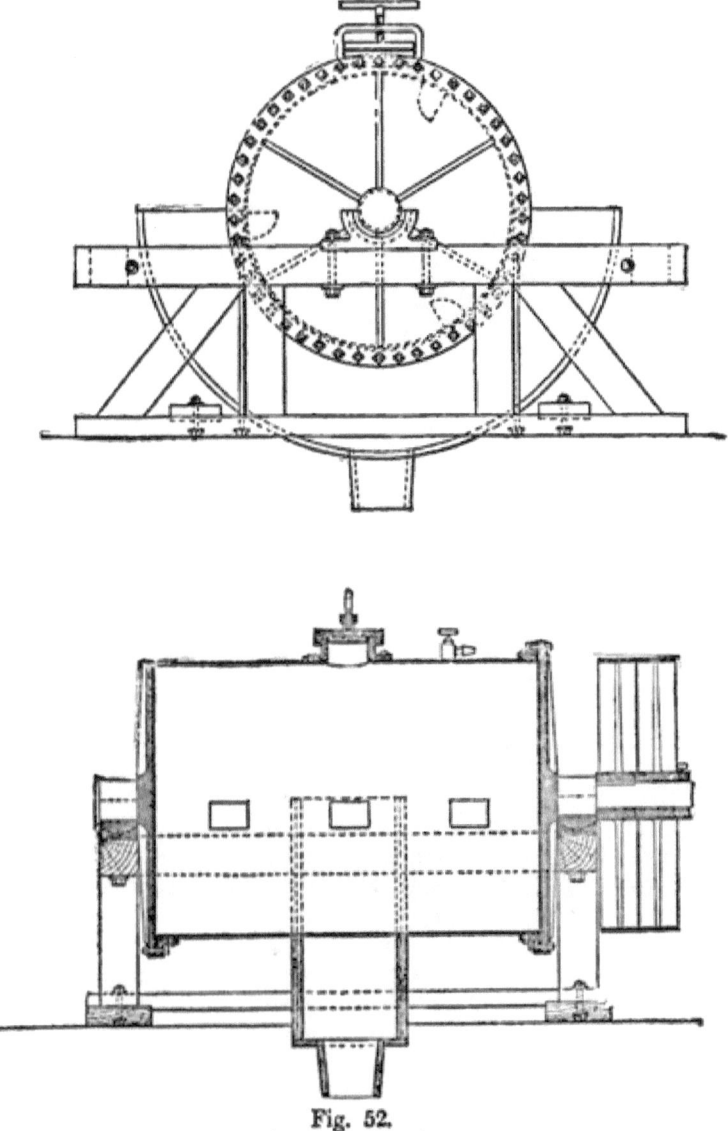

Fig. 52.
Scale, 1 in. = 2 ft. 6 ins.

charge of the Mears process at the Bunker Hill Mine for nearly four years before he made the improvements, in the year 1881, which have been associated with his name. He found that the hollow trunnions, the gooseneck, and the pressure pumps could all be dispensed with, and the chlorine gas generated inside the barrel itself to any required extent by the use of bleaching powder and sulphuric acid. This method had been mentioned by Mears in one of his earlier patents, but had been abandoned in favour of pumping the gas into the barrel. Thies proved it to be cheaper and better, as all joints liable to leakage are

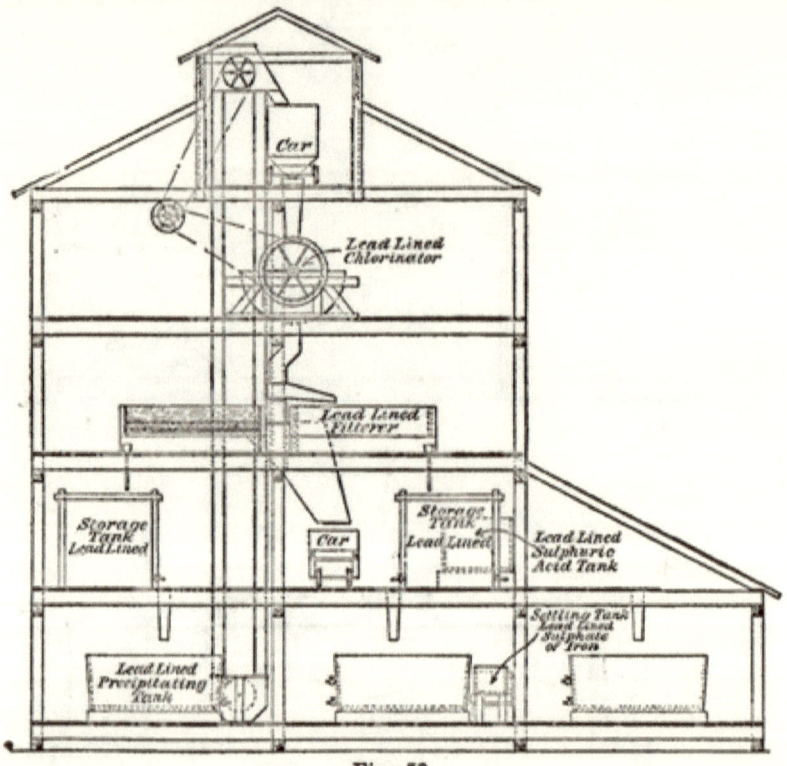

Fig. 53.

dispensed with, and the machinery is simplified and rendered much cheaper. Although there is no doubt that chlorine under very high pressure acts more efficiently and rapidly than at the ordinary pressure, still Thies showed that very good results could be obtained at a reduced cost by using a moderate pressure of chlorine of only a few pounds per square inch. The barrel used is similar to that already described as suitable to the Mears process, except that the trunnions are solid. It is shown in Fig. 52 in transverse and longitudinal section, while a complete Thies plant is shown in Figs. 53 and 53a.

CHLORINATION: THE BARREL PROCESS. 267

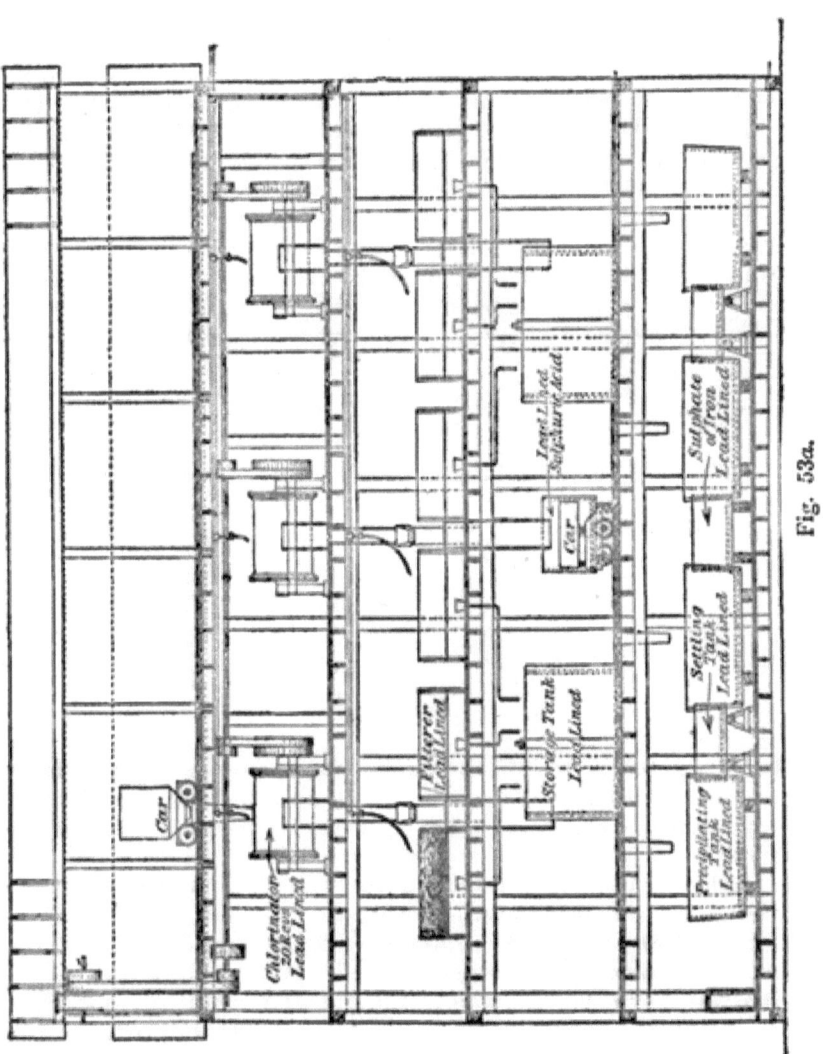

Fig. 53a.

An enlargement of part of the end of the most modern type of chlorination barrel is shown in A, Fig. 54; in B, the construction of an older form of barrel, which is stated to have been in use as late as the year 1891 at the Golden Reward Mill, is shown for purposes of comparison. The shaded parts of A and B are made of iron (the barrel-head and the flange being cast-iron, and the cylinder, boiler-iron); the lead lining is not shaded. In A the lead lining is burnt-on to the flanged cylinder and to the head respectively, and the head then firmly bolted to the flange. The lead-joint C is made tight by the blows of a hammer on a blunt chisel directed on the surface of the lead, in the direction shown by the arrow. When repairs are needed inside the barrel, the cast-iron head is removed and the work thus easily done. This barrel never suffers by leakage, and the cost for repairs is usually nominal. In the other barrel the lead lining of the end is burnt-on to the lining of the cylinder at D, leaving a hollow space, E. In consequence of the existence

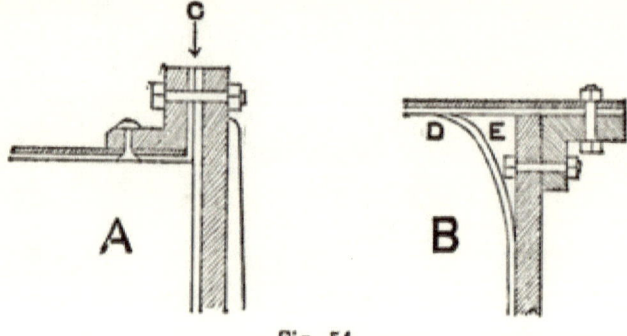

Fig. 54.

of this space, the lead-joint is often broken when the pressure in the barrel is great, and leakage then occurs; moreover, repairs must be effected by a workman crouching inside the barrel.

In charging-in, the water is added first, being run in by a hose through the manhole, until it reaches a certain mark on the inside of the barrel. The amount of water to be used varies with the nature of the ore—roasted pyrites absorbing much more water than siliceous ores; the quantity required is usually from 40 to 60 per cent. of the weight of the ore—*i.e.*, from 80 to 120 gallons of water per 2,000 lbs. of ore. This is enough to make an easily flowing pulp. It is much more than that employed in the Plattner process, where it is necessary to keep the mass porous, so as to enable the gas to pass through it. In the barrel, however, this reason for limiting the amount of water does not exist, and the ore is really treated by a solution of chlorine water. If too small a quantity of water is used, so that the pulp is not quite free-flowing, lumps are formed which are not broken by the revolution of the barrel, and these lumps are not perfectly chlorinated.

The ore is let fall into the barrel down a shoot from an overhead hopper, which may conveniently be made to contain the exact quantity of ore required for a barrel charge. The ore should be perfectly dry (not cooled after roasting by too much "wetting down"), as otherwise it sticks in the hopper, instead of sliding freely down the shoot. The latter may be conveniently made of canvas, so that it can be looped-up out of the way when not in use. The chemicals (bleaching powder and sulphuric acid) may be added in one of two ways. Either the lime is thrown into the water before the ore is added, and the acid subsequently poured upon the upper dry surface of the latter, just before the manhole is closed; or the acid is poured into the water, through which it sinks to the bottom without mixing, the ore let fall next, and the lime added on the top. This latter system is perhaps the better, as, if it is used, the generation of chlorine is never begun before the barrel commences to revolve, provided ordinary care is taken.

The amounts of bleaching powder and sulphuric acid to be added depend on the nature of the ore, and the care with which it has been roasted. If there is nothing in the ore which is open to attack except gold, the amount of chlorine actually absorbed in the course of the chemical reactions is extremely small, 1 oz. of gold requiring only 0·54 oz. of chlorine to convert it into the soluble trichloride.

The bleaching powder used should be of the finest quality obtainable, to prevent the introduction of an unnecessarily large amount of sulphate of lime into the charge. Bleaching powder usually has assigned to it the formula $Ca(OCl)Cl$, or $Ca(OCl)_2 + CaCl_2$, and may be, as the latter formula would indicate, a mixture of hypochlorite and chloride of lime. The reaction with acid is usually expressed thus—

$$CaCl_2 + Ca(OCl)_2 + 2H_2SO_4 = 2CaSO_4 + 2H_2O + 2Cl_2$$

This equation certainly does not accurately represent what happens, as much less chlorine is liberated than is indicated by it. The amount of "available" chlorine (*i.e.*, that which is liberated by the action of acids) contained in commercial bleaching powder varies from 20 to 35 per cent. Bleaching powder is gradually decomposed by the carbonic anhydride in atmospheric air (chlorine being liberated), and must consequently be kept separated from it as completely as possible. Even if preserved in hermetically sealed vessels, however, it suffers a slow change, by which some chlorate of lime is formed and the amount of available chlorine reduced. Under these circumstances, it is necessary to re-determine the value of the bleaching powder at short intervals of a few days. The shortest and best method of effecting this is to grind a sample in a stoneware mortar under water, and add the emulsion to an excess of a solution of potassic iodide. The whole of the available chlorine instantly displaces an equivalent quantity of iodine, which is set free and may be readily estimated by a standard solution of hyposulphite of soda in the usual manner. From this the amount of available chlorine in the sample is calculated. The whole operation can be performed in from ten to fifteen minutes when the standard solution has been prepared.

In calculating the quantity of lime to be added to a barrel charge, there are several considerations to be taken into account. A saturated solution of chlorine acts more rapidly than non-saturated solutions. At the ordinary temperature and pressure, water dissolves about $2\frac{1}{3}$ volumes of the gas, and it is desirable that the amount added to the barrel should be enough for this saturated solution to be formed. Now, suppose the barrel to have a capacity of 40 cubic feet, and that the charge consists of say 125 gallons of water and 2,500 pounds of ore. The volume of water added is 20 cubic feet, and if the particles of ore have a mean density of 3·0, which is approximately true in the case of many samples of roasted pyritic ore, the volume of ore and water will be, together, $33\frac{2}{3}$ cubic feet, leaving $6\frac{1}{3}$ cubic feet of air space inside the barrel. There must be, therefore, $46\frac{2}{3}$ cubic feet of chlorine dissolved in the water to make a saturated solution, and $6\frac{1}{3}$ cubic feet of chlorine in the air space, in order to keep that amount in solution. The weight of 53 cubic feet of chlorine is about 10·4 pounds, which could be supplied by the decomposition of about 30 pounds of bleaching powder or 24 pounds per short ton of ore. This quantity may be taken as one well adapted for the efficient chlorination of ordinary ores, the chlorine acting rapidly at that pressure. Nevertheless, much smaller quantities are in use with complete success on certain ores, and, as has been already stated, no general rule, applicable to all cases, can be laid down. When much oxide of copper or other metallic oxides capable of absorbing chlorine are present, or if the ore is insufficiently roasted, the pressure rapidly falls off, and a further addition of chemicals may become necessary, as at the Phœnix Mine. The total pressure exercised by the air and chlorine, if the solution is saturated by the last-named gas, is equal to two atmospheres—*i.e.*, one atmosphere in excess of the normal—and this may be taken as the greatest pressure which it is advisable to maintain inside a barrel. If greater pressures are used, either very expensive barrels are required, or else the valves, manhole, &c., soon begin to leak.

It must not be forgotten that, at higher temperatures, water dissolves less chlorine than the amount mentioned in the last paragraph, and, consequently, a pressure of chlorine equal to that of the atmosphere will be obtained with the use of smaller quantities of chemicals. Thus, at 40° C., with the other conditions identical with those given above, the amount of chlorine required to give the same pressure will be only 6·5 pounds instead of 10·4 pounds, and, as the temperature rises above this, the quantity of chlorine necessary falls off rapidly.*

The relative amounts of bleaching powder and acid used is varied with different ores. Theoretically, according to the equation given above, 7 parts of chloride of lime require 6 parts of sulphuric acid for complete decomposition, but practically a little more sulphuric acid is always added, because it is desirable to maintain an excess of the acid in the charge. This is done in order to prevent lead and lime from getting into solution as chlorides, which would entail a loss of chlorine, and also to assist in dissolving any oxide of copper that may be present, which otherwise would also absorb chlorine. The proportions added are usually 6 parts of bleaching powder to 7 or 8 parts of sulphuric acid of 66° B.

The time occupied in chlorinating usually varies from three to six hours. The continued presence of an excess of chlorine gas should be tested from time to time by opening a small valve momentarily. If the presence of the free gas cannot be detected, the barrel must be opened and further supplies of chemicals added. The amount of pressure in the barrel is not alone sufficient to prove the presence or absence of an excess of chlorine, as other gases may be generated and exert considerable pressure. When the chlorination is finished, the excess of chlorine is discharged by a hose-pipe outside the building, the barrel is filled up with water, again revolved, and the liquid decanted on to large, shallow filter-beds. The barrel is again filled up, revolved, and decanted as before, and, finally, the whole charge is emptied out and another wash-water given to the charge on the filter. In comparing this method of washing by decantation with that of direct filtration, usually adopted, Mr. Thies found that, with similar charges, the amount of water used in the latter case was nearly double, and the time occupied much longer, while the tailings contained $5\frac{1}{2}$ dwts. of gold per ton against about 1 dwt. when washed by decantation. He also found that there is little difficulty in filtering through a bed of fine ore from 3 to $4\frac{1}{2}$ inches thick, but if the thickness of the bed is greater, direct leaching becomes very tedious and ineffective, and decantation is much better.

The leaching vats are usually constructed of wood, which is either lined with lead or coated with tar and pitch. At Bunker

---

* For the amount of chlorine dissolved by water at different temperatures, vide Schönfeld, *Ann. Chem. Pharm.*, vol. xciii., p. 25; vol. xcvi., p. 8. The results are quoted in Roscoe & Schorlemmer's *Treatise on Chemistry*, vol. i. p. 123, 1881.

Hill the filter tanks are rectangular, and measure 6 feet by 18 feet, and 18 inches deep. They are lined with lead, and incline towards the drain hole, where the bottom is one inch lower than it is at the other side of the tank. The filter-bed, as is usual in California, is made of quartz-pebbles, gravel, and fine sand. In other mills the leaching vats are usually round.

The Thies barrel process has been greatly altered and improved during the last five years. The modern barrel chlorination process, as practised in Dakota, differs from it in several essential particulars; it is described in Chapter xiv., pp. 291-304.

Mechanical Difficulties of Leaching.—The difficulties of leaching vary enormously with the character of the ore and the treatment to which it has been subjected. Concentrates are among the best leaching ores, as, even if they were originally in a state of extremely fine division, the oxidising roasting in many cases appears to cause an agglomeration of the particles into porous granules, which do not pack down readily, and do not resist the passage of liquids. A small quantity of red oxide of iron in a very fine state of division is often present in roasted concentrates, and this is carried away by the water, passing into and partly through the filter-bed and appearing in the precipitating vat. This material, on settling to the bottom of the gold solution, often appears to carry down with it a large proportion of the gold, forming a layer of slimes which are extremely rich, and very difficult to treat except by smelting. In some cases, these iron oxide slimes are present in roasted pyrites in such quantities that leaching is greatly interfered with, and made very tedious. Siliceous ores, if properly pulverised, usually present no difficulty in leaching, but aluminous ores are exceedingly troublesome. At Mount Morgan, Queensland, the ore consists chiefly of hydrated oxides of iron, which offer the greatest possible resistance to leaching if treated raw, but, if roasted, the loss of their water of hydration is found to be accompanied by a remarkable agglomeration of the particles of ore, and ordinary gravitation leaching is thus rendered possible. The roasting is in this case performed merely for the purpose of facilitating the leaching, as it is stated that in the raw ore there are no constituents, except gold, which are readily acted on by chlorine.

In order to quicken the process of leaching various appliances have been suggested. Different forms of vacuum pumps have been used, the amount of air in the space below the filter-bed being reduced by them, and the liquid thus forced through by atmospheric pressure. Such methods have not apparently been attended with any conspicuous degree of success, as the increased packing of the ore tends to neutralise the effects of the pressure on the liquid. In 1889, at the Colorado Gold and Silver Extraction Company's Mill at Denver, the effect of the use of increased pressure of air applied directly on the surface of the liquid was tried. A special cast-iron vat was constructed capable of sustaining an internal pressure of 100 pounds per square inch, and furnished with valves, so that air and water could be simultaneously pumped into it. It was found that ores, which entirely prevented the passage of water through them, even after a vacuum of 20 inches of mercury had been established beneath the filter-bed, could be leached with great speed under a pressure of from 30 to 50 pounds per square inch. Moreover, when the leaching was complete, the ore could be freed from the water more completely by the passage of a current of air through it than by gravitation alone. This method, which was suggested by Mr. Dennes, the Company's engineer, has since been adopted at the Rapid City chlorination mill, and also, in a modified form, at the Golden Reward Mill, Dakota. The chief objection to it seems to lie in the great additional expense incurred in the construction of the leaching vats and in working the pumps.

Mr. Riotte has suggested * that the wash-water should be thoroughly mixed with the ore by agitation, and then removed as completely as possible by squeezing in a filter-press. To effect this, the ore, together with the necessary amount of water, is passed successively through two revolving barrels, entering and leaving them by means of hollow trunnions. The mixing is accomplished inside the barrels by means of projecting internal ribs, and the charge passes continuously through, and is received into large filter-presses. These are set to work as soon as they are full, and squeeze out all the liquid, retaining the tailings, the pressure used being from 25 to 30 pounds per square inch. Mr. Riotte finds that the average amount of moisture retained in the ore after being squeezed is about 6 per cent. As this would cause the retention in the tailings of from 3 to 6 per cent. of the soluble gold, according to the amount of wash-water used, the method is obviously inapplicable to rich ores, which would require to be subjected to treatment twice. Centrifugal leaching has also been proposed, and it is stated that the Mount Morgan ore can be leached in this way without previous roasting.

THE PRECIPITATION OF GOLD.

In adopting a system of precipitation suitable for application on a large scale, there are several points to be taken into consideration. (1) The precipitant should be capable of decomposing gold chloride rapidly and completely, even when the latter is present in extremely dilute solutions. (2) All other substances likely to be present in the solution should be left unprecipitated. (3) The excess of free chlorine in the solution should be removed by the precipitant by being converted into hydrochloric acid or into chlorides, so as to avoid running any risk of the precipitated

* *Eng. and Mng. Journ.*, March 31, 1888.

gold being partially redissolved. (4) The precipitate of gold must also be in such a form that it is easily separable from the liquid. It is possible to obtain it in such a finely divided form that it will not settle in water at all, and will pass through the closest filter.

The precipitants which have been proposed or used may be conveniently divided into two classes.

1. Soluble precipitants and gases, by which solid particles of gold are formed, suspended in the liquid. These are either allowed to subside and are separated by decantation, or they are removed by filtration.

2. Insoluble solid precipitants, on which the gold forms as a deposit, and from which it has to be separated by subsequent operations.

1. SOLUBLE PRECIPITANTS.

Ferrous Sulphate.—This is the best known precipitant, having been used by Plattner, see p. 217, and being still employed more frequently than any other. Some account of its use has already been given in the section on precipitation in the description of the vat process, p. 259. It is made by dissolving iron in sulphuric acid, with the aid of heat, and the crude solution thus prepared, which is in most cases added direct to the gold solution, always contains some free sulphuric acid. Precipitation takes place according to the equation—

$$2AuCl_3 + 6FeSO_4 = Au_2 + Fe_2Cl_6 + 2Fe_2(SO_4)_3$$

From this it might be inferred that one part by weight of iron, dissolved in sulphuric acid, would precipitate $1\frac{1}{6}$ parts of gold, but the oxidation of the ferrous salt is effected in other ways, notably by the excess of free chlorine present in the solution, so that much more sulphate of iron is required than is indicated by the equation. The difficulty of collecting and saving the precipitated gold has already been dwelt on. The gold settles better if it is well stirred, and Aaron recommends an addition of more sulphuric acid and vigorous stirring, two hours after the precipitation is complete, as a means of assisting the settling. It was proposed by Mr. Vautin to collect the precipitated gold by the use of centrifugal force. For this purpose the liquid, containing finely-divided gold, was placed in a circular vessel, capable of being rotated at great speed. On rotating the cylinder, the liquid in it was also gradually put in motion, and the finely-divided gold then moved outwards by centrifugal force, and was pressed against and adhered to the inner wall of the vessel, while the clear liquid could then be siphoned off. Although stated to be successful on a small scale, this device has not yet been adopted in practice. Besides gold, the only other metals precipitated by ferrous sulphate are those which form insoluble sulphates—viz., lead, calcium, strontium and barium. The last two of these are rarely present, and the others are dealt with in the manner already described above. Basic iron salts are not precipitated if enough free sulphuric acid is present, and, when precipitated, they may be removed from the gold by treatment with acids, or by slagging them off in the furnace.

Other proto-salts of iron are equally efficacious, but are never used in practice.

Organic Substances.—Many organic substances, such as oxalic acid, formic acid, ether, &c., also decompose chloride of gold, but are unsuitable as precipitants in practice, owing to their slowness of action and high cost, or to the extremely fine state of division in which the gold is thrown down, so that it is made difficult or impossible to collect. Nevertheless, it is probable that some of the more easily oxidisable organic compounds may at some future time be found to be suitable for the precipitation of gold on a large scale.

Sulphuretted Hydrogen.—This was formerly made at Deloro by heating paraffin and sulphur together, and the gas diluted with air was forced through the solution by means of a small air-pump. The use of the air was to keep the solution agitated, and to expel part of the chlorine mechanically, and so economise the sulphuretted hydrogen, which is decomposed by chlorine. The equation for this decomposition is given by Mr. W. Langguth [*] as follows:—

$$H_2S + 4H_2O + 8Cl = H_2SO_4 + 8HCl$$

but it is more likely that the greater part of the change is represented by the well-known equation—

$$H_2S + Cl_2 = 2HCl + S$$

although a small amount of sulphuric acid may be formed at the same time. The precipitate of sulphur thus formed before and during the precipitation of the gold, which begins before the whole of the chlorine has been destroyed, is to be avoided, and Mr. Langguth has, therefore, suggested the use of sulphur dioxide, generated by burning sulphur or pyrites, or by heating sulphuric acid with charcoal, to destroy the free chlorine, the reaction being as follows—

$$Cl_2 + SO_2 + 2H_2O = H_2SO_4 + 2HCl$$

When almost all the chlorine has been thus converted into hydrochloric acid, the passage of SO_2 is stopped, and sulphuretted hydrogen, now generated by the action of sulphuric acid on iron matte, is forced into the solution, destroying the last traces of chlorine and precipitating the gold. This system was introduced at the Golden Reward Chlorination Works in 1891, and has been subsequently adopted at other mills in Dakota with conspicuous success.

[*] *Eng. and Mng. Journ.*, Feb. 14, 1891.

The gold is precipitated as a sulphide mixed with more or less sulphur, the reaction being represented by the following equation—

$$2AuCl_3 + 3H_2S = Au_2S_3 + 6HCl$$

It is said to take less than an hour at the Golden Reward Works to precipitate the gold from 5,000 gallons of solution (resulting from the lixiviation of from 25 to 50 tons of ore). The liquid is quite cold, but the precipitate is in a collected, voluminous and flocculent form, that settles quickly. It is left undisturbed for two hours, and the liquid is then drawn off to within 4 inches of the bottom of the vat, and passed through a Johnson filter-press, provided with a set of heavy, canton-flannel filter-cloths. The head of liquid used for filtering is 25 feet, and the filtration is said to occupy from three to four hours, according to the amount of sulphides already contained in the filter. When the latter is full, a small air-pump is connected with it and a current of air passed through it for an hour to dry the mass of sulphides into hard cakes, which are easily handled and removed. The precipitate is then roasted in a muffle furnace, the filter-cloths being burnt with it. It is then melted down with a little borax and nitre, the total loss in handling being very small. The bullion is about 900 to 950 fine in gold, the remainder consisting chiefly of silver, copper, lead and arsenic. The bulk of the precipitate remains at the bottom of the vat. It is allowed to accumulate for a fortnight, and then treated, together with the material from the filter presses.

Of course, all the lead, copper, and silver contained in the liquids are precipitated with the gold. If much copper is present, the bullion may be very base, and Langguth suggests the removal of the copper from the precipitate by dilute nitric acid. It might also be removed with less cost from the roasted precipitate by dilute sulphuric acid; but, when much copper is present, it would probably be better to use some other agent for precipitation than sulphuretted hydrogen.

Sulphurous Acid, which has been already mentioned as a cheap agent for destroying free chlorine, precipitates gold very completely from dilute solutions, but its action in the cold is slow until the liquid is almost saturated with the gas (water absorbs 39 volumes of the gas at 20°), and the gold settles slowly. At higher temperatures the action of the gas is much more rapid and satisfactory, and very little is wasted in saturating the liquid. The experiment was made at the Portland Mining Company's Mill, Deadwood, Dakota, of using sulphurous acid gas instead of sulphuretted hydrogen, but it was found that the precipitate obtained was so finely divided that it was not retained in a Johnson filter-press, much of it passing through all the filter-cloths which were tried, and being lost. As a precipitant, the gas does not appear to present any advantages over ferrous sulphate.

2. SOLID PRECIPITANTS.

Charcoal.—The first experiments on the reducing action of charcoal on chloride of gold in solution were apparently made by Percy. In his laboratory, some sticks of wood-charcoal were immersed in water, and 32·50 grains of gold in the form of chloride were added on August 7, 1869. 85 more grains of gold, in the form of chloride, were added on November 3, 1869, and the bottle was left to stand. This bottle is in the Percy collection, at South Kensington, and the surface of the charcoal is now coated over with metallic gold. In the same collection is a model in gold of the surface of the end of a stick of charcoal left immersed in a strong solution of chloride of gold. The fibres and vessels of the stem are all shown on the surface of the metal.

Vegetable charcoal was first employed as a precipitant for solutions of gold chloride on the large scale by Mr. W. M. Davis, in a chlorination mill in Carolina, in the year 1880. Its use was discontinued, however, and nothing further was heard of it until it was adopted by Messrs. Newbery & Vautin, at the Mount Morgan Mine, Queensland, in 1887. The method adopted there is as follows :—The solution is heated to boiling, the free chlorine being thus expelled ; the liquid is then made to run slowly through large shallow tanks filled with pieces of charcoal, varying from the size of a small hazel nut to less than that of a pea. The tanks are about a foot deep, and the overflow from one is allowed to run through others, until the solution is found, on testing, to be free from gold, which is deposited on the surface of the charcoal. When the latter is coated sufficiently with the precious metal, it is burnt in small furnaces, furnished with large dust chambers or apparatus for condensing the fumes, and the ashes melted with borax, the gold being thus obtained in a remarkable state of purity. Animal charcoal cannot be used owing to the difficulty of burning it afterwards.

The exact action of the charcoal has not been fully demonstrated. It acts slowly on cold solutions, and its action is not rapid even at boiling point. It is under the disadvantage that it does not destroy free chlorine, which must therefore be expelled by boiling or by passing a current of air through the liquid before the precipitation of the gold is begun. Mr. Davis states that 240 parts of charcoal are required for the precipitation of $19\frac{1}{4}$ parts of gold. The prevailing opinion is that the hydrogen and hydrocarbons remaining in the charcoal are the active agents in the precipitation, hydrochloric acid and free gold being formed. Charcoal is not now used in Carolina, the only place where it is still employed being the Mount Morgan Mine.

Insoluble Sulphides.—The use of sulphide of copper as a precipitant for gold chloride was covered by a British patent, taken out many years ago, but the discovery was not applied in practice, and had been forgotten when Mr. C. H. Aaron described his method of application.* Experiments were subsequently conducted at the various mills engaged in working the Newbery-Vautin process, with the object of finding if the sulphides of copper or of other metals could be used in practice on a large scale. It was found in Denver that the best method of preparing the precipitated sulphide of copper is to add a boiling saturated solution of sulphate of copper to a boiling saturated solution of a mixture of the polysulphides of sodium, prepared by adding sulphur to boiling caustic-soda solution, and to stir the mixture vigorously. The sulphide of copper is precipitated in a granular form, which settles quickly in water, and allows liquids to filter through it readily, but offers a large surface, from the porosity of the granules. It is moreover easily decomposed, and so is especially active in precipitating gold. The reaction that occurs may be expressed thus:

$$3CuS + 2AuCl_3 = 3CuCl_2 + Au_2S_3$$

Some metallic gold may possibly be formed at the same time, according to the equation:

$$3CuS + 8AuCl_3 + 12H_2O = 8Au + 24HCl + 3CuSO_4$$

The cupric chloride is removed in solution and the sulphide of gold precipitated on the surface of the granules of copper sulphide. The precipitant may conveniently be contained in two or three small vessels, through which the gold-solution passes successively, either flowing in at the top and out at the bottom, or in the reverse direction. The sulphide of copper and the precipitated gold are partly carried off in suspension in the liquid, which must, therefore, be passed through a filter of flannel or other material. In order to quicken both the precipitation and the rate of filtration through the filter-cloths, the liquid may be heated, while a head of 10 or 15 feet makes filtration very rapid. The free chlorine need not be expelled from the liquid, as it is destroyed at once by the sulphide.

Mr. Blomfield found that the subsulphide of copper, Cu_2S, prepared by fusing together sulphur and copper, is more easily handled than the precipitated sulphide, besides being cheaper.† The Cu_2S is prepared for use by crushing it and retaining only that part which remains on a sieve having 100 holes to the linear inch, after having passed through a 60-mesh sieve. The precipitation vessels which he recommends are stout glass

---

* *Eighth Report Cal. State Min.*, 1888, p. 839.
† *Eng. and Mng. Journ.*, Jan. 24, 1891.

cylinders, 6 inches in diameter by 12 inches long, closed at each end by glass-lined cast-iron caps held together by two bolts. Each vessel has a false bottom to support the filter-cloth, and is charged with 4 to 8 inches of the precipitant. Mr. Blomfield recommends downward filtration of cold solutions, thus saving the expense of heating them. It is stated that this precipitant is now being used on the large scale at a chlorination mill in South America.

Whether precipitated or fused sulphide of copper is used, it is advisable to defer the clean-up as long as possible, until the first filter has had the greater part of its copper replaced by gold. It is then cleared out, and the contents of the second filter transferred to the first, and so on, fresh sulphide being added to the lowest filter. The mixture of gold and copper sulphides from the first filter is carefully dried, mixed with borax, and melted in plumbago crucibles. If the bullion produced is too base, some of the copper may be removed by roasting the sulphides and treating them with dilute sulphuric acid before fusion.

It has been proved by laboratory experiments* that both fused and precipitated sulphide of iron are more rapid in their action than the corresponding copper compounds. The only apparent disadvantage in their use lies in the fact, that some of the copper contained in the solution is precipitated with the gold. When copper is not present in the ore this drawback would not be felt.

Metals.—In recent experiments† the author has found that, in the laboratory, iron turnings constitute the quickest and most trustworthy precipitant for gold chloride, its superiority over the insoluble sulphides being more marked, however, at 60° or 80° C. than at ordinary temperatures. The separation of the gold from the iron may be readily effected by methods similar to those used in the separation from zinc, which has to be done in the MacArthur-Forrest cyanide process, as described below, p. 318.

The order of the rate of precipitation of gold from moderately dilute solutions of its chloride (one part in 20,000) appears to be: metallic iron (quickest), sulphide of iron, sulphide of copper, charcoal, ferrous sulphate, metallic copper (slowest). This order is that noted near the boiling point of water. The relative rate of precipitation by sulphurous acid and sulphuretted hydrogen was difficult to compare closely with that of the above reagents, but they appeared to act more slowly than iron and the sulphides. It does not necessarily follow from the results of these experiments that metallic iron is the best precipitant on a large scale, and it has not been tried in practice as yet. Possibly the order given above would not be preserved unchanged at ordinary temperatures.

* *Mining Journal*, March 11, 1893.
† *Loc. cit.*

MODERN PATENT PROCESSES OF CHLORINATION.

The Newbery-Vautin Process.—In this process an effort is made to combine the advantages of the Mears and Thies processes. Since a high pressure of chlorine increases the rapidity with which gold is dissolved, it is desirable to use pressure, but Thies had shown that the economy effected by reducing the quantity of chlorine employed outweighed the advantages gained by the high pressure. Messrs. Cosmo Newbery and C. Vautin proposed to obtain the desired pressure by pumping air into the barrel, keeping the amount of chlorine as low as possible. It was stated by them that the chlorine would be kept in a liquid state or, at any rate, that it would be entirely dissolved in the water if the air in the barrel were at a pressure of 60 to 80 pounds to the square inch. This is contradicted by theoretical considerations, and no attempt appears to have been made to prove it in practice. The only effect of the increase of air pressure would be an increase in the amount of air, but not in that of chlorine, dissolved in the water. The latter can only be affected by variation of chlorine-pressure. According to experiments made by the author at Denver, using the Newbery-Vautin barrels, air-pressure does not seem to exercise any influence on the rate of solution of the gold in ores, or upon the percentage of extraction. The other proposal made by Messrs. Newbery & Vautin was to leach by means of a single-acting vacuum pump. Vacuum leaching has already been discussed on p. 272. The particular form of pump proposed by Messrs. Newbery & Vautin does not seem to possess special advantages. Neither of these methods appear likely to pass into permanent use in gold chlorination.

It has often been asserted that the Newbery-Vautin process is identical with that formerly in use, with conspicuous success, at the Mount Morgan Mine, Queensland. Nevertheless, as far as the author is aware, neither air-pressure in the barrel, nor vacuum-leaching by means of a single-acting pump ever formed part of the method of procedure at Mount Morgan. Mills designed to use the Newbery-Vautin patents were erected about the year 1888 in the Transvaal, New Zealand, Colorado, and Hungary, but it appears that none of these mills are now at work.

The Pollok Hydraulic-pressure Process.—In this process also, there seems to have been a confusion of thought in the mind of the inventor, who does not clearly distinguish between the pressure of chlorine gas, which increases its chemical activity, and air- or water-pressure, neither of which has any effect on the condition of the chlorine or the rate of solution of gold. Mr. Pollok ensures the solution in water of the whole of the chlorine present in the barrel by filling the latter completely with water.

When the barrel is closed and all the air has been suffered to escape, more water is pumped in, until the pressure inside the barrel rises to at least 100 pounds per square inch. The chlorine is generated inside the barrel by the action of bisulphate of soda on chloride of lime, but it is doubtful if there is any economy on any gold-field in the use of bisulphate of soda in place of sulphuric acid. Pollok claims that by this high pressure the solution of chlorine is forced rapidly into the pores of the ore. The process in other respects presents no novel features, and it does not seem likely to come into general use.

The Swedish Chlorination Process.—This method was devised by Mr. Munktell, who seems to have worked it out without having visited either the vat or the barrel chlorination works already established in other parts of the world. The process enjoys the distinction, according to the published accounts, of having been worked at a profit. It was first used at the Fahlun Copper Works, Sweden, and has since been introduced into Hungary, where it is in successful operation at Brade and at Boitzas. The process as worked at Fahlun may be briefly described as follows:—The ore is roasted at a low temperature with the view of obtaining the copper in the form of sulphates. If silver is present, salt is added in the roasting furnace. The calcined ore is then placed in false-bottomed, wooden vats and leached with hot water, followed, if necessary, by hot, dilute sulphuric or hydrochloric acids, by which all the copper and much of the iron are removed from the ore. The solutions thus obtained are run through tanks containing scrap-iron, by which the copper and any silver in solution are precipitated. The solution of ferrous sulphate may be saved and used subsequently to precipitate the gold. The residues in the tubs are now in a condition to yield up their gold readily, and are accordingly treated by a solution of 0·6 to 0·7 per cent. of chloride of lime (bleaching powder) in water, mixed with an equal volume of dilute hydrochloric acid of specific gravity 1·002 or 1·003, or of dilute sulphuric acid. These solutions are mixed in troughs just before they flow into the ore-vat. Chlorine is slowly generated by the action of the acid on the hypochlorite of lime, and, being partly present in the nascent state, attacks the gold vigorously in spite of the extreme dilution of the solution, while there is very little smell of the gas above the vats. The liquid is passed through the ore until gold ceases to be dissolved, after which the tailings are thrown away. The solution is heated to 160° F. by steam, and precipitated by ferrous sulphate. The collection of the gold is expedited and ensured by adding acetate of lead to the solution; by this device lead sulphate is precipitated with the gold, and, in settling to the bottom, carries the particles of precious metal with it. This process was in continuous operation at the Fahlun Copper Works from 1885 to 1888. In that period,

1,500 tons of gold ore and the tailings from 29,000 tons of copper ore were subjected to treatment. It is stated that in the year 1886, the tailings from 14,000 tons of copper were treated with the following results:—

AVERAGE AMOUNT OF GOLD IN TAILINGS.

Before treatment, . . .	41·82 grains per ton.
After ,, . . .	4·04 ,,

COST PER TON.

Chloride of lime (6 lbs.), . . .	5d.
Sulphuric acid (8·37 lbs.), . . .	1d.
Plumbic acetate and other reagents, . . .	¾d.
Fuel for steam, . . .	1¼d.
Labour, . . .	1d.
Total, . . .	9d.

The low cost assigned to labour is very remarkable. These tailings had, of course, been previously crushed and roasted before being treated as above.

In the same year 960 tons of gold ore were treated at a cost of 12s. 2¾d. per ton, the ore containing 523·62 grains of gold per ton before treatment, and 6·02 grains per ton after treatment, so that no less than 98·85 per cent. of the gold was extracted.

The following details concerning the practice at Brade, Hungary, may be added.* In addition to gold, the ores contain iron pyrites, barytes, zinc-blende, antimonial minerals and argentiferous galena, and also in some cases calcite and carbonate of magnesia. The concentrates, which are subjected to treatment by the Munktell process, contain 0·86 oz. of gold and 4 ozs. of silver per ton, and 36 to 40 per cent. of sulphur. It is obvious that such concentrates could not be treated advantageously by the modern barrel process as described on pp. 291-304.

The oxidising roasting takes twenty-eight hours, after which 5 per cent. of salt is thrown upon the ore and mixed thoroughly with it, and four hours later the charge is drawn from the furnace and allowed to cool slowly in a covered-in brick pit. The loss by volatilisation is, however, considerable, and it is proposed to build seven-story furnaces. The leaching vats are made of wood, lined with lead, and are 3 metres broad, 5 metres long, and 0·75 metre deep, and hold 10 tons, the charge being 0·5 metre deep. There is a false bottom and filter-bed of quartz as usual. The ore is leached with five different solutions in succession, viz.:—(1) Warm water, 2,500 gallons being required for the charge: Chlorides of copper and zinc and about 25 per cent. of the silver are removed in solution. (2) A solution containing 2 per cent. of hyposulphite of soda, 1,600 gallons

* *Proc. Inst. Civil Eng.*, Session 1891-92.

being used; this dissolves the rest of the silver. (3) Dilute sulphuric acid, 2,700 gallons being used to remove the oxides of iron. (4) Weak solutions of bleaching powder and sulphuric acid to dissolve the gold; and (5) Cold water to wash the ore. The solutions are all preserved separately. The leaching takes in all nine or ten days, the chlorination alone occupying three days. The gold and silver are precipitated from their solution by sodium sulphide, and the precipitate is dried, pressed, roasted, and melted down.

The expenditure per ton of concentrates is as follows:—

	s.	d.
Salt (110 lbs.),	1	7
Sulphuric acid (102 lbs.),	6	0
Bleaching powder (25 lbs.),	3	7
Hyposulphite of soda (14·7 lbs.),	1	10
Fuel,	8	0
Labour,	3	8
Amortisation,	1	7
Total,	26	3

The percentage of extraction is not given.

Cassel's Process.—In this process a solution of salt in contact with the ore is decomposed electrolytically, and the chlorine thus set free attacks and dissolves the gold, which is deposited in a hollow iron shaft in the centre of the vessel. The apparatus is complicated and ill-adapted for its purpose, and the process has not been a practical success. It has now been completely abandoned, but is interesting as being the first attempt to generate chlorine by electrolysis for attacking gold.

The Julian Process.—This was devised by Mr. Julian, of Johannesburg, South Africa. After chlorination of the ore in a barrel, mercury is added to the charge and the barrel again revolved. The result of this is to amalgamate the coarse gold in the ore and to reduce the gold chloride, chloride of mercury and metallic gold being formed, the latter amalgamating with the excess of mercury. The charge is then emptied out and passed over a set of amalgamated copper-plate tables, by which the gold amalgam is partly retained. The tailings are then passed through a series of "electrolytic cells," each of which has a bath of mercury lying at the bottom, while an electric current is passed through the water. In these cells, the compounds of gold and silver, soluble in the water, and all floured mercury and amalgam are supposed to be collected. The process has not yet been proved to be a success in practice. The idea of decomposing gold chloride by mercury is not new, and previous experience has shown the action to be very slow and partial. The complicated nature of the whole process is against it.

Greenwood Process.—In this process gold is dissolved in barrels by chlorine produced by the electrolytic decomposition of a solution of common salt. The problem of producing chlorine economically by electrolysis is one of enormous importance, as one of the chief causes of expense in chlorination lies in the cost of the chemicals required. If power is alone required, then wherever water power is available, the whole process may be much cheapened. In the Greenwood patents, the chlorine so obtained is not well applied, and the machinery need not be described here.*

CHAPTER XIV.

CHLORINATION: PRACTICE IN PARTICULAR MILLS.

THE following description of the practice actually employed at the works named will serve to show how far the general descriptions given are applicable to particular cases :—

THE VAT PROCESS.

1. Butters' Ore Milling Works, Kennel, California.†—The practice varies with the nature of the ore. The Idaho sulphides from the Grass Valley constitute a case of extreme difficulty. "They contain considerable quantities of lime and magnesia, and some manganese, zinc-blende and chalcopyrite, with 4 ozs. of gold and 8 to 12 ozs. of silver per ton. Roasting with salt is prohibited by the heavy volatilisation losses thereby incurred. The roasting is performed at a black heat, the charge withdrawn, allowed to grow quite cold, and then wetted down and screened immediately, in order to prevent the agglomeration of the ore, as the presence of the anhydrous sulphates makes the mass a natural cement. The ore is then leached in a vat for from 48 to 72 hours with cold water, and the soluble sulphates of copper, iron and zinc thus removed. The solution thus obtained contains small quantities of silver and gold, perhaps dissolved by the agency of the metallic salts. These are recovered in the copper-precipitating tank. The alkaline bases and carbonates are next neutralised with sulphuric acid (60 to 100 lbs. of 66° Baumé acid being added per ton of ore), and 20 lbs. (diluted with water) of the same acid per ton are then added and circulated for 24 hours to dissolve the oxide of copper, and the charge washed for 48 hours with cold water, shovelled

---

* A detailed description of the Greenwood process is given in Eissler's *Metallurgy of Gold*, pp. 351-369.
† *Eng. and Mng. Journ.*, Dec. 20, 1890.

out, partially dried, screened back into the tank and impregnated with chlorine. After leaching, the tailings contain from $4 to $6 in gold and all the silver. Five per cent. of salt are added, and the charge dried and roasted at as high a temperature and as fast as possible, when little loss by volatilisation is experienced. The silver is then removed by leaching with hyposulphite solution."

It seems difficult to believe that this complicated system (which, it will be observed, bears a strong resemblance to Mr. Munktell's methods in several important particulars), can really be the most economical one that can be devised for this ore.

2. **Plymouth Consolidated Gold Mining Company, Amador County, California.***—The ore contains 11 dwts. of gold, mostly in the free state. It is crushed in a stamp battery, no. 8 screens (40-mesh) being used, passed over a length of 20 feet of amalgamated plates (the upper one of which is copper and the rest silver-plated) and then concentrated on Frue vanners. The concentrates amount to from $1\frac{1}{4}$ to $1\frac{1}{2}$ per cent. of the weight of the ore, and contain from 5 to 10 ozs. of gold per ton. They are treated at the rate of 100 tons per month, being kept damp until charged into the roasting furnace, to prevent the formation of lumps. A "Fortschaufelungsofen" is used for roasting, 80 feet long by 12 feet wide, its hearth consisting of a long continuous plane, holding three charges at one time, which are kept separate. The three stages are called the "drying," "burning," and "cooking" stages. In the (middle) burning stage, the bed of ore is kept thin and occupies double the space of each of the other charges. The furnace is worked by three eight-hour shifts of one man. The charges weigh 2,400 lbs., including 10 per cent. of moisture, and on an average contain 20 per cent. of sulphur. Just before the sulphur ceases to flame, $\frac{3}{4}$ per cent.—*i.e.*, 18 lbs. of salt—are added.

The chloridising vat is 9 feet in diameter and 3 feet deep: it holds 4 tons of ore. The filter-bed is 6 inches deep, and consists of, (*a*) at the bottom, wooden strips, $\frac{3}{4}$ inch wide, placed 1 inch apart; (*b*) above this, 6-inch boards placed 1 inch apart and laid across the strips; (*c*) coarse lumps of quartz, diminishing upwards to fine stuff; and (*d*) at the top, a cover of 6-inch boards placed similarly to the lower layer, but crosswise to them. Chlorine is introduced on both sides of the vat, which is left luted-up for two days, and then leached for four or five hours, the tank being kept full of water during the operation. A gunny-sack protects the surface of the ore from the direct impact of the water from the hose. The ore in the impregnation vat contains about 6 per cent. of water (crumbling after it has

* For a more complete description, see that given in the *Eighth Report Cal. Stat. Min.*, 1888, of which the account appended is an abstract.

been sieved), and is sieved into the vat through a screen of ½-inch mesh.

The gold solution is acidulated with sulphuric acid to precipitate the lead, and ferrous sulphate is then added to it in another tank, after which it is allowed to settle for two days before the supernatant liquor is siphoned off. The gold is left to accumulate for fourteen days and the wet gold is then filtered and fused in graphite pots. The average extraction is from 95 to 96 per cent. of the gold, and 7 dwts. per ton are left in the tailings. All wood is protected from the action of the acids by paraffin-paint. The cost of milling is said to be 39 cents per ton of ore, and the cost of concentration, roasting, and chlorination, $13·40 per ton of concentrates.

3. **The Alaska Treadwell Mine.***—These works turned out more gold than any other chlorination mill in the United States, until the Golden Reward and other Dakota mills were started. The old-fashioned vat process in use has been adopted after thousands of dollars have been spent in testing newer methods. The works have been in active operation since the year 1884. The material treated consists of the sulphides collected by Frue vanners from a stamp battery, and contains 40 per cent. of sulphur, mostly as iron pyrites, although the percentage of copper is increasing. The gangue is quartzose, containing from 2 to 5 per cent. of calcite, which necessitates the addition of salt in the roasting furnace.

Roasting.—This was first effected in Brückner cylinders, which were abandoned owing to the large amount of fuel consumed and the enormous losses by dusting and volatilisation (which are said to have amounted to 30 per cent. of the gold). The automatic Spence furnace was then tried and proved to be useless until it was used as a reverberatory. Six were erected, and the cost of roasting was reduced by one-half, but the capacity was small (10 tons per day in the six furnaces) and the consumption of fuel great, and a reverberatory furnace being erected, was found to be more satisfactory. The Spence furnaces were accordingly discarded, and 20 tons of concentrates are now roasted per day in four reverberatory furnaces of 13 by 65 feet, inside measurement, at a low cost. The ore is not roasted quite dead, Mr. Burfeind, the superintendent, having decided that when salt is added the best results, with least volatilisation, are obtained by leaving some sulphates undecomposed. He gives it as his opinion that improper roasting is responsible for all losses of gold in the tailings and by volatilisation.

The roasted ore is spread on the cooling floor, wetted down and sifted carefully into vats, each of which holds 4½ tons, and is then impregnated with the gas. This operation occupies four hours, after which the vat is left untouched for thirty hours,

* *Eng. & Mng. Journ.*, April 11, 1891.

fresh gas being forced in a few hours before leaching. The leaching usually requires twelve hours. The tailings are sampled and assayed, and, if found sufficiently poor, are sluiced into the sea.

The solution is run into collecting tanks and thence to the precipitating vats, which already contain the necessary amount of ferrous sulphate in solution. The precipitation is complete when all the solution has been run in, or when the vat is full. It is then stirred briskly for a few minutes and left to settle for from eighteen to twenty-four hours, when the supernatant liquor is siphoned off and passed through a large filter. The supernatant liquor usually contains from 23 to 25 cents of gold per ton of material treated, and this is all saved in the filter.

The clean-up is made twice a month, the drying and melting of about 600 ozs. of gold being done by one man in one day. The gold is washed into a small tub, allowed to settle, and the supernatant liquor returned to the precipitating vat. The gold is dried in an iron pan without filtering, and melted with a little borax.

Each one of the chlorination vats, holding $4\frac{1}{2}$ tons of ore, cost $50, and lasts for three years without any repairs. The filter in it costs only the price of a few gunny-sacks, and lasts for six months without any attention. The other vats are expected to last a lifetime. The ferrous sulphate is prepared on the works from sulphuric acid and scrap-iron. During the six months ending November 31, 1892, 120,002 tons of ore were mined and milled, the total cost being at the rate of $1·32 per ton. The concentration of the ore yielded 2,703 tons of sulphides, which were chlorinated at a cost of $8·42 per ton. This is very low for the Plattner process, although perhaps high when compared with barrel chlorination. In 1890, the cost was at the rate of about $10 per ton, the yield in gold being over $40 per ton. Later results are as follows :—

YEAR.	Tons Chlorinated.	Yield per Ton.	COST PER TON.		
			Labour.	Supplies.	Total.
June, 1890, to May, 1891,	5869	2·02 oz.	$5·03	$3·99	$9·02
June, 1891, to May, 1892,	6177	1·52 ,,	4·80	2·81	7·61

THE BARREL PROCESS.

1. Mears' Process at Deloro, Canada.—The following description of the barrel process adopted at these works is taken from the account given by the manager, Mr. John G. Rothwell, M.E. The ore is hand-sorted, crushed in rock-breakers and Cornish rolls so as to pass a 16-mesh screen, then sized and concentrated by jigs. The concentrates are roasted in two White-Howell cylinders, being passed through one, where nearly all the sulphides and arsenides are burnt, and then into the other, where dead-roasting is performed; the ore falls thence direct on to the cooling floor, from which it is loaded into cars and sent to the barrels. The first cylinder is 30 feet long and 5 feet in diameter, and has eight brick shelves to assist in stirring the ore. The second cylinder is 20 feet long and 4 feet in diameter, with six shelves. The air supplied to the first cylinder is heated by the escaping gases of the second. Both cylinders are jacketted by air-spaces and a covering of mineral-wool and paper. Ten tons of concentrates are roasted in twenty-four hours, yielding over 4 tons of arsenious oxide, which is condensed in brick chambers. These cylinders are specially arranged so as to receive an unusually large supply of air, by apertures in the fire-bridge and around the fire-box; they are said to treat more ore with less loss by dusting than ordinary White-Howell furnaces.

The chlorinators are cylinders of $\frac{1}{4}$-inch boiler-iron, having cast-iron heads rivetted or bolted on; each head has a long trunnion turned true and cored-out, with a shafting-box and gland on it. The cylinders are lined with 20-lb. (to the square foot) sheet-lead, the joints being burned. The barrels are of two sizes. The smaller size is 40 inches in diameter and 44 inches long (inside measurement); it has a capacity of $32\frac{1}{2}$ cubic feet, and holds 1 ton of ore. The larger size is 60 inches in diameter and 72 inches long, has a capacity of 118 cubic feet, and takes a charge of 3 tons of ore.

A manhole in the centre of the shell enables a man to get inside the barrel for repairs; in the centre of the manhole cover is an aperture for charging and discharging. Inside the cylinder, and extending from the charging aperture in the direction in which the cylinder turns, is a pocket covered and lined with lead, which is burnt on to the lining of the cylinder. The pocket in the 1-ton barrel holds 100 lbs. of sulphuric acid, in the 3-ton barrel, 300 lbs. There are several lead-covered shelves bolted to the inside of the cylinder, their use being to stir the pulp thoroughly. Through the trunnion passes the "gooseneck," (viz., an iron pipe covered and lined by lead), which is turned in the direction in which the barrel revolves. The gooseneck is held stationary in its upright position by clamps.

About 120 gallons (*i.e.*, 60 per cent.) of water are added to the ton (of 2,000 lbs.) of ore. The proportions of chemicals used are 6 parts of bleaching powder to 7 parts of sulphuric acid, the absolute amounts varying with the quality of the ore and the degree of perfection of the roasting. The usual amounts are from 36 to 54 lbs. of bleaching powder, and from 42 to 63 lbs. of acid per ton. The lime is mixed with the ore and water, and the acid poured into the pocket. The covering plate is then screwed on and the barrel is set in motion, when the acid is immediately emptied out of the pocket and is mixed gradually with the rest of the charge. The cylinder revolves for from one and a-half to three hours, after which the excess of gas is drawn off and the barrel is partly exhausted through the gooseneck. Then the barrel is opened and dumped into the filtering vat, which has a loosely fitting cover placed on it, and an exhaust-pump connected with the space below the filter-bed to quicken the leaching. Another method adopted is to decant the liquid contained in the barrel and conduct it to settling tanks; then to add more water, revolve for a few minutes and decant again. In this way a charge of three tons can be washed in two or three hours, while by exhaust-leaching, 1 ton takes from six to eight hours.

The clear solution is precipitated by sulphuretted hydrogen made from paraffin and sulphur, thus—The generator is a small cylinder of heavy boiler-iron with cast-iron heads rivetted in, and the seams caulked. In the head is a hole big enough to admit a man's hand for the purpose of cleaning out the vessel. In the shell is a charging hole and an outlet for the gas, consisting of a 2-inch gas-pipe. The generator is built in a small brick fire-place with a charging hole or pipe on the top, and a hand-hole outside. A small fire is used, and the charge consists of one part of single-pressed paraffin to two parts of common brimstone.

The precipitating tanks are fitted with tight covers (in which are well-secured manholes), and the only outlet from them is a wide exhaust-pipe leading out of the building. In front of these tanks are lines of steam- and gas-pipes connected separately with each one. The gas, after being generated, passes into a receiver to condense the oils, &c., and an air-pump forces it from this through a perforated lead pipe, nearly to the bottom of the precipitating vats. An air-hole left in the receiver prevents the formation of a vacuum there, and enables the pump to pass a mixture of sulphuretted hydrogen and air through the liquid, an arrangement which economises the sulphuretted hydrogen. When precipitation is complete, a short time elapses and the liquid is then allowed to flow through several small filters.* Three of these filters will empty a 1,500-gallon tank in from three or four hours. The precipitate is then pressed into cakes, dried, roasted, and melted with borax.

The ore is an arsenical sulphuret of iron in a gangue of quartz and calcspar. The "mineral" contains about 42 per cent. of arsenic, 20 per cent. of sulphur, and 38 per cent. of iron. The

---

* For a description of these filters, see *Eng. and Mag. Journ.*, Mar. 25, 1885.

roasted concentrates average $78·67 per ton in gold, and the tailings about $3, the extraction being at the rate of 96·16 per cent. of gold.

2. **The Thies Process at the Phœnix Mine, Concord, North Carolina, and the Hailo Gold Mine, South Carolina.**[*]
—The barrel process is used at these mines, the methods of procedure being similar at the two places. Chlorine was formerly forced in through a gooseneck, passing through the trunnion (Mears' process), but it was found that the gooseneck wore out quickly, and that its inner lead lining collapsed when the gas was exhausted. It was, therefore, discarded after two years' experience, and the chlorine is made inside the barrel by means of sulphuric acid and bleaching powder. The ore is roasted in a double-bedded reverberatory furnace, having a capacity of 3 tons of concentrates per day. The chlorinator is 42 inches in diameter and 60 inches long, the capacity being 48 cubic feet, and the charging-hole is 6 inches in diameter. The lining consists of 10-lb. to 12-lb. sheet-lead. The weight of the charge is from 2,000 to 2,500 lbs., to which from 100 to 125 gallons of water is added—i.e., enough to make an easily flowing pulp. At the Phœnix Mine, where the ore contains copper, one-half of the acid and bleaching lime is added first and the barrel turned at the rate of 15 revolutions per minute for 3 to 4 hours. For the Phœnix ores, this half charge of chemicals consists of 20 lbs. of lime and 25 lbs. of acid. After this time has elapsed the barrel is opened, and the other half of the chemicals added, after which the barrel is rotated for from two to three hours longer. The charge is then let fall on to a filter-bed, 6 feet long and 8 feet wide, the depth of the pulp on it being 4 inches, while the filter-bed is 5 inches thick. The first solution is allowed to run through until the surface of the ore begins to be exposed, when it is again covered with water to the depth of 3 to 4 inches, and when the ore surface appears again, the whole space above it, 11 inches deep, is filled, which is enough for the Phœnix ores. In all, 2 tons of water are used to leach 1 ton of ore.

The bottom of the filter-bed consists of perforated glazed tiles or the substance known as *mineraline*, and stones, gravel and sand are successively piled on these as usual. The tanks for precipitation are made less than 3 feet deep. The precipitation is effected by ferrous sulphate, and care is taken that no soluble chloride of calcium shall remain in the solution, or a bulky precipitate of sulphate of lime will come down on the addition of the precipitant. It is for this reason that an excess of sulphuric acid is added to the charge, and more is added to the solutions if they do not already contain some free acid. The large amount of lime and acid added to this ore is necessitated by the presence of copper. The solutions are allowed to settle

[*] *Eighth Report Cal. State Min.*, 1888, p. 836.

for three days and then allowed to run through tanks filled with scrap-iron, by which the copper is precipitated.

At the Haile Mine, where iron pyrites, free from copper, is treated, only 10 lbs. of lime and 15 lbs. of acid per ton of ore are used. Ten tons of roasted ore are treated in the large barrels per day of ten hours, and 94 per cent. of the gold is extracted.

The cost at the Haile Mine is $4·65 per ton, without reckoning the expenses of superintendence and assaying, the cost of roasting being $2·70, and of chlorination, &c., $1·95. The following details are given on the authority of Mr. Adolph Thies, the inventor of the process used and the manager of the mill:—

Fire-assay value of ore, delivered to the stamps, $4·50, of which $1·45, in free gold, is caught on the plates.

Average assay value of raw concentrates,	$30 per ton.
Sulphur in raw concentrates,	40 to 45 per cent.
Value of roasted concentrates,	$50 per ton.
Average value of tailings from chlorination,	$2 per ton.

Cost of Roasting.

½ Cord of wood at $1·40,	$0·70
Two labourers at $1·00,	2 00
Per ton of roasted ore,	$2·70

Cost of Chlorinating.

10 lbs. chloride of lime at 3 cents,	$0·30
15 lbs. commercial sulphuric acid at 2 cents,	·30
Two labourers, ½ day at 90 cents,	·45
Chlorinator, ¼ day at $2·00,	·50
Power,	·12½
Sulphuric acid for ferrous sulphate,	·12½
Wear and tear,	·10
Superintendence,	·05
Per ton of roasted ore,	$1·95

Total Cost.

Per ton of roasted ore,	$4·65
Per ton of raw concentrates,	3·50

A lead-lined barrel which had been in use for five years at this mill showed no signs of wear, the abrasion of the lining by the ore being evidently almost *nil*.

3. **The Modern Barrel Process at the Golden Reward Chlorination Works, Deadwood, Dakota.**—Mr. John E. Rothwell, in exchanging his position at Deloro for that of manager of these works, adopted a system differing considerably from the above, as may be seen from the following description which is based on his writings,[*] and on accounts furnished

---

[*] *Eng. and Mng. Journ.*, Feb. 7, 1891, and *Mineral Industry for 1892*, p. 233.

privately to the author by a late foreman of the mill and by others:—

The ore is crushed in a large Gates crusher and two sets of Krom rolls, roasted partly in four Brückner furnaces of 3 tons capacity each, partly in a large White-Howell furnace, and chlorinated in barrels of 3 and 4 tons capacity. After being passed through the rock-breaker, which delivers no pieces larger than $1\frac{1}{2}$ inch in diameter, the ore is dried by passing through a revolving cylinder, 18 feet long and 5 feet in diameter, the capacity of which is somewhat increased by being divided into four longitudinal compartments, by the same number of partitions, made of plate-iron, which terminate at a distance of 2 feet from each end of the cylinder. The fire-place is arranged to heat a large quantity of air, which is not entirely admitted through the grate, but partly through channels by its side. The revolving screens are hexagonal in shape, and are made double with a coarse mesh inside the fine mesh, the object being to protect the latter from undue wear. The mesh-frames are slid into grooves in the screen-frame, and are interchangeable and readily replaced when worn. The partially oxidised ores ("red ores"), coming from workings near the surface, are roasted in the White-Howell furnace, while the undecomposed pyritic ores ("blue ores"), derived from deeper levels, and containing much more sulphur than the first-named class, are roasted in Brückner furnaces. The ore from the roasting furnaces is spread out in a thin, furrowed layer on the cooling floors, and, when partially cooled, it is sprinkled with water and raked over, the water drying out.

The actual capacity of the barrels used at the Golden Reward Mill is $3\frac{1}{2}$ tons of roasted ore per charge. The following dimensions apply to a 5-ton barrel. This consists of a $\frac{1}{2}$-inch steel shell, 9 feet long and 5 feet in diameter inside, lined with ($\frac{1}{2}$-inch) lead of 24 lbs. to the square foot. The cast-iron heads are $2\frac{1}{4}$ inches thick and heavily ribbed, and are inserted inside the shell and bolted to it through a flange. A 3-inch iron rod, covered with lead, passes through the trunnions and the entire length of the barrel. The trunnions are 12 inches in diameter, and the charging-holes are oval, 11 by 16 inches, the cast-iron covers swinging off on levers when required. There are two charging-holes placed in the cover-plate of the manhole. The chlorination barrel is also the washing and leaching vessel; this is arranged by placing a supporting diaphragm, for a filtering medium, to form the chord of an arc of the circle of the barrel.

The filter or diaphragm, as it is called, is made of asbestos cloth, resting on a framework consisting of oaken planks, each 11 inches wide and as long as the barrel, and 2 inches thick. The area of the filter is nearly 30 square feet. In the planks are grooves running longitudinally, each groove being $\frac{3}{8}$ inch wide and deep, and the same distance apart. Transverse grooves, a little deeper than the longitudinal ones, are also cut at intervals of from 4 to 6 inches, and ⅜-inch auger holes are bored through the planks from the bottom of the transverse grooves at short intervals. These planks are supported on wooden cross-pieces placed transversely to the barrel, which rest on longitudinal strips bolted to the shell. The liquid passing through the asbestos filter-cloth collects in the grooves, and is drained off by the filter-holes. This oaken filter-plate, which was devised by Mr. D. Dennes, costs little, and is very effective and durable. Before his suggestion was adopted, an iron grating covered with lead, which was burnt-on, had been used to support the filter-cloth. In the grating on one barrel there were no less than 28,000 holes, through each of which the lead coating had to be carried and burnt-on separately. The grating thus made with great labour only lasted for a short time, as a faulty joint in the lead in any one of these holes allowed the iron to be corroded, and caused the grating to go to pieces. Boards with auger holes were tried, but the area of the filter was thus restricted to the sum of the areas of the apertures, and the speed of filtration greatly reduced. With the oaken-plates, however, the filter-cloth rests on the sharp ridges between the grooves, the surface being almost entirely available for filtration. Of course, when the pressure is applied, the cloth sags down into the grooves, but this only increases the area available for filtration. On the top of the corrugated plates is placed the filtering medium, an open-woven asbestos cloth. It is nearly as coarse as the ordinary gunny-sack, but the warp and woof are of much heavier thread. Over this is placed an open wooden grating, and the whole is held in place by cross-pieces, the ends of which rest under straps bolted to the inside of the shell; although the filter, when made in this way, is rigidly held in place, it can be very easily and quickly removed when the changing of the asbestos cloth becomes necessary. The time occupied in changing the cloth is about one hour and a-half, and, under ordinary conditions, it will last for from 15 to 18 charges or from 2¼ to 3 days. The wear is due to the scouring action of the pulp, from which the filter-cloth is very imperfectly protected by the wooden grating. This is made of 1½-inch wooden pieces, so that its apertures are 1½ inches square.* The woodwork of the supporting plates and gratings lasts much longer than the filter-cloths, but requires replacing at somewhat frequent intervals. One carpenter is continually employed in making the woodwork for the filters of the 4 barrels, and in fixing the new frames in position when necessary. The lead lining of the barrels, in which several thousand charges have

* At the Independence Mill, which stands near the Golden Reward Mill, the asbestos cloth is protected by a secret method, and lasts over three months.

been treated, show little signs of wear yet. Two valves on each side of the shell of the barrel, above and below the filter, are for the inlet and outlet of the wash-water and solution respectively.

The barrel is charged by first filling the space under the filter with water, which at the same time is allowed to pass through the filtering medium, and wash it; then the required quantity of water is put in above the filter. The sulphuric acid is then poured into the water, through which it sinks in a mass to the bottom, without mixing with it; the ore is then charged in, as follows:—The hoppers are furnished with 2 shoots, one for each charging-hole. The ore is let fall through these shoots alternately, the hole through which ore is not being passed serving as an air-vent. Meanwhile the bleaching powder has been weighed-out and placed in two small kegs. When the ore has all been introduced into the barrel, a workman, stationed at each charging-hole, hollows out a space in the surface of the dry ore with his hands, and, emptying one of the kegs into the barrel, closes up the charging-hole as quickly as possible. If all these operations have been conducted rapidly without a hitch, there is no immediate evolution of chlorine, but, if some time is suffered to elapse after charging-in the ore, the acid liquid, thoroughly stirred-up and mixed by the fall of the ore into it, gradually rises through and wets the charge, and the bleaching powder, falling on ore which has been wetted with acid, gives off copious fumes of chlorine before the cover-plate can be screwed on.

After the chlorination is complete—viz., in about one and a-half to two hours—the barrel is stopped, so that the filter assumes a horizontal position; the hose is attached to one of the outlet pipes and conducts the solution to the settling tanks. A hose is also attached to the inlet pipe, and water is pumped in under a pressure seldom exceeding 40 lbs. per square inch, and the leaching commences. The air in the top part of the barrel is compressed and forms an elastic cushion, which gives the wash-water perfect freedom to circulate evenly over the whole surface of the charge, and wash every portion of it thoroughly. By washing in this manner, no chlorine is allowed to escape into the building, as it is all absorbed by the water. At intervals the leaching is suspended, and the barrel is again revolved for a few minutes, so that its contents are thoroughly mixed-up together again. In this way the formation of channels in the ore are prevented, and perfect leaching is ensured. The wash-water coming from the barrel is tested for gold with sulphuretted hydrogen.

The length of time required for the leaching varies with the leaching quality of the ore treated—charges having been leached in forty minutes with a pressure of from 30 to 40 lbs. per square inch. With higher pressures the time can be materially shortened. In order to facilitate the leaching of charges carrying an excess of dust or slime, the following method, now not in use, is stated by Mr. Rothwell to have been formerly employed at this mill. A valve placed in the casting of the head, on a level with the surface of the pulp, is opened just after the barrel is stopped, and the dust and slimes which remain in suspension are run into an outside washing filter-press, or to the slime-filter where it can be treated separately, and the charge washed in the usual way.*

The tailings are discharged into a car which will hold the whole charge of ore and water, and then run out of the building; or, if water is abundant, they are discharged into a sluice, and washed away. The filter-cloth is washed clean by a jet of water under pressure directed successively to all parts of it. This water is discharged by revolving the barrel.

"For leaching purposes, the amount of water necessary to wash a charge varies very little with the richness of the ore, which goes to show the perfect leaching condition of the ore in the barrel. The amount required is about 120 gallons per ton over and above the quantity used in the barrel for chlorination, which is about 100 gallons per ton. The water from the final washings, which only contain small quantities of gold, is forced up into an overhead tank, and used for the succeeding charge in the barrel; the quantity of solution to be precipitated is thus reduced to about 120 gallons per ton of ore treated." †

The solution coming from the barrel, passes through a 2-inch acid-proof hose-pipe into a large settling tank, one of a considerable number. These are round, wooden vats, lined with lead, and are about 10 feet in diameter and $7\frac{1}{2}$ feet deep. When the first tank is full, it overflows into the next one, which is placed at a somewhat lower level, and this arrangement is repeated through the whole series, the lowest vat overflowing into a sump, whence the liquid, now nearly free from suspended particles of ore which have settled to the bottom of the vats, is pumped up into the precipitating vats, situated above the cooling floor near the top of the building. The force-pump used for this purpose is perhaps the weakest point in the whole plant as it is rapidly corroded by the acid liquid, and requires frequent renewal and repairs.

At the Independence Mill, the slimes are separated by being passed through slime-presses, similar in shape to a Johnson filter-press.‡ Asbestos cloth is used for the partitions, with the addition of a layer of Canton-flannel if the slimes are very

---

* This device was first used by the late Mr. J. T. Blomfield, at the Newbery-Vautin Chlorination Testing Works, in London, in 1889, and was afterwards adopted by Mr. Rothwell in 1890.

† *Eng. and Mng. Journ.*, Feb. 7, 1891, p. 166.

‡ For description of Johnson's filter press, *vide* Stetefeldt's *Lixiviation of Silver Ores*, p. 126.

finely divided. The same pressure which affects the leaching in the barrel forces the solution through the slime-presses. When these are full of slimes, they are washed out by water forced in the opposite direction to that in which the solution had moved. Slime-filters were formerly in use at the Golden Reward Mill also; they consisted of lead-lined cast-iron cylinders of 18 inches deep and 30 inches in diameter, in which were arranged filter-beds of sand and asbestos cloth. The chief difficulty encountered in the use of these slime-filters is stated to have been that the solution was squirted out in all directions in small jets, and part of the valuable liquid thus lost. At the Independence Mill, the filter-press is placed above a large, wide, shallow, lead tank, while it is completely surrounded by a leaden hood, against which all the jets of liquid strike and then run down into the tank.

At the Golden Reward Mill there are two precipitating tanks for each barrel. They are $6\frac{1}{2}$ feet in diameter and $10\frac{1}{2}$ feet deep, are fitted with covers, and are constructed of wood and lined with lead of 8 lbs. to the square foot. The filter press, which is situated on the ground-floor, over 25 feet below the precipitating vats, has twelve chambers of 19 inches square, and has a filtering area of 57 square feet. Two lead pipes connect the precipitating tanks with the press, one leading from the bottom of the tank and the other from a point in the side about 4 inches above the bottom. The precipitating gas generators are cast-iron cylinders of the same size as the slime-filters; only the sulphuretted hydrogen generator is lined with lead.

Before the precipitating tank is full of solution, the sulphurous acid gas generator is started, and the gas forced through a 2-inch leaden pipe leading to the bottom of the tank. The gas is made from sulphur burned in the generator, with the aid of a current of air forced in at the bottom and deflected over the surface of the burning mass. The mixture of sulphur dioxide and air, which is present in excess, is passed into the solution, where the former is absorbed, while the air, bubbling through, serves to stir the liquid. When the free chlorine has nearly all been converted to hydrochloric acid, sulphuretted hydrogen, generated from sulphide of iron and dilute sulphuric acid, is forced in, together with a large excess of air, by which the liquid is kept in a state of agitation, and the free chlorine expelled, thus effecting a saving of SH_2. The precipitate is flocculent, and settles quickly.

After the precipitation is complete the "tank is allowed to stand for two or three hours, when it has settled enough to draw the supernatant liquor off through a filter-press by the pipes in the side of the tank. There is little danger of precipitating arsenic or antimony that may be in the solution when it is worked cold, as they do not commence to come down till some time after the gold has precipitated and collected. Of course, any copper or lead in solution will be precipitated with the gold, but if small quantities only are present, they can be removed subsequently. The loss in gold is considerably less if the precipitate is allowed to accumulate in the tanks, and a clean-up made after six to ten precipitations, than if it were filtered through a press and collected after every precipitation. Continuous filtration causes the filters to be coated and clogged with the sulphides, and retards the operation unless high pressure is used, which is sure to increase the loss, while a similar result attends the handling of a large number of filter-cloths." * Rothwell has stated more recenly that gold can be fractionally precipitated from solutions containing copper if free acid is present. The sulphide cakes in the presses are compressed by blowing air through them, and are turned out in the form of lumps, which break up into powder when they are touched; they are then dried, roasted, and fused with borax, nitre, carbonate of soda and sand.

The amount of precipitants required to precipitate a tank of 2,500 gallons of solution are—sulphur, 2 lbs.; iron sulphide, 4 to 5 lbs.; sulphuric acid, 16 lbs.; water, 9 gallons. The chemicals required for chlorination are—bleaching powder, 8 lbs., and sulphuric acid, 15 lbs. per ton of roasted ore. The power is furnished by an engine of 125 H.P., consuming six or seven cords of wood, or about 4 tons of bituminous coal. Steam is also required for the air-compressor, for steam-pumps, and for heating the building. No salt is used in roasting, as the ore treated contains little copper, &c., and no attempt is made to extract the silver; the fuel used (chiefly pine-wood) amounts to a little more than that needed for power. The men employed on the mill number about forty. "One man of ordinary intelligence and a helper are able to take care of three barrels—that is, to look after the charging, leaching, and discharging. If the tailings are sluiced out they can also attend to that; but where they have to be trammed out, one more man is necessary."†

There are four barrels, each of which treats 6 charges in 24 hours, or a total of over 90 tons per day. The cost of treatment is given below. When the ore is crushed through an 8-mesh screen, the tailings usually contain about 3½ dwts. of gold per ton, or 20 per cent. of the total contents; but it is stated that 92 per cent. was being extracted in 1894. ‡

The disadvantages of the process used are due to the construction, but do not seriously interfere with the successful working of the barrel; they are, among others, the amount of space taken up by the filter and the portion of the barrel under the filter, and the fact that whenever the barrel is charged and running, it is not in a perfectly balanced condition. These disadvantages could be minimised, according to Mr. Rothwell, by using a filter

---

* *Loc. cit.*   † *Ibid.*   ‡ *Eng. and Mng. Journ.*, Jan. 27, 1894.

placed close to the shell, and separated from it by a space only just enough to allow of free circulation, and reaching to the same height on the sides as the horizontal filter; then, by using compressed air to displace the solution and wash-water, an equally good result could be obtained. The mechanical difficulties in constructing a curved filter-bed, however, prevent its introduction.

According to Mr. Dennes, it appears that one of the disadvantages in the use of the diaphragm is that it is difficult to keep it in place. The ore and water wash to and fro through the filter-cloth, and there is a strong tendency for the latter to become displaced. If one plank in the somewhat complex framework, by which it is held in place, becomes loosened, the whole soon gives way as the barrel revolves, and the mishap is not detected until leaching begins, when ore and water come out together. The asbestos cloth is expensive, and only lasts for about three days, and all repairs must be done by a man getting inside the barrel, owing to its construction. It has been more recently suggested that the asbestos cloth could be replaced without difficulty by a sand filter.

The cost of barrel chlorination in 1891, at the Golden Reward Mill, is given by Mr. John E. Rothwell, as follows:—*

	July.	August.	September.
Amount of ore treated,	1,430 tons.	1,195 tons.	1,513 tons.
,, treated per day,	46·1 ,,	38·5 ,,	50·4 ,,
Cost per ton—			
Milling,	$1·49	$1·40	$1·44
Roasting,	1·53	1·35	1·37
Chlorination,	1·76	1·67	1·65
Office, salaries,	0·40	0·47	0·37
Construction and repairs,	0·36	0·93	0·21
	$5·54	$5·82	$5·04

The total cost given above includes all the working expenses, but is exclusive of interest on capital, taxes, insurance and amortisation (*i.e.*, extinction of capital by a sinking fund, which covers the depreciation of plant, &c.).

The roasting was at that time done in Brückner cylinders, and its cost was afterwards reduced to about 70 cents per ton by the adoption of a modified White-Howell furnace, and in January, 1894, the total cost of treatment per ton was stated to be only $3·77.

The amount of chemicals used in the chlorination and precipitation departments, in 1891, is given on the next page.†

* *Eng. and Mng. Journ.*, Mar. 25, 1893, p. 269. † *Loc. cit.*

	July, 1891.		August, 1891.		September, 1891.	
	Lbs. per ton of Ore.	Cost in Cents.	Lbs. per ton of Ore.	Cost in Cents.	Lbs. per ton of Ore.	Cost in Cents.
Sulphuric acid,	26·82	45·6	25·14	44·0	24·77	43·3
Chloride of lime,	11·04	35·8	10·17	33·1	10·09	32·8
Crude sulphur,	0·37	1·14	0·31	0·9	0·29	1·2
Iron sulphide,	0·77	4·54	0·79	4·0	0·84	4·2
Total cost,	...	87·1	...	82·05	...	81·5

Method employed at the Bromination Mill, Rapid City, Dakota.—The following details were supplied to the author by Mr. D. Dennes, the late foreman of the works. The method of crushing has already been described, p. 222. The ore which was treated in 1892 contained 6 or 7 per cent. of sulphur, a similar amount of arsenic, and some iron, but little or no copper. It was roasted in two White-Howell furnaces, each 30 feet long and 60 inches in diameter inside the iron, lined with firebrick 5 inches thick, shaped so as to fit the curve of the cylinder. There are six shelves in each furnace. The inclination of the cylinder was varied with the ore, and the speed of revolution was also varied from one turn in two and a-half minutes to three turns per minute. A large Brückner cylinder was also tried, 24 feet long and 9 feet in diameter, holding a charge of 15 tons of ore, the duration of the roast being about six hours. This gave a better product than the White-Howell furnaces, but the cost of fuel was greater. The ore was cooled by being spread out in a thin layer, and was sprinkled with water when very hot. The roasted ore was then charged into lead-lined revolving barrels of the same dimensions as those used at the Golden Reward Mill—viz., 8 feet long and $4\frac{1}{2}$ feet in diameter, and having a capacity of $3\frac{1}{2}$ tons. Hot water was introduced into the barrel before charging-in the ore, only 66 gallons per ton or 33 per cent. of the weight of the ore being used. This seems too small a quantity for efficient chlorination. The barrels had cast-iron trunnions 12 inches in diameter, and were revolved at the rate of twelve turns per minute by a worm and wheel. They were made of boiler iron, $\frac{3}{8}$ inch thick, and were lined with thick lead of 24 lbs. to the square foot on the ends, and 18 lbs. to the foot on the cylinder.

After adding the ore and water, the required amount of bromine was poured in, and the barrel closed at once. Little inconvenience is stated to have been caused to the workmen by the bromine fumes, which were evolved owing to this crude method of manipulation; the men's eyes only were slightly affected. From 6 to 13 lbs. of bromine were added to the charge of 4 tons, the amount being adjusted so that a very slight excess of free bromine remained at the end of the operation. Chlorine generated from sulphuric acid, and bleaching powder was used for several months before the introduction of bromine, but the cost was greater and the tailings only $\frac{1}{4}$ dwt. poorer in the former case. Moreover, the filter cloths lasted longer, the health of the workmen suffered less injury, and the gold precipitate was purer when bromine was used. The barrel was revolved for

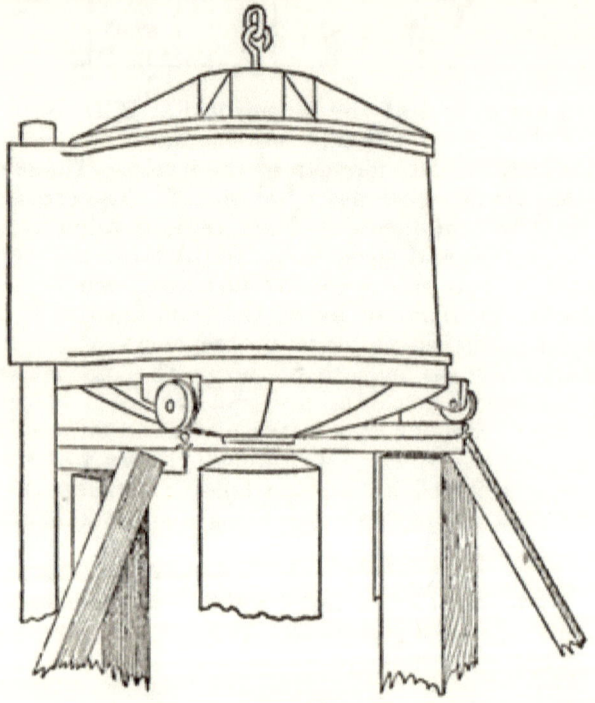

Fig. 55.

from thirty minutes to one and a-half hours, and then discharged into the leaching vat below, shown at Fig. 55. This is $7\frac{1}{2}$ feet in diameter and 3 feet deep, made of cast-iron lined with lead, and has a strongly ribbed cover, so that it is capable of withstanding an internal pressure of 100 lbs. to the square inch. Internally, it tapers slightly upwards. The sides of the vat do not rest on the bottom, but are supported by columns direct from the floor. The bottom is not connected with the rest, but is supported by a hydraulic ram, shown in the figure. On this true bottom there is a filter-bed and false bottom. The filter-cloth consisted of gunny-sacking. The vat was worked as follows:—When a charge was about to be introduced, the bottom was raised by the ram and pressed tightly against the sides of the vat, a good joint being made by a thick, rubber ring. The hydraulic pressure was more than enough to overcome the air pressure subsequently applied. The charge was then introduced, the cover replaced and fastened down tightly by screw-bolts, and air was pumped into the vat above the surface of the charge. Water was also introduced by means of a lead pipe extending round the vat at about the surface of the charge. This pipe was pierced with a number of small holes, by means of which jets of water were thrown upon all parts of the ore surface. The air pressure, which never exceeded 60 lbs. to the square inch, forced the wash-water through the charge very quickly. The solution was of a strong ruby-red colour, due to the presence of free bromine, and of the bromide of gold, the colour of which in solution is very intense. There was no need to test the issuing liquid, as its colour was found to be a perfect guide to an experienced eye. When it had become colourless, the water supply was shut off and the charge dried by pumping air through it. This was necessary, as the company was not allowed to sluice the tailings into the river. When no more water could be driven out, the air-pressure was let off, and the bottom of the vat lowered until the top of the filter-bed was just clear of the sides of the vat. The bottom was then drawn sideways from below the vat by means of a second hydraulic ram, and the charge fell into a large ore-car below, having a capacity of 4 tons. This ore-car then ran by gravitation to the dump, and was emptied and drawn back by a wire rope winding on a drum actuated by steam power. It is stated that the leaching occupied only twenty minutes, and the vat was used to leach the charges from the two barrels, which discharged into it alternately. The vat was so rapid that it was always waiting for the barrels, and 100 tons per day could be leached by it. It was designed by Mr. D. Dennes, and constructed by Messrs. Fraser & Chalmers, and, though of high initial cost, was found to be of such great efficiency that its introduction was thought to be amply justified.

The liquid was forced into large, lead-lined wooden precipitation vats, each 20 feet × 10 feet × 6½ feet, which were placed outside the mill, buried under 2 feet of manure to keep them warm in winter. The precipitation was effected by the successive use of sulphur dioxide and sulphuretted hydrogen, applied in exactly the same manner as at the Golden Reward Mill. The collection of the sulphides was also effected in a similar manner. When dried, the sulphides were put, together with the necessary amount of borax, into a small barrel, revolving by machinery, and were mixed thoroughly. The mixture was then transferred by a scoop to a red-hot clay crucible (size No. 100) in the furnace, additions being made at intervals until the crucible was full of bullion and molten slag. The latter was very rich and full of shots of metal. It was stored and eventually sold to the smelters, but owing to the difficulty in sampling it, it would have been better to fuse it with lead and to sell the latter.

The air-compressor used was 12 inches in diameter, and had an 18-inch stroke; its maximum velocity was eighty revolutions per minute. A large air receiver, 16 feet long and 4 feet in diameter, was used between the pump and the leaching vat. The maximum air pressure in the receiver was 80 lbs. per square inch. The dimensions of the bullion smelting furnace were as follows:—Depth 3 feet 3 inches. Inside diameter—at bottom 2 feet 8 inches, at top 2 feet 2 inches. The flue measured 8 inches by 12 inches, and was situated 10 inches below the top of the fire-box.

From 60 to 75 per cent. of the gold in the ore was extracted, the tailings usually containing over 5 dwts. of gold. The reason for this unsatisfactory result is said by Dr. L. D. Godshall (*Eng. and Mng. Journ.*, Jan. 6, 1894) to lie in the coarseness of the crushing, a 10-mesh sieve being used. When a 30-mesh sieve was used, only about 2½ dwts. were left in the tailings, and the product was easily leached. It seems possible that inefficient leaching, due to the formation of channels in the ore, may have been another cause of loss.

Cost of Treatment.—The cost of treatment of the ore by chlorination at the Rapid City Mill for 40 tons per day, everything running double shift, was as follows:—

Crushing Mill.

2 men (1 each shift), running ore from bins to Blake, at $2·00 ⅌ day	=	$4·00
2 ,, ,, attending to the Blake, . . 2·00 ,,		4·00
2 ,, ,, ,, ,, Rolls, . . 3·00 ,,		6·00
2 ,, ,, oiling in the mill, . . 2·50 ,,		5·00
Coal used by rotary dryer, ½ ton, at . . . . 5·75 ,,		2·87
Oil used by the mill, 2 galls., at . . . . ·40 ,,		·80
⅔ of the whole power used (see Power Account), . . ,,		19·65
Consumption of roll-shells and Blake jaws, . . . ,,		4·80
		$47·12

or $1·177 per ton.

Roasting.

2 men (1 each shift), attending to White-Howells, at . $3·00 ⅌ day	=	$6·00
2 ,, ,, helping ,, at . 2·00 ,,		4·00
2 ,, ,, cooling ore from ,, at . 2·00 ,,		4·00
4 ,, (2 each shift), wheeling ore to barrels, at . 2·00 ,,		8·00
2 ,, (1 each shift), getting wood for roasters, at . 2·00 ,,		4·00
Cordwood consumed per 24 hours by 2 roasters = 6½ cords, at $3·50 per cord, ,,		22·75
Oil for this department, including lighting, and broken lamp chimneys, ... ,,		1·05
1/12 of the whole engine and boiler power, . . . ,,		1·95
		$51·76

or $1·294 per ton.

Chlorinating and Leaching.

2 men (1 each shift), running the barrels, at $3.00 ⅌ day	=	$6.00
2 ,, ,, ,, leaching apparatus, at 3.00 ,,		6.00
2 ,, ,, ,, tailings out on to dump, at 2.00 ,,		4.00
Gunny-sacking for filtering table of leaching apparatus, ,,		.10
¼ of the whole boiler and engine power, ,,		7.85
Chloride of lime, 14 lbs. per ton, or 560 lbs. per day (at 2½ cents),	=	14.00
Sulphuric acid, 24 lbs. per ton, or 960 lbs. per day (at 3½ cents),	=	33.60
Oil for this department, at 40 cents per day,	=	.40
		$71.96

or $1.780 per ton.

Precipitating and Recovering.

2 men (1 each shift), precipitating and handling solution, and drying and emptying filter presses, at $3.00 ⅌ day	=	$6.00
Sulphur for generating SO_2 and H_2S, 20 lbs., ,,		.40
Flannel cloth for filter presses, $10.00 per clean-up, or $2.00 per day (clean-up say every 5 days), ,,		2.00
$\frac{1}{16}$ of the whole boiler and engine power, ,,		1.96
Borax glass, carbonate of soda, crucibles, and coke, ,,		4.50
		$14.86

or 37.1 cents per ton.

General Expenses—Power.

2 Engineers, each $3.25	=	$6.50
2 Firemen, ,, 2.50	=	5.00
Coal, 7 tons at 2.75	=	19.25
Oil,	=	.70
		$31.45

or 78.5 cents per ton.
(This is included in the other items.)

Offices, &c.

1 Assayer, at	$5.00
1 Assistant, at	2.00
1 Clerk, at	2.00
1 Office boy, at	1.00
	$10.00
1 Blacksmith,	$3.50
1 Machinist,	3.50
1 Helper,	2.00
Blacksmith's coal and iron per day, average,	1.75
Oil, waste, and general stores consumed per day,	2.00
	$12.75

= $22.75 or 56.9 cents per ton.

Cost of bromination in place of chlorination as follows :—1·5 to 3·25 lbs. of bromine at 17 cents per lb. for each ton of ore = 25·5 to 55·25 cents per ton, labour and other things being equal.

Total cost of chlorination as above—

Crushing,	$1·177 per ton.
Roasting,	1·294 ,,
Chlorinating and leaching,	1·780 ,,
Precipitating, &c.,	·371 ,,
General expenses,	·569 ,,
	$5·191 ,,

Superintendence and amortisation are not included in this total.

It will be noticed that a saving of at least 60 cents per ton is effected by substituting bromine for chlorine.

The Future of Chlorination.—At the present day chlorination is the most favoured method of treating auriferous concentrates obtained from the tailings of stamp batteries, except in districts where smelting works, producing mattes or base bullion, exist. It is probable that, in general, such concentrates may be more economically dealt with by the cyanide process, but it seems doubtful if the high percentage of extraction attained by the use of chlorine can be equalled by the newer method without roasting. During the last few years the chief extension given to chlorination has been due to the perfecting of the barrel process; by its use certain virgin ores, especially those of Mount Morgan, of Carolina, and of South Dakota, have been successfully treated. The latest methods introduced in Dakota, especially at the Golden Reward Mill, certainly offer great advantages over the original practice in Carolina and Queensland, and improvements in detail will probably continue to be made. For example, the recent increase in size of the barrel in Western America from one containing 3 tons of ore to one holding 10 tons is of considerable importance. Nevertheless, a doubt may be expressed whether the complicated and costly machinery necessitated by these methods will not stand in the way of their adoption in other gold-fields. It is at least possible that American pioneers of chlorination are working in the wrong direction. Dr. Munktell appears to have shown that a weak solution of chlorine in water, acting on ores not subjected to agitation, is, if sufficient time is allowed, almost as efficacious in dissolving gold as a strong solution combined with agitation, or as chlorine gas acting on moistened ore. If this should prove to be the case with most auriferous pyrites, a plan of operations in chlorination, much simpler than those hitherto in vogue, can be pursued. The roasted ore might be charged into large round tanks of wood or masonry holding hundreds of tons of material, such as those now in use with the cyanide process in the Transvaal. It would then be sufficient to leach with a weak solution of chlorine, allowing it to run through by gravity, and, as it issues at the bottom, continually raising it by means of a pump (preferably of the injector type). The solution could then be poured on the surface of the charge again, strengthened by the addition of chemicals if necessary, and the whole operation repeated until the dissolution of the gold is complete. There can be no doubt as to the cheapness of the plant required in this case, and it is to be remembered that similar methods have been selected as most advantageous in the old hyposulphite, the Russell and the cyanide processes, in all of which the progress that has been made is in the direction of greater simplicity. In consideration of these facts, it seems likely that, in spite of the great technical success of the operations in Dakota, it is in the results of the Munktell process in Central Europe that indications may be found of a possible extension of chlorination to the treatment of large quantities of low-grade refractory ores.

As an instance of the direction in which the industry is progressing, the operations in Queensland may be quoted.* Here a few years ago barrels were almost exclusively used in chlorination. At Mount Morgan, the largest chlorination mill in the world, wooden barrels were used in which chlorine was generated by bleaching powder and acid. However, both at Mount Morgan and generally throughout Queensland, a return to the vat process has been made, with a view to reduce the working costs. The vats are, however, made larger, those at Mount Morgan holding 25 tons of ore. The chlorine is now applied there as chlorine water, and it is claimed that the consumption of the solvent when used in this form can be better controlled and kept within the absolute requirements for the extraction of the gold. The gas is generated in stills by means of manganese dioxide, salt and sulphuric acid, and the use of chloride of lime has been completely abandoned. The chlorine is absorbed by water in scrubbing towers. The ore contains 1·66 oz. of gold per ton, and 95 per cent. of this is extracted at a cost of about 12s. per ton, the amount treated being about 200 tons of ore per day. It will be observed that all the American methods have now been abandoned at this mill, and the simple method of procedure now adopted seems to have much to recommend it.

* See article by E. A. Weinberg in *Report for* 1894 *on Queensland Mines.*

CHAPTER XV.

THE CYANIDE PROCESS.

The Cyanide Process, like most others, gradually grew up from small beginnings. The power possessed by potassic cyanide solutions of dissolving metallic gold and silver has long been known, as is pointed out, p. 333, but it was formerly believed that the use of an electric current in conjunction was needed to quicken the action, which was otherwise too slow to be of any practical value. The use of cyanide of potassium in the stamp battery has already been referred to, p. 124. It is probable that its effect in dissolving gold was entirely overlooked when it was being used in this way, the effect which it was desired to produce being the removal or prevention of formation of the crusts of carbonate of copper on the plates. In California, prior to the year 1870, gold ores were sometimes digested in tubs with water, to which a lump of cyanide had been added; after digestion, the ores were amalgamated by stirring with mercury, to which a little sodium-amalgam was sometimes added. It is obvious that, when no sodium was present, only the undissolved gold was amalgamated, the dissolved gold being lost, not being precipitated by mercury. These operations, however, can hardly be said to foreshadow the cyanide leaching process, as they were conducted in ignorance of the principles on which it is based, and partly in defiance of them.

The first attempt at the direct extraction of gold by the use of cyanide was made by J. H. Rae, who took out a patent in the United States in 1867 (No. 61,866) for a process dependent on the removal of gold and silver from their ores by the combined action of a "current of electricity and of suitable solvents or chemicals," such, for instance, as cyanide of potassium, the gold being simultaneously precipitated on copper plates by the electric current. This patent has now lapsed, and, although other United States patents claiming the use of cyanide in the treatment of gold ores were issued in 1880 and 1881, no real progress seems to have been made.

In 1885, J. W. Simpson, of New Jersey, obtained a patent in the United States (No. 323,222) which was of greater interest. He proposed to treat ores containing gold, silver and copper by a solution containing 3·0 per cent. of potassium cyanide and

0.19 per cent. of ammonium carbonate. Copper was to be dissolved at the same time as the gold; if silver was present also, an addition of common salt was made. The inventor appears to have believed that, by using carbonate of ammonia, the necessity was obviated of employing an electric current, in conjunction with cyanide of potassium, in order to dissolve the gold. Of course the carbonate of ammonia would have no effect on the amount of gold dissolved, whilst its action in inducing the hydrolytic decomposition of cyanide, described in the sequel, is a decided disadvantage. The precipitation of gold was effected by "a piece or plate of zinc." The process does not appear to have been tried on a large scale, nor did any results follow from the later experiments made on the subject by Endlich and Mühlenberg,[*] and by Louis Janin, Jr.,[†] in America.

In 1886, however, a series of experiments on wet processes of treating gold ores were begun by J. S. MacArthur and R. W. and Wm. Forrest in Glasgow, and it was entirely owing to their energy and skill that cyanide of potassium was successfully applied in practice to the treatment of gold ores. Patents were taken out by them in Great Britain in 1887 (No. 14,174), in the United States in 1889 (No. 403,202), and in other parts of the world. Their process consists essentially in attacking gold and silver ores by dilute solutions containing less than one per cent. of KCy, caustic soda or lime being added to ores rendered acid by the oxidation of pyrites, and then in precipitating the precious metals by zinc shavings. This process has now passed into use in all parts of the world. Its success is complete on many ores, and its extension will probably be very great, partly at the expense of the chlorination process. The chief advantage of the cyanide process over the chlorination process is that roasting is not necessary, even if sulphides are present; this is a most important point in the treatment of low-grade ores, especially in places where fuel and labour are costly. Moreover, silver as well as gold is extracted by cyanide solutions. Messrs. MacArthur and Forrest, in the course of their investigations on wet methods, became dissatisfied with chlorine as a solvent, owing to its energetic and preferential action on sulphides of base metals and other bodies, and its inapplicability to ores containing silver. They desired to find a solvent which would exercise a selective action in favour of the precious metals, instead of other substances. With this object in view, they experimented with a number of solvents (such as ferric bromide, ferric chloride, &c.), and finally decided that potassium cyanide possessed advantages over all other substances. These chemists lay especial stress on the advantages to be obtained by using very dilute solutions, as they believe that, in this way,

[*] *Eng. and Mng. Journ.*, Jan. 17, 1891, p. 86.
[†] *Ibid.*, Dec. 29, 1890.

the waste of chemicals, caused by reactions with the base minerals present, is almost entirely avoided, while the dissolution of gold and silver is not retarded.

The MacArthur-Forrest Process.—The following general description of the working of the MacArthur-Forrest cyanide process is based partly on accounts furnished by Mr. J. S. MacArthur, to whom the author is much indebted for a large amount of information, and partly on the actual work being done in South Africa, in America, and elsewhere. The mechanical treatment of the ore—viz., crushing, charging into vats, leaching, and discharging the tailings—is based on the same principles, and could be effected by the same general methods as are adopted in chlorination or in any other leaching process. Many details will, therefore, be omitted as having been already covered in the course of the descriptions given in Chapters xi. to xiv.

The process comprises four distinct operations, viz. :—(1) Preparation of the ore for treatment. (2) Solution of the gold in potassium cyanide. (3) Precipitation of the dissolved gold by means of zinc; and (4) Conversion of the precipitated gold into bullion by fusion.

1. Preparation of the Ore.—The ore is crushed to a suitable degree of fineness, which depends partly on the condition of the gold and partly on the nature of the gangue. In the Transvaal, where tailings are treated, the ore is crushed in stamp batteries previous to amalgamation, the screens being usually about 25 or 30-mesh. In this case, sizing is performed to some extent during the operation of catching the tailings, the slimes in part passing off suspended in the water used in the stamp-mill. Nevertheless, masses of slimes occasionally accumulate and resist the passage of the cyanide solution or retain the dissolved gold. If mixed with sand, the slimes are less harmful, but they tend to separate from the sand again when the leaching begins.

At the Mercur Gold Mine, Fairfield, Utah, the ore is a silicious limestone,[*] carrying magnetic oxide of iron, and in places containing much silt, one of the proofs of its aqueous origin. Any attempt at fine crushing results in the sliming of the greater portion, and renders leaching impossible. The course of procedure adopted is consequently as follows:—The ore is passed through a Dodge rock-breaker, set to break the ore to a maximum size of 1 inch; it is then reduced to $\frac{1}{2}$ inch by a pair of 20-inch corrugated rolls; thence it is passed over a $\frac{1}{2}$-inch grizzly to a similar pair of 12-inch rolls, which reduce it to a maximum size of $\frac{1}{4}$ inch. After this treatment, 61 per cent. remains on a No. 12 screen, while 26 per cent. passes through a No. 30 screen, this part consisting almost entirely of impalpable

[*] *Eng. and Mng. Journ.*, Nov. 5, 1892.

powder. This material can be leached, although somewhat more slowly than is desirable.

As a rule, ores are crushed so as to pass through screens varying from sixteen to thirty holes to the linear inch; in the product thus obtained there is usually from 90 to 95 per cent. of the gold which can be reached by the solution, and, if rolls have been used for crushing, it is not difficult to leach. Stetefeldt gives an instance* in which leaching by hyposulphite solutions was successfully performed on the Bremen tailings, at Silver City, New Mexico, when the average fineness was such that 87·8 per cent. of the ore would pass through a screen with 150 holes to the linear inch. The rate of leaching in this case was about $\frac{1}{2}$ inch per hour, with a vacuum of 14 inches of mercury, a degree of slowness which would militate strongly against the success of any works. For the separation of slimes from crushed ore, see p. 311.

2. Solution of the Gold.—This is effected in wooden, iron, or cement vats, with false bottoms similar in construction to those used in the Plattner process (see p. 250), or to those employed in the Russell process described by Stetefeldt.† The vats should be round, and may be discharged by sluicing through a door in the side or bottom, or by shovelling out. Those in use in Utah have sheet-iron sides with a wooden bottom, and have a capacity of 14 tons of ore. In South Africa the vats are often of 2,000 cubic feet capacity, and hold 75 tons, but the Langlaagte tanks contain 450 tons each, and the Simmer and Jack vats 600 tons each.

The use of large vats in preference to small ones is to be strongly recommended, as the attention and superintendence needed by a 10-ton vat costs as much as for a 600-ton vat. Stetefeldt considers‡ that constructive difficulties limit the diameter of wooden vats to 16 feet, but the Transvaal vats of 40 feet diameter give no trouble. The inside depth, Stetefeldt continues, should not be more than 5 feet if the tailings are shovelled out, and 7 feet if they are removed by sluicing. The depth has usually little influence on the rate of leaching, as the greater "head" of liquid tends to compensate for the increase of compactness at the bottom of the vat. At Cusi, in treating silver ores by the Russell process,§ it was found that the rate of leaching was greater for a charge 6 feet deep than for one which was only 2 feet deep. In South Africa the vats are often 10 feet deep.

In calculating the capacity of a leaching vat, the volume of a ton of raw ore may be taken at from 22 to 28 cubic feet when dry, and from 20 to 26 cubic feet when wetted down.

The false bottom is usually a wooden framework, constructed

---

* *Lixiviation of Silver Ores*, p. 149. † *Ibid.*, pp. 114-119.
‡ *Ibid.*, p. 114. § *Ibid.*, p. 149.

of boards pierced with numerous auger holes, or of wooden slats crossing each other and covered by canvas or cocoa-nut matting, which are not rapidly destroyed by the solution, as they are in the chlorination process. Below this a layer of coarse sand and pebbles, through which the solution percolates, is sometimes placed.

An excellent false bottom recommended by the Cassel Gold Extracting Company, the owners of the MacArthur-Forest patents, consists of angle-irons with the angles uppermost, and the free edges nearly touching. The irons are supported on wooden slats 1 inch thick, which give a space of that depth below the filter-bed, in which the liquid may accumulate. The triangular channels between the irons are filled with sand. When the vat is being cleared out, the workmen's shovels slide along the top of the angles and leave the filter-bed undisturbed. Triangular pieces of wood may be used instead of angle-iron. Thick canvas duck, resting on matting, forms a trustworthy filter-cloth.

As in the case of chlorination, a layer of perforated glazed porcelain tiles to support the filter-cloth is better than wood, as no absorption of the solution takes place when the former is used. If wood is used it is covered by a coating of paraffin paint, or by a mixture of asphaltum and coal-tar. It was found by Mr. Julian of the Salisbury Cyanide Works, South Africa, that pine wood, lying 34 hours in a solution containing 0·3 per cent. of cyanide reduced the strength to 0·005 per cent., while glass reduced it to 0·2 per cent. and cement to 0·24 per cent. Cement tanks have been used at the Langlaagte Estate since May, 1892.

An iron pipe communicates with the space below the false bottom, and conveys the liquid to the pumps or to the zinc boxes. The solution does not attack wood or iron: brass and bronzes are attacked and corroded rapidly.

The crushed ore is conveyed in cars pushed by hand, running on an overhead railway, and emptied into the vats by tipping. The supports of the railway must not touch the sides or any part of the vats or their supports, as the jarring caused by the running of the cars makes the ore settle down, and so renders the leaching more difficult and tedious.

At the Langlaagte Works the tailings from the stamp batteries are hauled for 600 yards up an incline of 1 in 15 to the tramway above the vats, at a cost of 2d. per ton. At the Salisbury Cyanide plant, a tailings wheel lifts the discharge from the plates into a flume, which carries it to a hydraulic separator, where the slimes are separated from the coarse tailings. The latter are then run directly into the cyanide vats, thus saving handling; this method of direct filling is now adopted in several works on the Rand, but the separation of the slimes from the sand is not complete, and this has a prejudicial effect on the percentage of gold extracted. The separation of the slimes is beneficial in expediting the filtration in the vats.

A widely-adopted form of slime separator is that known as Butters & Mein's collecting vat. It consists of a large round vat, 20 to 30 feet in diameter and 20 feet deep. The tailings from the stamp battery are run direct into a revolving distributor, which spreads the pulp evenly over the surface of the water with which the vat is filled. The sand settles to the bottom and accumulates there, while the slimes overflow at the top. When the accumulation of sand in the vat reaches to within 5 feet of the top, the water is drained off and the ore discharged by shovelling through an aperture in the bottom into ore-cars below. This sand is now in a suitable condition for leaching, containing only small quantities of slimes. At the Jumpers mine the pulp collected in this way is treated directly without removal. At the May Consolidated and the Croesus mills, where the method of "double treatment" of the tailings is employed, the first solution of cyanide is directly applied to these vats, and, after draining, the pulp, wetted with cyanide solution, is transferred to the second treatment vats (Hatch and Chalmers). As already stated, the usual method is to remove the collected sands to the cyanide vats for treatment.

The vats are filled to within a few inches of the top, and the charge is levelled by means of hoes. The amount of ore charged-in is such that, after the solutions have been applied, the surface of the charge may stand at about 12 inches below the rim. In levelling the ore, the labourer must not step into the vat or forcibly press down the ore, as irregular filtering is produced by this cause. The shrinkage of the charge on the addition of liquid is from 10 to 18 per cent. The ore should be charged-in as dry as possible, but a few per cent. of moisture makes very little difference to the subsequent leaching.

The charge is now ready for lixiviation. If soluble salts are present in it, a preliminary washing with water is required, while, if decomposed pyrites are present, treatment with an alkaline solution is also necessary, for reasons to be explained when the chemistry of the process is considered. In each case, from 7 to 9 cubic feet of liquid per ton of ore are added and allowed to remain on the charge for some hours before being displaced. The liquids are run on to the top of the charge through pipes of $2\frac{1}{2}$ to 4 inches diameter, and are distributed by a large box with a perforated bottom, by which the force of the discharge is broken. The orifice below the filter-bed is closed before beginning to add the liquids, and not opened until the solution stands about 3 inches above the surface of the ore, when the leaching is begun. After the caustic alkali (if it is used) has been allowed to remain in contact with the ore for a sufficient length of time, it is displaced by wash-water, which is added above, while the alkaline solution is allowed to run out at the bottom. A dilute solution of cyanide is then run in, the stopcock below the filter-bed being closed as soon as all the wash-water has been displaced. In leaching, the charge is always kept completely submerged, and the surface of the liquid is never allowed to sink below the surface of the ore for reasons which will be given subsequently.

The cyanide solution is often allowed to stand in contact with the ore for from 12 to 24 hours undisturbed before the stopcock below the filter-bed is opened and the solution drawn off. It is usually conveyed at once to the zinc boxes, but it was formerly preferred either to raise it and pass it through the charge again (circulation method), or to transfer it to a second or even a third charged vat before precipitating the gold. The advantage of these "circulation" and "transference" methods is that the solutions become much richer in gold than if they are only allowed to percolate through a single charge of ore, and consequently they give a cleaner deposit on the zinc with much less consumption of cyanide, the volume of solution passing through the precipitation boxes being less. At the Mercur Mine the circulation is kept up for from 24 to 240 hours, according to the speed of leaching, the usual time being about sixty hours. At the Robinson Mine 20 tons of solution cover the ore in a 75-ton vat, and are continually pumped backed into the same vat for thirty-six hours, and then passed to the zinc boxes.

The "Simmer and Jack" works, South Africa, are shown in Fig. 55a.* Here it was necessary to raise the tailings, and this was done by hauling them up inclined platforms carried on trestles. A double gangway runs over the vats for filling. The leaching vats are the largest on the Rand gold-fields, each holding 600 tons, being 40 feet in diameter with 14-foot staves, the staves, as well as the bottom, being of 9-inch by 3-inch material. There are eight discharge doors, 16 inches in diameter, in the bottom of each vat. The three solution vats are 26 feet in diameter, with 15-foot staves, and the liquid is returned to them from the sumps by three 2-inch centrifugal pumps.

It has been thought desirable to stir the solution and ore together, in certain cases, but the loss by decomposition of the cyanide thus caused, and the cost of the power, are usually more important than the additional amount of gold recovered. The Cassel Gold Extracting Company recommend the following method of procedure when agitation is used :—The solution-vat or agitator (A, Fig. 56) is 4 feet in diameter and 4½ feet deep. This is filled to a depth of from 18 to 20 inches with the solution, from the vat C, and the mixer-shaft set going at about sixty

* Reproduced by permission from the paper by Butters and Smart in the *Proceedings of the Inst. of Civil Engineers*, vol. cxx., pt. ii., 1895.

THE CYANIDE PROCESS. 313

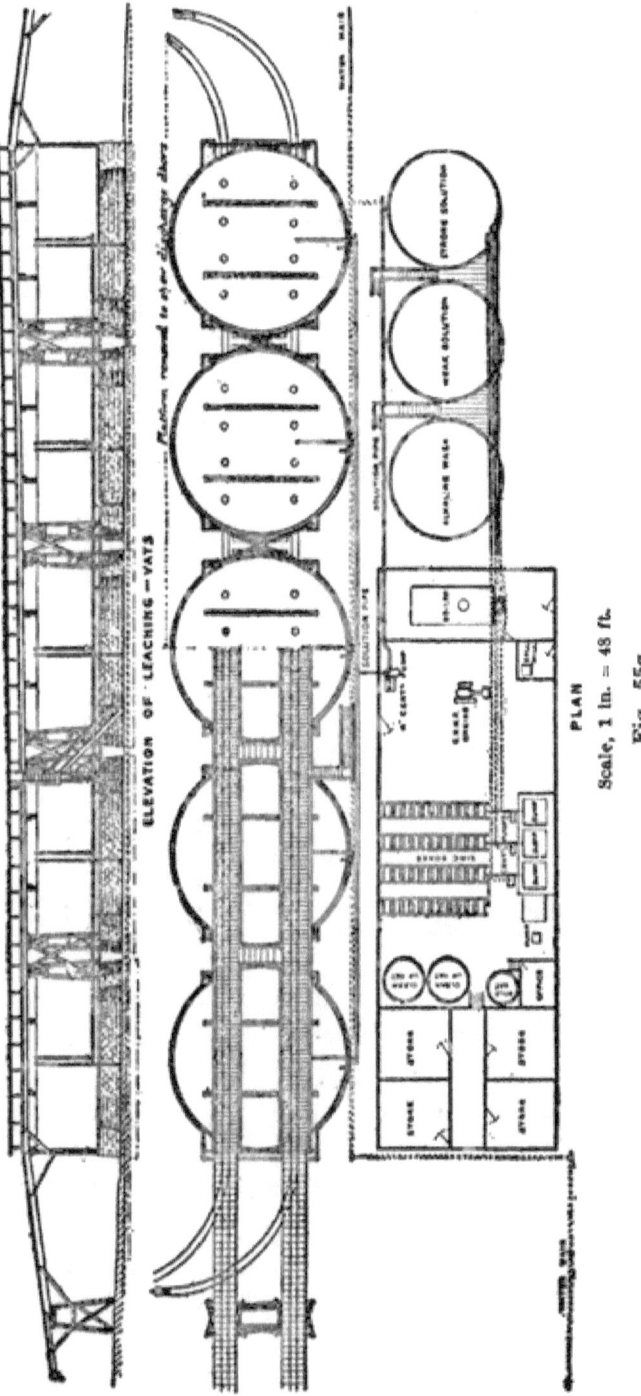

Fig. 55a.

revolutions per minute. The ore (1¾ tons, or double the weight of the liquid) is then added gradually from a hopper, B, overhead, in which it has been stored, and agitation is continued until the gold is dissolved. The pulp is then discharged through a 2-inch iron pipe into D, the large filtering vat, which is 12 feet in diameter and 4 feet deep, having a capacity of about 20 tons; here the leaching is performed as usual. This treatment is recommended for some highly pyritic ores, and in all cases where

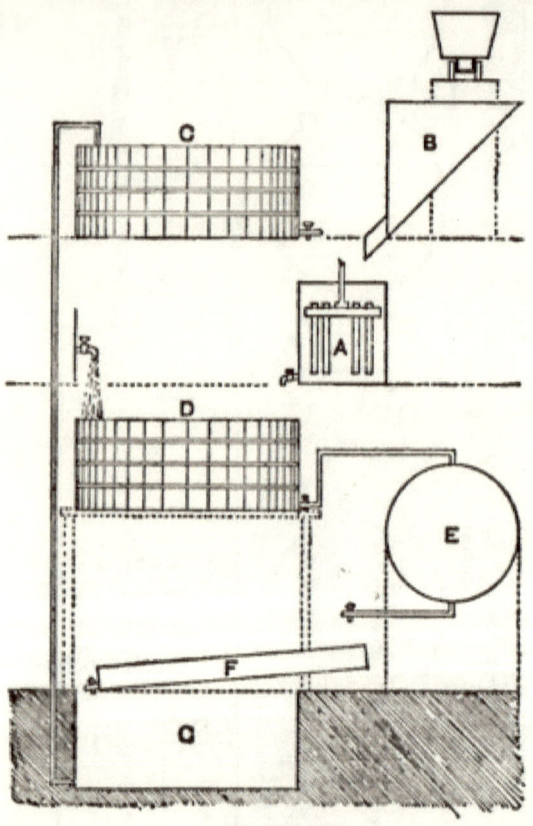

Fig. 53.

leaching is slow owing to the sliming of the ore, which is especially liable to occur when much galena is present.

The power required to agitate the pulp is said to be at least one horse-power per ton, and it is also stated that the ore cannot be completely freed from the solution, which does not drain readily from the ore after the agitation is finished, although the supernatant liquid, forming about half of the solution, can of course be removed by decantation. In Fig. 56, E is the vacuum boiler, F the zinc box, and G the sump.

The rate of leaching may be increased by passing the liquid

upwards and downwards alternately, or by creating a vacuum below the filter-bed. The latter may be done in several ways, as has already been pointed out in Chapter xiii., p. 272. The use of a large boiler, in which a vacuum is created by a Westinghouse or other pump, is perhaps the most advantageous course; as soon as the pressure in the boiler falls to about half an atmosphere, it is connected with the aperture of the vat below the filter-bed. Korting's injector is recommended by Stetefeldt for the production of a vacuum. The rate of leaching is often doubled by the diminution of the pressure, below the filter-bed, to half an atmosphere, and in some cases it is increased from $\frac{1}{2}$ or 1 inch to 7 or 8 inches of liquid (in the leaching vat) per hour. In some cases, as already remarked (p. 272), the creation of a high vacuum is rather a cause of delay than otherwise, the ore packing down and making leaching quite impossible. The author has seen an attempt to leach a roasted ore, containing much slimes, by means of a vacuum of 20 inches of mercury, result in the production of a compact mass which could scarcely be penetrated by a pick-axe, while water stood above and could not be drawn through in any way.

Strength of the Solution.—In the Transvaal it is customary to use two different strengths, viz., a "strong" solution, containing from 0·25 to 0·35 per cent. of cyanide of potassium, and a "weak" solution, containing from 0·05 to 0·15 per cent. of cyanide. The strong solution is allowed to filter through during a period of from eight to forty-eight hours, and the weak one is then run through for twenty-four or forty-eight hours. The amount of each solution used is, in general, about $\frac{1}{2}$ ton per ton of ore, but, as before stated, only about $5\frac{1}{4}$ cwts. of solution per ton of ore is used at the Robinson Mine with the "circulation system.". At the Mercur Mine the strength of the solution in use is 0·25 per cent.

When leaching is complete, so that no more gold is being dissolved, a point which can be determined by testing the escaping solution with bright zinc shavings, the addition of fresh solution is stopped, and the surface of the ore is laid bare by the sinking of the liquid. There still remains in the vat, however, about 6 or 8 cwts. of solution per ton of ore. This might be reduced to 3 or 4 cwts. by draining for some time, but the charge would then shrink and crack and separate from the side walls, so that the succeeding operations would be less effectual. To displace the solution, therefore, water is run on to the ore surface as soon as the latter emerges, and is kept running until the quantity added is equal to that of the solution originally retained by the charge. Up to this point the discharge from below the filter-bed is allowed to run into the "stock" solution, although somewhat more dilute than the main mass, owing to slight admixture with the water. The operation is now stopped and the tailings may be drained and discharged, but, as they still contain a little cyanide, a slight saving is effected by displacing the first wash-water by a second. In this case the first wash-water is run into a separate vat, and used again to displace the stock solution, which is thus reduced in strength less than if fresh water were used. More cyanide is left in the tailings if the solution be drained off as completely as possible before being displaced, as the wash-water finds its way down the cracks, leaving the unbroken masses of ore less efficiently freed from cyanide.

By this practice the volume of the solution is kept constant, but its strength falls off considerably, owing to decomposition in the ore and in the zinc boxes. It is tested in the "sump" to which it flows from the zinc boxes, and enough cyanide added to raise it to the required strength. The cyanide is usually dissolved in a little water before being added to the stock solution, as the amount of KCy present is more easily determined in a strong solution than in any other form. At the Robinson Mine lumps of cyanide are added to the liquid which is circulating through the ore, as the insoluble impurities ("carbides of iron") are in this way left on the surface of the charge, and are dumped with the tailings instead of accumulating in the sumps.

Messrs. MacArthur & Forrest recommend the concentrated solution of cyanide to be made and kept in a small vat standing at a higher level than the large vats used for the storage of the stock solution. Cyanide of potassium is supplied in zinc-lined boxes holding 190 to 195 lbs. of cyanide, containing about 78 per cent. of available KCy. The contents of these boxes are broken into lumps and placed in wire-gauze trays, which are immersed in water contained in the small vat. The soluble salts are soon dissolved, potassium cyanide in particular being very rapidly removed, and the tray is then lifted out, still containing the insoluble material, which is thrown away. The strength of the concentrated solution, which is kept to raise the stock solution to its normal strength, may be from 10 to 25 per cent. It should not be prepared until it is wanted, as it undergoes somewhat rapid decomposition.

Disposal of the Tailings.—These are sometimes sluiced out of the tanks by water under pressure. The sluice gates may be of about 1 or 2 square feet in area, and are placed so that their lower edges are level with the surface of the filter. Where the supply of water is not large, or the fall of the ground is insufficient for sluicing out the tailings, they are removed by shovelling out, the cost being about 6d. per ton with European labour. Before the vats are emptied, samples are taken, usually by means of a long iron semicircular probe, shaped like a cheese-taster, which is thrust to the bottom of the vat, then revolved by means of the handle, and withdrawn with the tailings adhering to it. If the results are high, samples from various parts of the vat should be assayed separately, so as to find out if the percolation has been uneven. At the Roodepoort Works, round vats have been constructed, 40 feet in diameter and 8 feet deep, with a capacity of 360 tons. They are placed on firm concrete foundations, raised 6 feet from the ground, and the tailings are discharged through four openings in the bottom communicating with tunnels. It is said that these vats show no signs of leakage, and that they can be discharged in six hours by twenty-eight natives at a cost of 1½d. per ton.*

At the Langlaagte Estate, the five large 450-ton vats, whose upper edges are level with the surface of the ground, are discharged by a steam crane by which truck loads of tailings are elevated and placed on a tram line. The trucks are lowered into the vats and loaded up by Kaffirs, and the total cost of discharge is said to be 2d. per ton, including maintenance. The vats and sumps are excavated in the solid ground, the walls being built of brick and lined with cement.

3. Precipitation of the Gold.—The precipitation of the gold is effected by shavings of zinc freshly turned on a lathe. In South Africa, the zinc linings to the boxes in which the cyanide is imported are worked up for the purpose, a light spongy mass easily traversed by the solution and presenting a large surface for precipitation being formed. The shavings are placed in wooden troughs, the "zinc-boxes;" they are divided into compartments by partitions, which cause the solution to flow alternately upward and downward. Each alternate compartment is empty, in order that, in passing through the shavings, the solution may invariably flow upwards. "It is arranged so, because, if the solution were to flow down, the gold slimes would collect on the upper surface of the zinc, and impede further flow, but by upward flowing the gold slimes are precipitated on the under surface of the zinc, from which they continually drop off, and permit free passage of the solution."

At the Robinson Mine, the shavings are supported on wire-screen trays, so that the finely-divided, precipitated gold falls through to the bottom of the trough, while the unaltered zinc remains on the sieve. At this mine the boxes are 20 feet long and 2 feet square in cross-section; there are ten compartments in each, and about 40 lbs. of shavings in each compartment, except those at the head and foot, which are left empty in order to allow the slimes to settle. About 30 tons of solution pass through each box in a day of nine hours, so that every part of the solution is in contact with the zinc for about twenty-five minutes. The solution contains from 1 to 3 ozs. of gold per ton, and, after passing through the boxes, contains from ½ to 2 dwts. per ton. There are two sets of boxes, one for the "strong" solution, the other for the "weak." In the latter, less zinc is consumed, but the gold-slimes are poorer, less gold being con-

* *Eng. and Mng. Journ.*, Mar. 25, 1893, p. 273.

tained in the "weak" solution. The total amount of zinc consumed is 100 lbs. per day, the gold extracted being about 120 ozs. At the Mercur Mine, the boxes are 40 feet long and 12 inches square in cross-section, with compartments 3 feet long, and contain zinc enough to maintain the contents of the last compartment quite bright. The consumption here is about 1 lb. of zinc for 1 oz. of gold recovered.

When the solution comes in contact with bright zinc, the latter turns black at once, owing to the deposition on it of finely divided gold. The zinc is gradually dissolved, and the shavings fall to pieces, those in the first compartment being consumed most rapidly. As the precipitation proceeds, the zinc is transferred from the lower compartments to the upper ones, and fresh zinc is added at the foot of the box.

Clean-up.—At the Robinson Mine, the clean-up takes place once or twice a month. The screens containing the undissolved shavings are lifted out of the zinc boxes, and the zinc-gold slimes, remaining at the bottom of the boxes, are allowed some time to settle. The clear liquor is then siphoned off and the slimes allowed to run out by withdrawing a plug in the bottom of the box, and drained through a 40-mesh screen, which retains a small part only. After the residue has been rubbed on the screen with sticks tipped with indiarubber, and well washed, it is put back again into the first compartment of the box, on the top of fresh shavings, as it consists mainly of small pieces of unconsumed zinc. The subsequent fineness of the bullion greatly depends on the care with which these operations are conducted. By careful manipulation most of the undissolved zinc can be separated in this way. The slimes proper are now transferred to enamelled iron pans and carefully and slowly dried. The richness of the dried precipitate depends on the strength of the cyanide solution, and on the time of contact as well as on the quantities of base metals which are present in the solution. By the prolonged action of the cyanide, the zinc shavings become partly corroded and disintegrated, so that the precipitated gold is mixed with zinc debris. Ordinary commercial zinc contains a considerable proportion, generally over 1 per cent., of lead, and a small quantity of carbon, besides other impurities, such as arsenic and antimony. All these accumulate and are collected with the gold. The precipitated gold and the carbon from the zinc are invariably in a state of fine division, and, by using a fine sieve, the gold can be very closely separated from the zinc, but, as the moderately fine particles of zinc still retain a considerable proportion of gold entangled, it is not always desirable to use a very fine sieve. If the cyanide solution contains copper, it is found with the gold.

According to Messrs. Butters and Clennel,* the pans in use

* *Eng. and Mng. Journ.*, Oct. 15, 1892, p. 365.

at the Robinson Works contain about 5 or 6 gallons of dried precipitate (often called "gold slimes"), which may contain as much as 150 ozs., or as little as 20 ozs., of gold. A little silver is also contained in it, the remainder being chiefly zinc and lead, with smaller quantities of tin, antimony, organic matter, &c. The average in South Africa is said to be approximately:—

Gold and silver,	20 to 50 per cent.
Zinc,	30 to 60 ,,
Lead,	about 10 ,,
Carbon,	,, 10 ,,

A carefully prepared sample of gold slimes which was obtained at one of the African works contains:—

Gold,	60 per cent.
Silver,	10 ,,
Base metals and carbon,	30 ,,

This is by no means average, but is exceptionally good, and must not be taken as showing what is usual. The Hanauer Smelting Company found the precipitate from the Mercur Mine to contain the following substances:—*

Zinc,	39·1 per cent.
Carbonate of lime,	36·7 ,,
Gold,	4·4 ,,
Cyanogen,	3·5 ,,
Sulphur,	2·6 ,,
Iron,	2·4 ,,
Residue,	6·0 ,,
	94·7 ,,

The carbonate of lime present is probably deposited in the boxes chiefly from suspension in the solution, the gangue of the ore mainly consisting of limestone.

4. **Production of Bullion from the Precipitate.**—In the United States, most mines will, no doubt, find it convenient to follow the example of the Mercur Mine and sell the dried precipitate to smelting companies, which pay about $20 per oz. for the gold contents. In South Africa this course cannot be pursued, and various methods have been suggested for effecting the elimination of the zinc and other base metals. The usual course is to melt the mass in graphite or clay crucibles, using as fluxes sand, borax and bicarbonate of soda. The slag, which consists of silicates of zinc, soda, &c., corrodes the pots rapidly. Large quantities of zinc oxide are given off as fumes, forming thick crusts in the flues, and doubtless taking some gold with them, and evil-smelling products of decomposition of the cyanides are also evolved. The bullion produced varies in colour from a pale yellow to a brownish linnet-green, and is about 650 fine, but

* *Eng. and Mng. Journ.*, Nov. 5, 1892, p. 440.

cannot be obtained uniform in composition, so that accurate assays are difficult to obtain.

The results of analyses made on three ingots of bullion produced by the MacArthur-Forrest process in South Africa, and shipped to London, are appended :—

	I.	II.	III.
Gold,	60·3	61·7	72·6
Silver,	7·0	8·1	9·2
Zinc,	15·0	9·5	7·1
Lead,	7·0	16·4	4·9
Copper,	6·5	4·0	4·8
Iron,	2·2	0·3	1·4
Nickel,	2·0	...	.
	100·0	100·0	100·0

The slags obtained in this way are always rich in gold, part of which is sometimes in the form of shots, and may be recovered by crushing and panning. The crushed slag is then fused again with the addition of granulated lead, or better still, of litharge, when all the gold is concentrated in the lead. If the lead thus obtained is granulated, it can be used again until rich enough to be worth cupelling. It has been stated that the infusibility of the slag and its richness in gold is caused by the presence of a large amount of finely divided carbon, which had been originally taken up by molten zinc in the course of its manufacture, and left as residue when the zinc was dissolved.

If the slimes are roasted in a muffle furnace, some of the zinc is volatilised, and a bullion 800 fine can thus be obtained, according to Butters and Clennel. The separation of the zinc by solution in sulphuric or hydrochloric acid, or in acid sodium sulphate, has been stated to be impracticable, owing to the difficulty of filtering the slimes. Moreover, the vessels in which the solution is effected boil over from the copious evolution of hydrocyanic acid, and there are other drawbacks to this method. J. S. MacArthur, however, points out that this is due to the presence of soluble cyanide salts containing zinc. By the action of acids, cyanide of zinc ($ZnCy_2$) is set free and precipitated, forming a gelatinous mass which cannot be filtered, and is not readily dissolved by solvents other than potassic cyanide. This substance is also very troublesome in the crucible, rendering the slags pasty and rich. If, however, the slimes in their original condition, while still alkaline, are treated in a filter-press such as Johnson's (described and figured in Stetefeldt's *Lixiviation of Silver Ores*, p. 126), the soluble cyanide salts may readily be separated by washing, and the greater part of the zinc thus removed in solution. The residue is then spread in a thin layer on iron trays, and dried and heated to a barely-perceptible red heat in the flue of a furnace. The carbon, zinc, arsenic, &c., ignite readily, being in a very fine state of division, and roasting proceeds regularly through the mass to the bottom without any stirring, leaving a completely oxidised porous mass, aggregated more or less into granules. This residue contains no carbon and little arsenic, and consists chiefly of oxides of zinc and lead, with metallic gold and silver. It is rapidly acted on when heated with dilute sulphuric acid, which dissolves the zinc, and forms insoluble sulphate of lead. There is no tendency for the mixture to boil over, as is the case when cyanides are present, and the residue is easy to filter, the granular sulphate of lead offering little resistance. When the residue is washed and dried, it is easy to fuse, not requiring large amounts of fluxes, and forming liquid slags in which the greater part of the remaining impurities are contained. The bullion thus prepared is often worth £3 10s. per oz., the average value being over £3 per oz. At the Robinson Mine, over three-fifths of the foreign metals present in the gold bullion consists of silver, the remainder being chiefly zinc. However, the presence of the last-named metal does not diminish the price given for the bullion, and, if the latter is more than 500 fine, no abatement is made in the payments for the gold.

Molloy's Method of Precipitation. — Instead of using zinc, Molloy proposes to pass the gold solution through some kind of amalgamator, such as the hydrogen amalgamator described at p. 164. A "weak" electric current is sent through the solution, the mercury being used as the negative pole. The solution is decomposed and metallic potassium is released on the surface of the mercury, but instantly exchanges itself for the gold in the solution, gold amalgam being formed and potassium cyanide regenerated. In another modification, sodium is liberated by the decomposition of sodium carbonate and replaces the gold in solution, an amalgam of the latter being formed. The sodium cyanide in the solution is as efficacious as potassium cyanide for the purpose of dissolving more gold, and it is said that there is much less decomposition of the cyanide than if zinc is used. Moreover the gold amalgam is much more easily handled than the zinc slimes, and considerable losses are thus avoided, while a saving of skilled labour is effected. These methods have never got beyond the experimental stage, and it is not likely that they will prove of practical value. It is certain, from the results obtained by Dr. Gore's experiments[*] that mercury itself will not precipitate gold from the cyanide solution, and therefore the action depends entirely on the electric current.

Electro-deposition of the gold is used in the Siemens-Halske

[*] See p. 334.

process (see p. 329) and the conditions necessary for success have been worked out and stated by the inventors of that method. One of the most important provisions is that the layers of liquid between the anodes and cathodes must be very thin in order that the ions may not be compelled to travel far before they reach the poles. Hence, when great quantities of liquid have to be treated, it is necessary for the electrodes to have a very large surface. Moreover they should stand in a vertical position so that they may not become coated with slime settling from the liquids, which are usually more or less muddy. It follows that mercury cannot be used in practice for the cathodes. Von Gernet has calculated that 80 tons of quicksilver would be required for the treatment of 100 tons of solution per day, and the initial cost of the mercury would, therefore, be some £20,000. Besides this, about 3,000 ounces of gold, worth nearly £13,000, would be retained by the bath after the whole had been subjected to filtration. Thus the produce of, say, the first four months' work would be permanently locked up, until the mercury was re-distilled.

Messrs. MacArthur & Forrest experimented on the action of sodium amalgam, using a small tower filled with pieces of sodium amalgam, through which the solution of gold in potassium cyanide was allowed to trickle slowly. They came to the conclusion that sodium was not more efficacious than zinc, but was much more expensive. The saving of expense in the production of the bullion from the precipitate was outweighed by the cost of the sodium required to form the precipitate.

Plant Required.—For the treatment of 2,000 tons of tailings per month, Messrs. MacArthur & Forrest consider that six circular vats each 19 feet in diameter and 4 feet deep are sufficient. Each vat holds about 40 tons of ore, and must be charged, leached, and emptied in three days. The other requirements of this plant are two large vats, each of 2,000 cubic feet capacity, to hold the strong and weak cyanide solutions, and a similar vat to hold the soda solution. If lime in the solid state is mixed with the ore before it is charged into the leaching vat, this last tank may be dispensed with. Zinc boxes, sumps, pumps, pipes, and launders are also required, and furnaces to fuse the zinc residues. The amount of water required for this plant is about 2,000 to 3,000 gallons per day, or 30 to 45 gallons per ton of ore. The labour needed consists of eight men for a shift of twelve hours. A zinc box 14 feet long, 2 feet deep, and 1 foot wide, divided into six or eight compartments, and holding 100 lbs. of zinc, should suffice to treat in about nine hours the gold solution obtained each day.

A plant capable of treating 20 tons of pyritic ore by means of agitation, consists of six agitator-vats (each holding a charge of $1\frac{1}{4}$ tons of ore), three large filtering vats, and two vacuum boilers to assist in the filtering. The discharge from the agitators to the filters is by means of pipes which are at least 2 inches in diameter. Pumps are required to raise the liquid from the sump to the stock solution vat, and to create a vacuum in the boiler. The power required for the agitators and pumps is about 6 H.P. The consumption of water is from 600 to 1,000 gallons per day, or from 30 to 50 gallons per ton of ore. Labour consists of three men for a shift of eight or twelve hours.

Treatment of Ore Slimes.—Ore slimes, consisting of fine particles of clay or of ferric hydrates, sometimes exhibit a curious action in withdrawing gold from almost any of its solutions. This has been noticed both in chlorination and cyanide mills. As the slimes settle, so that the muddy solution becomes clear, the gold appears to go to the bottom with them, so that they often contain many ounces of gold to the ton after treatment, although not especially rich at first. This fact must be remembered in the treatment of ores which slime badly. It is possible that the precipitated gelatinous ferric hydrate formed in the vats by the action of alkalies on oxidised salts of iron may have some similar effect. This substance should be collected and assayed occasionally.

On the Rand, there have been accumulated hundreds of thousands of tons of slimes which cannot be made to yield any of its gold to cyanide on the large scale, owing to its impermeability, although the gold contained in it is readily soluble in the laboratory. The average value of the Robinson slimes is between 7 and 8 dwts. of gold per ton, and the fineness is such that it would pass through a 225-mesh screen. It has been proposed by Bettel to treat this material in Johnson's filter-presses. The slimes are mixed with a very dilute (0·01 per cent.) cyanide solution and thoroughly agitated, then filtered, and water forced through under a pressure of 100 lbs. to the square inch. This pressure leaching is similar to that suggested by E. N. Riotte in connection with the hyposulphite lixiviation process (see p. 273). It is stated that about 98 per cent. of the assay value of the slimes can be extracted in this way, but the method has not yet been applied on the large scale.

Filter press separation was in use at the Crown Mines, New Zealand, as early as the year 1889. Separation of the liquid by decantation has also been used with even greater economical success.

Testing of Ores.—Experiments with small quantities of material in the laboratory will usually determine the maximum extraction that can be looked for. The weight of ore taken may be from 100 to 200 grammes. It should be digested in a beaker, with sufficient solution to form a thin mud, with occasional stirring, or better still, in a funnel or lamp glass, through which the solution slowly passes. The solution is then separated, and the residue thoroughly washed by filtration, and assayed to find how much of the precious metals has been removed. The consumption of cyanide may be determined by titrating the solution before and after use. In this way, the maximum extraction of gold and silver from an ore which has been crushed to different degrees of fineness, with solutions at different strengths, acting at various temperatures and for different lengths of time, may be determined. Such data will be of great value in determining the degree of suitability of the process to any given ore, but the results obtained cannot usually be repeated on the large scale. Mechanical difficulties preclude fine crushing beyond a certain point: the relative amount of solution used in practice must be less than the large quantities giving the best results on a small scale. Moreover, the washing and filtering is more perfectly performed in the laboratory than in the mill, but, on the other hand, the consumption of cyanide may be less in the latter case than in the former.

Preliminary tests as to the amount of alkali which it is necessary to add, and the amount of cyanide decomposed are made as follows:—(a) *Alkali.*—200 grammes of ore are made into a paste with water in a porcelain basin, and tested with litmus paper. If an acid reaction is observed, a titrated solution of caustic soda is run in, little by little, until the mixture is neutral. It is convenient to make the alkaline solution of such a strength that each c.c. corresponds to one pound of caustic soda per ton of the ore which is being tested. (b) *Decomposition of Cyanide.*—100 grammes of the ore is shaken in an 8-oz. stoppered bottle with 50 to 100 c.c. of a solution of KCy for fifteen minutes, then left undisturbed for twelve hours, and finally shaken again for five minutes. The available cyanide in the separated solution is then estimated and compared with that in the original solution. By merely shaking for a few minutes, most of the decomposition of cyanide is made to take place, but not the whole. The strength of solution used may be from 0·1 to 0·5 per cent. of KCy according to the strength required for dissolving the gold. In general, the stronger the solution the greater the amount of cyanide decomposed. If soluble salts are present in the ore, they should be removed by washing with water before applying the cyanide, and the effect of a soda solution in saving cyanide should also be tested.

The Cassel Gold Extracting Company supply a small testing plant capable of treating 3 cwts. of ore at one time, under conditions closely resembling those obtaining in a good mill. The plant consists of a small revolving barrel for solution, a filtering vat and vacuum boiler, and a zinc box for precipitation. The time of treatment in the barrel varies from six hours upwards. Such a barrel is at the Royal College of Science.

Direct Treatment of Rand Ore.—Several attempts have

been made to treat auriferous banket by the cyanide process alone, without first amalgamating it. In general, however, these have been unsuccessful. The coarser particles of gold, usually removed on the copper plates, require far too much time for their complete dissolution to admit of a high percentage of extraction being attained. The pyritic ores have not yet been proved to be amenable to this method, but with certain oxidised ores from near the surface of the ground, all the gold is in a very finely-divided state, and 75 to 80 per cent. is extracted by cyanide in two or three days. At the George and May Mining Company's works, the oxidised banket is very friable and easily disintegrated, and this was crushed dry, and treated direct with complete success in 1895. Here nearly half the ore was converted into intractable slimes if it was crushed in the battery, and in the last few months of 1894 only 55 per cent. of the gold had been extracted by amalgamation, followed by treatment with cyanide. In the method now adopted the ore is crushed dry by a Gates crusher, and dumped at once into the vats, where it is treated as follows (Hatch & Chalmers) :— The ore is saturated with a solution containing 0·06 per cent. of KCy, and this is displaced by a "strong solution" containing 0·28 per cent.; a "weak" solution containing 0·10 per cent. of cyanide follows, and is washed out with one containing 0·05 per cent. "About sixty hours are required for the whole treatment. Some lime is mixed with the ore, and on an average 0·89 lbs. of cyanide of potassium are used per ton of ore treated." The ore is so friable that it is reduced to barren pebbles and loose sand by merely passing through the rock breaker. About 75 per cent. of the gold is extracted, the ore containing about $5\frac{1}{4}$ dwts. per ton. The total cost of mining and treatment is about 12s. per ton.

Use of Cyanide in the Stamp Battery.—Efforts have been made in South Africa and the United States to treat the ore direct from the battery instead of first passing it over amalgamated plates. Ore from the May Consolidated Mine, Johannesburg, was crushed by the African Gold Recovery Company in the latter part of the year 1892,[*] with cyanide solution instead of water, and led at once into the filtering tanks. The results are stated to have shown that the coarse gold resisted the attack of the cyanide for so long a time as to render the process uneconomical. No doubt this system will meet with great success in cases where all the gold is in a finely divided condition.

At the Stewart Mine, Bingham, Utah, a combination of the amalgamation and cyanide processes is said to have been applied successfully to ore containing a mixture of coarse free gold and fine rebellious particles.[†] The ore is crushed and amalgamated

---

[*] *Eng. and Min. Journ.*, Oct. 8, 18.2, p. 342.
[†] *Ibid.*, April 15, 1893, p. 339.

in Huntington and Crawford mills successively, the water used being a solution of cyanide of potassium. After leaving the mills the solution and pulp are run into settling tanks and the liquid drawn off. It would appear likely that the decomposition of the cyanide in the mills would be excessive, but no details of the working have been published.

Method of Treatment at the Robinson Mine.—At the Robinson Mine* there is a 60-stamp battery, pulverising 280 tons per day through a 40-mesh sieve, the ore containing nearly 2 ozs. gold per ton. Of this amount 71·04 per cent. is caught on the amalgamated copper plates, 5·13 per cent. is extracted from the concentrates at the chlorination works, and 18·55 per cent. from the tailings at the cyanide works, while 5·28 per cent. is lost. The pulp from the battery is concentrated, and the tailings caught in pits and shovelled into cars and conveyed to the cyanide works, while the concentrates go to the chlorination works. There the concentrates are placed in heaps, assayed for sulphur, and mixed so as to make a uniform product for roasting. They are roasted in three furnaces, each 60 feet long, treating 600 to 800 tons per month. When roasted, the material is charged into ten circular chlorination vats with bottom discharge and luted rings and lids. The chlorination occupies four days, 6 lbs. of chlorine being used for each ton of concentrates. The gold is precipitated by a solution of ferrous sulphate, the mass being kept agitated by means of a jet of compressed air. The consumption of ferrous sulphate is from 10 to 15 lbs. per ton of concentrates. The gold is allowed to settle for some time in a series of large vats, and a complete clean-up takes place only four times a year.

At the cyanide works there are twelve vats, each holding 75 tons. The treatment here also occupies four days, the output being 225 tons per day. The whole of the tailings are subjected to treatment, the coarse gold having been removed on the plates, and the pyrites by concentration, so that almost clean sand, containing from 8 to 10 dwts. of finely divided gold, is left for the action of the cyanide. Of this from 1½ to 2 dwts. are left in the tailings, the extraction being over 70 per cent.

The Cyanide Process at the Sylvia Gold and Silver Mining Company, Thames Valley, New Zealand.†—The ore from the deeper levels of the mine contains a high percentage of complex minerals, consisting of galena, zinc-blende, copper, and iron pyrites, and does not, as a rule, show any visible gold. It is crushed in a battery and passed over amalgamated Muntz-metal plates, by which the coarse gold is extracted, the amount thus saved being about 5 dwts. per ton. The ore is then

---

* B. H. Brough, *Journ. of the Soc. of Arts*, vol. xli., 1893, p. 173.
† This description is taken chiefly from the reports of the manager, Dr. A. Scheidel.

carefully sized and concentrated by a complicated system of jigs, rotary tables, buddles, and sizers. The concentrates obtained are classed as follows:—jigger concentrates, first-class slime concentrates, second-class slime concentrates, and buddle concentrates. The jigger concentrates contain 4 ozs. 5 dwts. of gold and 20 ozs. of silver per ton, and consist chiefly of iron and copper pyrites and zinc blende, with a little galena and quartz. The galena in the ore carries greater amounts of gold and silver than the other minerals. It forms slimes for the most part, and is found in the slime concentrates. Of these the first-class contain about 20 per cent. of galena, and an average of 10 ozs. 6 dwts. of gold and 44 ozs. of silver per ton: the second-class concentrates contain very little lead, and assay 4 ozs. 5 dwts. of gold and 20 ozs. of silver per ton. The buddle concentrates are midway in richness between the two slime-concentrates. About 80 per cent. of the total values in the ore are saved on the tables and in the concentrates, and the latter can all be treated by the cyanide process.

It was originally intended to dispose of the concentrates by sale, but the prices realised, after deducting the expenses of bagging, carting, shipping, insurance and treatment, were so small as to render treatment on the spot an imperative necessity. The system ultimately adopted was the MacArthur-Forrest process, after exhaustive trials, and a plant was erected capable of treating 20 tons per day. "It consists of three large agitators, three vacuum filters, a grinding-pan, cyanide solution tank, tanks for gold solution, vacuum and other pumps, and some minor appliances. The whole plant is of local manufacture." The filtration of the concentrates after agitation is difficult to accomplish, and necessitated the introduction of a vacuum filter patented by Dr. Scheidel. "The time of agitation, and the strength of solution applied, vary in accordance with the quality of the material. The quantity of cyanide used for the highest grade of ore amounted to less than 1 per cent., and for low-grade material to considerably under 0·5 per cent. The time of agitation varied between five and twenty-four hours."

The Sylvia Company acquired the right of using the MacArthur-Forrest process on payment of a royalty of $7\frac{1}{2}$ per cent. on the bullion extracted. The patentees did not interfere with the construction of the plant, which was left in Dr. Scheidel's hands. He states that "the results of extraction have varied in accordance with the quality of the material, the slimes generally giving better results than the other products, and the (richer) first-class slimes returned a higher percentage of gold and silver than the lower-grade materials." The extraction was as follows:—

	Gold.	Silver.
First-class slimes: (1) average extraction,	86·11 per cent.	67 per cent.
,, ,, (2) highest ,,	96·45 ,,	94·59 ,,
Second- ,, average ,,	85·34 ,,	68·7 ,,
Jigger concentrates: ,, ,,	80·32 ,,	50 ,,
Buddle ,, ,, ,,	77·9 ,,	54·45 ,,

The average extraction on the whole was 82·67 per cent. of the assay value; the tailings are stacked for a possible future course of re-treatment, as they contain large percentages of lead and copper, besides precious metals to the value of about £5 per ton.

The bullion produced by amalgamation and lixiviation combined is thus seen to be about 66 per cent. of the total value of gold and silver contained in the ore. It is obvious that, if smelting works are established in the neighbourhood at any future time, the concentrates will probably be sent to them, as the tailings after lixiviation are still valuable. Nevertheless, it is clear from the results obtained that in places far removed from smelting works, certain complex ores, including those rich in galena, may be treated by the cyanide process successfully. The cost of treatment, consumption of cyanide, &c., are not given by Dr. Scheidel.

New Primrose Cyanide Works.*—At these works the so-called "double treatment" is used, the tailings being treated with cyanide in two vats in succession. The pulp from the battery is passed into hydraulic classifiers (*spitzluten*), and coarse sands and pyrites, amounting to about 10 per cent. of the whole product of the stamps, are separated from the finer material. The coarser materials, which contain 15 dwts. of gold, are drained and treated for three or four days with a solution containing 0·3 per cent. of cyanide. They are then transferred to other vats and treated for a further period of three days with weaker solutions, which are finally displaced by water. The extraction amounts to 85 to 93 per cent., the consumption of cyanide being about 1½ lbs. per ton. The fine material, overflowing from the classifiers, is fed into a distributing tank and thence to other tanks in which the sand settles, while the slimes overflow and are carried away. The collected sands are then drained dry, the operation being expedited by the use of vacuum pumps, a vacuum of 15 inches of mercury being employed. A solution containing 0·04 per

---

* The account given here is mainly based on the notes supplied by Mr. W. Bettel, consulting chemist of the New Primrose Mine, to Messrs. Hatch & Chalmers' excellent work on *Gold Mines of the Rand.*

cent. of cyanide is then run on, and replaced by a stronger solution, which is allowed to stand for twelve hours, and then drained off and the sand removed to the ordinary leaching tanks. Here a solution containing 0·2 per cent. of cyanide is added, and allowed to stand for twelve hours; it is then replaced by weaker solutions, 0·08 and 0·04 per cent. being used successively, and finally water, rendered alkaline by milk of lime. An average of 78 per cent. of the gold is extracted, while, before the "double treatment" was introduced, only 55 per cent. was obtained. The capacity of the plant is 17,000 tons per month, and the working expenses average 4s. 1d. per ton. The consumption of chemicals per ton of ore is as follows:—cyanide, 0·98 lb., zinc, 0·22 lb., and caustic soda, 0·22 lb. The bullion recovered is worth £3, 5s. per oz.

Siemens-Halske Process.—In this process the gold is deposited from solution by the passage through the liquid of a current of electricity. Moreover, as the precipitation is equally complete and as readily obtained in extremely dilute cyanide solutions as in those containing 0·1 per cent. or over, the strength of the solutions used in dissolving the gold from the ores is made less when electrical precipitation is employed than if zinc is the precipitant. Hence the whole method forms an interesting variation of the ordinary MacArthur-Forrest process, and bids fair to assume great importance in the future. Gold can be extracted from its ores as completely by a solution containing 0·03 per cent. of cyanide as by one containing 0·3 per cent., the only difference being that the time required is considerably longer. On the other hand, the advantage in using the more dilute solution is that the selective action in favour of the gold is increased, and the amount of cyanide decomposed by "cyanicides" in the ore is diminished. In addition to this, some cyanide solution is invariably left in the ore, and if the "weak" solution used to finish the dissolution of the gold contains only 0·01 per cent. of cyanide, instead of 0·1 per cent., the amount of cyanide lost in this way by mechanical means is also reduced.

The process was first adopted on a large scale at the Worcester mill in the Transvaal. Here the vats are 20 feet in diameter and have their sides formed of staves 10 feet long; the five vats hold 135 tons each. The battery pulp, after passing over Frue vanners, is classified into four products by hydraulic classifiers. The first series, which consists of spitzluten, removes the coarse sand and pyrites, amounting to 15 per cent. of the pulp and containing 15 dwts. of gold. This product is treated for nine days with solutions of 0·08 per cent. of cyanide, and, after being washed with 0·01 per cent. solutions, gives residues assaying 1½ to 2 dwts., so that from 87 to 90 per cent. of the gold is extracted. The second product yielded by the hydraulic classifiers comprises 50 per cent. of the pulp, contains 6 dwts. of gold per ton, and, after five days' treatment with solutions ranging from 0·05 per cent. downwards, yields residues containing from 1 to 1·25 dwts., showing an extraction of 80 to 84 per cent. The finest sand is separated from the slimes by pointed boxes; the slimes amount to 25 per cent. of the whole pulp, and are not treated, being unleachable, though they assay 4½ dwts. The fine sands, constituting 10 per cent. of the pulp, contain 4½ dwts. of gold, and after treatment yield residues assaying 1 dwt. per ton. The consumption of cyanide averages ¼ lb. per ton of the tailings, of which 3,000 tons are treated per month.

The precipitation plant consists of four boxes, each 18 feet long, 7 feet wide, and 4 feet deep. Copper wires are fixed along the tops of the sides of the boxes and convey the electric current from the dynamo to the electrodes. The anodes are made of iron, and the cathodes, on which the gold is deposited, of lead. These metals appear at present to be the most suitable for the purpose, but it is possible that other substances may be found equally good. It is necessary to have a very large surface on which to deposit the gold, so that a bath of mercury is out of the question, von Gernet calculating that it would require 80 tons of the metal to give the 10,000 square feet of surface necessary to deal with 100 tons of solution in twenty-four hours. Amalgamated copper plates were tried but abandoned, as the mercury penetrated the copper under the influence of the current, and a dry amalgam resulted which did not adhere to the plate. Lead answers all the requirements of the cathode laid down by von Gernet, which are:—(1) that the precipitated gold must adhere to it, (2) that it must be capable of being rolled out into very thin sheets to avoid unnecessary expense, (3) that it must be easy to recover the gold from it, and (4) that it must not be electro-positive to the anode, in order to prevent return currents being generated when the depositing current is stopped. A fifth requirement might be added, that the gold should be separable from the cathode without destroying the latter. This requirement is not fulfilled by lead.

At the Worcester mill the anodes are iron plates, 7 feet long, 3 feet wide, and ¼ inch thick; they are supported in a vertical position by wooden strips nailed to the box, and are covered with canvas to retain the small quantity of Prussian-blue produced. The sheet-lead cathodes are stretched on wires fixed in light wooden frames which are suspended between the iron plates. There are in all 3,000 square feet of cathode surface. The solution is made to circulate between the plates passing alternately over and under the edges of the anodes. The boxes are kept locked except when cleaning up is necessary, when the cathodes are removed and replaced with fresh sheets of lead.

The lead, which contains from 2 to 12 per cent. of gold, is then melted into bars and cupelled. The consumption of lead is 750 lbs. per month, its cost being equal to 1½d. per ton of tailings. The gold is comparatively free from base metals. The total cost of treatment at the Worcester mill is under 3s. per ton of tailings. The electrodes are placed 1½ inches apart, and a current of 4 volts is employed.

Results of Process.—The success of the process in South Africa, where a new industry has been created, has been complete. The rapid progress which has been made there may be judged from the production of gold by cyanide in each year since its introduction in 1890:—

Year.	Total Output at Rand.	Production by Cyanide.	Percentage of Total Output.
	Ozs.	Ozs.	Per cent.
1890,	494,817	300	0·06
1891,	729,238	35,000	4·8
1892,	1,208,928	176,231	14·6
1893,	1,476,502	304,498	20·6
1894,	2,023,198	587,388	29·0
1895,	2,278,110	677,435	29·7

There were at first large accumulations of tailings to be dealt with, but practically all of these have now been treated, and for the past year little but tailings or concentrates coming direct from the stamps have contributed to swell the output. In the future, cyanide may be expected to continue to yield about 30 per cent. of the total output on the Rand. Out of 677,435 ozs. of gold produced in 1895, 38,708 ozs. were from concentrates, and 638,727 ozs. from tailings. The average amount of gold contained in the concentrates was about 2 ozs. per ton, but one parcel of concentrates from the Village Main Reef yielded 19·5 ozs. per ton. In the same period, concentrates, treated by chlorination, yielded 65,833 ozs., the relative amounts of gold from these materials produced by the two processes being about the same as for the previous two or three years. It would therefore appear that there is no striking advantage to be gained by using either system instead of the other in treating concentrates on this field.

The cost of treating tailings by cyanide in the Transvaal varies from less than 3s. per ton to 10s. per ton, the average being probably about 4s. The cost of the cyanide consumed is from 1s. to 2s. per ton, and with this exception the biggest item is that for handling the pulp, in charging and discharging the vats, which amounts to 1s. per ton and upwards. The royalty generally amounts to about 1s. per ton of the tailings treated. The amount of tailings now being treated is about 8,000 tons per day, and the average yield is about 4·8 dwts., which is equal to about 70 per cent. of the gold contained in the material treated. It is evident from these results that the minimum value of tailings capable of being treated at a profit in the Transvaal ranges from about 1½ to 4 dwts. of gold per ton. The cost of the treatment of concentrates is usually over 20s. per ton, while the amount extracted is about 90 per cent.

The importance of the process to the mining industry is not entirely represented by the output given above, as few of the mines could work at a profit if they depended merely on the gold extracted on the plates. The profit of say 15s. per ton derived from the tailings converts the mine from a losing into a paying concern in a large number of cases, and this profit could not be earned without the cyanide process.

In other parts of the world, the gold-mining industry is less dependent on the cyanide process than in South Africa, but nevertheless it is gradually making its way on every quartz-mining gold-field of any importance. In Australia many mines have adopted it, and a large customs mill is at work in Queensland where over 14,000 ozs. of gold bullion were produced in 1894. In New Zealand the process is being worked at the Waihi Mine, the Crown Mine, and a large number of others. The progress in this colony may be judged from the fact that in 1893, 17,271 ozs. of bullion of the value of £28,657 were produced by cyanide, while in 1895 the amount had increased to 133,162 ozs. of gold. The process has only lately been introduced into India, where in 1895, 3,614 ozs. of gold were produced, but several works are in course of erection, and the output may be expected to be greatly increased in the near future. In the United States the progress of the MacArthur-Forrest process was for some years somewhat slow when compared with that in South Africa and New Zealand. It was adopted at the Mercur Mine, Utah, as early as the year 1891, and was a great success from the start. It is now at work in many of the States, the production in Colorado from its use being considerable. In 1895 the production of gold in the whole of the United States by cyanide was valued at over £300,000. The process will certainly continue to grow in importance for some time to come in all parts of the world, and in 1895 the total output of gold by its use could not have fallen far short of £3,000,000.

CHAPTER XVI.

CHEMISTRY OF THE CYANIDE PROCESS.

Action of Potassium Cyanide on Gold and other Metals.—It has long been known that metallic gold is soluble in potassium cyanide. Elkington, in 1840, in a patent specification, speaks of dissolving finely-divided metallic gold in this solvent, and Bagration, in 1843,[*] studied the action of cyanide on plates of gold, and announced that they were slowly dissolved. Faraday, in 1857,[†] pointed out that gold-leaf is dissolved by a dilute solution of the salt, and also showed that if the gold floats on the surface of the liquid, so that one side of the leaf is in contact with the air, while the other is bathed by the solvent, the action is much more rapid than if the metal is completely submerged. Elsner had previously proved [‡] that the presence of oxygen is required for the solution of the gold. On evaporating the solution, colourless octahedral crystals of auro-potassium cyanide, $KAuCy_2$, are formed, which may be viewed as being a double cyanide, produced as follows :—

$$4Au + 8KCy + O_2 + 2H_2O = 4KAuCy_2 + 4KOH$$

This equation is exothermic. In *calories*, it may be expressed—

$$4(+82\cdot3 + x + y - 64\cdot7) = 4(17\cdot6 + x + y)$$

Here, $+82\cdot3$ is evolved by the formation of potash from potassium, oxygen, and water, x is evolved by the union of Au and Cy, and is probably positive (see below, p. 335), y is evolved by the union of AuCy and KCy, and is also a positive quantity, and $-64\cdot7$ is absorbed by the decomposition of potassium cyanide.

The equation for the solution of silver is

$$2Ag + 4KCy + O + H_2O = 2KAgCy_2 + 2KOH$$

This may be expressed in stages thus—

$$2Ag + 2KCy + O + H_2O = 2AgCy + 2KOH$$
$$AgCy + KCy = KAgCy_2$$

Theoretically, therefore, 130 parts by weight of KCy in the presence of eight parts of oxygen suffice for the solution of 196·8 parts of gold. This has been recently proved by Mr. J. S. Maclaurin § to be the case in all carefully conducted experi-

---

[*] *Bull. de l'Acad. des Sciences de St. Petersbourg* (1843), vol. ii., p. 136.
[†] *Roy. Inst. Proc.*, vol. ii., p. 308.
[‡] *Erdm. Journ. Prak. Chem.*, vol. xxxvii. (1846), pp. 441-446.
§ *Journ. Chem. Soc.* (1893), vol. lxiii., p. 724.

ments. The amount of oxygen dissolved in liquids not specially prepared, to say nothing of that contained in a porous mass of pulverised ore, is consequently enough for the solution of great quantities of gold.

The voltaic order of the metals in different solutions of cyanide of potassium is given in the following table:—

1 part KCy in 8 parts water.*	1 part KCy in 160 parts water.†		1 part KCy in 3½ parts water.†	
	At 50° F.	At 100° F.	At 50° F.	At 100° F.
+ Zn-amalgam.	+ Al.	+ Mg.	+ Mg.	+ Mg.
Zn.	Mg.	Zn.	Zn.	Al.
Cu.	Zn.	Cd.	Cu.	Zn.
Cd.	Cu.	Al.	Al.	Cu.
Sn.	Cd.	Co.	Cd.	Cd.
Ag.	Sn.	Cu.	Au.	Sn.
Ni.	Co.	Ni.	Ag.	Au.
Sb.	Ni.	Sn.	Ni.	Ag.
Pb.	Ag.	Au.	Sn.	Ni.
Hg.	Au.	Ag.	Hg.	Hg.
Pd.	Hg.	Pb.	Pb.	Pb.
Bi.	Pb.	Hg.	Co.	Co.
Fe.	Fe.	Sb.	Sb.	Sb.
...	...	...	Te.	Te.
Pt.	Pt.	Bi.	Bi.	Bi.
Cast iron.	Sb.	Fe.	Fe.	Fe.
...	...	Te.	...	...
− Coke.	Bi.	Pt.	Pt.	Pt.
...	Te.	...	...	...
...	− C.	− C.	− C.	− C.

The position of the metals in the tables denotes also their relative initial tendency to dissolve under the given conditions, one of which is, of course, that they form part of a complete electric circuit. It may be deduced that mercury is inapplicable as a precipitant for gold from solution, and that aluminium would perhaps act similarly to zinc in cold dilute solutions. Moldenhauer proposed to apply this in practice.

The following table ‡ shows the actual amounts of metals (stated in grains per square inch of surface per hour) dissolved by a solution containing 1·86 per cent. of KCy:—

* Poggendorf, vide Gore's *Electrolytic Sep. of Metals*, p. 57.
† Gore in *Proc. Roy. Soc.*, vol. xxx., p. 38.
‡ Dr. Gore, *Proc. Roy. Soc.*, vol. xxxvii., p. 283.

	At 60° F.		At 160° F.	
	Wt. dissolved.	Amounts dissolved, Gold being taken as 1.	Wt. dissolved.	Amount dissolved, Gold being taken as 1.
Al,	·0095	10·5	·4940	17·5
Zn,	·0054	6	·0495	1·75
Cu,	·0053	5·9	·1258	4·5
Ag,	·0016	1·8	·0070	0·25
Cd,	·0014	1·6	·0046	0·16
Au,	·0009	1	·0282	1·0
Sn,	·0006	·66	·0042	0·15
Ni,	·0005	·55	·0027	0·10
Fe,	Trace.	...	·0010	0·035
Pt,	None.	...	None.	...
Pb,	...	...	·0031	0·11

It will be observed that the rate of dissolution of gold is increased thirty-one times, and that the "selective action" in favour of it is also increased (except as regards aluminium) by the increase of temperature. It must be noted, however, that Dr. Gore's table refers to metals in the form of sheets: it must not be assumed that the relative rates of dissolution given above will hold good when ores are being treated. (See also p. 341). The position of zinc on the table is instructive. Dr. Gore has also shown that the rate of solution of gold is increased by 50 per cent. if it is placed in contact with iron, and that it is then five times more soluble than the iron.

The heat disengaged in the formation of metallic and other cyanides is shown in the following table, the authorities for which are Berthelot and Thomsen:—

Components.	Compound Formed.	Heat Disengaged.	
		Compound in Solid State.	Compound Dissolved in Solution of KCy.
H + Cy	HCy	...	+ 11·4
K + Cy	KCy	+ 68·3	+ 65·4
Na + Cy	NaCy	+ 60·4	+ 59·9
½Ca + Cy	½(CaCy$_2$)	...	+ 57·7
½Zn + Cy	½(ZnCy$_2$)	+ 26·7	+ 31·1
½Cd + Cy	½(CdCy$_2$)	+ 17·7	...
½Hg + Cy	½(HgCy$_2$)	+ 9·5	+ 15·3
Ag + Cy	AgCy	+ 4·3	+ 10·8

The heat disengaged in the union of equivalent quantities of the metals in the solid state (except Hg) with cyanogen is given.

Messrs. MacArthur and Forrest have stated that no increase in the rate of solubility of gold and silver takes place on increasing the concentration of the cyanide solution. As exception has been taken to this statement by other authorities, the author conducted the following experiments in the laboratory at the Royal Mint, in order to increase the available data on the subject as far as metallic gold is concerned.

Gold cornets, which had all been subjected to the same treatment, being cupelled, boiled, and annealed together, were submerged in 30 c.c. of KCy solution. Cornets offer a large surface for action, as they are spongy in texture. After some time the cornets were taken out of the solution, well washed, dried, ignited, and weighed.

The results obtained were as follows:—

Weight of Gold taken.	Strength of Solution.	Temperature.	Duration of Time of Action of Solvent.	Gold Dissolved.	
				Weight in Grammes.	Per Cent.
·4997 gramme,	1 p.c.	50°C.	1¼ hours.	·0023	0·46
·4990 ,,	1 ,,	15°	1¼ ,,	·00095	0·19
·4991 ,,	5 ,,	,,	1¼ ,,	·0009	0·18
·4988 ,,	25 ,,	,,	1¼ ,,	·0010	0·20
·5001 ,,	1 ,,	,,	5 ,,	·00305	0·61
·5001 ,,	1 ,,	100°	5 ,,	·0065	1·30
·5000 ,,	1 ,,	60°	1¼ ,,	·00285	0·57

These results tend to show that the rate of solution of nearly pure gold is a function of time and temperature, but the degree of concentration of the solution is less important. At 100° the cornet was blackened, and a black precipitate was formed and remained floating in the liquid. For the relative rate of action of potassium cyanide, chlorine and bromine, compare the results given above with those on p. 255.

Mr. J. S. MacArthur remarks on the subject of temperature:—
"Bagration * states that the solution of metallic gold in cyanide of potassium is facilitated by a rising temperature. We have not found this so in the case of dissolving gold from its ores. Up to a temperature of about 100° to 110° F. (the highest temperature one is likely to encounter in the Tropics) the solution from ores is unaffected, but beyond that the consumption of cyanide is seriously increased and the selective solvent action retarded till, at the boiling point, it ceases, and probably is to a certain extent reversed, as we have found that gold, which had been dissolved

* *Journal für Praktische Chemie*, Leipzic, 1844, vol. xlvii., p. 367.

from an ore by a cold solution of cyanide, was re-precipitated on the ore by heating towards the boiling point."

Mr. J. S. Maclaurin of Auckland University, New Zealand, has recently demonstrated * that the rate of dissolution of pure gold, in the form of plates, in potassium cyanide solutions passes through a maximum when proceeding from dilute to concentrated solutions. The maximum is reached when the solution contains 0·25 per cent. of KCy. The solubility of gold is very slight in solutions containing less than 0·005 per cent., but increases rapidly as the strength rises to 0·01 per cent., when the rate of dissolution is ten times as great as in the 0·005 per cent. solution, and about half as great as that in the 0·25 per cent. The rate increases slowly as the strength rises to 0·25 per cent., and thereafter decreases much more slowly, until in 15 per cent. solutions the rate of dissolution is about equal to that in 0·01 per cent. solutions. Higher strengths show a gradual diminution in the rate of dissolution up to saturation point. Silver is also dissolved at a maximum rate in solutions containing 0·25 per cent. of cyanide, and the changes in the rate are similar to those noted above in the case of gold, the rates for silver being always about two-thirds of the corresponding rate for gold, or, roughly, in the same ratio as the atomic weights of the two metals. In both cases there is hardly any change in the rate of solubility as the strength rises from 0·1 per cent. to 0·25 per cent. From these results, it is seen that the most active solutions are now used in the ordinary practice of the MacArthur-Forrest process in South Africa, and that the solutions used in the Siemens-Halske practice (0·01 per cent. to 0·1 per cent.) are little inferior to them in activity. It is remarkable that the solubility of oxygen in cyanide solutions undergoes similar changes as the concentration increases, and it is to this fact that Maclaurin is disposed to attribute the variations in the rate of dissolution of the gold. The oxygen which unites with the cyanide, converting it into cyanate, is thereby made inactive, as the presence of cyanate of potassium has no effect on the rate of dissolution of the gold.

The table on p. 338 gives the results of some experiments made by Mr. Louis Janin, jun.,† on the solubility of metallic silver in potassium cyanide. In this case it appears that a maximum rate of dissolution is reached when the strength of the solution is only about 1 per cent.; the rate then diminishes gradually as the strength increases from 1 to 3 per cent., and continues to diminish, although much more slowly, until the strength reaches 15 per cent., after which an increase in concentration has no effect on the rate of dissolution. The influence of the volume of solution seems to be small, and that of time not very great, after the first few hours.

* *Journ. Chem. Soc.*, vol. lxiii. (1893), p. 731; and vol. lxvii. (1895), p. 199.
† *Eng. and Mng. Journ.*, Dec. 29, 1888.

Weight of Cement Silver. Mgs.	Strength of Solution Per Cent.	Mgs. of Silver Dissolved.	Percentage Dissolved.	Volume of Solution in c.c.	Time of Standing in Hours.
500	0·50	123·8	24·7	500	12
,,	0·50	126·5	25·2	1000	12
,,	1·00	138·2	27·6	500	12
,,	1·00	109·8	21·9	500	6
,,	1·00	179·5	35·9	1000	96
,,	2·00	112·4	22·4	500	12
,,	2·00	170·0	34·0	1000	96
,,	3·00	100·0	20·0	500	12
,,	3·00	145·5	29·0	1000	96
,,	4·00	100·0	20·0	500	12
,,	5·00	100·0	20·0	500	12
,,	6·00	92·1	18·4	500	12
,,	10·00	76·0	15·2	500	12
,,	15·00	88·0	17·6	500	12
,,	20·00	88·0	17·6	500	12
,,	25·00	88·0	17·6	500	12

Decomposition of Potassium Cyanide.—Hydrocyanic acid is one of the weakest acids known, and is expelled from its salts by all mineral acids and many organic acids. Carbonic acid decomposes potassium cyanide in presence of water thus :—

$$2KCy + CO_2 + H_2O = 2HCy + K_2CO_3$$
(Exothermic: $+ 10·5 + 82·3 + 10·1 - 64·7 - 34·5 = + 3·7$)

The smell of hydrocyanic acid, noticeable whenever KCy or its solutions are exposed to the air, is accounted for by this reaction.

In the presence of air, potassium cyanide takes up oxygen, and is converted first to cyanate and then to carbonate—

$$KCy + O = KCyO (+ 69·7 \text{ cal.})$$
$$2KCyO + 3H_2O = K_2CO_3 + CO_2 + 2NH_3 (+ 3·7).$$

These reactions are much more rapid if heat is applied. Strong solutions turn brown in the air. In dilute solutions, potassium cyanide appears to be changed into HCy and KOH, since the passage of a stream of any neutral gas such as nitrogen through the solution causes an evolution of hydrocyanic acid, while the solution becomes alkaline. The equation is—

$$KCy + H_2O = HCy + KOH$$
(Endothermic: $+ 10·5 + 82·3 - 64·7 - 34·5 = - 6·4$)

If the solution is boiled with acids or alkalies, hydrolysis of the cyanide occurs rapidly, ammonia and formic acid being formed, thus—

$$KCy + 2H_2O = NH_3 + HCO_2K (+ 9·5)$$

This equation does not represent the whole effect, as acetates and other organic substances are also formed. The reactions proceed slowly even in the cold, especially if free alkali is present. It is obvious, from the observed facts of the decomposition of cyanide solutions mentioned above, that (1) the solution must be kept from contact with the air as far as possible, by having closely fitting lids to the storage and leaching vats, the zinc boxes, &c., and by transferring the solution from one to the other by means of iron pipes (not open launders); and (2) the solution must be kept free from acids, and this can be effected by the addition of a little alkali.

Decomposition in the Zinc Boxes.—Pure zinc has only a very slow action on solutions of potassium cyanide, but the action of the gold-zinc couple, formed by the black deposit of gold (which may be really a compound of gold and zinc), and the unaltered zinc, is much more vigorous. This gold-zinc couple probably develops enough electromotive force to decompose water thus—

$$Zn + 2H_2O = Zn(OH)_2 + H_2 (41\cdot 8 - 34\cdot 5 = +7\cdot 3)$$

The positive element, zinc, is thus oxidised, and subsequently dissolved by the cyanide solution. As a matter of fact, Messrs. Butters & Clennel* observed a vigorous evolution of small bubbles, which proved to be mainly hydrogen, in the zinc boxes at the Robinson Mine. The hydrogen thus evolved doubtless carries off mechanically some hydrocyanic acid, and so the odour of the latter, noticeable above the zinc boxes, may be accounted for. The nascent hydrogen also unites directly with hydrocyanic acid, forming methylamine, thus—

$$HCN + 2H_2 = CH_3 . NH_2$$

The presence of the methylamine may in part account for the ammoniacal odour sometimes occurring above the zinc boxes, which may, however, be also caused by the hydrolytic decomposition. The hydrate of zinc, formed as shown above, dissolves at once in excess of the cyanide—

$$Zn(OH)_2 + 4KCy = K_2ZnCy_4 + 2KOH,$$

and the increase in alkalinity of the solution is thus explained.

It is observed that when strong solutions of caustic soda have been used for neutralising the acid salts of the ore, a white deposit is formed on the zinc, and Messrs. Butters & Clennel suggest that this may be cyanide of zinc, $ZnCy_2$, formed by the action of the sodic zincate, $Zn(ONa)_2$, on the double cyanide of zinc and potassium, the equations being—

$$Zn + 2NaOH = Zn(ONa)_2 + H_2$$
$$Zn(ONa)_2 + K_2ZnCy_4 + 2H_2O = 2ZnCy_2 + 2NaOH + 2KOH$$

In this way the solution would continually become more alkaline.

* *Eng. and Mng. Journ.*, 1892, p. 416.

By the reactions in the zinc boxes, not only is the potassium cyanide solution weakened by decomposition, as above, but a large amount of zinc is dissolved. The gold is precipitated thus—

$$2KAuCy_2 + Zn = K_2ZnCy_4 + 2Au,$$

the silver in the ore being dissolved and precipitated by means of precisely similar reactions to those given for gold. The double cyanide of zinc and potassium is not available for the solution of the precious metals, and as long as an excess of zinc is present, no gold will be dissolved by a solution of potassium cyanide flowing through the boxes. The presence of large quantities of the double cyanide of zinc and potassium in the solutions is not prejudicial to the solvent action of the simple cyanides. At the Mercur Mine, the stock solution was apparently as efficacious after nine months' use as at the start, although it must have contained large quantities of zinc cyanide. Feldtmann showed that gold in ores can be dissolved by zinc-potassium cyanide, but J. S. C. Wells points out that the double cyanide remains undecomposed by gold so long as any simple cyanide is present. An accumulation of base heavy metals in the solution can be got rid of by the addition of soluble sulphides.

On the influence of temperature on precipitation, Mr. J. S. MacArthur writes :—"We have hardly any data regarding the influence of temperature on the precipitation by zinc. The few experiments which have been done did not indicate any appreciable difference in precipitation by a rise of temperature, and, on the other hand, they have clearly shown an enormous waste of cyanide by the formation of urea, which manifested itself by its strong unpleasant odour."

Action of Potassium Cyanide on Metallic Salts and Minerals Occurring in Ores.—The ordinary gangue of most ores (viz., silica and silicates of the alkalies and alkaline earths) exercises no direct influence on the cyanide solution. The carbonates of the alkaline earths are also probably without influence. The decomposing effects of sulphides of the heavy metals vary with the physical state of the sulphides.

In 1892, Kedzie[*] made a large number of experiments, using different samples of pyritic ores and concentrates, and solutions containing from 0·5 to 1·5 per cent. of pure potassium cyanide, and found that the consumption (decomposition) of the cyanide varied from 3 to 50 lbs. per ton of ore. He found that the consumption increased with the time, the volume of the solution, and the degree of concentration, the effect of the latter being especially marked. An increase of concentration from 0·5 per cent. to 0·9 per cent. caused the consumption to be doubled, and a further increase to 1·5 per cent., doubled the consumption

[*] *Eng. and Mng. Journ.*, p. 606, 1892.

again. Sulphide of iron, he found, acts very little on potassium cyanide, but sulphides of copper and zinc are rapidly dissolved.

Messrs. MacArthur and Forrest find that dilute cyanide solutions exercise a "selective action" in dissolving gold and silver in whatever form they may be present, in preference to sulphides or other salts of the base metals. There are exceptions to this rule, some of which are noted in the sequel. "For instance, cyanide of potassium solution has a strong tendency to dissolve precipitated sulphide of zinc, but its action on the natural sulphide of zinc, blende, is almost *nil*. The same holds for compounds of iron, and thus we prove selective action by the average result of a series of experiments on ores. Let us suppose a pyritic ore containing about 7 per cent. of iron and 8 per cent. of sulphur with about 1 oz. of gold to the ton. After grinding, this ore is treated with a solution containing about 1 per cent. of cyanide of potassium. The most of the gold will be dissolved and the rest of the ore left practically untouched. It is obvious that the amount of cyanogen contained in the solution is insufficient to combine with the iron present in the ore, yet, notwithstanding the much greater mass of iron sulphide present and open to attack, it is the gold that is selected for action by the cyanide solution. Taking the average result of our work we find that a higher percentage of gold than of silver is extracted, which justifies us in concluding that the selective action is greater on the former than on the latter. One of the ores on which our early investigations were done was composed as under:—

Copper,	0·15 per cent.
Arsenic,	15·09 ,,
Antimony,	Traces.
Sulphur,	14·65 per cent.
Iron,	18·77 ,,
Silica,	36·20 ,,
Lead,	2·66 ,,
Zinc,	4·00 ,,
Alumina,	4·20 ,,
Gold, per ton,	2 ozs. 2 dwts. 16 grs.
Silver, ,,	2 ozs. 13 dwts. 8 grs.

In this ore we had an extraction of gold 85 per cent., silver 50 per cent., for a consumption of cyanide of about 0·45 per cent., and investigations showed that the action was directed in the order, gold, silver, iron, zinc, copper. From the amount of cyanide consumed it is obvious that the amount of base metals dissolved must have been very slight.

"The consumption of cyanide on fresh concentrates varies naturally with the composition of the concentrates. In many cases it is less than 0·2 per cent. of their weight. When the concentrates contain marcasite there is a greater consumption of cyanide than when the pyrites is entirely of the ordinary yellow cubical description. The presence of compounds of copper, physically soft, also tends to increase the consumption."

It has been laid down as a general rule that oxides, hydrates, carbonates, sulphates and sulphides of those metals which are electro-positive to gold in cyanide solutions are dissolved more rapidly than the last named metal, whether it is present in the metallic form or contained in its commonly occurring salts.

This rule certainly applies to the precipitated salts commonly occurring in the laboratory, but J. S. MacArthur has shown that the case may be quite different when the naturally occurring minerals are concerned. Thus, not only is precipitated sulphide of copper rapidly dissolved, but also a sooty form of the same substance occasionally met with as a mineral occurring in ores. On the other hand, fused copper matte is scarcely acted on at all, and in the majority of cases the same may be said of the hard dense sulphides of copper usually found in nature. Sulphide of zinc exhibits the same differences of behaviour: the "black-jack" concentrates of the Ravenswood Mine, Queensland, can be treated with good results, little zinc being dissolved. Again, oxide of copper, if freshly precipitated, is strongly acted on by the cyanide, but if it is heated to dull redness in a muffle it becomes insoluble, and a large excess of this material added to a gold ore makes no difference in the percentage of extraction, while the consumption of cyanide is not increased by its presence.

The action of cyanide solutions on sulphide of silver is similarly dependent almost entirely on its physical state. Experiments conducted by Fresenius showed that if a weak solution of silver nitrate is precipitated by a weak solution of sulphide of ammonium, the resulting sulphide of silver is soluble in a weak solution of cyanide of potassium. On the other hand, if strong solutions of silver nitrate and ammonium sulphide are mixed together, the precipitated silver sulphide can only be dissolved by concentrated solutions of potassic cyanide, and if this solution is subsequently diluted with water the sulphide of silver is reprecipitated. These results tend to show that sulphide of silver is not decomposed by cyanide of potassium, but is held in solution by it as a hydrate. Similar peculiarities in the behaviour of sulphide of silver are observed in practice when ores are being treated, and in this case an increase in the strength of the solution quickens the action of the potassium cyanide even though dilute solutions may be eventually efficacious if enough time is allowed.

A number of experiments made by Louis Janin, Jun.,[*] on various salts of silver point to the following conclusions:—Silver chloride is readily soluble in cyanide, and the arseniate is also rapidly dissolved. Silver sulphide and antimonide are less easily acted on, but are not so refractory as metallic (cement) silver, for the solubility of which see the table on p. 338. The

[*] *Eng. and Mng. Journ.*, Dec. 29, 1888.

presence of copper salts appears to exercise a detrimental action on the solubility of silver sulphide.

Action of Potassium Cyanide on Oxidised Pyrites.—When the pyrites occur in tailings which have been subjected to the action of the weather for some time before treatment, compounds are formed which are more prejudicial to the solution than the sulphides. Sulphide of iron, FeS_2, is oxidised by air and water, ferrous sulphate and free sulphuric acid being formed, thus—

$$FeS_2 + H_2O + 7O = FeSO_4 + H_2SO_4$$

The protosulphate suffers further oxidation, and normal ferric sulphate ($Fe_2.3SO_4$) is produced, which eventually loses acid and becomes a soluble basic sulphate, $Fe_2O_3.2SO_3$. Other basic salts of complex and unknown composition appear to be formed also. In the presence of such oxidised copper and iron pyrites, the following reactions take place:—

(1) The free sulphuric acid liberates hydrocyanic acid.

$$H_2SO_4 + 2KCy = K_2SO_4 + 2HCy.$$

(2) Ferrous sulphate reacts on the cyanide, forming ferrous cyanide, which dissolves in the excess of potassium cyanide, so that it does not appear in the free state.

$$FeSO_4 + 2KCy = FeCy_2 + K_2SO_4$$
$$FeCy_2 + 4KCy = K_4FeCy_6$$

The potassic ferrocyanide, if sufficient acid be present, reacts with fresh ferrous sulphate forming a bluish-white precipitate.

$$FeSO_4 + K_4FeCy_6 = K_2Fe_2Cy_6 + K_2SO_4$$

This precipitate oxidises in the air to Prussian blue if free acid is present *—

$$4K_2Fe_2Cy_6 + O_2 + 2H_2SO_4 = 3FeCy_2.2Fe_2Cy_6 \text{ (Prussian blue)} + K_4FeCy_6 + 2K_2SO_4 + 2H_2O$$

Both these precipitates are decomposed by potash or soda and cannot therefore be formed in their presence. The reactions may be represented as follows:—

$$K_2Fe_2Cy_6 + 2KOH = K_4FeCy_6 + Fe(OH)_2$$
$$3FeCy_2.2Fe_2Cy_6 + 12NaOH = 3Na_4FeCy_6 + 2Fe_2(OH)_6$$

Consequently, if free acid is not present Prussian blue is hardly formed at all, as the solution soon becomes alkaline, and the precipitate is decomposed as fast as it is formed.

It follows from these reactions that if the blue colour of Prussian blue is visible in the vats or on the surface of the tailings heaps, an enormous waste of cyanide must have taken place, and the matter should be at once investigated.

(3) Ferric sulphates are decomposed by potassium cyanide, hydrocyanic acid being evolved and ferric hydrate precipitated.

* Valentine's *Chemical Analysis*, p. 42.

(4) A mixture of ferrous and ferric sulphates produce Prussian blue by reacting with potassium cyanide, ferrocyanide of potassium being formed at first as above; the equation is—

$$3K_4FeCy_6 + 2Fe_2(SO_4)_3 = 3FeCy_2 . 2Fe_2Cy_6 + 6K_2SO_4$$

(5) Sulphate of copper, $CuSO_4$, acts differently from $FeSO_4$, cuprous cyanide, Cu_2Cy_2, being formed, soluble in excess of KCy to $K_2Cu_2Cy_4$, a compound very prone to decomposition. Copper sulphate also gives a precipitate with potassic ferrocyanide, thus—

$$K_4FeCy_6 + CuSO_4 = K_2CuFeCy_6 + K_2SO_4$$

(6) Ferrous hydrate, when formed as above, is instantly dissolved in KCy, thus—

$$Fe(OH)_2 + 6KCy = K_4FeCy_6 + 2KOH(+175·6)$$

Ferric hydrate, however formed, does not act on potassium cyanide, its only action is mechanical, as it collects in a gelatinous mass on the filters and checks the flow of liquid.

Copper and zinc in the condition of hydrates or carbonates are quickly dissolved in preference to the precious metals. If sulphates of these metals are formed in an ore containing limestone or clay, double decomposition occurs with the production of sulphate of lime or alumina, and oxides or carbonates of the heavy metals, which are dissolved by the cyanide, thus—

$$ZnSO_4 + CaCO_3 = ZnCO_3 + CaSO_4$$
$$ZnCO_3 + 2KCy = ZnCy_2 + K_2CO_3$$

The Soda Solution.—Since acidity of the ore causes decomposition of the cyanide, an obvious method of reducing the loss is to add alkali in some form. Before doing this, the free sulphuric acid and soluble salts may be removed by leaching with water, and then a solution of caustic soda or lime is run on to the ore, and after standing for some time is drained off and followed by the cyanide solution. The insoluble basic salts are thus converted into ferric hydrate and soluble sulphates—

$$Fe_2O_3 . 2SO_3 + 4NaOH + OH_2 = Fe_2(OH)_6 + 2Na_2SO_4$$
$$2Fe_2O_3 . SO_3 + 4NaOH + 4 . OH_2 = 2Fe_2(OH)_6 + 2Na_2SO_4$$

Leaching with water then removes the excess of alkali; but, as this cannot be done completely, except with considerable expenditure of time, it is usual at the Robinson Mine to use lime instead of soda. Although the action on the iron salts is slower, an excess of lime is less detrimental than soda to the cyanide solution, and does not attack the zinc. It is found that, even after treatment of the oxidised pyrites by alkalies, the loss of cyanide is much greater than in the case of free milling ores. The reason for this may, in part at least, be attributed to the action of soda on the protosalts (such as sulphates or carbonates) of copper, zinc, &c., by which these metals are precipitated as hydrates, readily soluble in KCy. The preliminary washing with water must always be carefully performed, until no coloration is obtained with ammonium sulphide, so as to remove the soluble salts as far as possible, but some always remain and are converted into hydrates by the alkali.

It is now usually regarded as more advantageous to add lime as a dry powder to the ore before it is charged into the vats, instead of an alkaline solution. The necessary amount is added to each truck load of tailings, and is intimately mixed with it by the time it is charged into the vat.

The amount of alkali to be mixed with a charge of ore is determined by laboratory experiments, adding little by little an alkaline solution of known strength to a given weight of the ore, until the whole is neutral to litmus paper.

Re-Precipitation of Gold and Silver in the Leaching Vats.—If the solution is acid there is a precipitation of gold previously dissolved, insoluble aurous cyanide being thrown down, according to the equation—

$$KAuCy_2 + HCl = KCl + HCy + AuCy$$

This, however, need not be feared as long as there is an excess of KCy, which must all be destroyed by the acid before the aurous cyanide can be precipitated. There is danger in transferring a solution containing gold to a vat containing pyritic material. If the latter should contain any soluble salts of the heavy metals, insoluble salts are thrown down, *e.g.* :—

$$2KAgCy_2 + ZnSO_4 = K_2SO_4 + ZnAg_2Cy_4$$

The salt $ZnAg_2Cy_4$ is probably a true double salt, but the opinion has been expressed that it is merely a mixture of simple cyanides.

Testing the Strength of the Solution.—The ordinary method of estimating the amount of potassic cyanide present in a liquid is by titration with a standard solution of silver nitrate. Silver cyanide is formed, and re-dissolves in the excess of potassic cyanide, until one-half of the latter has been decomposed. The equations are as follows :—

$$AgNO_3 + KCy = AgCy + KNO_3$$
$$AgCy + KCy = KAgCy_2$$

When one-half of the KCy present has been converted to AgCy, an additional drop of $AgNO_3$ solution causes the formation of a permanent white precipitate of AgCy. The amount of silver solution added is then read off, and the percentage of cyanide calculated. The equation of the end reaction is—

$$KAgCy_2 + AgNO_3 = 2AgCy + KNO_3$$

A few drops of a solution of potassium iodide are often added to make the end reaction sharper.

This method is difficult to apply when solutions containing soluble cyanides of zinc and other metals require to be titrated. "A white flocculent precipitate occurs at a certain stage, probably consisting of simple (insoluble) cyanide of zinc, formed by decomposition of the soluble double cyanide—

$$K_2ZnCy_4 + AgNO_3 = KAgCy_2 + ZnCy_2 + KNO_3$$

This precipitation occurs long before the whole amount of potassium has been converted into the soluble double salt of silver ($KAgCy_2$), for the solution, after the appearance of the flocculent precipitate, still gives the Prussian blue precipitate with acidulated ferrous sulphate."*

Mr. Bettel has devised the following methods† of testing solutions for free cyanide, hydrocyanic acid, and double cyanides respectively.

(1) *Free Cyanide.*—50 c.c. of solution is taken and titrated with silver nitrate to faint opalescence, or first indication of a flocculent precipitate. This will indicate (if sufficient ferrocyanide be present to form a flocculent precipitate of zinc ferrocyanide) the free cyanide, together with cyanide equal to 7·9 per cent. of the potassic zinc cyanide present.

(2) *Hydrocyanic Acid.*—To 50 c.c. of the solution add a solution of bicarbonate of potash or soda, free from carbonate or excess of carbonic acid. Titrate as for free cyanide. Deduct the first from the second result, and the percentage of free hydrocyanic acid is obtained.

(3) *Double Cyanides.*—Add excess of caustic normal soda to 50 c.c. of solution, and a few drops of a 10 per cent. solution of KI, and titrate to opalescence with $AgNO_3$. The zinc potassic cyanide is decomposed by the silver nitrate, and the $ZnCy_2$ thus formed instead of being precipitated is acted on by the soda, sodium zincate being formed and some of the double cyanide regenerated. This method gives the whole of the combined CN present, whence the separate results can now be calculated.

According to Watts‡ and Fresenius,§ the total amount of cyanogen in a solution, whether present as simple or double cyanides, may be estimated by boiling with an excess of oxide of mercury and water, when all the cyanogen is obtained as cyanide of mercury and the metals pass into oxides. The cyanide of mercury is then precipitated by nitrate of silver, with the precautions recommended by H. Rose and Finkener.∥

* *Eng. and Mng. Journ.*, 1892, p. 417.
† *Chem. News*, 1895, vol. lxxii., p. 287.
‡ *Dict. of Chem.*, 1864, vol. i., p. 202.
§ *Quant. Chem. Anal.*, 7th edition, 1876, vol. i., p. 376.
∥ *Loc. cit.*

The amount of gold in the solution is usually determined by evaporating a known bulk to dryness with litharge, reducing the lead by fusion in a crucible with charcoal, and cupelling the lead button. The amount of zinc and other heavy metals present may be determined by concentrating the solution by evaporation, adding an excess of sulphuric acid, and heating until "almost all the sulphuric acid has been expelled. The residual mass is then free from cyanogen. It is dissolved in water, if necessary with the addition of hydrochloric acid, and the oxides determined by the usual methods. This way is not adapted for cyanide of mercury, as a little of the metal would escape with the fumes of the sulphuric acid."

Strength of Solution Required.—Below is a table, given by Mr. J. S. MacArthur, showing the relative effect of weak solutions up to 1 per cent., from which it appears that on certain ores an extremely weak solution does practically the same work in gold extraction as one eight times as strong. While there is a slight tendency to raise the extraction of silver by the increased strength of solution, the greater tendency of the stronger solutions to attack the base metals is shown by the fact that, where one of the stronger solutions is used, the amount of cyanide consumed is equal to the whole amount present in the weaker solutions.

	Strength of Cyanide Solution.	Consumption of Cyanide.	Extraction.	
			Au.	Ag.
	Per cent.	Per cent.	Per cent.	Per cent.
Decomposed pyritic ore containing— Au. 1 oz. 8 dwts. 3 grs. Ag. 1 ,, 3 ,, 3 ,,	0·125 0·250 0·500 1·000	0·090 0·110 0·150 0·230	93 91 93 91	70 72 73 76
Pyritic ore containing— Au. 3 ozs. 11 dwts. 1 gr. Ag. 1 ,, 9 ,, 17 grs.	0·125 0·250 0·500 1·000	0·040 0·050 0·050 0·150	86 85 89 88	76 74 82 79

The Consumption of Cyanide varies with the ore. From 0·1 to 0·2 lb. per ton of ore is left mechanically mixed with the tailings unless special means are taken to remove it by washing with water. The decomposition due to exposure to the air, to contact with the ore, and to the reactions in the zinc boxes may not be more than ¼ lb. to 1 lb. per ton. In the treatment of pyritic ores it is usually much greater, amounting to from 3 to 50 lbs. per ton on Ouray ores.* Agitation of the ore with the solution, by means of mixer shafts, is said, by Mr. Charles

* *Eng. and Mng. Journ.*, 1892, p. 606.

Butters, to increase enormously the decomposition of the cyanide, but Mr. MacArthur declares that this increase is not more than 0·05 per cent. of the ore, or little more than 1 lb. per ton.

Methods of increasing the Speed of Action of Potassium Cyanide on Gold.—These methods are described here for convenience, as being more intelligible after the chemistry of the process has been discussed. The necessity of the presence of oxygen has already been dwelt on above. It has, however, been frequently pointed out that in the interior of a mass of ore undergoing treatment the conditions are not favourable for the maintenance of a sufficient quantity of oxygen in a free state. Both the cyanides and the pyrites of the ore tend to unite with it, and further absorption of free oxygen from the air is extremely slow. Hence the time required for the treatment of a charge is many hours, or even days, although under favourable conditions the gold could be dissolved in a few minutes, or at most in two or three hours. To supply the oxygen, various oxidising substances have been tried. For example, Crosse passed a current of air through the solution, and the addition of potassium ferrocyanide, of bleaching powder, of hydrogen peroxide, of manganese dioxide, and of other substances has been made. Most of these oxidisers were tried by Dixon in 1877 * in his unsuccessful attempt to find a process for treating refractory ores. These substances hasten the solvent action of cyanides on metallic gold, but are not used in practice, as they act as "cyanicides," destroying large quantities of the solvent by direct or indirect oxidising effects.

Dr. N. S. Keith suggests † that the action of oxygen is due to its strong electro-negative relation to gold in cyanide solutions, and has experimented with various materials which are electro-negative to gold in solutions of cyanide of potassium. In the list of such materials, given by Dr. Gore, are carbon, iron, lead, and mercury (see p. 334). Keith tried the effect of finely-powdered carbon mixed with the ore, and continually agitated with it in a cyanide solution. He found that the gold was more rapidly dissolved than if no carbon had been present, and supposed that some of the particles of gold came into contact with the carbon and formed galvanic couples, in which the electro-positive element, gold, was quickly acted on. Such contact, however, could not be attained in practice except to a small extent, and the method is therefore useless, whilst the use of lead or iron in this way is *a fortiori* impossible. On the other hand, mercury can be more readily subdivided and distributed through the ore, but, as it amalgamates with the gold, the conditions are changed, and, as a matter of fact, the gold in pasty amalgam is only slightly more rapidly dissolved by cyanide than is pure gold.

* *Chemical News*, December 20, 1878, p. 293.
† *Engineering*, vol. lix. (1895), p. 379.

From the fact that mercury is electro-negative to gold in cyanide solutions, Skey in 1876 * concluded that metallic gold in contact with a solution of mercury cyanide would rapidly dissolve and mercury be reduced. He found this to be the case alike with gold and silver, which dissolved with almost equal readiness. In 1895, Keith proposed † to add a small quantity (2 ozs. to 12 ozs. per ton of liquid) of potassium mercuric cyanide, $HgCy_2.2KCy$, to ordinary cyanide solutions to quicken their action by enabling the presence of free oxygen to be dispensed with. The gold displaces the mercury from solution, and so is dissolved, whilst the mercury is precipitated on the surface of the particles of gold and amalgamates them. Keith supposed that the whole of the gold would thus be rapidly dissolved, and the precipitated mercury would then be redissolved in the cyanide, and thus be ready to react as before. This view seems to be incorrect. According to the author's experiments, the gold is at first rapidly dissolved, and the mercury precipitated. As the action proceeds, however, the dissolution of the gold becomes slower and slower, the mercury appearing to protect it more and more as the percentage of gold in the amalgam is diminished. If the particles of gold are only moderately fine (*e.g.*, gold precipitated from the solution of the chloride by sulphurous acid) the action becomes extremely slow after about 85 per cent. of the gold has been dissolved, the amalgam then consisting of about three parts of mercury to one of gold. If, on the other hand, very finely-divided gold is used, such as gold leaf, the action is fairly rapid until about 95 per cent. of the gold is dissolved, and in one case 98 per cent. of such gold was dissolved in four days by a solution containing 1·5 per cent. of $HgCy_2.2KCy$. In the case of gold leaf, however, which contains both silver and copper, the rate of dissolution is higher than it would be for pure gold in a similar state of subdivision, as the presence in the alloy of either silver or copper favours the dissolution. The retarding effect exercised by metallic mercury when amalgamated with the gold is exemplified by the results of some experiments in which the solutions contained 1·5 per cent. of $HgCy_2.2KCy$, the time of treatment was thirty-six hours, and the weights and state of aggregation of the metal treated were approximately the same. Under these conditions, about 86 per cent. by weight of some samples of pure gold were dissolved, and only 14 per cent. of the gold contained in amalgams consisting of two parts of mercury and one of gold. Dissolution of gold by solutions containing mercury cyanide is greatly expedited by heat. In the author's experiments on ores, the quickening effect of mercury cyanide on solutions of cyanide was very slight.

The Hood Process.‡—Careful consideration of the known facts

---

* *Transactions and Proceedings of the New Zealand Institute*, vol. viii., p. 334. † *Engineering, loc. cit.*

‡ This section is inserted by permission of Dr. Hood.

regarding the dissolution of gold by cyanide solutions, led Dr. J. J. Hood, A.R.S.M., to the conclusion that the generally accepted theories on the subject are only partly true. He suggests that gold can only be dissolved by cyanide when some other metal is present in the solution, which is displaceable by gold. Thus when gold is digested with solutions containing an alkaline cyanide, together with certain compounds of mercury or lead, it is dissolved, and an equivalent quantity of mercury or lead is precipitated.

If, for example, gold is digested with an aqueous solution of potassium cyanide and one or other of the chlorides of mercury, the action is represented, so far as weights are concerned, by the following equations:—

$$2Au + HgCl_2 = Hg + 2AuCl$$
$$2Au + Hg_2Cl_2 = Hg_2 + 2AuCl.$$

These equations do not, of course, represent the whole of the interchanges. The presence of the alkaline cyanide doubtless plays some part in the dissolution of the gold, and it would seem to be obvious that the cyanide is instrumental in keeping it in solution, although there are no published experiments proving that the dissolved gold exists either wholly or in part in the form of the double cyanide, $KAuCy_2$. By using an excess of gold, the whole of the mercury can be removed from solution, and equivalents of gold dissolved as represented by the equations given above, $2 \times 196\cdot8$ parts of gold being dissolved when 200 parts of mercury are added as mercuric chloride or 2×200 parts as mercurous chloride. In some experiments made by the author employing 0·5 gramme mercuric chloride and 1·0 gramme potassium cyanide, the amount of gold dissolved by the hot, strong (5 per cent.) solution in thirty minutes was 0·724 gramme, whilst theory requires 0·726 gramme. Under similar conditions, the amount of gold dissolved by potassium cyanide alone was only 0·010 gramme.

Dr. Hood maintains that pure alkaline cyanides could not dissolve gold, and that they act by means of impurities contained in them. Suppose, for example, a trace of a metal were present which could be displaced by gold; the impurity would be precipitated like copper by zinc, or reduced like ferric chloride dissolving tin or zinc, and an equivalent quantity of gold dissolved. The precipitated metal could not of course be redissolved by cyanide while there remained undissolved any portion of the particle of gold on which it had been precipitated. If, however, the impurity were oxidised by contact with air or other means, it might be redissolved, and would then be available for the dissolution of a further quantity of gold. The reduced compound might be similarly regenerated. In this way a small amount of impurity might suffice for the dissolution of a comparatively large amount of gold. It is to such indirect action that Dr. Hood attributes the efficacy of oxygen in promoting the solvent action of potassium cyanide. No doubt, he suggests, ores frequently contain soluble substances which would increase the solvent action of cyanide on gold—for example, a trace of an oxidised compound of lead present in the ore—but nevertheless to depend on such fortuitous circumstances is unwise, when a solution efficient in itself at the start can be used.

One of the solutions used by Dr. Hood in the treatment of gold ores is obtained by dissolving in water 0·03 per cent. of mercuric chloride and 0·06 per cent. of alkaline cyanide. He finds that the mixture in this ratio is so far stable that the decomposition of cyanide due to the presence of acid sulphates in the ore is much less than if potassium cyanide alone were used. He also adds caustic soda or carbonate of soda to the solution. In the Hood process the presence of oxygen is quite unnecessary, and the treatment of concentrates is thus stated to be more rapid and cheaper than it is by the MacArthur-Forrest process. The recovery of the gold from solution is effected by precipitation by means of the copper-zinc couple, which was described by Gladstone and Tribe. This couple on a small scale appears to be far more rapid in its action than zinc alone. The excess of mercury left in solution is precipitated with the gold, and can be recovered.

Many successful trials on gold ores have been made, and the process is about to be tried on a working scale in Australia. Experiments made upon some of the Australian gold bearing iron oxides and pyritic ores ranging from a few dwts. to 3 ozs., as well as upon some Mexican auriferous sands carrying 700 ozs. of gold to the ton are said to have given a very high percentage of extraction. According, however, to some experiments made by the author, it would appear that the process is inapplicable to ores containing coarse particles of gold.

It is proposed to prepare the double compounds of mercury from the solutions of the alkaline cyanides obtained in the manufacture of cyanogen compounds through ammonia and carbon bisulphide which is now being worked on a large scale in England. Such double compounds are readily crystallised, very stable, and easy of transport.

The Sulman-Teed Process.—Among other suggestions for rendering the presence of oxygen unnecessary, may be mentioned that made by Messrs. Sulman and Teed,[*] who use cyanogen bromide, CNBr. The addition of this substance to a solution of potassium cyanide makes it three or four times more rapid in dissolving gold. They put forward the equation

$$CyBr + 3KCy + 2Au = 2K.AuCy_2 + KBr$$

[*] *Proc. Institute of Mining and Metallurgy*, Feb., 1895.

in explanation of the action of their solvent, which is inoperative except in the presence of an alkaline cyanide. Cyanogen chloride and iodide give equally good results as far as rate of solution is concerned, but are not convenient for use on a large scale. In tests on concentrates, and on various complex ores, the results obtained by Sulman, Teed, and others were remarkably good, high percentages of extraction being obtained in a few hours from ores which yielded little or no gold to ordinary cyanide solutions in the same time.

Description of Ores suitable to the Cyanide Process.—Up to the present the ores on which the most striking success has been obtained on the large scale have been the tailings of the free milling ores of the Witwatersrand Gold Field, in which the gold is finely divided, and the amount of pyrites present is very small. In a large number of cases in South Africa, pyritic ores have been treated with great success as far as the solution of the gold is concerned, over 90 per cent. having been extracted, but the consumption of cyanide is considerable. Ores containing sulphide of silver, mixed with base sulphides, often yield no silver at all. It is, however, no longer doubtful that concentrates and other highly pyritic materials can often be treated more cheaply by cyanide solutions than by roasting and chlorination.

The presence of decomposing marcasite, especially if some copper is contained in the sulphides, is frequently fatal to the process, as great quantities of cyanide are destroyed by contact with such ores, the amount often exceeding 50 lbs. per ton, even after careful washing with water and treatment with alkali. However, the concentrates of many mines can be treated successfully, even if they consist chiefly of sulphides of iron, lead, zinc, &c. The time of treatment is in these cases often as much as three or four weeks, and the charges in the vats are sometimes drained and stirred up or transferred to other vats. Nevertheless, high percentages of extraction are obtainable from such materials.

The process is particularly applicable to low grade ores, containing only finely-divided gold and small quantities of base metals. Owing to the necessity for comparatively coarse crushing which exists with all wet processes, and the difficulty of handling great quantities of dilute solutions of the precious metals without loss (which absolutely precludes the use of sufficient solution and wash water to remove the whole of the soluble gold from any ore), the percentage of extraction possible with the majority of ores only amounts to from 70 to 90 per cent. When coarse particles of gold are present, they must be removed by amalgamation, before treatment with cyanide. Ores containing coarse gold cannot be treated by any wet process, no solution being sufficiently rapid in its action.

CHAPTER XVII.

PYRITIC SMELTING.

The best known and most extensively practised smelting processes for the treatment of gold and silver ores—viz., lead smelting, copper matte smelting, and smelting for the direct production of copper bottoms, in which the precious metals are concentrated—may be best dealt with in the volumes in this series devoted to Lead, Copper, and Silver, and will not be described here. A brief account of iron matte smelting is appended, however, as its main object is the treatment of purely gold ores. This system is said by Eissler[*] to have originated in Hungary. It consists in fusing auriferous iron pyrites in a blast furnace, with the object of obtaining a regulus of iron, in which the gold is concentrated. The richness of the regulus, under the original system, is increased by repeatedly fusing it with fresh ores, or by alternately roasting and fusing it until the percentage of gold has risen to a certain limit, which varies in practice, but never exceeds 50 ozs. per ton, after which the gold is extracted from this product by some other method, either by roasting and chlorination, or by lead or copper smelting. In the United States, the production of a rich matte in one operation is effected by mixing the ore judiciously, and burning out part of the sulphur in the furnace. It has been proposed by Mr. H. Lang[†] to restrict the use of the term "pyritic smelting" to the reduction of gold and silver ores in blast furnaces, with the formation of a rich matte in one operation, the distinguishing feature of the work being the use of the sulphur in the ore as a fuel. This description applies to the system invented by Dr. W. L. Austin, of Denver, and now in use in several localities in Western America. The definition, however, is somewhat narrow, and, in this chapter, pyritic smelting is taken as meaning iron matte smelting. The method is especially applicable in districts where no lead ores are to be obtained, where fuel is cheap, and where there are available large quantities of iron pyrites containing a small quantity of gold, with which purely quartzose ores can be mixed if it is desirable.

Iron pyrites, as is well known, on being heated with a limited supply of air, may be made to lose about half its sulphur, and is then converted mainly into FeS, which is readily fusible. By

---

[*] *Metallurgy of Gold*, London, 1891, p. 378.
[†] *Eng. and Mng. Journ.*, Dec. 26, 1891, p. 721.

increasing the supply of air, the amount of matte produced may be reduced to any required extent, the iron being oxidised and slagged off. If auriferous pyritic ores containing a quartzose matrix are in course of treatment, a flux must be added to slag off the quartz. At the Deadwood and Delaware pyritic smelting works in South Dakota, nearly pure quartzose ores are treated, and here the flux (in this case pure limestone) is said to cost about 25 cents per ton of ore.* A matte is thus formed beneath the layer of slag, and is drawn off for further treatment.

On account of the comparatively high temperature at which the matte solidifies, the ordinary blast furnace, with a deep crucible, used in lead smelting, is thought in Hungary to be unsuitable for pyritic smelting, as the matte chills as soon as it sinks below the smelting zone and freezes up the tap-hole, so that tapping is rendered a difficult operation. Consequently, in Hungary, the "Spur-Ofen" has been substituted, in which there is no crucible, the bottom of the furnace sloping to a tap-hole immediately below the tuyers. The tap-hole is kept open continually, and the matte, as fast as it is melted, flows through it by a narrow channel into wells placed outside the furnaces, where it is separated from the slag by gravity, and is tapped into moulds, while the slag overflows into ordinary slag-pots on wheels. There seems no reason, however, why the methods adopted for keeping metallic copper molten and in a fit condition for tapping at Terrazas, Chihuahua, Mexico, by Mr. H. F. Collins,† should not be equally applicable to pyritic smelting.

The matte thus obtained is sometimes subjected to partial roasting and then mixed with further quantities of crude auriferous pyrites and smelted again, until the product is sufficiently rich. An iron matte can be advantageously made richer than lead without undue losses in the slag, but the limit is reached when it is worth about £200 per ton. The losses in the slag at Mineral City, Idaho, and at Deadwood are said to be about $1 per ton in gold and silver together, with mattes of nearly this value. These mattes are produced by the Austin process, which was introduced at Toston, Montana, in 1890, and is now in use at Leadville, Colorado, and at Mineral, Idaho. Details of the work at these places are given below.

If the sulphides treated contain a high percentage of sulphur, this is of great value as fuel, and the supply of coke may be diminished to a corresponding extent. At the Bimetallic Smelting Company's works, at Leadville, no fuel at all other than sulphur is used, except an occasional charge of coke when some irregularity occurs.‡ Raw sulphide ores, direct from the

---

* *Eng. and Mng. Journ.*, Jan. 14, 1893, p. 28.
† Smelting Gold and Silver Ores, *Proc. Inst. Civil Eng.*, vol. cxii. (1893), part ii.
‡ *Eng. and Mng. Journ.*, Feb. 4, 1893, p. 99.

mine, are fed into the furnace, with the necessary proportions of quartzose ore and limestone, to form slag. The furnace is of peculiar construction, and was designed and built by the Colorado Iron Works, Denver. The following results, however, were obtained by the use of an old blast furnace, which had been previously used for lead smelting. The internal dimensions of this furnace were 36 inches by 80 inches, and it was altered and adapted for the Austin system of pyritic smelting. A hot-blast stove was erected, capable of delivering the required quantity of air heated to 400°C., and the other machinery consisted of one 100-H.P. Buckeye engine, two 40-H.P. boilers, and heavy line shafting extending through the works conveying power to the blowers, rock-breakers, slag hoist, &c. This machinery is stated to have been enough to satisfy the requirements of six blast-furnaces, but nevertheless it was found that one 40-H.P. boiler was not quite sufficient when one furnace was at work. In a trial run in this furnace, in March, 1892, 1,206 tons of ore were smelted with 216 tons of limestone as flux in twenty-five days. The amount of coke burnt was $6\frac{1}{4}$ per cent. of the weight of the ore, its use being mainly to support the fine ores. The hot-blast stove was heated by oil, which was also employed to generate steam. The following is a summary of the cost per ton of ore smelted in this run:—

Coke, at \$7·50 per ton,	\$0·47
Oil, at \$1·10 per ton,	0·49
Limestone, at \$1·90 per ton (using 18 per cent.),	0·34
Labour and sundries,	2·62
Total cost per ton,	\$3·92

The ores used were stated to contain about 12 per cent. of sulphur and 1 ton of matte was produced from every $9\frac{1}{2}$ tons of ore. The richest matte obtained contained gold 0·33 oz., silver 258 ozs., copper 14·2 per cent., the loss being, gold 3·43 per cent., and silver 4·15 per cent. of that contained in the ore. It appears that, as a result of this trial, continuous work on a large scale has been begun and is meeting with considerable success.

At Mineral, ores containing approximately silica 30, calcite 30, iron 10, sulphur 12·5, and arsenic 2·5, with some zinc, are run down in one operation into a very rich matte, without the use of fluxes, and without admixture of other ore, using 7 per cent. of coke. Most of the sulphur and arsenic is burnt off and the corresponding proportion of iron and zinc allowed to go into the slag, thus effecting a desirable concentration of the matte, and at the same time utilising the heat of combustion of the elements named.* Mr. Lang, the manager of these works, also states † that baryta is not disadvantageous in pyritic smelting, and he

* *Eng. and Mng. Journ.*, March 18, 1893, p. 244.
† *Ibid.*, April 22, 1893.

does not consider that the presence of even 25 per cent. of heavy spar would render an ore unsuitable, although it would be almost hopeless to attempt to treat such an ore by lead smelting. One of the main reasons for this difference is the fact that sulphates do not increase the percentage of matte formed in pyritic furnaces as they do in lead smelting, their acid being volatilised unchanged in the former case, but reduced by the coke in lead smelting. Mr. Lang has found in practice, at Mineral, that the best smelting mixture contains approximately silica 30 per cent., sulphur from 10 to 15 per cent. (the larger the percentage of sulphur the less fuel is required), iron 10 per cent., lime, magnesia, baryta, &c., 30 per cent. A few per cent. of zinc, lead, copper, &c., do no harm, and the lead and copper will be retained in the matte. Arsenic is advantageous as it economises the fuel. The cost of smelting such a mixture, and refining the matte is about $3 per ton in Western America, at points conveniently situated on railroads, within a moderate distance of a coal-field.

Pyritic smelting, for treating gold ores, is as yet in its infancy, and few details of working have been published by the managers of the various works. It may possibly be found applicable to the deep-level pyritic ores in South Africa and elsewhere.

CHAPTER XVIII.

THE REFINING AND PARTING OF GOLD BULLION.

General Considerations.—By whatever process gold may have been extracted from its ores, it is necessary to melt the crude bullion and cast it into bars so that its value may be ascertained, and that it may be put into a form convenient for transportation and sale. The name "bullion" may be conveniently restricted to the precious metals, refined or unrefined, in bars, ingots, or any other uncoined condition, whether contaminated by admixture with base metals or not. It is, however, often applied to coin, and the appellation "base-bullion" is given to the pig-lead or to copper bottoms or pig-copper, which have been obtained in smelting operations, and which may only contain a few parts per thousand of gold and silver, the main portion consisting of base metals. The treatment of base-bullion, however, properly belongs to the metallurgy of argentiferous lead, and copper, and the descriptions given in this chapter apply only to bullion which consists chiefly of gold and silver. Refining operations which involve cupellation on a large scale may also be more conveniently considered under the heading of the Metallurgy of Silver.

The operations to which the retorted metal, gold precipitate or bars from the chlorination mills, &c., are subjected may be summarised as follows:—

1. The bullion is melted in crucibles (a rough refining operation being usually effected at the same time) and cast in ingot-moulds.

2. Assay-pieces are cut from the cast ingots or dipped from the molten metal before pouring, and assays are made on these, by which the value and composition of the bars are ascertained.

3. The bars are then usually sold to the refineries, where the base metals are eliminated and the gold and silver separated by "parting," and cast into bars separately. Both before and after the parting it is sometimes necessary to subject the bullion to further refining operations. The bars of gold and silver thus obtained, being of a high degree of purity, are in a condition to be used for minting, or for the various industrial purposes to which they are applied.

Rough unrefined gold is frequently sold to the refineries attached to the American and Australian mints, in the state of retorted metal, &c., without being previously melted and assayed, the producing mills relying on the good faith of the officials at these establishments.

REFINING.

Composition of Bullion.—Bullion varies greatly in composition, and gold may be present in any proportion from zero up to nearly 100 per cent. Native gold always contains more or less silver, but silver quite free from gold is not uncommon. The Mount Morgan gold is the finest gold which has yet been found; this is 997 fine in gold, the alloying metals being chiefly copper with a trace of iron. The gold obtained in most chlorination mills is of a high degree of purity and rarely contains much silver. This precipitated gold, however, generally makes brittle bars owing to the presence of a few parts per thousand of lead, bismuth, antimony, and other metals of high atomic volume. From some chlorination mills the gold is far from pure, owing to various causes, which include lack of care. If ferrous sulphate is used as the precipitant, the precipitate may contain large quantities of ferric hydrate from which some iron is reduced in the crucible, and if sulphuretted hydrogen is used and the gold precipitated as sulphide, it is contaminated with all the heavy metals contained in the solution, copper, iron, and lead being most often encountered. These may amount to several per cent.

Retorted metal is of very different degrees of fineness, according to the nature of the ore and the course of treatment. Placer gold is usually finer than that derived from lodes, containing a smaller percentage of silver, while the nature of the material treated and the methods used in placer operations are not favourable to the contamination of the bullion with base metals, which vary in amount only from 0 to 20 parts per thousand and seldom approach the latter figure. The average fineness of the placer gold obtained in the United States is given in the following table, which contains details concerning the chief producing States :—*

STATE.	AVERAGE COMPOSITION.		
	Gold.	Silver.	Base Metals.
California,	883·6	112·4	4·0
Idaho,	780·6	213·4	6·0
Montana,	895·1	100·9	4·0
Oregon,	872·7	123·3	4·0

The finest placer gold from California is about 980 fine, viz., that from El Dorado County, and the basest is about 720 fine.

Australian placer gold is in general finer than Californian, averaging about 950 fine, the remainder being chiefly silver; the silver in gold from batteries has been increasing in amount of late years. Gold from New South Wales contains from 15 to 350 parts of silver per 1,000; that from Victoria averages 940 fine in gold, containing only 45 parts of silver and 15 of base metals per 1,000. The Queensland gold, except that from Mount Morgan, is less fine, containing about 120 parts of silver per 1,000 on an average. Australian gold is generally brittle before refining, owing to the presence of small quantities of lead and antimony.

Battery retorted gold is usually much less pure than placer gold, the percentage of base metals being in particular much higher, a state of things due in great part to the difference in the method of treatment. In California, the bullion is usually from 750 to 800 fine in gold, seldom falling below 600 nor rising above 850, though in rare cases it is as much as 920. The impurities are principally silver, copper, and iron, the greater part being copper and iron in many cases.

The battery gold from the Transvaal is on an average about 850 fine in gold, a large part of the remainder being silver.

The bullion from pan-amalgamation is less fine than battery gold, containing less gold, more silver, and more base metals.

* *Report of the U.S. Census*, 1880, vol. xiii., p. 352.

Some of that produced in California by the Reese River and Washoe processes, which are not described in this volume, is only 500 or 600 fine in silver or even lower, with a few parts of gold per 1,000, and the remainder consisting chiefly of lead and antimony, or copper.

The Furnace.—The furnace used for melting the bullion is of simple construction. It is usually square, with walls about 12 inches thick, consisting of an outer layer of ordinary brick and an inner layer, at least 4 inches thick, of the best fire-brick. There is often a complete outer casing of iron, which is useful in keeping the furnace from falling to pieces, but radiates more heat than the bricks. The fire-box is about 1 foot square and from 14 to 18 inches deep; below it is an ashpit, provided with a working iron door, through which the air-supply of the furnace is made to pass, and by which it can be regulated. The fire-bars are movable, and their ends rest loosely on iron supports. The top of the furnace may be made flat or sloping up towards the back at an angle of about $30°$. In this case a wide flat ledge should be provided at the front, on which crucibles and moulds can rest. The top is always made of a cast-iron flanged plate, with an opening of the same area as the fire-box. This opening is closed by a cast-iron sliding door made in one or two pieces, and preferably lined with fire-brick and running on rollers. The flue is placed at the back of the fire-box near the top; its cross-section should have an area of about 16 or 18 square inches—6 by 3 inches, and 4 inches square are both convenient sizes. The flue communicates with a chimney, which must be of brick for the distance of 2 or 3 feet from the furnace, but may be of wrought-iron tubing in its upper part. The height of the chimney will depend on the position of the furnace, and should be as great as possible, 60 feet giving better results than any less amount. Some authorities consider that a height of 30 feet is the minimum that can be allowed in order to ensure a good draught, but very satisfactory results can be obtained with a chimney only 16 feet high. The furnace can be built by any bricklayer acting under directions. No mortar is used in its construction; clay, mixed with an equal bulk of sand, being substituted for it. A sliding damper in the flue at a convenient height above the ground is necessary, so as to regulate the draught. The fuel used in such a furnace may be anthracite, charcoal, or good coke, made in ovens, not in gas-retorts, and broken into pieces of about the size of a hen's egg. If the coke is of high quality, it is the most satisfactory fuel, making a hot fire and lasting for a long time, so that it does not require very frequent replenishing. Neither dust, nor very small, nor very large pieces must be used. Charcoal is preferred in the United States Mints for small charges, and anthracite for large ones.

The Crucibles.—The bullion is melted in either graphite or clay crucibles. If nitre is used to refine the metal, the graphite pots are sometimes coated inside with clay. The amount of refining that can conveniently be done in this way is limited by the fact that the molten oxides produced rapidly corrode the crucibles, and may perforate them in course of time, thus causing loss. The Salamander crucibles, manufactured by the Battersea Crucible Company, are the best graphite pots, as they require very little annealing, and will stand frost without being disintegrated. A No. 20 crucible, of 9 inches in height, and holding 400 to 500 ozs. of bullion, can be used in the furnace described above. The size of the crucibles and the weight of the charges of bullion vary greatly, but in extraction mills, as a general rule, a gold-charge does not exceed 400 ozs., and a silver-charge 1,200 ozs. in weight. In mints and refineries, much larger crucibles are employed, holding different amounts up to 6,000 ozs. of metal.

Melting the Bullion.—All crucibles must be thoroughly annealed before being used; otherwise, the contained moisture being suddenly converted into steam when the crucible is heated rapidly, the pots fly in pieces. The crucible is kept on a shelf near the flue, for as many days or weeks as convenient, before being used. It is then placed on the top of the furnace or in the ashpit for a few hours, when it will probably be safe to hold it over the open furnace by means of the crucible tongs, until it becomes gradually warm. After a few minutes, the crucible being turned round at intervals, it can be lowered rim downwards upon the burning fuel, and as soon as the rim becomes red-hot, the crucible is quite safe, and may be turned over and placed in position for the reception of the gold. With Salamander crucibles, a less degree of care in annealing will suffice, as they are well annealed before being sold. The crucible rests on a fire-brick about 3 inches thick, which is laid on the bars of the grate. If the fire-brick were omitted, the bottom of the pot, resting directly on the fire-bars, would be too cold, while a layer of fuel, if placed below it, would soon be burnt out, and could not readily be replaced, so that the pot would sink down to the bars. The fuel is built up round the pot until it reaches to its rim, and the fire urged slightly until the whole pot is at a full red heat. One or two spoonfuls of borax are then thrown into the crucible by means of a scoop or wrapped in paper; this flux not only assists the metal to fuse, but slags off the earthy impurities, and makes the metallic oxides more liquid. As soon as the borax is melted, the introduction of the bullion is commenced. The safest way to do this is to use the shoot shown in fig. 57, which is held in position, its lower edge being inside the crucible, with the left hand, while the metal is transferred to it in a scoop by the right hand. In this way the melter avoids all danger of loss which might be encountered if the metal scrap were wrapped in paper and added by the tongs. Large pieces of metal are added by the crucible tongs. The cover, which must also have been previously well annealed, is kept on the crucible as much as possible, and the fuel pushed down with the poker to avoid scaffolding, and fresh pieces of coke added when required. The crucible is not allowed to become more than two-thirds full at any time, but more metal is added when the first supply has been melted down, and the operation repeated until the pot is sufficiently full of molten material.

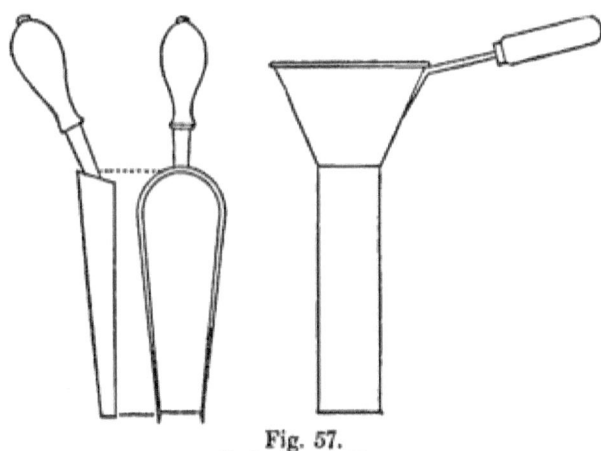

Fig. 57.
Scale, 1 in. = 9 ins.

Refining the Bullion.—If the bullion is of a high degree of purity, containing but little dirt or base metals, not much flux is added, a spoonful or so of carbonate of soda and nitre being enough. In this case the slag is not skimmed off but poured with the metal. If the bullion is very base, however, it is usual to refine it partially by adding nitre and borax, a little at a time, and skimming off the slag when all action has ceased. The nitre exercises a powerfully oxidising effect on the base metals in the bullion, and the resulting oxides form a liquid slag with the borax. When graphite crucibles are employed, the nitre must be prevented from coming in contact with the sides, as in that case the carbon would be oxidised and the pot rapidly corroded. On the other hand, clay pots do not withstand the action of molten oxides slagged with borax. For these reasons a favourite practice is to use graphite pots, covering the surface of the molten metal with bone-ash sprinkled on, the layer being thickest round the sides. Holes are made near the centre of this cover with an iron rod, and nitre introduced through them, in small amounts at a time. As the fusible oxides are formed they are absorbed by the bone-ash and prevented to some extent from attacking the crucible. When sufficient nitre has been added, a point which is judged by an experienced melter from the appearance of the surface of the molten metal, the slag should be of a moderately pasty consistency suitable for skimming. If it is too liquid it is difficult to skim and must be thickened with bone-ash; if it is too pasty, shots of gold may become entangled in it, and it must be thinned with borax.

Mr. Hanks, the late State mineralogist, describes the operation of skimming, at the San Francisco Refinery, as follows:—* " A skimmer is prepared by bending the end of an iron rod of a $\frac{1}{4}$ inch diameter into a spiral of about $1\frac{1}{2}$ inches in diameter, shaped so that when the skimmer is let down vertically into the crucible, the spiral will lie flat upon the surface of its contents. When the slag and metal are perfectly fluid, the surface of the former is touched by the skimmer, to which some slag adheres. It is then withdrawn, quenched in a bucket of water, and at once replaced in the crucible, thus causing a further small portion of the slag to solidify. This operation is repeated until the greater part of the slag has been removed from the crucible. Care must be taken not to allow the bare iron to come in contact with the molten gold, as in that case some of the latter would adhere to it, and for this reason the slag is left adhering to the skimmer during the latter part of the operation. The wet skimmer must not be plunged below the surface of the molten metal or an explosion would ensue. If too much slag accumulates on the skimming tool it is detached by quenching and hammering."

If the bullion is very base the addition of bone-ash, nitre, and borax, followed by skimming, may have to be repeated two or three times. The method was formerly known as the "Poussee" process. Lead is not readily removed from bullion by the action of nitre, which is best adapted for the oxidation of copper. When much lead is present, alternate additions of sal-ammoniac and nitre are made, by which the lead is rapidly oxidised. It is probable that the sal-ammoniac acts by decomposing basic compounds of lead which resist the action of nitre. In obstinate cases, a blast of air is directed upon the surface of the molten metal, and the lead is in this way rapidly oxidised and slagged off. Sometimes sal-ammoniac is sprinkled on, to remove lead, after skimming.

Other fluxes which are sometimes used are sand, pearlashes, and metallic iron. Sand is added to assist in forming a liquid slag, and to protect the crucible from corrosion by the oxides, especially when iron is present. Pearlashes are sometimes added when the bullion contains tin.

Bars which contain antimony or arsenic can be rapidly refined by stirring briskly with an iron bar, a little nitre being added to

* *Californian State Mineralogist's Report*, 1884.

the charge. After three or four minutes' stirring, the greater part of the antimony will have been removed as antimonide of iron.

The refined gold should now be of a brilliant green colour, and its surface should remain quiet, without showing any iridescent films or other signs of continued oxidation.

Toughening the Bullion.—After melting with nitre, bullion is sometimes toughened before being poured, as small quantities of lead, antimony, arsenic, and bismuth are still retained and render it brittle. Nearly fine gold, which is the product of chlorination mills or of parting operations, is similarly treated. The toughening is usually done in one of three ways, viz. :—

1. Sal-ammoniac and corrosive sublimate are added to the molten metal.
2. The metal is melted with oxide of copper.
3. Chlorine is passed through the molten metal.

The method of procedure in each case is as follows :—

1. Sal-ammoniac is sprinkled on to eliminate the lead and tin, after which repeated small additions of powdered corrosive sublimate (mercuric chloride) are made. After each addition the door of the furnace must be at once closed, as dense poisonous fumes arise and must not be breathed by the workers. Volatile chlorides of zinc, copper, antimony, bismuth, &c., are formed and pass off, carrying with them some gold, of which there is an appreciable loss. A little corrosive sublimate sprinkled on the surface of molten gold will completely toughen every part of it without being mixed with it by stirring, even although the crucible contains several hundred ounces of the metal.

When the metal is supposed to be tough, a small sample is dipped out and made into a thin ingot, which, after it has been cooled in water, is doubled up by hammering and its degree of toughness thus tested. It is then often remelted with copper to make up the standard alloy of the country, and again cast and hammered or cut in two with a shearing machine. The reason for doing this is that impure gold, although it may be tough when unalloyed with copper, may make brittle standard bars. If the gold is still found to be brittle, the main bulk of it left in the crucible is subjected to a repetition of its former treatment as often as is necessary, and as soon as the toughening is complete, the gold is covered with a layer of charcoal in the form of powder or lumps and thoroughly stirred before being poured.

The melting under charcoal is sometimes necessary to render silver bars fit for coinage when they have been treated for a long time by nitre. When silver has been raised to a high degree of fineness, it is affected by a peculiar bubbling which may be due to the evolution of oxygen previously absorbed from the nitre. If this continues it is necessary to stir continuously with a graphite rod, keeping the surface covered with charcoal powder, until the bubbling ceases. If the metal, while still effervescing, is poured into a mould, it sprouts at the surface, and a shower of extremely minute particles are projected, often to some distance from the mould, requiring to be swept up; if the crucible is covered by a lid, very heavy effervescence ensues when the lid is lifted. In this case the silver ingots formed are not marketable, being brittle, of low density, and covered by heavy efflorescences.

2. Black oxide of copper is less frequently employed. It is stirred in with the molten metal, and the whole then allowed to remain in the furnace for about half an hour, with occasional stirring. Antimony, arsenic, bismuth, &c., are oxidised at the expense of the oxide of copper, and volatilised or slagged off with borax. It is stated that 2 per cent. of antimony can thus be removed. The process is efficacious, but the pot is rapidly corroded, and the gold is of course contaminated with the reduced copper, a matter which, however, is of little consequence, for the reason that the pure metal is seldom required.

3. The use of gaseous chlorine is described under the heading of Miller's process of Parting, p. 383.

The time occupied in refining and toughening a crucible full of gold of course varies greatly, but often occupies from one to three hours after complete fusion has been effected.

Casting the Ingots.—It is necessary to stir the charge thoroughly before pouring, as the bar must be as homogeneous as possible to insure a correct assay. Since segregation may occur on cooling, assay pieces are often dipped out immediately after stirring. The subject of taking assay pieces in further considered in Chapter xx., p. 432. The stirring is usually done with a peculiarly shaped graphite rod, made expressly for the purpose. It is annealed carefully and raised to a full red heat before being introduced into the crucible, and is held firmly by a pair of tongs with special concave curved faces to its jaws, so as to fit the round rod. In the case of very small meltings it is sufficient to lift the crucible out of the furnace with the tongs and to give it a rotary motion just previous to pouring. In doing this, the metal must not be allowed to cool too much or the casting will be defective. It is advisable to close the dampers wholly or in part, so as to check the draught, when stirring is being done.

Meanwhile the ingot-mould in which the gold is to be cast has been prepared. It is cleaned thoroughly inside by rubbing with emery paper and oil, or with pumice stone, and wiping with an oily rag; it may also be blackleaded inside, as this prevents contact between the gold and the iron of the mould. It is then warmed by being placed on the top of the furnace; its temperature must not be sufficiently high to ignite the oil, but it should be too hot to touch with the hand. When the bullion is ready to pour, the mould is placed on a level surface, such as an iron stool, at a height above the floor of about 12 or 18 inches and oil poured into it. Any cheap non-volatile oil will do, whether animal or vegetable. The mould is filled to a depth of about a quarter of an inch with oil, which is made to flow over all parts of the interior.

The crucible is then carefully lifted from the furnace, usually with basket tongs, and the contents poured rapidly but steadily into the mould, the crucible being moved to and fro so that the stream of molten metal is directed to all parts of the mould in succession. The crucible is then held in the inverted position for a short time, and jarred once or twice to cause the last portion of the metal to flow from it. The oil is ignited, and burns on the top of the cast metal, thus keeping it from tarnishing. In small castings, the slag is allowed to flow out and remain on the top of the metal in the mould; in large castings, the slag is usually skimmed off before pouring. Beads of metal are caught in a large iron tray with raised edges, in the centre of which the mould is placed. If the mould is clean and has been hot enough, and if enough oil has been used, a clean untarnished bar is produced. It is turned out of the mould by inversion of the latter, while still too hot to be handled, and the slag is separated by one or two light taps with a hammer. The bar is then, in many establishments, momentarily dipped into water to assist in the complete removal of the last fragments of the slag, and it is also a favourite practice to dip the bar, first into dilute sulphuric acid, and then into clean water, the bar retaining warmth enough after removal to expel all moisture. This treatment removes all tarnish, and any adherent particles of slag are then chipped off and assay pieces cut from the bar.

In some refineries large crucibles are used, and 3,000 or 4,000 ozs. of metal are refined at once. In this case, several ingot moulds are filled successively from one pot, the weight of the gold bars manufactured being usually either 200 or 400 ozs. each. Silver bars, on the other hand, are made much larger, often weighing 1,000 or 1,200 ozs. Mixed bars of bullion (containing both gold and silver) are seldom cast of a greater weight than 600 ozs. in mills.

The method of refining which has been described above is seldom attempted in extraction mills, and is often unnecessary in refineries, before parting the silver from the gold. The object at a mill is merely to melt down the bullion obtained, so as to bring it to a marketable form with as little loss as possible. With this object in view, no nitre is used, but the metal is kept covered by a layer of charcoal to prevent the formation of oxides, and the crucible is poured as soon as the charge has been fused and stirred.

In conducting all these furnace operations the use of a thick pair of mittens, made of sacking or rubber, is to be recommended for the protection of the hands.

Losses of Bullion Incurred in Melting.—The losses sustained in melting vary according to the composition of the metal, but are usually very small. At the New York Assay Office they are found to vary from 0·5 to 1·5 per 1,000. The losses may be divided into mechanical loss, and loss by volatilisation. The mechanical loss is reduced by care in the conduct of the operation; it may be due to a number of causes. The crucible may break and its contents fall into the fire, or be scattered over the floor of the melting house when on the point of being poured. To avoid loss in this way, the ashpit may be constructed of a cast-iron tray, which can be easily scraped out. The floor of the room is made of carefully-laid flagstones, or, better still, of iron plates, in which there are no cracks or crannies capable of hiding metal beads. Projection by spirting out of the crucible may be occasioned if certain impurities, such as tellurium or antimony, are present in the bullion, and if the nitre is added in large quantities at a time, the pot may "boil over," and part be lost. When flux is to be added, the surface of the metal must be at least 6 inches below the rim of the pot. Recovery of any metal lost in the ashes is effected by panning.

The slags formed in the course of refining frequently contain some small shots of metal, which may be recovered by grinding the slag finely and washing down the product in a pan or on a vanning shovel. At the San Francisco Mint, the residue from the vannings are allowed to accumulate, and at intervals dried and fused with borax in an old graphite crucible at a high temperature. The crucible is left in the furnace over night to cool, and is then broken up, and the shots of metal at the bottom picked out. They are chiefly silver, very little gold being thus recovered. Shots of metal can be recovered with greater certainty from the slags by fusion with lead.

The crucibles, stirrers, lids, &c., also contain a certain quantity of gold and silver. After each melting they are scraped, and the scrapings panned, or, better still, calcined and fused with lead; but, in spite of this treatment, precious metals accumulate in the pots, and, when worn out, they are ground-up in a Chilian mill or other grinder, and panned, in order to separate the shots of metal. The tailings from this treatment will often pay for fusion with lead, and subsequent cupellation. Certain refineries treat large quantities of such residues, with which may be included sweepings of the floor of the melting house.

At the Philadelphia Mint it is found worth while to recover the gold from the iron tools used in stirring, dipping, &c. For this purpose they are melted down in a graphite crucible with a little charcoal to make grey-iron, and kept at a white heat for some time, after which the charge is allowed to cool slowly. Under this treatment the gold and silver separate out (an alloy containing three or four parts of silver to one part of gold being better for the purpose than pure gold), and are found at the bottom of the crucible sharply marked off from the surface of the iron, which is now quite freed from the precious metals.*

The mechanical loss in melting at the San Francisco mint, where crude bullion is bought and valued, is stated to be only about 0·005 oz. in a melt of 100 ozs. This is only one part in twenty thousand, but the total loss of gold is usually much greater, and, as the crude bullion cannot be accurately valued before it is melted-up, it is always difficult to determine how much loss has actually occurred.

The loss by volatilisation is often more serious than the mechanical loss. It is increased by the passage of a rapid current of air over the surface of the molten metal; to avoid this, the fire-box is made deep enough to allow the top of the pot to be sunk some distance below the flue, and the exposure to the draught coming from the opening in the top of the furnace is thus diminished. The metal is also sheltered by being kept covered by the crucible lid as much as possible, and covers of charcoal powder, of fragments of charcoal, or of bone-ash are in general use. For the effect of different alloying metals on the volatilisation of gold, see p. 7. Napier found † that much volatilisation took place during pouring, so that if a wet beaker is held inverted over the mould during pouring, gold is condensed in it and can be subsequently detected by the ordinary tests. As for other conditions, it appears from the results on p. 6 that the longer the gold is kept melted, the higher the temperature employed, and the smaller the mass of gold, the greater will be the percentage loss by volatilisation. These results are in accordance with the experience of those engaged in the operations on a large scale.

The loss by volatilisation when the gold is toughened by the use of corrosive sublimate is particularly high; the total loss of gold incurred when bars are subjected to this treatment is stated to be on an average about 0·85 per 1,000 or £850 per million sterling, while the ordinary loss on melting, both mechanical and by volatilisation, only amounts to about 0·15 or 0·17 per 1,000, part of which is recoverable.

In order to condense the gold and silver which are carried off by the volatilised copper, Mr. J. Feix, foreman of the refining department of the San Francisco Mint, devised an ingenious flue-dust chamber, which is used there as an adjunct to the melting furnaces.‡ The chambers are built of brick, and the

---

* Egleston's *Metallurgy of Silver, Gold, and Mercury*, vol. ii., p. 728.
† *Chem. Soc. Journ.*, vol. x., p. 229.
‡ *Ninth Report Cal. State Min.*, 1889, p. 66.

horizontal flues, coming from the fire-box, open almost directly into them. The chambers are 12 feet high, 1 foot wide, and of considerable length; the flues from them are placed within 1 foot of the top and communicate with the stack. Iron doors, placed at the end of the chamber, enable the sweepings and dust to be readily withdrawn. It is stated that almost all the precious metals which have been volatilised are condensed in this way. The sweepings are heated with borax to a high temperature in an old graphite crucible, when the metal accumulates at the bottom as a regulus. It is not stated how much gold and silver is recovered in this way, and whether the amount has paid for the erection of the chambers and the extra labour and fuel required.

The loss of weight in melting will of course be frequently much greater than the actual loss of gold. Thus mercury, if present, will be expelled by volatilisation, and this metal usually forms from 0·5 to 1 part per 1,000 of retorted gold. Zinc is not so readily expelled, but, if any is present, some will be volatilised, the amount depending on the temperature and duration of the melting. Moreover, the earthy material and all the base metals which have been oxidised by the nitre will be slagged off by the borax, and the total diminution in weight may thus amount to several per cent.

Refining by Means of Sulphur is said to be practised in the United States Mints;* the following is a brief account :—It is effected in plumbago crucibles, and has for its main object the elimination from retorted metal of iron, when, as sometimes happens, it is present in large quantities. The metal is kept just above its melting point, the temperature being as low as possible in order to avoid unnecessary waste of sulphur by volatilisation. Sulphur is sprinkled round the edges of the molten mass, and stirred in with a graphite stirrer. If sulphur is added near the centre, particles of gold are lost by projection. Sulphide of iron is formed with great energy, and sulphide of silver also, but the latter is not produced rapidly until nearly all the iron has been already converted into sulphide. The gold is unaffected by the sulphur and subsides to the bottom. It is not usually cast by pouring, but allowed to solidify in the pot, a better separation between the gold and the matte being thus effected. The pot is turned out as soon as solidification has taken place, and the matte is broken off by a hammer, the gold being remelted and cast into a bar. The small quantity of gold taken up by the matte is separated by melting with metallic iron.

Osmiridium in Gold Bars.—Gold from some districts of California and from the Fraser River district of British Columbia contains some platinum and palladium, and frequently some osmium and iridium. As these metals remain together during

* *Ninth Report Cal. State Min.*, 1889, p. 64.

the treatment, the mixture is commonly called osmiridium. If this is present in perceptible quantities, a fact which is not usually detected until after the bars have been parted, the gold is remelted in a clean crucible and kept fused at a high temperature for about half an hour, when the osmiridium will settle to the bottom of the crucible. This is due to the fact that osmiridium does not seem to form a true alloy with gold, and, being of high density and very infusible, the particles, unfused or partially fused, settle through the liquid gold. The crucible is then gently lifted out of the furnace, and the greater portion carefully but rapidly poured into a mould; the remainder, which contains almost all the osmiridium, is allowed to cool in the crucible until it has solidified, and then assayed for osmiridium. An alternative plan is to allow the whole charge to solidify in the crucible, and then to cut off the lowest portion, which is set aside. The osmiridium settles better from an alloy chiefly consisting of silver than from pure gold, and the rich bottoms are consequently melted several times with silver, the lowest part being cut off each time. The gold is thus gradually replaced by silver, which eventually forms by far the greater part of the mass. It is then granulated and parted, and the resulting powder of gold and osmiridium is treated with aqua regia, by which the gold is dissolved, and the osmiridium separated as a black powder. Such osmiridium is worth from eight to twenty shillings per ounce, and selected grains are used to make the "diamond points" of gold pens. Iridium is, however, not invariably separated from the gold bars in which it is contained, and traces can be observed in a considerable proportion of the refined commercial bars met with in London. Platinum and palladium are, in great part, extracted by the nitre, and enter the slag.

PARTING.

Parting is the separation of silver from gold. During the course of the operation the base metals are separated from both, but, as the presence of these base metals is injurious to the successful conduct of the processes which are chiefly in use, a preliminary refining by one of the methods already described is usually necessary. Only about 10 per cent. of base metals is permissible in the alloys when sulphuric acid is used, although a somewhat larger quantity does no harm if nitric acid is employed.

The processes of parting may be tabulated as follows:—
 1. Cementation.
 2. Melting with sulphide of antimony.
 3. Melting with sulphur, and precipitation of the gold from the regulus by silver, iron, or litharge.

4. Parting by nitric acid.
5. Parting by sulphuric acid.
6. A combination of these last two methods.
7. The Gutzkow process.
8. The new Gutzkow process.
9. Parting by chlorine gas.
10. Parting by electrolysis.

The first two of these methods were known to the ancients, and the third was described as early as the beginning of the 11th century.

Cementation.—In this ancient and obsolete process, gold was freed from small quantities of silver, copper, &c., contained in it. The method was mentioned by Pliny and described by Geber, who wrote in Arabic, probably in the eighth or ninth century; it is possibly still in use in Japan and in some parts of South America. It consists in heating granulations of argentiferous gold mixed with a cement, consisting of two parts of brick-dust, or some similar material, and one of common salt, in pots of porous earthenware. The temperature used is a cherry-red heat, which is insufficient to melt the granulations. After about thirty-six hours' treatment, the greater part of the silver is converted into the state of chloride, and this, together with the cement, can be removed from association with the granulations by washing with water. The gold can in this way be raised to a fineness of about 850 or 900. The silver is recovered from the cement by amalgamation with mercury.*

Parting by Means of Sulphide of Antimony.—This process was also used to purify gold which contained only small quantities of silver. The alloy was repeatedly melted with sulphide of antimony, upon which the gold became alloyed with the antimony and sank to the bottom of the mass, while the silver was converted into sulphide and floated on the top, mixed with the excess of antimony sulphide added. The gold was subsequently refined by a blast of air directed upon it, the antimony being thus oxidised and volatilised. The method is now obsolete, but was in use at the Dresden Mint up to the year 1846, and gold of the fineness 993 was said to be produced in this way.

Parting by Means of Sulphur.—This method was formerly used for the purpose of concentrating the gold contained in auriferous silver in order to obtain a richer alloy. The granulated alloy was melted with sulphur and some of the silver was thus converted into a matte. The gold was then precipitated from the matte and collected in a smaller quantity of silver by fusion with pure silver, or with iron, or litharge. No attempt

---

* For a full account of this interesting process, as well as of the next succeeding two methods, the student is referred to Percy's *Metallurgy of Silver and Gold*, pp. 356-402.

was made to obtain pure gold in this way, and the enriched alloy of gold and silver was parted by nitric acid. The silver was recovered from the matte by fusion with iron. The method was in use in several refineries in Europe at the beginning of the present century. The employment of sulphur in refining at the United States Mints has been already noticed, p. 368.

Parting by Nitric Acid.—The first clear mention of the use of nitric acid for parting silver from gold is made by Albertus Magnus, who wrote in the thirteenth century, but the process does not appear to have been employed on a large scale until two centuries later in Venice. Here, according to an old tradition,* some Germans were employed in separating gold from Spanish silver in the fifteenth and sixteenth centuries, the art being kept secret. These refiners were not inaptly named "gold makers" by those who were unacquainted with their methods. The process was fully described by Biringuccio in his treatise,† published in 1540, and by Agricola‡ in 1556. It was first used in the Paris Mint about the year 1514, and in London at least as early as 1594, but for a long period the operations were conducted in secret in both countries, and it is supposed that this method of refining was not fully practised in England until about the middle of the eighteenth century.

Parting by means of nitric acid is conducted on the large scale in the same general manner as in the assaying of gold bullion. It consists of the following operations:—

1. Granulation of the alloys.
2. Solution of the silver in nitric acid.
3. Treatment of the gold residues, viz.:—Sweetening by washing with water, drying, melting, and casting into bars.
4. Precipitation of the silver as chloride by salt solution.
5. Reduction of the silver chloride by zinc and sulphuric acid.

Granulation of the Alloys.—The gold to be parted must be approximately free from base metals, particularly from those which are not soluble in nitric acid, such as tin, arsenic, antimony, &c. If these were present they would form insoluble oxides, which would remain with the gold, so that further refining operations would be necessary; they would, moreover, cause a great increase in the consumption of nitric acid, so, if they are present, the gold is freed from them as far as possible by melting with nitre, &c. Copper, lead, and other metals which are readily soluble in nitric acid are less obnoxious, and small percentages of these are allowed to remain, as they are not difficult to separate from the silver when in solution with it, while the presence of copper in particular is advantageous in promoting rapid dissolution of the alloy. If present in large

---

* Beckmann's *History of Inventions*, vol. iv., p. 578.
† *De la Pirotechnia*, Florence, 1540.
‡ *De re Metallica.*

quantities, however, even these metals would create difficulties and expense, increasing the consumption of acid.

The bars are melted together to form an alloy which, it was formerly believed, must contain one part of gold to three parts of silver (hence the term "inquartation" applied to this process).* This proportion is still adhered to in many English and European establishments, and at the Philadelphia Mint. In some refineries, however, the proportion of silver used is less, the minimum being two parts to one part of gold. Doré bars containing small quantities of gold are, of course, preferred to bars of fine silver for the purpose of alloying with the argentiferous gold bars. After the "inquarted" alloy has been thoroughly mixed by being stirred while still in the furnace, the crucible containing it is lifted out, and the metal is poured into copper tanks filled with cold water, which is sometimes kept cool by ice, while, in some refineries, a stream of water is kept constantly flowing through the tank. The metal is poured with a circular and wavy motion in a thin stream to prevent the formation of lumps; leafy granules and small hollow spheres are thus formed. The pouring is done either directly from a crucible or from a dipper, the vessel being held in either case about 3 feet above the surface of the water. In the tank is a perforated copper pan, which is lifted out when the pouring is completed, and the granulations allowed to drain.

Dissolving the Granulations.—The granulated metal is heated with nitric acid in vessels of earthenware, porcelain, or platinum. The earthenware vessels are usually cylindrical. Those in use at the Philadelphia Mint are 21 inches in diameter and 22 inches deep, and contain 1,500 ozs. of granulations: they are placed on a lattice work of wood, which is laid on the bottom of lead tanks, and are surrounded by water 10 inches deep kept at the boiling temperature by means of steam. The earthenware vessels are covered by a closely-fitting lid provided with a delivery tube to carry away the fumes. Messrs. Johnson & Matthey's platinum vessels, which hold 800 ozs. of metal, are heated by separate furnaces, the fuel being coke or coal gas.

The strength of the acid used varies from that of spec. gr. 1·33 (38° B.) to that of spec. gr. 1·2. When nitric acid alone is used, about 3 lbs. of acid (spec. gr. 1·2) are used to dissolve each pound of granulations, but of this quantity the amount used in the last boiling (about 20 per cent. of the whole) is available for

---

* A smaller proportion of silver, however, was used at least as early as the year 1627 in Paris. Thus Savot observes in his *Discours sur les Medalles Antiques*, Paris, 1627, chap. vii., p. 72:—"S'il n'y a beaucoup plus d'argent que d'or, l'eau n'agira aucunement : de sorte qu'il faut qu'il y ayt au moins les deux tiers d'argent, et un autre tiers d'or, et encore que l'eau soit tres-bonne : car, si elle est foible, elle n'operara point." Savot did not seem to regard this proportion of 1 to 2 as of recent introduction.

further use; more acid is required if there is much copper present. The first addition of acid is of about 1½ lbs. to each pound of metal; it is kept boiling gently for about five or six hours, by which time most of the silver will have been dissolved. The solution is allowed time to settle, and the hot supernatant liquid is siphoned off by a gold or glass siphon, and diluted with water to prevent the formation of crystals on cooling. The second addition consists, in some establishments, of strong acid (specific gravity 1·414), and in others of acid of the same strength as before. The second boiling is for two or three hours only, and the third boiling for only one or two hours, the liquid being siphoned off after each boiling.

The vessels are provided with hoods and small chambers in the delivery tubes, in order to effect a partial condensation of the acid, and also to recover the small amount of silver nitrate which is carried over mechanically, owing to the violence of the disengagement of gas bubbles. The fumes are conducted to the melting furnace where they are consumed, giving up their oxygen to the fuel.

The reactions that occur are partially expressed by the following equations:—

$$6Ag + 8HNO_3 = 6AgNO_3 + 2NO + 4H_2O$$
$$3Cu + 8HNO_3 = 3Cu(NO_3)_2 + 2NO + 4H_2O$$
$$4Zn + 10HNO_3 = 4Zn(NO_3)_2 + N_2O + 5H_2O$$

and similar reactions for other metals. The amount of nitrous oxide evolved increases towards the end of the operation. It is seen that silver decomposes less than its own weight of nitric acid, while copper and zinc destroy nearly three times their weight of the acid. Nitric acid of specific gravity 1·2 contains about 32 per cent. of anhydrous HNO_3, so that the quantity of acid of this strength theoretically required to dissolve 1 lb. of silver, copper and zinc is about 2·4 lbs., 8·3 lbs. and 7·6 lbs. respectively.

Treatment of the Gold Residue.—The pulverulent gold is "sweetened" by being washed thoroughly in perforated earthenware dishes with boiling distilled water, stirring being performed with a spatula of wood, platinum, or porcelain. The gold is thus freed from nitric acid and nitrate of silver, the operation being continued until the washings show no signs of turbidity on the addition of salt. The washings are added to the first silver solutions, serving to dilute them, the dilution, as has already been observed, being necessary to prevent crystallisation on cooling. The sweetened gold is generally pressed, dried, melted, and cast into bars, which are now made of a weight of either 200 or 400 ozs. The gold thus obtained is usually of a fineness of about 997 or 998, the remainder being chiefly silver, which would not pay for extraction, although part of it could be separated with a further expenditure of time, fuel, and acid. The gold is pressed by a hydraulic ram, the pressure exerted being about 800 lbs. to the square inch. The cakes of metal are dried at a cherry-red heat, and then broken up for melting.

Treatment of the Silver Solution.—The solution of nitrate of silver is diluted with water, allowed to cool, and then treated with a strong solution of salt which is regulated so as not to be in excess, a constant agitation being kept up by revolving wooden agitators driven by steam power, or by hand paddles. When all the silver has been precipitated as chloride, the whole is allowed to settle overnight, and, in the morning, the clear solution of nitrate of soda, containing most of the base metals originally present in the alloys, is drawn off and filtered. The precipitated chloride is washed several times by decantation and agitation, and finally sweetened in wooden filters by boiling water, which incidentally dissolves out the chloride of lead. The filters are usually lined with linen or some similar material.

Reduction of the Silver Chloride.—The silver chloride is then reduced in lead-lined tanks by means of granulated zinc and water acidulated with sulphuric acid. Thirty-three pounds of commercial granulated zinc are stated to be enough to reduce 100 lbs. of silver from the chloride.* The reactions involved are as follows:—

(1) $2AgCl + Zn = ZnCl_2 + 2Ag$

(2) $Zn + H_2SO_4 = ZnSO_4 + H_2$

(3) $H_2 + 2AgCl = 2Ag + 2HCl$

(4) $2HCl + Zn = ZnCl_2 + H_2$

These reactions explain the fact that, while zinc slowly reduces silver chloride in the presence of water only, the action is quickened by the addition of free acid, by which the zinc is attacked and hydrogen evolved. Nascent hydrogen is a powerful reducing agent, and decomposes silver chloride much more rapidly than zinc does, hydrochloric acid being formed and rendered available for the production of more hydrogen. The result of this is that the action, which is at first slow, becomes more and more rapid as hydrochloric acid accumulates in the solution. The sulphuric acid is only needed to start the reaction, but, of course, the more that is added, the more quickly will the operation proceed, and if much is added, the chemical action is more violent at first than afterwards, the amount of free acid present in this case falling off. At the San Francisco Mint 1 lb. of acid of 60°B. is added for every 2 lbs. of silver to be reduced. Hydrogen is evolved copiously and is carried off by a hood and flue. Energetic stirring with wooden paddles is desirable to prevent the formation of lumps of chloride, protected by a layer of silver powder.

* *Ninth Report Cal. State Min.*, 1889, p. 70.

The white chloride of silver gradually turns black as the silver is reduced. To test whether the reaction is complete, some of the silver is taken out, washed well with ammonia, filtered, and the clear solution acidified with nitric acid. A white precipitate signifies that undecomposed silver chloride is still present in the vat.

When the reduction is complete, the vats are allowed to settle, the solution drawn off, and a little sulphuric acid added to dissolve any residue of zinc that may be present. The dark grey pulverulent silver is then washed by decantation, after which it is removed to a wooden filter, and sweetened by washing with boiling water, and finally pressed, dried, and melted into bars, which are about 998 fine. The zinc and sulphuric acid used in this process are lost, and a considerable quantity of undecomposed nitric acid is also run to waste, being contained in the solution from which the silver chloride is precipitated.

The cost of refining and parting by the nitric acid process at the United States Mints in Philadelphia and New York is somewhat less than 2 cents per oz. of the parting alloy, and in San Francisco it is nearly 3 cents. The cost for doré silver is considerably lower. In Europe, the cost is less than in the United States.

Parting by Sulphuric Acid.—This process has now, in the majority of refineries, superseded the nitric acid method, which is much more expensive, owing to the higher cost of the acid used and of the plant required. The German chemist, Kunckel, who lived in the seventeenth century, is said to have been the first to employ sulphuric acid in parting, but it was not used on the large scale until the year 1802, when it was introduced into France by C. D'Arcet, and worked in a refinery built in Paris for the purpose. It was established in London at the Mint Refinery in 1829 by Mr. Mathison, and has been in almost continuous use there ever since, with little change, having been leased to a member of the Rothschild family since 1852.

The method used varies considerably in different refineries, but essentially consists of the following operations—

1. Mixing and granulating the alloys.
2. Dissolving the silver from the granulations by sulphuric acid.
3. Washing and melting the gold residue.
4. Precipitating the silver from its solution by means of copper.
5. Recovering the copper sulphate by crystallisation.

The account given below is a general view of the operations in various refineries, the modifications adopted not being described in most cases.

Mixing and Granulating.—The alloys must be carefully prepared so as to be of suitable composition, as otherwise difficulties are encountered. The most suitable proportion of gold in the alloy is said by Dr. Percy* to be from 18 to 25 per cent., including whatever copper there may be present; but some American authorities consider the proportion of one part of gold to two and a-half parts of silver to be the most desirable, whilst at a refinery at San Francisco the alloy consists of two parts of gold to three parts of silver. This proportion was instituted when alloying silver was scarce in California and has never been abandoned, but the gold thus separated is only 990 fine, containing ten parts of silver, the maximum allowed by law in the gold coins of the United States. If the ordinary proportion of three parts of silver to one of gold is used, however, the gold can be obtained about 996 fine, and the fineness of the gold can always be increased to about 998 or 999 by fusing it, first with bisulphate of potash and subsequently with nitre. If there are only a few parts of gold per 1,000 of the alloy, it has been stated that the silver left in the gold amounts to as much as 3 or 4 per cent.; nevertheless, such an alloy, when subjected in the form of bars to the action of the acid, instead of being granulated, yields gold at San Francisco of no less than 996 fine, after one boiling only.

The amount of base metals present in the alloy must be carefully regulated, as their sulphates are little soluble in concentrated sulphuric acid, and consequently are precipitated and interfere with the progress of the operation. Bars consisting in great part of copper are often received at the San Francisco works. These are melted with fine bars so as to reduce the proportion of copper, which must not be more than about 10 or 12 per cent. of the whole; a small amount of copper facilitates the solution of the silver. A small quantity of lead is said to assist in the solution of the copper, which is somewhat slowly attacked by concentrated sulphuric acid, and a maximum amount of 5 per cent. of lead does not interfere with the operation. From the economy with which this system of parting can be practised, silver containing only 0·5 part of gold per 1,000 can be separated from it at a profit, while the nitric acid process is unremunerative if applied to an auriferous silver alloy containing one part of gold in a thousand. At the Vienna Mint, bars are parted containing 0·9 part of gold per 1,000, and at Freiberg bars containing only 0·4 part per 1,000 are profitably treated.

In England, silver bars are passed through the parting operation, if they contain at least 2 grains of gold per troy pound, or 0·35 part per 1,000.

The parting alloy is usually granulated, but at the San Francisco Refinery the doré silver is not granulated but melted and cast into bars ¾ inch thick, 9 inches wide, and 15 inches long.

Solution of the Silver.—This is usually effected in cast-iron kettles, platinum having been abandoned on account of its

* *Metallurgy of Silver and Gold*, p. 471.

high cost. The iron used is fine-grained compact white iron, preferably containing 3 or 4 per cent. of phosphorus, which increases the durability, although 2 per cent. only of phosphorus is considered enough by some refiners. The kettle is slowly dissolved by the acid, ferrous sulphate being formed, and, in the course of about two years, the thickness of the vessel is reduced from about 2 inches to from $\frac{1}{4}$ to $\frac{1}{2}$ inch, when it is discarded. The perfect exclusion of air from the interior increases the length of life, and dilute acid must not be allowed to come in contact with the iron, as the latter is freely dissolved by it. The vessels are rectangular or cylindrical, with flat or hemispherical bottoms, the latter being preferred in Europe and the former in America. They are covered with cast-iron lids, about $\frac{1}{2}$ inch in thickness, which are bolted tightly to the vessels, and have bent leaden pipes fitted to them for carrying off the fumes, which consist largely of SO_2. This is sometimes reconverted into sulphuric acid in leaden chambers arranged for the purpose. The cover has also an opening (supplied with a lid made air-tight by a water-joint) through which the alloys and acids are added and the operation watched. The heating is usually done by a wood fire.

The charge for the pots varies from 200 to 1,000 lbs. of alloy, and the amount of acid required varies from 2 to $2\frac{1}{2}$ times the weight of the alloy, depending on the composition of the latter. About one-half of the acid, which is strong commercial acid of 66° Beaumé (specific gravity 1·85) is added at first, and the temperature cautiously raised to boiling point, when the pot is closely watched, and, if the ebullition becomes too violent, the temperature is lowered by regulating the fire and by adding cold acid a little at a time. The charge is stirred occasionally with an iron tool, particularly towards the end of the operation, when the undissolved granules of metal must be freed from the surrounding sediment, consisting of sulphates of the base metals, and exposed to the action of the acid. The ebullition gradually subsides and action ceases in about five or six hours, the presence of a greater proportion of base metals increasing the length of time required. The reactions are as follows:—

(1) $2H_2SO_4 + Ag_2 = Ag_2SO_4 + SO_2 + 2H_2O$
(2) $2H_2SO_4 + Cu = CuSO_4 + SO_2 + 2H_2O$

and similar reactions with tin and lead. The reactions with antimony, bismuth, zinc, and iron are more complicated. It is obvious that 63 parts of copper decompose as much sulphuric acid as 216 parts of silver. It is clear, therefore, that an increase in the percentage of copper present necessitates an increase in the amount of sulphuric acid required.

One part of sulphate of silver is soluble in $\frac{1}{4}$ part of boiling concentrated sulphuric acid, but the solubility rapidly falls off as the temperature and concentration diminish, so that 180 parts of cold acid of specific gravity 1·08 are required for the same purpose. Sulphate of copper dissolves slightly in the boiling concentrated acid, but is almost all precipitated in the form of the white anhydrous salt on cooling. Tin and zinc behave similarly, and lead makes the solution turbid and milky. The iron would not be so much attacked if it were not for the increasing dilution of the acid during the process, owing to the formation of water, which is, however, in great part boiled off as fast as it forms, or taken up by the anhydrous sulphates. The presence of copper checks the dissolution of the iron.

When the dissolution is complete, the fire is withdrawn and a few pounds of cold acid of 55° B. are added to the charge, by which the acid is cooled and diluted, and some crystals of silver sulphate are formed. These, falling to the bottom, carry down with them the suspended fine particles of gold, and so clarify the solution. If much copper is present, however, this is not necessary, as the slight cooling of the acid, caused by the withdrawal of the fire, is enough to precipitate some sulphate of copper, which falls to the bottom and adheres to it very firmly, thus clarifying the liquid and enabling it to be poured off or ladled out very closely. The clear silver solution is then ladled out with iron ladles into lead-lined rectangular wooden vats already partly filled with hot water, in which the precipitation is subsequently effected.

Washing and Melting the Gold Residue.—The residue in the dissolving pot, if the amount of base metals present is not large, is then boiled twice more with fresh concentrated sulphuric acid added hot, after which the gold residue is hard and heavy and rapidly subsides to the bottom, and the liquors are ladled into the precipitating vat. The gold is dipped out with an iron-strainer and transferred to a lead-lined filter-box where it is thoroughly washed, first with hot dilute sulphuric acid and subsequently with boiling water, after which it is pressed, dried, and melted. It is almost always brittle, from the occurrence in it of traces of lead or tin which are difficult to separate by sulphuric acid owing to their insolubility. These metals are eliminated by fusion with nitre or by a blast of air, and the bars thus toughened.

If the amount of base metals present is very large, the gold residues are ladled into a vessel of hot dilute sulphuric acid and boiled with it by means of steam. In this way, most of the sulphate of silver and the whole of the copper, zinc, iron, &c., remaining with the gold are rapidly dissolved. Care must be taken, however, to add the residues a little at a time, as otherwise the anhydrous sulphate of copper will form lumps, which are only slowly dissolved. The gold is then allowed to settle, and, after the solution has been drawn off, is boiled again with acid if necessary, or if it is already pure enough it is at once washed, dried, and melted.

In Europe it is not customary to attempt to obtain pure gold from auriferous silver in one operation, but the gold is concentrated in a small quantity of silver and then mixed with other alloys rich in gold and parted again. The product of gold thus obtained is purified by heating in a furnace in small iron pots with about half its weight of bisulphate of potash, by which some additional silver is converted into sulphate. The temperature is not raised much above the fusion point of the salt. The fused mass is then boiled in sulphuric acid, and again washed, dried, and melted. In the United States these methods are not used, auriferous silver being cast into slabs and parted in one operation by boiling with sulphuric acid; fusion with bisulphate of potash is rarely resorted to.

Precipitation of the Silver.—On pouring the sulphuric acid solution into water, most of the silver sulphate is precipitated at once in the form of small crystals, consisting of bisulphate, and the liquid must then be raised to boiling, by means of steam, in order to redissolve them. When the original alloys contain much lead this is not redissolved, and it is, therefore, necessary to let the solution settle and transfer the clear liquid to another vessel. Some particles of gold are usually found in the precipitate thus formed.

The reduction and precipitation of the silver is effected by means of copper, which takes its place in solution. The copper is usually added in the form of scrap while the liquid is being heated up by steam. The precipitation is assisted by constant stirring by means of wooden paddles. In San Francisco, however, the copper is cast into slabs, which are suspended side by side in the solution in a vertical position. The solution should be of about 24° B.; if it is much more concentrated than this, the precipitation of the silver is imperfect. The end of the reaction is detected by testing with salt solution, and when complete, the stirring is stopped, the solution allowed to settle for two hours, and the clear liquid tapped into lead-lined vessels, where further settling of the suspended particles of silver takes place. The precipitate of silver is thoroughly washed with boiling water in wooden filters lined with lead, until the reaction for copper can no longer be obtained with ammonia. Care must be taken that no fragments of metallic copper remain with the silver. The metal is then pressed, dried, and melted, and is usually from 998 to 999 fine, even without fusion with nitre, when the copper plates are used for reduction. At the London refineries, the silver produced is only about 996 fine.

Crystallisation of the Sulphate of Copper.—This is effected by alternate evaporation and crystallisation in lead-lined wooden tanks. The solution, which is still of 24° B., is run from the precipitating tank into the evaporating pan and concentrated to 40° B. by heating with steam; thence it is transferred to the crystallising tanks, where it is allowed to cool and remain for from ten to twelve days. The mother liquor of 36° B. is then run off and reconcentrated to 45° B., after which it is again allowed to crystallise, reconcentrated to 55° B., and a third crop of crystals obtained, which contain much iron. The clear acid mother liquor can now be used to dilute the solution of sulphate of silver in the dissolving pot as already described. The excess of acid in surplus liquids is neutralised with oxide of copper, more copper sulphate being thus formed.

The crystals of bluestone are found adhering to the sides and bottom of the tanks. They are detached with copper chisels, redissolved in pure water and recrystallised, the mother liquors being eventually added to the first liquor from the precipitating vats. When the liquors become over-charged with iron, the copper in them is precipitated by means of metallic iron, and they are thrown away or evaporated to get the crystals of sulphate of iron. The bluestone crystals are packed in barrels for the market. One pound of metallic copper with 1·5 lbs. of sulphuric acid of 66° B. will make 4·5 lbs. of crystallised sulphate of copper.

The cost of the process of parting by sulphuric acid in Europe is about one farthing per ounce troy of the parting alloy.

Combined Process.—At the Philadelphia Mint a combined process is used, nitric acid and sulphuric acid being employed in succession. The alloys are granulated and digested with nitric acid of 39° B. for four or five hours in the same manner as has already been described; the solution is then siphoned off, and the gold washed two or three times with distilled water, by decantation, and subsequently sweetened in lead-lined filters with boiling water. The gold is then introduced into cast-iron cylindrical kettles and boiled for five hours with sulphuric acid of 60° B., the gold being stirred up with an iron rod every ten or fifteen minutes to prevent agglomeration, and the solution is then ladled out and treated as already described, p. 379.

The gold is again boiled in the same kettles with concentrated sulphuric acid for two hours, after which it is washed thoroughly and sweetened in wooden filters, boiling distilled water being poured through it until the washings will no longer redden blue litmus paper. The silver is precipitated from these washings as chloride by the addition of salt. The gold is then pressed, dried, melted, and cast into bars, which are from 996 to 998 fine. By a third boiling in sulphuric acid it is said that they can be raised to 999·5 fine, but according to English refiners this can only be done by fusion with potassic bisulphate.

This process is much cheaper than the nitric acid process, costing 20 per cent. less for acids, and saving some fuel. The granulations contain 100 parts of gold in 285 of the alloy. After the boiling in nitric acid only 6 per cent. of silver is left with the gold. The cost of refining is a little over one cent. per oz.

The Gutzkow Process.—This process of parting by sulphuric acid was invented and patented by Mr. F. Gutzkow in 1867. It has been extensively worked in Germany and in San Francisco, and up to the year 1891 had been instrumental, on the authority of Mr. Gutzkow, in refining one hundred million dollars' worth of silver. It is fully described in Percy's *Metallurgy of Silver and Gold*, p. 479, and only a brief account will be given here. When the patent had expired, Mr. Gutzkow introduced and patented several improvements on it, which will be described at greater length. This improved process is now successfully at work at the Consolidated Kansas City Smelting and Refining Company's Works at Argentine, Kansas, where it was established in the spring of the year 1892.

The original Gutzkow process, as employed at the San Francisco Assaying and Refining Works for many years, may be summarised as follows:—The bullion treated is of three kinds, viz., (1) Gold bars from retorted metal, containing about 900 parts of gold, 10 to 20 of base metals, and the remainder silver; (2) Comstock silver bars or doré bars, usually containing 20 to 100 parts of gold per 1,000; (3) base bars from the Reese River district and from pan-amalgamation of tailings, containing from 100 to 800 parts of silver, and the remainder chiefly copper, with sometimes a little gold. The gold bars (1) are alloyed with silver and granulated, but the others are cast into bars, and parted in that form. The doré bars, when prepared for solution in the acid, weigh about 100 lbs. each, and are 12 inches long, 6 inches broad, and 5 inches thick. The base ingots are melted with fine bars to reduce the average copper contents to 12 per cent., and are cast into bars 1 inch thick, the gold from which is only about 992 fine.

The boiling is done in flat-bottomed thin cast-iron kettles (A, Fig. 58), of which the bottom is only $\frac{3}{8}$ inch thick when new, and $\frac{1}{4}$ inch when worn out. The solution can be rapidly heated, owing to the thinness of the iron kettles, and 200 lbs. of alloy are dissolved in four hours by means of 300 lbs. of sulphuric acid, which comes from the tank C, and is forced into the kettle through the pipe f by the plunger d. The solution is then siphoned off through the pipe m into the tank E, and diluted with a large quantity of hot mother liquor from a previous crystallisation, which is mainly sulphuric acid of about 58° B.; some water is also added, and the solution partially cooled, so that some crystals of silver sulphate are enabled to separate out and carry down with them the milky precipitate of lead sulphate and any suspended particles of gold; green basic sulphate of iron also settles firmly. The clear solution is then

382 THE METALLURGY OF GOLD.

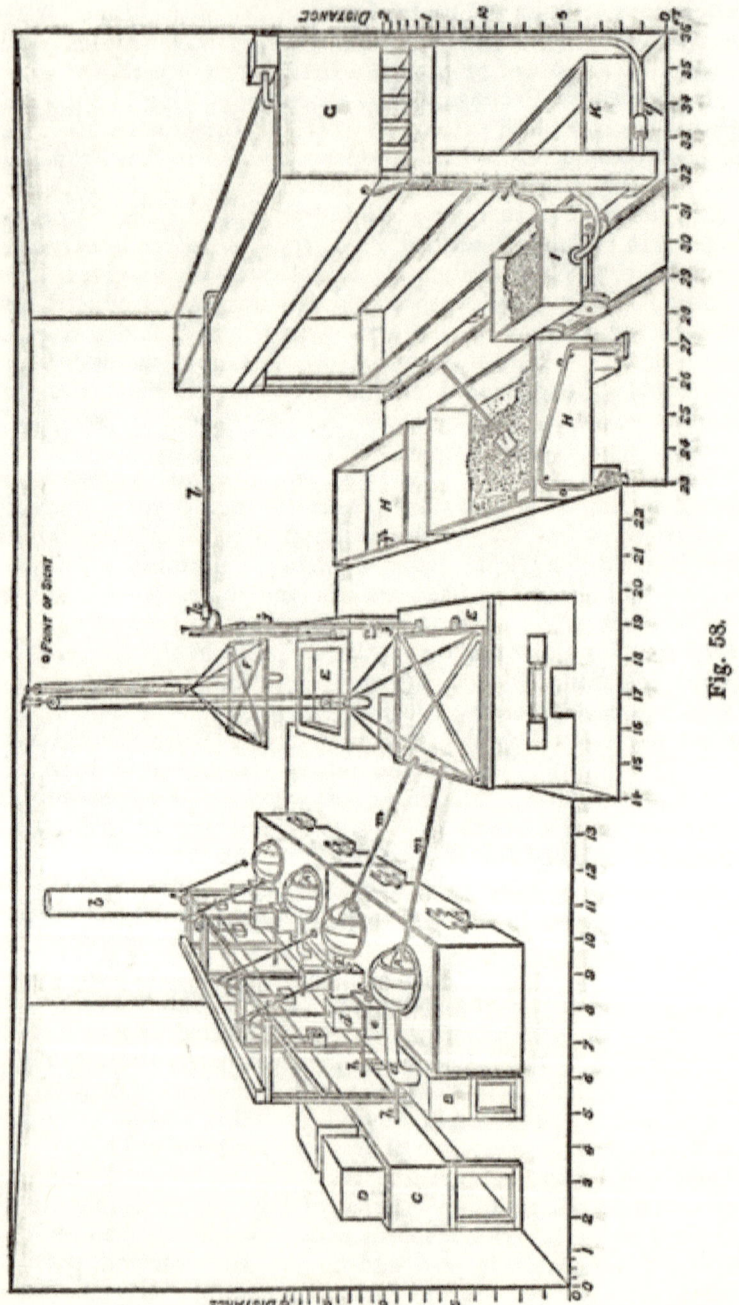

Fig. 58.

siphoned off into H and cooled to 80° F., and almost all the silver sulphate thus crystallised out. If the acid is concentrated, white soft crystals of bisulphate are formed, which is not desired; if, however, the acid is only at about 58° B., large hard yellow crystals of monosulphate, free from acid but contaminated with copper, are deposited. The mother liquor is pumped back into the tank E or to the original acid tank, the device employed for this purpose being to exhaust them of air, so that the acid is sucked up without passing through any valves, which would soon wear out. The crystals of sulphate of silver are transferred to the filtering box I by iron shovels, and a hot solution of green vitriol of 25° B. run on to them from G. This is at first mainly occupied in dissolving the sulphate of copper, and the first portion of the solution, after passing through the filtering box, is run into a storing vat, where the silver, incidentally dissolved, is precipitated by copper, and the latter subsequently recovered by means of iron. After a time, the copper being dissolved, the silver begins to be reduced, the green solution of iron turning coffee-brown; the reaction is as follows:—

$$2FeSO_4 + Ag_2SO_4 = Fe_2O_3 . 3SO_3 + 2Ag$$

The reduction may also be effected by sheets of metallic iron, which is first converted into ferrous sulphate and then into ferric sulphate, the silver being simultaneously reduced to the metallic state. The brown solution of ferric sulphate is boiled with metallic iron in K, in order to regenerate the ferrous salt. The silver is washed, pressed, dried, and melted. The gold from the original dissolving kettle is also washed in a filter, pressed, dried, and melted.

Such was the original Gutzkow process as employed in treating doré bars. Its chief advantage over the ordinary sulphuric acid process was the saving of acid. In the ordinary process, none of the acid used is saved, so that it is reduced in amount as much as possible, but does not fall below twice the weight of the silver dissolved. This reduction in the amount of acid used makes the finishing of the dissolving a difficult and delicate operation. In the Gutzkow process, however, only the acid decomposed by the silver is lost; the weight of this is about equal to that of the metal, the rest of the acid being all recovered and used over again in the boiling. Moreover, the long and tedious crystallisation of copper sulphate is avoided, and the space required for the crystallising vats saved. However, several large lead-lined vessels are required for the storage of the various solutions, and the expense of these, as well as the space required, is greatly reduced by the recent improvements described below.

The New Gutzkow Process.[*]—Mr. Gutzkow has lately

---

[*] This is fully described by Mr. Gutzkow in the *Eng. and Mng. Journ.*, Feb. 28, 1891, p. 257, and May 7, 1892, p. 497, from which this account is summarised.

pointed out that if a large amount of acid is used for the boiling, not only is the silver more completely dissolved and the operation greatly expedited, but the presence of a high percentage of copper does not hinder the parting, as it is kept in solution by the excess of free acid. Thus, for ordinary doré silver, he uses four parts of acid to one of bullion; for bars containing 20 per cent. of copper he uses six parts of acid; for still baser bullion, more acid, and so on, never losing more than one part of acid for one of bullion, and recovering the remainder.

The charge for a pot 4 feet in diameter and 3 feet in depth is 400 lbs. of doré silver: the pot is flat-bottomed, with a basin-shaped pocket or well in the centre which is useful for the collection of the gold. The bullion is first attacked by fresh acid of 66° B., run in by gravity from a large tank, and, when most of the silver has been dissolved, mother liquor from a former operation is added, a pitcher-full at a time, until the charge is completely dissolved, which takes from four to six hours. The fire is then moderated, and the pot filled with mother liquor to within 1 or 2 inches of the top, when the temperature of the acid will have been so far reduced that only faint fumes are discernible. If no fumes are visible the acid is too cold and some silver sulphate will be precipitated, but otherwise the large excess of acid will keep it in solution. The well-stirred charge is now allowed to settle, which is perfectly accomplished in ten minutes, as the yellowish slowly-subsiding persulphate of iron is transformed to a greenish flocculent compound by the water in the mother liquor, and this settles quickly and carries all suspended matter to the bottom. More iron is dissolved from the kettle than in the ordinary process, owing to the greater dilution of the acid used in boiling.

The solution is now siphoned from the kettle by means of a $\frac{3}{4}$-inch gas pipe into a large cast-iron vessel, only about 1 foot deep, standing in a larger vessel which can be filled with water for cooling the charge. Steam is blown into the still hot acid solution through a lead nozzle, $\frac{1}{8}$ inch in diameter, pointing vertically downwards. This both dilutes and warms the solution, the heating being necessary in order to prevent crystallisation of the silver consequent on the dilution. As soon as the dilution has proceeded sufficiently far to ensure the crystallisation of the hard yellow monosulphate instead of the soft white bisulphate of silver, a point which is found by dipping out small quantities at intervals, and observing their behaviour on cooling, the steam is shut off and the vat cooled with water and left all night. The silver crystals form a coating of about 1 inch thick, which is contaminated with copper sulphate if the mother liquor, by repeated use, has become saturated with it. The mother liquor is now pumped back into the acid storage tank by the creation of a vacuum, and the crystals of sulphate of silver are detached with an iron shovel and thrown into a filtering-box provided with a false bottom. Cold distilled water is sprinkled on the charge, and is allowed to filter through it and flow back into the crystallising vat, until the greater part of the free acid has been removed. The stream is then deflected into a "silver filter" where any silver is precipitated that may have been dissolved at the same time as the sulphates of iron and copper. The silver filter is a lead-lined box, partly filled with precipitated copper and provided with a false bottom. The silver separates on the top of the copper as a spongy sheet, a corresponding amount of copper being dissolved. When the crystals of silver sulphate in the first-named filtering box have been completely freed from acid, and from copper and iron sulphates, the stream of water is discontinued. The spongy sheet of silver is then removed from the "silver filter" box and treated with hot water and a few crystals of silver sulphate to dissolve the copper still retained by the sheet. During this whole operation of sweetening the crystals of silver sulphate, only about 3 per cent. of it is dissolved, as it is little soluble in cold water. The copper solution, after passing through the "silver filter," is either run to waste or precipitated by scrap iron in wooden tanks at a nearly boiling temperature.

The crystals are now dried in an iron pan which is placed above a furnace, and, after being mixed with about 5 per cent. of charcoal, they are at once charged into a hot crucible in a melting furnace. The silver sulphate is reduced at a low red heat to metallic silver, carbonic and sulphurous acid gases being evolved. By the time the temperature of melting silver is reached, these gases will have all passed away. The silver is toughened by adding nitre and borax until the so-called "boiling" indicates that the sulphur has all been eliminated, and the metal is then cast into bars.

The gold residue in the dissolving kettle contains insoluble sulphates of lead, iron, antimony, mercury, and often some copper and silver. It is ladled out and boiled with water to dissolve out the sulphates of silver, copper, iron, &c., and, after thorough filtering, it is stirred in a dish with hot water, and decanted on to a filter-cloth until the insoluble sulphates of lead, &c., have all been washed off, and the gold is left bright and clean. The gold is stored until enough is collected to make a 200-oz. bar, which is usually brittle. The material collected on the filter-cloth is re-washed once or twice to recover the particles of gold from it, and can then be reduced with charcoal and cupelled. If lead is present in the original alloy, part remains with the gold, and is dealt with in the manner which has been already described, but the greater part is carried off with the silver solution, and is deposited both while the steam is being passed in, and also subsequently during crystallisation of the sulphate of silver, which is coated with it. The sulphate of lead is removed from the crystals by stirring them well in a stream of cold water, by which the light insoluble particles of lead and antimony sulphate are carried away; it can then be collected, reduced, and cupelled. Any silver that may be dissolved in the course of this washing is precipitated by copper as before.

The process is seen to differ from the original one in three essential particulars:—1. The solution is diluted with steam instead of with mother liquor, the amount of liquid in use, and consequently the number of lead-lined vats required being thus reduced. This is an important item, especially in the United States where lead-burning is expensive, owing to the existence of a powerful union. 2. The weak silver solution is precipitated at once, instead of being stored in tanks to be used again or to be precipitated at leisure. 3. The silver sulphate is reduced directly with charcoal in a crucible in the furnace. This saves the pressing of the silver and, what is of greater importance, avoids the use of the solution of sulphate of iron, which needs to be stored. The reduction in the crucible and subsequent melting requires scarcely more fuel than would be used to melt the pressed silver. One of the minor advantages of the process is said to be that no stirring is required during the boiling, owing to the large amount of acid used. This saves labour and enables the acid fumes to be more easily condensed, as they are not mixed with air, which in the ordinary way would enter through the aperture left for stirring. The exclusion of air also helps to prolong the life of the iron kettles by checking the attack on them by sulphuric acid. Mr. Gutzkow also declares that, owing to the excess of acid present, it is not necessary to specially prepare the alloys for dissolution. Bars, retorted metal of any shape, scrap, &c., may be added just as they are, provided that the amount of gold in them is so small that they can be fairly called doré bars. Finally, it is stated that all the silver and gold charged into the kettles in the morning can be melted into bars and made ready for assay before night. The cost of this process was stated by the general manager of the Kansas City Works in April, 1892, to be 0·35 cent per oz. of doré. The wages at these works are from $3 to $4 per day, and sulphuric acid costs $1\frac{1}{2}$ cents per lb. The refining charges in the Eastern States average 1 cent per oz. of metal, and in California about 2 cents per oz. It is evident that these charges can be greatly reduced by the new process.

Miller's Chlorine Process.—The use of chlorine gas for the purification of molten gold was first proposed by Mr. L. Thompson in 1838, and the results of his investigations were published in the *Journal of the Society of Arts** two years later. He stated that "it has long been known to chemists, that not only has gold no affinity to chlorine at red heat, but it actually parts with

* Vol. liii., part i., p. 17.

it at that temperature, although previously combined. . . . This, however, is not the case with those metals with which it is usually alloyed. It offers, therefore, at once an easy and certain means of separation."

In 1867, Mr. F. B. Miller, Assayer of the Sydney Mint, applied this property of chlorine to the separation of gold from silver on the large scale, and his process has been in use at Sydney ever since, being particularly suitable for the purpose under the local conditions. Among these conditions may be mentioned the facts that acid is very costly, and that there is a scarcity of silver bullion containing small quantities of gold, while the gold produced in Australia contains but little silver. The result is that the ordinary parting processes would prove very expensive, but the chlorine process can be applied cheaply, as it requires very little acid, and is efficacious in removing small quantities of silver from gold bullion which has not been made up into alloys of definite composition. Before the introduction of the chlorine process no attempt was made to extract the silver from any of the native gold of Australia and New Zealand which was coined at the Sydney Mint. Sovereigns were manufactured containing several per cent. of silver, which replaced part of the copper used as the alloying metal. These sovereigns, some of which are still in existence, can be easily recognised by their pale tint, due to the presence of silver. Such sovereigns have not been manufactured since 1867. Besides separating the silver, the chlorine process removes the small quantities of lead, antimony, &c., which render most of the Australian retorted gold brittle, and so in one operation prepares the gold for coinage. Practically the whole of the gold produced in Australia is now deposited in the mints of Sydney and Melbourne, and refined by this process, the amount treated in 1892 having been 1,673,000 ozs.

The following description is abridged from that given by the late Mr. Miller,* and from later writings:—The furnace used is an ordinary melting furnace, such as has been already described. The tile cover of the furnace has a hole in the centre to allow the chlorine tubes to pass through. French clay crucibles are used, Nos. 17 and 18 being convenient sizes, holding about 600 or 700 ozs. of gold; they are placed inside graphite pots to prevent loss by cracking. They are glazed inside by melting borax in them to prevent them from absorbing molten chloride of silver. Graphite crucibles are said to be unsuitable, as silver chloride appears to be reduced, presumably by the hydrogen contained in them, as fast as it is formed. The crucibles are covered by loosely fitting lids, through which the clay pipe-stems of about $\frac{3}{16}$-inch bore are passed to the bottom of the crucible for

* *Journ. Chem. Soc.*, 1868; and *Trans. Roy. Soc. of New South Wales*, 1869.

the conveyance of chlorine. The pipe-stem is made red-hot before being introduced into the molten metal, as otherwise it would crack and break off. The chlorine generator consists of a stoneware jar furnished with three necks, and capable of holding from 10 to 15 gallons of liquid. The three openings are fitted with well-secured rubber plugs, through two of which two tubes are passed, viz., the safety tube, which is 8 or 10 feet high, with its open end bent over so as to deliver into a large jar, and the eduction tube, which is closed by a stopcock till it is required. The generator is partly filled with from 70 to 100 lbs. of manganese dioxide in small lumps, an amount which will suffice for many operations; hydrochloric acid is introduced through the safety tube when the gas is required. The generator is warmed by a steam jacket.

The chlorine gas is conveyed in leaden pipes to the furnaces. All joints and connections are made by well-wired india-rubber tubes, which must be protected from direct radiation from the furnace. Screw compression clamps on these rubber tubes enable the supply of chlorine to be regulated to a nicety. When the clamps are closed the gas accumulates and forces the acid up the safety tube into the vessel placed overhead, and so the further generation of gas is prevented. Two such generators and three melting furnaces are enough to refine 2,000 ozs. of gold, containing 10 per cent. of silver.

The generators being in readiness, the crucibles are slowly heated to redness, and the full charge of 600 or 700 ozs. of bullion introduced and melted, 2 or 3 ozs. of borax being sprinkled on its surface or poured on in a molten state. The chlorine is now allowed to pass slowly through the clay pipe to prevent metal from entering it, and the pipe is plunged to the bottom of the molten metal and kept there by means of a weight attached to it. The full stream of chlorine is now turned on and is heard to be bubbling into the molten metal, by which it is completely absorbed, so that no splashing and projection of the metal occurs. A height of 16 to 18 inches in the safety tube corresponds to and balances a height of 1 inch of gold in the refining crucible. The safety tube acts as an index of the pressure in the generator and of the rate of production of the gas: any leakage or the exhaustion of the acid is at once indicated by a fall of the liquid in the tube. Fresh acid is added at intervals as it is required.

When the chlorine is introduced, dense fumes at once arise from the surface of the metal owing to the formation of volatile chlorides of the base metals, which are the first to be attacked: lead gives especially dense fumes, which can be condensed on a cold object held in them. After a time these fumes cease and silver chloride is formed, very little chlorine escaping from the crucible, even if an extremely rapid current is passed into it; consequently the operation is expedited by every increase in the volume of the current. Towards the end of the operation splashing is more noticeable, and dark brownish-yellow fumes appear, consisting chiefly of free chlorine. The completion of the refining, however, is indicated by a peculiar reddish or brownish-yellow stain which is imparted to a piece of white tobacco-pipe when exposed to the action of the fumes for a moment. It is suggested by Prof. S. B. Christy that the stain contains gold. This stain appears in about one hour and a-half from the start, when 600 ozs. of gold, containing 10 per cent. of silver, are being subjected to treatment. The current of gas is then at once stopped, and the crucible lifted out of the furnace and allowed to cool sufficiently for the gold to solidify. Probably, if the operation were continued after the appearance of the brown stain, losses of gold by volatilisation would occur.

The chloride of silver, still molten, and floating on the top of the gold, is then poured off into iron moulds, and the crucible inverted on an iron table, when the red-hot cone of gold falls out. This is now fine, and after any adherent chloride of silver has been detached from it by scraping, it simply requires melting into ingots, 98 per cent. of the gold being thus at once rendered available for use. The remainder of the gold is contained in the chloride of silver, partly in the form of entangled shots of metal, but chiefly as a double chloride of silver and gold. It was formerly recovered by melting the chloride with about 10 per cent. of metallic silver, rolled to about $\frac{1}{8}$ inch in thickness. The gold is reduced by the silver and alloys with the excess, settling to the bottom of the pot where it solidifies after ten minutes cooling, so that the chloride of silver can be poured off into large iron moulds, slabs suitable for reduction being thus formed.

It was found at the Sydney Mint,[*] that the above method of separating the gold from the silver chloride was subject to several disadvantages. In particular, although on a small scale the amount of gold in the silver could be reduced to from 0·3 to 1·0 per 1,000 by careful and continuous stirring with silver foil for a great length of time, nevertheless in practice on a large scale the results varied greatly, and the silver bullion produced usually contained from 10 to 25 parts of gold per 1,000. Several reducing agents, such as argol, resin, hydrogen and coal gas were successively tried but were not found to give good results. Finally, the application of soda carbonate, which had been proposed by Leibius in 1868, was adopted, the method of procedure being originally as follows :— [†] "The argentic chloride is covered by a layer of fused borax, about $\frac{1}{4}$ inch thick, and when all is well fused, the powdered soda is sprinkled on the top of the borax, without stirring, as rapidly as the ensuing action will

---

[*] *Fourth Annual Report of the Royal Mint*, 1873. Report by A. Leibius, p. 63.

[†] *Loc. cit.*

admit. Occasionally the top layer is dipped with a stirrer slightly underneath the molten argentic chloride, without stirring the latter. When all the necessary soda is added and the action is nearly over, the pot is covered with a lid, and left for about ten to twenty minutes to increased heat, and, when the contents are quite liquid, the pot is lifted out of the fire without previous stirring, and allowed to cool, so as to enable the argentic chloride to be poured off from the gold button at the bottom of the pot.

"Although in several experiments all but 0·1 of gold per 1,000 was eliminated from the silver bullion produced, in no case is every trace of gold removed in one operation. To free the argentic chloride entirely from gold, producing therefore silver bullion free from gold, was, however, accomplished by subjecting the argentic chloride to a second treatment, with a small quantity of soda, in a separate boraxed clay pot, similar to the first operation.

"A convenient quantity of argentic chloride, to be treated in a No. 18 French clay pot, was found to be 230 ozs. The amount of soda required for 230 ozs. of chloride may range from 16 to 20 ozs. Less than 16 ozs. leaves too much gold in the silver, while more than 20 ozs. produces a very silvery gold button, and yet without completely freeing the argentic chloride from gold.

"The use of 18 ozs. of soda for 230 ozs. of chloride produces a gold button weighing between 30 and 35 ozs., assaying about 920 to 930, and leaves from 0·5 to 1·0 part of gold in 1,000 parts of silver bullion produced.

"With 20 ozs. of soda the results were:—Gold, about 35 ozs., assay 870-880; gold left in the silver bullion produced from 0·2 to 0·5 per 1,000.

"With 16 ozs. of soda:—Gold from 30 to 33 ozs., assay 940-950; gold left in the silver bullion, from 1·0 to 2·0 per 1,000, and sometimes as much as 6·0 per 1,000.

"To free the argentic chloride from gold, a second treatment with 3 ozs. of soda per pot of 200 ozs. chloride, containing but a minute quantity of gold, will always be found to answer, the only care required being gradual application of the soda and enough heat at the end of the operation."

The time required for the two operations is about half an hour.

"The presence of a large proportion of chloride of copper has been found to prolong the operation considerably on account of oxide of copper being formed on addition of soda, as a much greater heat is required in order to fuse the whole mass. The argentic chloride produced from base gold alloys would contain a large proportion of chloride of copper, &c., and it would be better, therefore, to reduce it direct, and dissolve the reduced metals in acid, to separate gold and silver therefrom."

The silver chloride may be assayed for gold by cupellation with lead foil and subsequent parting.

The method just described was adopted at the Sydney Mint in 1872, and at the Melbourne Mint in the following year, with excellent results.

The process of reduction of the silver chloride was devised by the late Mr. A. Leibius, fellow-assayer of Mr. Miller at the Sydney Mint, and was described in a paper communicated to the Royal Society of New South Wales.* In this process, 1,400 ozs. of argentic chloride are reduced in 24 hours by the apparatus, of which the following is a brief description:—Seven zinc plates, each 14 inches long, 12 inches wide and $\frac{1}{2}$ inch thick, are supported about $1\frac{1}{4}$ inches apart in a vertical position in slots in a wooden frame. Six slabs of argentic chloride, each 12 inches long, 10 inches wide and $\frac{3}{4}$ inch thick, are suspended by loops made of silver bands, in such a way that each slab is placed between two of the zinc plates and separated from them by spaces of about $\frac{1}{4}$ inch. The silver loops are connected with silver bands on which the zinc plates rest, so that there is metallic connection between the slabs of chloride and the zinc plates. The whole is now plunged into water, to which some of the liquor from a previous operation containing chloride of zinc in solution is added as an exciting agent. Galvanic action soon begins, the liquor gets gradually warmer and a strong current is discernible. The silver chloride is gradually reduced to metallic silver, the slabs undergoing no alteration of form, and the zinc is dissolved. The slabs of silver chloride are generally free from most of the base metals, but copper, if present in the original alloy, is not volatilised in the crucible, and its chloride remains mixed with that of the silver. The two metals are now reduced together. When all action has ceased, the slabs of cupreous silver are lifted out and boiled, first in acidulated water and then in pure water, while still suspended in their silver loops. The porous metal is now ready for melting. As no acid is used the amount of zinc consumed is the theoretical quantity required by the equations—

$$2AgCl + Zn = ZnCl_2 + 2Ag$$
$$CuCl_2 + Zn = ZnCl_2 + Cu$$

The weight of zinc consumed usually amounts to from 24 to 25 per cent. of the weight of the slabs of fused chloride. The zinc plates are used over again until worn too thin for safety, after which they are melted-up and cast into new plates. They suffer no loss if the apparatus is left untouched for any length of time after the whole of the silver has been reduced.

At the Melbourne Mint in the year 1889, the zinc plates

---

* *Trans. Roy. Soc. New South Wales*, 1869. The paper is given almost at full length in Percy's *Metallurgy of Silver and Gold*, p. 418.

employed as described above were replaced by sheets of iron with satisfactory results. "Upon the reducing bath being heated with steam, the chloride of copper dissolving, disengages itself freely from the slabs of chloride of silver, and coming into contact with the iron is reduced, and the metallic copper falls to the bottom of the bath in large quantities, leaving the reduced silver in a much cleaner state than when zinc was used. The noxious fumes which were formerly given off on the melting of the reduced silver sponge are also avoided." *

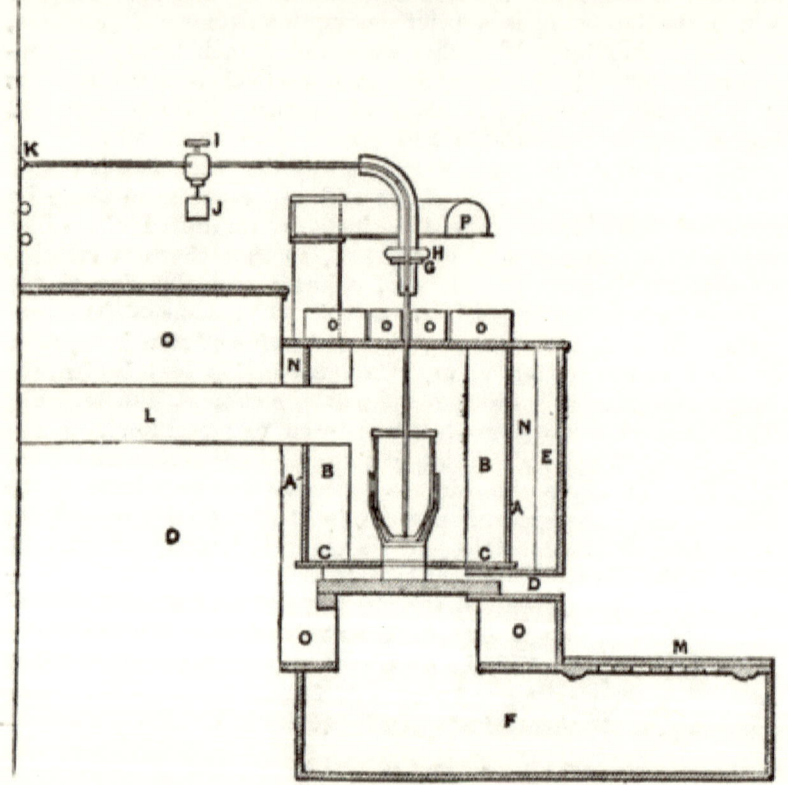

Fig. 58a.

The Chlorine Process as now practised at Melbourne.—The following description has been kindly supplied by Mr. Francis R. Power, the Assayer at the Royal Mint, Melbourne, by permission of the Deputy Master. It gives the exact methods and apparatus in use in the early part of 1896. As will be seen, these differ considerably from those described above, and from the practice at Sydney.

* *Twentieth Annual Report, Royal Mint*, 1889, p. 126.

Furnaces.—These are thirteen in number. They are built cylindrically (see Fig. 58a, in which one of these furnaces is shown in section, with crucible and pipe-stem in position), being more compact in this form, more easily cleaned from clinker, and more economical in fuel than the square ones. They are 12 inches in diameter and 21 inches deep. The five firebars, $1\frac{1}{2}$ inches square and 18 inches long, are set in a cast-iron box, D, 12 inches by 2 inches, which passes through the brickwork in front of the furnace, the other ends of the firebars resting on an iron bar set in the brickwork at the back of the furnace. The bars are 6 inches above the floor. The draught is obtained through a grating in the floor, which covers a portion of the ashpit, over which there slides a cast-iron plate, M, $\frac{1}{4}$ inch thick, for regulating the admission of air, and pivotted in one corner. The flue, L, is 6 inches square, and communicates with a series of five condensing chambers, 8 feet by 8 feet by 5 feet, running the length of the furnaces (42 feet), all communicating and leading to the stack, 80 feet high, common to refining and melting furnaces, which are twenty-one in all. There are three furnace covers, two of them 20 inches by $6\frac{1}{2}$ inches, the third a little smaller, and all are bound with iron. The middle one is perforated by a 1-inch hole, through which the chlorine delivery pipe passes. Glenboig arched firebricks, B, 9 inches by $4\frac{1}{2}$ inches, and tapering from $2\frac{3}{4}$ inches to 2 inches, are used for lining the furnaces, and are set with touching joints in an iron cylinder, A, $21\frac{1}{2}$ inches in diameter, and at least $\frac{1}{4}$ to $\frac{3}{8}$ of an inch thick, which is supported by a cast-iron plate, C, $\frac{5}{8}$ of an inch thick, and 22 inches in diameter, with a 12-inch hole in the centre. This plate is supported by the brickwork which forms the foundation. The ashpit is a cast-iron flanged box, easily cleaned in case of an accident. Round the iron cylinder concrete, N, is rammed, the front iron plate of the furnace being shifted 2 or 3 inches in, until this is set and then moved out, thus providing an air space, E, and keeping the plates cooler. The furnace top is a plate of cast iron and, so as to facilitate repairs, should be in two pieces for each furnace, halved into one another, the hole being slightly bossed at the edge so that the firetiles may run easily on them. One piece has a hole 6 inches in diameter over which the swing ventilating hood, P, is placed by which the pot is covered when removed from the fire. This hood communicates by a passage through the brickwork with the flue. The cylindrical furnace is calculated to last for three years, the square ones lasting only eighteen months and taking three hours to reline, while the cylindrical ones take one hour.

The Crucibles, &c.—The guard pot, placed for safety under the white pot and afterwards used for remelting the refined gold, is a plumbago crucible $8\frac{1}{4}$ inches high, 6 inches inside diameter,

$\frac{5}{8}$ of an inch thick at the top, and $\frac{3}{4}$ of an inch at the bottom, which is flat inside and stands on a cylindrical firebrick 5 inches in diameter and $2\frac{1}{2}$ inches deep. The white pots, fitting loosely into the guards, are $10\frac{1}{2}$ inches high, 5 inches in diameter, and $\frac{3}{8}$ of an inch thick at the top, tapering from 1 inch at the bottom. They are covered by a closely-fitting lid, dished at the top to catch any globules spirted out by too rapid a current of gas and perforated by two holes $\frac{3}{4}$ of an inch in diameter. A new pattern of lid to be introduced shortly will have a notch in the edge for the pipe-stem to pass through, the advantage of this being the easy removal of the lid without withdrawing the pipe-stem, as is necessary with the old lids.

Fig. 585.

The pipe-stem is 24 inches long, tapering from $\frac{3}{8}$ to $\frac{1}{2}$ an inch at the end inserted into the gold, and is wedge-shaped to facilitate the escape of the chlorine when resting on the bottom of the pot. The bore of the pipe-stem is $\frac{1}{8}$ of an inch in diameter. The thin end of the pipe-stem is attached to the branch delivery pipe by a piece of $\frac{1}{2}$-inch rubber about $2\frac{1}{2}$ inches long, which connects with an ebonite junction, G, 3 inches in length, with a bore of $\frac{1}{16}$ of an inch, turned with a ring round the middle, which acts as a rest for the 8-oz. weight, H, used as a sinker for the pipe-stem. One end of the ebonite junction is $\frac{1}{2}$ an inch in diameter, the other $\frac{3}{4}$ of an inch; the latter being connected by a stout rubber tube 3 or 4 inches long to a 14-inch lead pipe ($\frac{1}{2}$ an inch in diameter) which is attached [by a rubber junction] to a glass stopcock, I, from the spigot of which a $\frac{3}{4}$-lb. lead weight, J, is suspended to prevent the pressure of gas from blowing it out. The glass stopcocks have replaced the compressor clamps, which were not satisfactory owing to the rubber cutting through, and chlorine leaking past. The rubber joints are sufficiently flexible to allow the pipe-stem to bend down into the pot or to be laid horizontally on a rest when not in use. Each furnace is provided with one glass stopcock to control the flow of gas. The cock is far enough away at the back of the furnace, to be unaffected by the heat when the firetiles

Fig. 58c.

are removed, and is connected by a $\frac{1}{2}$-inch lead pipe with the main pipe running along the wall. Thinner pipe-stems are found to be as serviceable as the above and do not require such careful annealing. The tubes, stopcock, &c., are somewhat diagrammatic in Fig. 58a, and can be studied in Fig. 58b, which is from a photograph of the furnaces by Mr. R. Law. This also shows the crucibles, guard pots, ventilating hoods, and chloride cakes.

The generators are shown in Fig. 58c, which is also from a photograph by Mr. R. Law. The pressure regulator and the reduction tank are also shown. The generators, eight in number, are three-necked cylindrical stoneware vessels with domed tops, and having a flange round the middle by which they are supported on the stoneware steam jacket 16 inches high, 16½ inches in diameter, and ⅜ of an inch thick. The domed vessel is 2 feet high. The three necks have 1¾-, 1⅜-, and 1¾-inch holes, the first for charging-in the manganese ore and closed by an indiarubber plug, the second for the pipe leading to the main chlorine pipe, and the third for a branch acid supply tube, ½ an inch in diameter, fitted with a glass stopcock a foot above the neck, between which and the stopcock another ½-inch tube, the overflow, branches. The overflow tube, through which the hot waste of the generators has to pass, is provided with an ebonite stopcock which is turned off during refining. Stout combustion tubing is used. The bottom of the generator is covered by 4 inches of quartz pebbles, to prevent choking of the acid delivery pipe, which reaches to within 1 inch of the bottom of the vessel. 56 lbs. of manganese ore (about 73 per cent. peroxide), broken to ¼ to ½ inch square, is placed on the pebbles, and commercial hydrochloric acid, of specific gravity 1·16, is added as required through the acid delivery pipe by turning the glass stopcock. The acid pipe is of glass, and leads to the eight storage tanks 20 feet above the floor, which hold 320 lbs. of acid each, and are interconnected by glass tubes luted into the bottoms; the delivery of acid, however, being from one at a time. The gas delivery pipes from the generators all connect with a 1 inch lead pipe, which leads to a distributing vessel with two necks and partially filled with manganese chloride solution. A pressure guage of 1 inch glass tube and 15 feet high is luted into the bottom of this vessel, and is fixed to the wall by brackets, 10 to 11 feet of the solution being required to overcome the resistance of about 7 inches of metal in the pots. The pressure in refining is equal to 5 lbs. per square inch. A four-way tube of lead or pottery is passed through the second neck of the vessel, and each arm is connected by thick rubber to glass stopcocks to which ½-inch lead pipes are joined, these pipes leading to sets of four, four and five furnaces, so that the supply of gas can be delivered to a few or all the furnaces, as desired, the subdivision being made for safety in case of a leakage or for convenience if only a few furnaces are in use. All the generators are used whether the quantity of gold to be refined be large or small, the same quantity of acid being run into each. When the flow of chlorine through the gold is stopped the acid in the generators is forced back through the overflow pipe by opening the ebonite tap. It is found necessary to have the main pipe in communication with another two-necked earthenware vessel containing such a quantity of water that when the pressure of gas exceeds the working pressure required, the end of a glass tube, passing to the bottom of the vessel and connected above the neck with an upright 4-inch lead pipe 10 feet high, becomes unsealed, and the gas escapes through the water in large bubbles, escaping through a glass pipe, inclined at an angle at the top of the lead pipe, into the air. When sufficient gas has escaped to reduce the pressure to the working limit the pipe is sealed. Thus the pipe acts automatically in keeping the pressure below such an amount as would endanger the apparatus or cause joints to leak. It is found expedient to cover all the rubber junctions in the generating room with calico and then to paint it. Protected in this way it will last until stopped up by the action of the chlorine which fills it with lemon-yellow incrustation, at the same time reducing its thickness. All junctions are secured with copper wire where practicable.

Refining Operations.—The guard, with the white pot in it containing 2 or 3 ozs. of fused borax, is placed in the furnace, and is heated gradually until the bottom of the white pot is dull red. The ingots (of which the larger are slipper-shaped) to be refined, amounting in all to 650 to 720 ozs. in weight, are then placed loosely in the pot, the furnace filled with fuel, and the dampers opened. As soon as the gold is melted, which generally happens in about one and a-half hours, the boraxing of the pots being also effected at the same time, the perforated lid is put on, and the pipe-stem, previously brought carefully to a red heat to prevent cracking or flaking, is pushed to the bottom of the pot. As the pipe is being inserted, the chlorine is gently turned on to avoid stoppage of the passage through the stem by the solidification of metal in it. The supply of chlorine is controlled by the glass stopcock over the furnace, and the amount is adjusted so that the whole of the gas is absorbed and no globules of metal can be thrown up. This can usually be ascertained by feeling the pulsations of the gas through the indiarubber connections as it escapes in bubbles out of the bottom of the pipe-stem. When the gold contains much silver or base metals, the absorption of the chlorine takes place rapidly but gently, very little motion of the contents of the crucible being apparent, but when the gold to be refined is of high assay and also in all cases towards the end of refining, the gas is admitted only in a small stream, and requires careful watching to prevent spirting. When base metals are present in large quantities (over 2 per cent.) dense characteristic fumes of the chlorides of these are given off, and the metal or metals present may be generally identified by the fume or incrustation caused by the condensation of the base chlorides on the pipe or lid.

Gold containing 2 per cent. of silver and 0·5 per cent. of base metal is refined to about 995 fine in one and a-half hours, while that containing 3·5 per cent. of silver and 1·5 per cent. of base metals takes two hours. When larger percentages of silver or base metals are to be dealt with, the time taken is not proportionately longer, because, as mentioned above, a much greater stream of gas may with safety be admitted, though, in all cases, at the beginning the chlorine must be introduced gently on account of there being air in the chlorine mains, and, also, at the end of refining, the supply must be greatly reduced. When nearing completion, the "flame" issuing from the holes in the lid becomes altered in appearance, and much smaller; it now contains much chlorine mixed with small quantities of the volatile chlorides. The actual completion of the operation is generally known by the appearance of a very characteristic "flame," which is luminous, with a dark brown fringe. In case of doubt, a piece of clean pipe-stem is used as a test. It is placed, cold, for a few seconds in the issuing flame, and if the refining is finished, a clear reddish-brown stain, tending to yellow, is imparted to the test end. This stain consists of ferric oxide and chloride from the oxidation of ferrous chloride, and contains gold and sometimes chloride of silver, and is probably caused by small quantities of chloride of iron retained by the fused gold and non-volatile chlorides, from which it is freed by the unabsorbed chlorine bubbling through. Traces of copper and iron are always found in the refined gold, the bulk of the alloy being silver. As soon as the stain is found to be of the right colour, the current of gas is reduced, and is allowed to pass for a further fifteen minutes, and the pipe is then withdrawn and the clay pot lifted out of the guard. The pot is allowed to stand under a hood (to carry off the fumes) until the gold is set, which usually takes place in from five to seven minutes, the fact that solidification has taken place being observed by thrusting a piece of red-hot pipe-stem down through the fused chlorides. The chloride is then poured into a mould provided with a ventilating hood, which, in consequence of the high density of the fumes necessitating a sharp draught to remove them, is connected with the stack. Any borax poured off with the chlorides is allowed to remain, as it is required as a cover for the chlorides in the subsequent fusion for the separation of their gold contents. The pot is then broken, as the cone of gold will not fall out of it soon enough, and the cone of refined gold is remelted in the guard and cast into two flat ingots, 12 inches by 4 inches by $1\frac{1}{2}$ inches, which, when set and still red hot, are placed on a copper lift, dipped in dilute sulphuric acid and then in water, and after removal from the water are still sufficiently hot to dry by their own heat. The broken pots are ground in a small Chilian mill and panned off, and the gold obtained is added to the "end" that is returned at the end of the day. 9,000 ozs., containing up to 10 per cent. of silver and base metals, constitute a day's refining.

An improvement has recently been introduced, by which a

considerable saving of time and material is made. This is the dipping of the fused chlorides and borax from the pot while it is still in the fire, and without previous solidification of the gold, by a small clay crucible, from which they are poured into a covered mould projecting over the furnace, the drops falling back into the pot. This had previously been the practice when the percentage of silver was large, as the silver doubles its bulk on conversion to chloride, and would have overflowed. The last "dip" always contains some gold, and is poured into a separate mould, in which the metal sets at once. The chloride is thence poured into the larger mould, and the gold returned to the pot. The chloride remaining in the pot is then made into a paste with bone ash, after which the refined gold is stirred and cast into ingots, the pot being at once returned to the furnace to be used a second time.

The chlorides, which hold from 5 to 10 per cent. of gold in feathery particles, are remelted during the day in plumbago crucibles holding 300 ozs. of chloride. When fused, 7 per cent. of their weight of bicarbonate of soda is added, cautiously and without stirring, which produces a shower of globules of reduced silver, and these falling through the chlorides carry down nearly all the gold. As one addition of bicarbonate of soda does not entirely free the chlorides from gold, a second addition is made, without removing the crucible from the furnace, ten minutes being allowed after each addition. The pot is then lifted out and placed on one side to allow the metal to set, when the chlorides are poured into a mould 12 inches by 10 inches by 2 inches, practically free from gold and ready for reduction. The silvery button obtained contains from 40 to 60 per cent. of gold, and is refined on the following day.

The silvery ingot and the refined gold contain 99·85 per cent. of the gold issued in the morning for refining, the bulk of the deficiency being in the pots. The amount of gold which goes immediately into work after refining is 97·6 per cent. (an average of thirty days refining). The amount of gold left in the silver after reduction from the chloride reaches a maximum of 1 part in 10,000, but is usually from $\frac{1}{2}$ to $\frac{1}{3}$ of this quantity.

The cakes of impure chloride, weighing about 250 ozs. each, may be colourless and translucent to brown or chocolate colour and opaque, the colour depending on the amount of copper salt present. They consist of argentic and cuprous chlorides, with traces of other chlorides and 9 per cent. of chloride of sodium from the decomposition of silver chloride by bicarbonate of soda. When cool, each cake is sewn up in a coarse flannel bag to prevent loss of any silver which may become detached during reduction, and they are then boiled with water in a wooden vat for four or five hours. By exposure to air and moisture the cakes become coated with a green deposit, owing to the conversion of the cuprous chloride into cupric oxychloride and hydrated cupric chloride, and the successful removal of a large proportion of the cuprous salt in the vat is due to its solubility in a hot solution of common salt and of hydrated cupric chloride, from which it is redeposited on dilution.

The cakes for reduction are placed alternately with wrought-iron plates $\frac{1}{8}$ inch thick in a cast-iron tank lined with similar plates. The plates are prevented from touching the bags by laths of wood, otherwise the copper would be reduced in the bags, and would be difficult to separate from the silver. The reduction is slow in starting, unless either some liquor is left from a previous operation or some chloride of iron added. The bath is heated by the direct injection of steam (this is absolutely necessary), and the reduction is complete in from two to four days, though sometimes it takes longer. The time may be lessened to twenty-four hours by putting the chloride cakes in metallic connection with the iron plates. The completion of the reduction may be easily ascertained by feeling the cake, when, if any chloride be unreduced, it is felt as a hard lump. The reduced silver is taken out of the bags, washed in boiling water for about an hour, and then melted, no fluxes being necessary. The use of the flannel bags makes the reduced silver of high standard, as the reduced copper is thus prevented from adhering to the silver cake, from which it was found very hard to detach it without tearing off silver as well. A small percentage of reduced silver and silver chloride is found at the bottom of the tank, its presence being probably due to the solubility of the chloride in solutions of the chlorides of copper, iron, and sodium. In 1894, the mean standard of the refined silver was 982·2, the lowest ingot being 955 fine, and the highest 998. In 1895, the mean standard was 981·6, the lowest being 936, and the highest 995 fine.

The mean fineness was only about 930 prior to the year 1889. In 1889, after the introduction of the iron plates in place of zinc, the fineness rose to 948. The subsequent improvement was probably due to the introduction of the flannel bags.

The silver contained in the gold operated on is distributed in the following manner, the mean results of the last five years (1891-95) being given :—

Silver in ingots,	88·77 per cent.
,, left in refined gold,	7·62 ,,
,, in "sweeps,"	2·00 ,,
,, unaccounted for,	1·60 ,,

The "sweep" from the condensing chambers amounts to about 3 cwts. per annum, and contains an average of 41 ozs. fine gold and 157 ozs. fine silver, which are carried over as globules or volatilised as chloride and condensed.

The mean amount of gold refined per annum during five years (1891-95) was 949,527 ozs., containing gold 937·7, silver 49·6, and base metals (by difference) 12·7. The mean assay of the refined gold for the same period was 995·9, and the mean loss of gold in the refining operations for the same period was 0·175 per thousand.

The approximate cost of refining per ounce gross weight refined was as follows:—

	In 1894.	In 1895.
Material,	0·1397 of a penny.	0·1215 of a penny.
Wages,	0·1485 ,,	0·1439 ,,
	0·2882 ,,	0·2654 ,,

Half the cost for materials was for hydrochloric acid at £20, 15s. per ton.

The amount of gold refined in 1894 was 1,049,529 ozs., containing in parts per thousand—gold, 933·9; silver, 51·4; base metals, 14·7; and the gold refined in 1895 amounted to 1,083,243 ozs., containing gold, 932·0; silver, 52·3; and base metals, 15·7 parts per thousand.

In some experiments made by Mr. Barton, who took gold alloyed separately with $4\frac{1}{4}$ per cent. of copper, $4\frac{1}{4}$ per cent. of lead, 4 per cent. of iron, and $4\frac{1}{2}$ per cent. of tin, when the cost of hydrochloric acid was $2\frac{1}{2}$d. per lb. and manganese ore (70 per cent. peroxide) 1d. per lb., the following results were obtained, operating on quantities of gold containing 30 ozs. of each of these metals:—

Cost of extracting copper,	4·5 pence per oz. Troy.
,, lead,	1·4 ,,
,, iron,	3·9 ,,
,, tin,	2·5 ,,

The gold in each case was brought up to the usual fineness. These results give approximately the cost of extracting these metals in the quantities in which they ordinarily occur in Australasian gold.

The Miller Process at the Sydney Mint.—Mr. J. M'Cutcheon, the Assayer at the Sydney Mint, writes that the process of freeing the chlorides from gold now used by him is as follows:— The chlorides produced during the operation are separated into two classes, termed "balers" and "non-balers." The first is that portion baled, or rather ladled, out during the operation, to prevent overflow; this is re-melted in quantities of 350 ozs., and whilst in the molten state, half a pound of bicarbonate of soda is projected on the surface. This has the effect of reducing some of the chloride, and the metal in sinking to the bottom of the pot carries with it all the gold. The "non-balers," or that portion of chloride which is poured off the refined gold when it has set, is treated as above, but 7 ozs. of granulated zinc is used instead of the bicarbonate.

The chlorides are poured into slabs, and are now ready for the reduction process, in which the silver loops formerly used have been abandoned, iron plates being now used instead of zinc; the water is acidulated with hydrochloric acid.

The bullion treated at the Sydney Mint during 1895 contained— gold, 832·1; silver, 135·5; and base metals, 32·4 parts per 1,000.

The average fineness of the gold produced by this method at Sydney in the period of nine years from 1884 to 1892 was 995·9. The remainder is silver, which apparently cannot be profitably removed by chlorine or by any other method. The average fineness of the silver produced at Sydney was about 970 in 1893, the fineness of individual bars varying from 917 to 987. The alloying metal in the silver bars is almost all copper. The gold retained by the silver formerly amounted on an average to 1·3 parts per 1,000, but it has since been materially reduced and is now a mere trace. Analysis of the silver resulting from the refinage of gold, known originally to have contained, among the base metals in the alloy, copper, lead, antimony, arsenic, and iron, gave the following results:—

Silver,	972·3
Copper,	25·0
Gold	2·7
Zinc and iron,	Traces.
	1000·0

The losses of gold in the course of the process are very small, varying from 0·11 to 0·19 per 1,000; this is considerably less than would have been lost by merely toughening the gold with corrosive sublimate without parting it from the silver. The loss of silver at Sydney was about 4·25 per cent. in 1895. These losses are reduced if the amounts recovered from the flue-dust and from the ground-up crucibles are taken into account. The total cost of the process is about 1d. per oz. of crude gold at Sydney, and was about 0·65d. per oz. in Melbourne in 1873, but has since been reduced. The loss of gold by volatilisation is probably prevented from reaching the large amounts which might be expected from the results of Christy and of the author (see p. 21) by the fact that, during the whole time that the chlorine is being passed, silver and base metals are present, and, by absorbing the gas, protect the gold from its action.

Professor Thomas Price, who treated some of the Californian gold-bullion by this method on a working scale in his laboratory in San Francisco, states * that, with Californian gold, which generally contains more silver than Australian gold, the gold

* *Trans. Am. Inst. Mining Eng.*, 1888, vol. xvii., p. 30.

taken up by the chloride of silver amounted to 5 per cent., and even to 10 per cent. of the total weight of the gold. For this reason, and on account of the large amount of silver bullion in the San Francisco market requiring parting, Professor Price considers that the Miller process, while technically successful with Californian gold, is hardly able to compete commercially with the ordinary sulphuric acid process. He suggests that it might be well adapted for refining the nearly pure brittle gold produced at chlorination works, where chlorine is at hand and other methods of refining are not convenient. The idea of refining brittle gold in this way had previously been acted on by Professor Roberts-Austen in the year 1871 in performing the experiments and operations detailed below.

Refining Brittle Gold by Chlorine Gas.—In spite of the fact that brittle ingots of gold are not accepted for coinage at the Royal Mint, a number of such ingots formerly found their way into work there, their lack of ductility not being observable until after they had been alloyed with the copper requisite to make standard gold, which is 916·6 fine. In 1871, no less than 40,000 ozs. of such brittle standard gold had accumulated at the Mint, having been set aside as unfit for coinage. These bars contained as impurities traces of antimony, lead, bismuth, &c. In order to determine how far such impurities could be eliminated by chlorine gas, applied as in Miller's process, the following experiments were conducted by Professor Roberts-Austen at the Royal Mint:—*

1. A bar of standard gold, weighing 241·20 ozs., which broke with a slight tap from a hammer, was melted in a clay crucible, and chlorine passed for three minutes. The metal was found to be perfectly toughened and to have sustained a loss of weight of only 0·17 oz., which proved to consist entirely of base metal.

2. Gold bars containing 0·5 per 1,000 of antimony and 0·5 per 1,000 of arsenic, or a total of 1·0 per 1,000, were converted by a stream of chlorine into gold of excellent quality in three and a-half to four minutes.

3. A bar of standard gold, weighing 301 ozs., in which the impurities noted below were contained, was operated on. The base metals consisted of:—

Antimony,	3·0 per 1,000.
Lead,	2·0 ,,
Zinc,	2·0 ,,
Iron,	2·0 ,,
Tin,	2·0 ,,
Arsenic,	3·0 ,,
Bismuth,	1·0 ,,
	15·0 ,,

---

* *First Annual Report of the Mint*, 1870, p. 95.

This gold was as brittle as loaf-sugar, and could be broken with the fingers. After chlorine had been passed for eighteen minutes, the metal was perfectly ductile. The loss of metal was found to be—

Base metals (as above),	4·51 ozs.
Copper,	3·44 ,,
Gold,	·15 ,,
Total,	8·10 ,,

These experiments having been successful, the 40,000 ozs. of brittle gold were treated similarly in graphite crucibles, the charges being about 1,100 ozs. of gold in each. The time of passage of the chlorine varied from five to seven minutes, and the total loss of gold after recovery from the "sweeps," was returned as only $\frac{1}{10}$ oz. With this loss and at a very slight expenditure, the whole of the stock of brittle gold was toughened and rendered fit for coinage. Considering the success of this venture, it is surprising that the method has not yet found wider application in the treatment of brittle bars.

Parting by Electrolysis—*The Moebius Process.*—This process was patented by Mr. Bernard Moebius, in England, as long ago as the year 1884 (Dec. 16, No. 16,554), and is now in successful operation in several localities in the United States and Germany. It is said* to be specially suitable for refining copper bullion containing large proportions of silver and gold with small quantities of lead, platinum, and other metals, but is chiefly used in parting doré silver.

The apparatus required consists of a number of wooden vats coated inside with graphite paint, and filled with a solution containing 1 per cent. of nitric acid, which constitutes the electrolyte. The anodes consist of plates of bullion of about ½ inch thick and 14 inches square, which are hung in muslin bags destined to catch the insoluble impurities after the silver, copper, &c., have been dissolved. The cathodes consist of plates of pure silver, slightly oiled to prevent adhesion of the deposited metal. These plates are continually scrubbed by a mechanical arrangement of brushes by which their surfaces are kept free from loose crystals of electro-deposited silver. This loose silver falls on to trays placed below, which are removed at intervals, and the silver collected from them.

The current should have an electromotive force of from one to three volts for each vat. The copper is not deposited unless the solution becomes too weak in silver or too rich in copper, and even if some happens to be deposited, it falls into the cathode trays with the silver and does little harm, since it will be

* Gore's *Electrolytic Separation of Metals*, London, 1890, p. 240.

gradually redissolved if the conditions are corrected. When too much copper has accumulated in the solution the latter must be removed, the method being as follows :—The bullion anodes are replaced by carbon ones, and a weak current passed until all the silver is deposited. The silver cathodes are then replaced by copper ones, and a strong current passed so as to deposit the copper as rapidly as possible as a loose powder, which falls into a copper box. This box is connected with the cathode to prevent corrosion by the acid which is set free. The liquid thus regenerated is used again as the new electrolyte.

The process is stated to be the cheapest parting process known. If no copper were present in the bullion it is clear that there would be no consumption of acid, and it would never be necessary to change the electrolyte. Under the most favourable conditions, therefore, with water power available and the amount of copper in the bullion very small, the cost of parting doré silver would be merely nominal.

The following description of the Moebius plant which is in successful operation at the Pinos Altos Mine, Chihuahua, Mexico, is an abstract of that given by Mr. George Maynard.[*] The plant is capable of treating from 3,500 to 4,000 ozs. of doré silver in 24 hours, and consists of a tank 12 feet long, 2 feet wide and 20 inches deep, divided into seven compartments. Four silver plates (the cathodes) and six doré plates (the anodes) are placed in each compartment in such a way that an anode is opposite to each face of a cathode. The trays to catch the deposited silver have perforated bottoms and are covered with asbestos cloth, and in order to facilitate cleaning-up, the electrodes, frames, anode bags, trays and brushes can all be lifted up simultaneously by a hoisting arrangement worked with a crank and handle, so that the exciting liquid alone remains in the cells.

The bullion is from 800 to 900 fine in silver, and 25 to 50 in gold, the rest being chiefly copper. The exciting liquid, which at first contains 1 per cent. of nitric acid, is gradually converted into a solution of silver and copper nitrates. The silver nitrate is continually decomposed by deposition of the metal, but the copper nitrate accumulates, fresh acid being added at intervals to prevent the copper from being deposited. The silver dissolves from the anodes and is precipitated on the cathodes in the form of heavy tree-like crystals. The action of the brushes or scrapers is of vital importance to the success of the process, the advantages derived from their action being summarised as follows :—

1. The liquid is agitated and so kept homogeneous.
2. Polarisation is prevented.
3. The electrodes can be brought very near to one another, and the resistance thus reduced without any fear that short circuiting will ensue by the bridging over of the space between them by crystals of deposited silver.

[*] *Eng. and Mng. Journ.*, May 9, 1891, p. 556.

The lead (as peroxide), platinum metals, antimony and other impurities remain with the gold in the bag surrounding the anodes. When the exciting liquid becomes too highly charged with copper, the solution is used instead of bluestone in the amalgamating pans of the mill, but the copper could of course be recovered as usual if it were desirable. The manual labour is all performed by the assayer and his assistant; the cathode trays are hinged, and by letting them turn on their hinges the silver is let fall into a movable tank on castors furnished with a false bottom, on which it is washed and dried, and it is then ready to be melted into bars of about 999 fine. The gold slimes are similarly washed and filtered; when they are fused, the lead is almost all slagged off.

The electric current employed is 170 ampères, the electromotive force being 8 volts; this requires an expenditure of 2½ H.P. to refine from 3,500 to 4,000 ozs. per day. The cost of parting is said to be less than ⅓ cent per gross oz. of bullion, and the royalty is ⅛ cent per oz.

The original cost of the plant is said to have been about $6,000, including $1,000 for the silver in the cathode plates. This sum included the cost of conveying the plant to the mine, which was very high, as a journey lasting several weeks on mule-back had to be performed. The amount of silver contained at any one time in solution in the bath is about 300 ozs. The weight of the forty-two anode plates is about 4,200 ozs. when they are fresh, and this silver can be melted up and recovered about forty-eight hours after the operation has been started, so that in this plant only about 10,000 ozs. of bullion are necessarily locked up continuously in the apparatus. The clean-up takes place once a month.

The Moebius process has also been in successful operation at the works of the Pennsylvania Lead Company at Pittsburg since September, 1886; here it is said that from 30,000 to 40,000 ozs. of doré bullion are refined daily at a cost of three-fourths of a cent per oz. The silver produced is from 999 to 999·5 fine. A small plant was also erected at the Kansas City Smelting and Refining Company's works, and a large one in 1891 at St. Louis. The process is also worked by the Norddeutsche Refining Company at Hamburg.

CHAPTER XIX.

THE ASSAY OF GOLD ORES.

THE assay of gold ores is almost universally conducted in the *dry* way—*i.e.*, by furnace methods. Exceptions will be noted later. The plan of operation is to concentrate the precious metal in a button of lead in one of two ways, viz. :—(1) By fusion in a crucible; or, more rarely, (2) By scorification. The button of lead obtained by either method is then subjected to *cupellation*, by which the lead is oxidised and removed, and the resulting bead of precious metal is weighed. Since in these operations silver and the metals of the platinum group remain with the gold, they are subsequently separated by *inquartation* and *parting*, and in the case of platinum and its allies by further special methods.

The exact method to be used in any particular case varies with the richness of the ore, the nature of its gangue and the presence or absence of compounds of the base metals. As a general rule, poor ores—*i.e.*, those containing less than 2 ozs. gold per ton—are better assayed by the fusion process so that a comparatively large quantity of material may be operated on. Rich ores may be assayed either by fusion or scorification, the errors arising from the small amount of material used in the latter process being less important in their case. Telluride ores, arsenical and antimonial ores, and ores containing tin, nickel, or cobalt must all be scorified if possible, but in these cases it is better to make the ordinary crucible assay, and then to scorify the lead button obtained. Either method can, however, be used for any ore.

*Assay by means of the Blowpipe.**—This method, though less exact than that made in a furnace, is of importance, because in prospecting expeditions it is possible by its means not only to detect the gold and silver in any ore, but also to determine its amount quantitatively with fair accuracy. On such expeditions it is impossible to carry the cumbrous apparatus required to make an ordinary assay. The amount of powdered ore taken is usually 100 milligrammes, and this is mixed with borax and about 1 gramme of granulated lead. The whole is wrapped in paper and heated on charcoal in the reducing flame of a blowpipe until the fusion is complete, and then for a short time with the oxidising flame. The lead is then separated from the slag

---

* For a full description of this method, see Plattner's *Manual of Analysis with the Blowpipe*, pp. 360-406. London, 1875.

and heated on a bone-ash cupel until it is all converted into litharge. The diameter of the button of silver and gold thus obtained is carefully measured on an ivory scale, which at once gives the percentage amount in the ore. The gold is usually separated from the silver by parting in nitric acid, but Richards has recently stated * that the silver can be distilled off by the blowpipe, leaving a bead of pure gold, which can be measured.

Sampling the Ore and Preparation of the Sample for Assay.—The value of an assay depends largely on the care with which a sample of the ore is selected. The sampling is done by processes which are as far as possible automatic, and independent of the will or judgment of the assayer. The best and most widely applicable method is that of dividing and intercepting a part of a stream of ore falling vertically downwards or sliding down an inclined plane. This is done in many ways in mines and mills. A convenient method in use in Colorado may be described as follows :—The mineral, having been passed through a stone-crusher and one or two pairs of rolls so as to be reduced to the size of coffee beans, falls through a vertical tube, on to the apex of a cone. The surface of the cone is inclined at an angle of $45°$ to the vertical and has one or more openings or windows bounded by straight lines drawn from the apex to the base. The ore, falling on the cone, slides over its surface into the ore-bin, excepting a portion, usually one-sixteenth, which passes through the openings in the cone and is conveyed down a tube to another receptacle. The operation may be repeated on the sample thus taken, after it has been further crushed.

In the assay office the sample, however obtained, is further reduced in bulk by the implement known as the *sampling tin*. This consists of a series of troughs arranged side by side and fastened at equal distances from each other (the width of the spaces being equal to that of the troughs) by strips of metal soldered on to their ends. An even stream of ore being let fall from a shovel on to this sampler, half is retained in the troughs while half passes through. Careful experiments have proved that each half is representative of the whole. A repetition of the process reduces the sample to any required extent. A sampler with troughs 1 inch wide is suitable for treating materials which include lumps of not more than $\frac{1}{4}$ to $\frac{1}{3}$ inch in diameter. For finely ground materials a convenient width for the troughs is $\frac{3}{8}$ inch.

In taking a sample from a large bulk of ore, or where sampling tins cannot be procured, the method of dividing into quarters may be employed. The ore is piled into a heap, thoroughly mixed and divided into four by a shovel by two lines at right angles to each other. The two opposite quarters are removed, and the remainder again mixed and divided as

* *Journal of the Franklin Institute*, June, 1896.

before. When the sample is reduced to a few pounds in weight it is broken down to the size of coffee beans, reduced in bulk by quartering or the sampling tin, crushed finer, again sampled and so on, until finally, when reduced to a mass of from $\frac{1}{2}$ lb. to 1 lb., it is all crushed fine enough to pass through a 60-mesh or 80-mesh sieve (*i.e.*, one containing 80 holes per linear inch). Before this fine crushing can be done, it is usually necessary to dry the sample; the percentage of moisture should always be estimated at the same time by weighing the sample both before and after it is dried.

The implements employed for crushing samples of ore are various. A small rock-breaker, with reciprocating jaws, similar to those used on the large scale, and worked by hand or steam power, is useful for breaking down large lumps, which otherwise may be broken by a hammer. For finer pulverisation a small pair of steel-faced high-speed rolls may be used if steam power is available. This method is adopted at large smelting or sampling works, where great numbers of samples are crushed daily. In smaller works or offices the *buckboard* (Fig. 59) is most suitable.

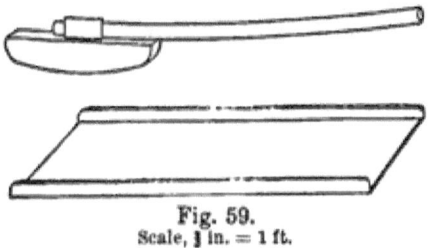

Fig. 59.
Scale, $\frac{1}{4}$ in. = 1 ft.

It is a smooth plate of iron about 2 feet square with a 1-inch rim surrounding it on two or three sides. On this a *bucking hammer* is worked—a heavy piece of iron 5 to 10 lbs. in weight, with a large smooth curved face and a handle 30 inches long. It is moved about on the iron plate (on which the ore is spread) with both hands, one holding the handle, the other pressing the head downwards, the curved face being below, while an oscillatory movement is imparted by the handle. The instrument is very effective if the ore is previously broken down to the size of coarse sand in a mortar. The pestle and mortar are of value in breaking down samples from the size of nuts to that of coarse sand. In grinding down siliceous material, so as to enable it to pass through an 80-mesh sieve, the pestle and mortar is far inferior to the buckboard.

"*Metallics.*"—In many ores, both gold and auriferous silver occur native in grains or threads. These "metallics" are not readily reducible to a fine state of division, and, though a part always passes through the sieve, some of the larger pieces which have resisted abrasion fail to do so. In some assay offices part of the pulverised ore is thrown back into the mortar with the metallics, and grinding is continued until everything passes the sieve. This is a dangerous practice, as it is impossible to ensure the even distribution of metallics through the sample. The smaller pieces which pass through the sieve in the first instance constitute an unavoidable evil which is increased by every piece of metal that follows them. The safer plan is to cupel by themselves the whole of the metallics left on the sieve, and calculate their value per ton of ore independently of the result obtained by the ordinary assay. The total value of the ore is found by adding these two results together.

The prepared sample may be stored in tin boxes or glass jars, which should be labelled by numbers, none of which are ever repeated. Before weighing out the powdered ore for assay, the whole sample should be turned out into a wide bowl or on to glazed paper, or, better still, rubber cloth (which does not crack and wear out like paper), and thoroughly mixed with a spatula. The sample should never be mixed by shaking, and care should be taken to avoid jarring it after mixing, as metallics in that case tend to settle to the bottom from their superior density, and a fair sample cannot then be easily obtained. For the same reason the part of the mixed sample which is taken for assay should not be hastily shovelled on to the balance pan from the top of the pile, but a vertical slice should be taken, some of the lowest layer being carefully scraped up from the rubber cloth.

FUSION OR CRUCIBLE PROCESS OF ASSAY.

This process is divided into three parts, viz. :—(1) Fusion; (2) Cupellation; (3) Parting.

(1) **Fusion.**—The object of this operation is to concentrate the precious metals in a button of lead, while the whole of the remainder of the ore forms a fusible slag with suitable reagents, in which lead sinks. The fusion is made in clay (or rarely iron) crucibles in a wind furnace, the fuel being coke, charcoal or anthracite. The size of the furnace depends on the amount of work to be done and on the fuel employed, charcoal requiring more space than coke. If only one fusion is to be made at once, the internal dimensions of the fire-box may be 8 inches square, while a furnace 14 inches square will hold nine crucibles. The furnace is exactly similar to that used for melting bullion described on p. 359.

In the United States it is usual to perform fusions in a muffle furnace, similar to that described below under cupellation, p. 406. The temperature required is about the same as that used in scorification. The advantages claimed are greater cleanliness and neatness and more uniformity in the conditions, the temperature of a muffle being more easily kept constant and uniform than that of an ordinary fusion furnace. Six or eight fusions can be performed at one time in a large muffle.

There are three shapes of clay crucibles in common use, namely—the French, the Battersea, and the Colorado. The French pot differs from the others in the thickness of its walls near the bottom, and consequently it requires very careful annealing. The Battersea pot is of coarser texture than the others, having a larger percentage of sand in its composition. The Colorado crucibles, manufactured by the Battersea Company, are specially made for fusion in the muffle; the "20 gram" size is the most generally useful.

In weighing the materials for a crucible charge the use of a set of *assay-ton* weights saves much labour in calculation. The assay-ton is a weight which contains as many milligrammes as a ton contains ounces. Thus an English ton of 2,240 lbs. contains 32,666 Troy ounces, so that the corresponding assay-ton must weigh 32,666 milligrammes or 32·666 grammes. Now suppose the weight of the resulting bead of gold (or silver) from an assay-ton of ore to be 1·5 milligrammes, it is obvious that the ore contains 1·5 ozs. gold per statute ton. If the value per ton of 2,000 lbs. is required, the weight of the assay-ton is 29·166 grammes, since there are 29,166 Troy ounces in 2,000 lbs. avoirdupois. If grain weights are preferred the weight of the assay-ton may conveniently be taken as 326·66 grains (or 291·66 grains for the short ton); the weight of the resulting bead of gold in hundredths of a grain then gives the value of the ore in ounces per ton. Sets of weights ranging from $\frac{1}{5}$ A.T. (assay-ton) to 4 A.T. can be bought, or they can be made up from an ordinary box of decimal weights. In the following pages this system is used in giving the weights of fluxes, &c., required: the grammes or grains of the various authors referred to are converted into the approximately corresponding number of assay-tons.

General Charges.—The charge for a fusion assay varies according to the nature of the ore. The following proportions are supposed by their respective authors to be suitable to all gold and silver ores—

Mitchell's Formula—[*]

Ore,					1 A.T.
Soda carbonate,	.	.	.	.	1 ,,
Litharge,	.	.	.	.	5 ,,
Borax glass,	.	.	.	.	1 ,,
Salt to cover,	.	.	.	.	1 ,,

and the amount of argol or of nitre necessary to ensure the formation of a button of lead weighing about 200 grains. These quantities can only be determined by a preliminary assay.

[*] Mitchell's *Manual of Assaying*, 1881, p. 546.

Aaron's Formula*—

Ore,	1 A.T.
Soda carbonate,	3 ,,
Litharge,	1 ,,
Borax,	½ ,,
Sulphur,	1/10 ,,
Flour,	1/10 ,,
Iron,	3 nails.
Glass.	
Salt to cover.	

"Melt and leave in a hot fire about twenty minutes after fusion."

When ores contain only small quantities of base metals the following formula is recommended by Brown & Griffiths †—

Ore,	1 A.T.
Soda carbonate,	1½ ,,
Litharge,	1½ ,,
Borax glass,	½ ,,
Carbonate of potash,	½ ,,
Silica,	1 ,,
Charcoal,	·6 gramme.
Salt to cover.	

Percy's Formula ‡ for ores containing only small proportions of base metals—

Ore,	¼ to 1 A.T.
Red lead,	1 ,,
Soda carbonate and borax,	1 ,, together.
Charcoal,	13 to 17 grains.

The charges given above are varied according to circumstances. One A.T. of the ore is a suitable quantity if the value in gold is from ·5 oz. to 10 ozs. or more. With very poor ores 2, 3, or 4 A.T. may be taken, and with very rich ores ½ A.T. may suffice. The fluxes are altered in proportion. Formerly the usual practice of assayers was not to reduce all the red lead or litharge; part was left in the slag, forming easily fusible silicates. Aaron was the first to describe the method of reducing the whole of the litharge employed; the gangue is in this case slagged off by soda carbonate, borax or silica, and the base metals separated as a matte. The advantages claimed for this method are that the button is never much contaminated by copper, and that the crucible is but little attacked. Brown & Griffiths observe that beginners find Aaron's method less easy than the other; but Aaron's method is now always employed at the Royal College of Science, London, except in rare cases, such as, for instance, when much copper is present.

The following remarks on the use of the fluxes may serve as an aid in making up charges in particular cases:—Litharge or

---

* Aaron's *Assaying*, 1884, p. 53.
† Brown and Griffiths' *Manual of Assaying*, p. 174.
‡ Percy's *Metallurgy of Silver and Gold*, p. 245.

red lead is added in the proportion of at least one part to two of ore; if too much litharge is used the slags are not clean, as a slag containing lead may mean a loss of silver and gold. Whatever method is used, the amount of lead to be reduced should be from 250 to 450 grains. Raw ores or regulus containing much sulphide of copper may be fused with from 4 to 6 A.T. of litharge to 1 A.T. of ore. In this case the other fluxes, except sand, may be omitted. Some assayers prefer to concentrate the copper as a regulus, and then to treat this over again: little more than the ordinary quantity of litharge is then used, and not much iron.

The amount of the charcoal added varies with the reducing power (percentage of ash, &c.) of the particular sample which is employed, as well as the degree of oxidation of the ore. In some highly basic oxidised ores as much as 30 to 40 grains of charcoal powder is required for 1 A.T. of ore. If there is much Fe_2O_3 in the ore (e.g., roasted pyrites) the slag is often rich. The oxide must be completely reduced to FeO, hence the need for the large quantity of charcoal. In such cases 3 or 4 per cent. of the gold may be lost.* The remedy usually adopted is to increase largely the quantity of soda carbonate, while sand must also be added to prevent the crucible from being perforated by the scouring action of the ferrous oxide. If ores contain much sulphur no charcoal is used, and it may be necessary to add nitre to burn off the excess of sulphur; otherwise a rich matte may be formed, or, if the old practice of adding much red lead is followed, the amount of lead reduced may become too great. The addition of much nitre is, however, to be deprecated since the pot is liable to boil over; with very large quantities of sulphides it is better to make a matte and treat the latter again.

Carbonate of soda is used to flux silica, while borax is valuable in basic ores to prevent corrosion of the crucible, and render the slag more liquid. The relative amounts required are judged from the appearance of the ore in the first place, and afterwards modified according to the success of the fusion. Even when the ore is entirely composed of silica, some borax is added. The most convenient form is borax glass.

Silica is used only for ores full of lime, baryta, compounds of the base metals, &c., or generally whenever the ore does not contain much quartz. It aids fusion in these cases, and protects the crucible from corrosion. Fluorspar is added to the charge when the ore contains sulphates of barium or calcium, and in the fusion of cupels. Like borax, it increases the fluidity of almost any charge, but it attacks the crucible, and care must be taken to avoid deficiency of silica when it is used. In general, it may be noted that for basic impurities an acid flux is used, and for an acid gangue a basic flux.

* C. & J. J. Beringer's *Assaying*, p. 125.

The ore is weighed accurately to $\frac{1}{10}$ grain unless very poor, when a less exact approximation is sufficient. The charcoal is always weighed with great care; the litharge is best measured by a ladle or shot measurer; the fluxes may also be measured out by a ladle. By practice the assayer becomes very rapid in measuring out reagents. The various ingredients are thoroughly mixed together on rubber cloth or in the crucible in which the fusion is to take place. Part of the borax is kept separate and used as a cover, being put on the top of the rest of the charge after transference to the crucible. A cover of common salt is used by German and American assayers; for its employment at the Royal College of Science see Dr. Ball's remarks on p. 416. Iron is added to the charge in the shape of large nails or hoop-iron, or even scrap, if sulphur or arsenic is present. Sulphur is thus kept out of the lead button.

The crucible is carefully annealed in the ashpit of the furnace before using. It is lowered into a hollow in the fuel of the furnace made by piling the coke round an old pot and then carefully withdrawing the latter. The tongs most used are shown at B, Fig. 60, A being those used when the fusion is performed in a muffle. The fire should be at a good red heat on charging-in, and care must be taken that no coke in the furnace is black. If the upper layer of coke is much colder than that below, the top of the charge in the crucible remains unmelted while the bottom begins to effervesce, and the crucible may froth over and part of its contents be lost. As the charge melts the fire is urged, and brisk effervescence occurs chiefly from the escape of carbonic acid from the soda carbonate as it unites with silica. After a lapse of from twenty to forty minutes the charge is in a state of tranquil fusion, with the exception perhaps of slight action round the sides or next the iron, if any is used. The crucible is then lifted out of the fire by the tongs, the nails withdrawn and any lead adhering to them shaken off into the pot. The pot may now be tapped on the floor to assist the lead to settle in the slag, allowed to cool and broken by a hammer to extract the lead button, or the charge may be at once poured into an iron mould. The mould must be cleaned, blackleaded and warmed before being used, and, after pouring, it is tapped on the bench to collect the lead at the bottom.

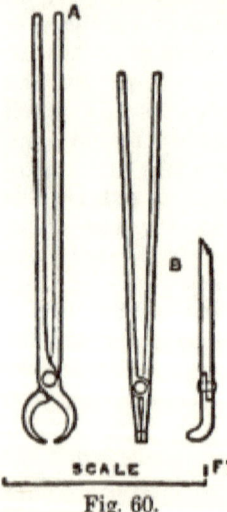

Fig. 60.

'If the charge does not fuse completely, so that the slag is pasty or has lumps in it, it is advisable to recommence the assay, making such alterations in the charge as experience suggests, having regard to the considerations mentioned above, p. 413.

When the mould is quite cold its contents are readily separated from it if the precautions mentioned above are taken, and the slag is detached from the lead button with a hammer on an anvil. The slag should be glassy and homogeneous; if it is streaked it is probable that the fusion has not been perfect. It is most often black and opaque, but is red owing to the presence of cuprous oxide if the ore contains much copper.

The lead button should be soft and malleable. If it is hard or brittle it may contain sulphur, arsenic, antimony, copper, &c. Sulphur and arsenic are kept from entering the lead by the addition of iron, and then form with the iron a regulus or speiss, which separates as hard blackish-grey or white layers found just above the lead. They are richer than the slag, and may often yield appreciable quantities of gold on further treatment by scorification or roasting and fusion. Antimony makes the lead hard, white, brittle and sonorous; it can be removed by adding nitre to the fusion charge, but, according to Rivot,* it is not a source of loss if forming less than 1 per cent. of the lead button (see under Cupellation below). The lead must be completely freed from the slag, very small quantities of the latter interfering seriously with the cupellation, forming a scoria and occasioning loss.

On the subject of fusing gold ores, Dr. E. J. Ball, who was until lately the Instructor in Assaying at the Royal College of Science, London, writes as follows:—

"I find a very good plan in assaying a gold (or a silver) ore to be as follows, noting the points:—

"(1) That the main aim at the beginning of the assay is to produce an intimate contact between the molten reduced lead and the gold freed by the original grinding of the ore.

"(2) That when this has been effected, the particles of ground-up ore must be, as it were, 'atomically' ground-up by solution, in order to liberate enclosed particles of gold.

"The first of these stages should not be accompanied by a fusion of the charge, because immediately that happens the lead sinks to the bottom of the crucible, and is at once removed from the possibility of contact with the gold floating about in the charge. To bring about this contact in the second stage, the bath must be as thin-fluid as possible, and convection currents should be induced by irregular heating of the crucible, in the hope that, in the course of one of their gyrations, the particles of gold liberated by the solution of the quartz particles may strike against the surface of the molten lead at the bottom of the crucible.

"The maximum quantity of lead which the ordinary cupels in

* *Traité de Docimasie.*

use at the Royal College of Science will take is about 450 grains. I therefore recommend (when treating fairly pure quartz, containing say 1 oz. of gold per ton) that the charge should be made up as follows :—

Ore (passed through an 80-sieve),	1,000 grains.
Red lead,	500 ,,
Charcoal,	25-30 ,,

The charcoal and red lead are first mixed together, and the ore is then carefully incorporated with them. Then 250 grains of sodium carbonate are roughly stirred in, so as to prevent the formation of a sort of sand-bottom, which would not dissolve in stage 2.

"The charge is then maintained at as high a temperature as possible, actual fusion being avoided for fifteen minutes, when another 1,000 or 1,200 grains of sodium carbonate are charged-in little by little, and the temperature raised to about 950°. At this stage, any necessary addition may be made in order to make the bath fluid, borax for instance, but borax seems to give low results if added at the beginning of the assay. I *always* add a piece of hoop iron to help to decompose any lead silicate or sulphide.

"I never recommend the use of salt, but sometimes when a large excess of sodium carbonate has been added, the 'boil' at the end seems never likely to stop, owing to the action of the acid crucible on the basic charge. In that case, a little salt stirred into the bath seems to volatilise, prevent the contact of the charge with the walls of the crucible for a moment or two, and to quiet the bath, enabling the charge to be poured properly."

Roasting before Fusion.—Ores containing large quantities of sulphur, arsenic or antimony may often be roasted with advantage as a preliminary to fusion. Roasting is effected in shallow circular clay dishes, in a muffle, or in the crucibles in which the fusion is afterwards performed. The temperature must be kept low at first and the ore frequently stirred with an iron wire or spatula, to prevent fritting, and to expose fresh surfaces to the air. The roasting takes place in two stages: at first, sulphur dioxide, arsenious oxide (As_2O_3) and antimonious oxide (Sb_2O_3) are formed and volatilised, the sulphur burning with a blue flame. The formation of lumps is most to be feared during the first few minutes of the operation, and can scarcely be prevented if much sulphide of antimony is present; in this case an equal weight of pure silver sand is mixed with the crushed ore before charging it into the muffle.

After a time the blue flame disappears, the odour becomes less strong, and sulphates, arseniates and antimoniates form. By raising the temperature sulphates are decomposed, but arseniates and antimoniates are stable at high temperatures and cause loss of silver in the fusion. To prevent their formation the ore should be roasted in a coke furnace, starting to heat it very gradually and admitting a limited supply of air. In all cases the roasting is nearly complete when the glow caused by stirring is shown only by a few specks of ore; the temperature may then be raised to a strong red heat without danger of fusion. The operation is complete when the ore remains of a uniform colour on stirring. The fusion of roasted ores requires less lead than raw ores and more charcoal powder, the amount of the latter needed being sometimes as much as 40 grains per A.T. of ore.

Cleaning the Slag.—The slags of very rich ores may retain enough gold to necessitate further treatment. The slag is roughly crushed and fused with 300 to 500 grains of litharge, 15 to 20 grains of charcoal and a little carbonate of soda, the same crucible being used again. If any regulus or speiss forms during the first fusion it must be preserved with special care and re-fused, charcoal being omitted. The button of lead from the second fusion is cupelled with the first button, or alone; the slag is almost always poor enough to be thrown away. It is not usually necessary to clean the slag of ores containing less than 5 ozs. gold per ton.

The Treatment of Base Ores.—The difficulty of roasting arsenical and antimonial ores may be avoided [*] by taking ore, 1 A.T.; red lead, 1,000 grains; sodic carbonate, 500 grains; potassic ferrocyanide, 550 grains, with a cover of salt or borax. The button is scorified, together with the matte formed, before cupellation. It is perhaps better to scorify these ores as well as those containing much copper if they are rich enough. If poor, copper ores may be treated in three ways,[†] so that each method may serve as a check on the others, viz.:—(*a*) Fusion with iron nails: the lead button becomes cupriferous, and should be scorified together with the matte; (*b*) roasting, followed by fusion and scorification; (*c*) treatment with nitric acid, by which all the sulphur and copper are removed. The silver dissolved in the liquid is then precipitated by a solution of common salt, of which a large excess should be avoided. The insoluble residue is dried, and can now be readily fused and cupelled. By treatment (*c*) the lead button is kept free from copper, the presence of which in the lead obtained by methods (*a*) and (*b*) renders cupellation difficult and unsatisfactory. As already observed, some assayers prefer to concentrate the copper as a regulus. The regulus first obtained still retains gold, and must be re-treated.

Cupellation.—This operation is conducted in a muffle furnace, the construction of which is shown in Figs. 62-64, p. 436. The

---

[*] Rickett's *Notes on Assaying*, New York, 1887, p. 77.
[†] Percy's *Metallurgy of Gold and Silver*, p. 247.

fire is lighted, a little bone-ash is sprinkled on the floor of the muffle to prevent its corrosion by litharge in case of the upsetting of a cupel, and the cupels introduced as soon as a bright red heat is attained. Cupels are little cups made of bone-ash, and are either round or square. In their manufacture the bone-ash is finely powdered so that it will pass a 40-mesh sieve, then slightly moistened with water (to which a little carbonate of potash is sometimes added), put into a mould (Fig. 65) and compressed by the blows of a mallet so as to cohere firmly. Cupels must be dried very carefully and slowly but completely, as otherwise cracks appear when they are heated, and loss is thereby occasioned. The cupels at the Royal Mint are made two years before being used, and are dried slowly on shelves at some distance from the furnaces. If too much water is used in mixing the bone-ash, the cupels lose part of their porosity: if too little is used they are too soft and crumble readily. About 1 oz. water to each troy pound of bone-ash answers very well.

The cupels having attained the same temperature as the muffle, the lead buttons obtained as described on p. 415 are charged in by the tongs (A, Fig. 61). The buttons collapse

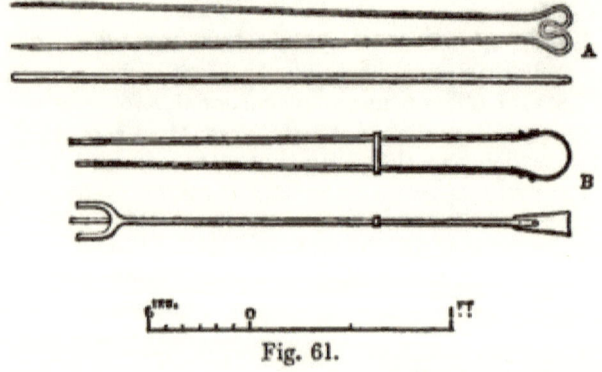

Fig. 61.

and lose their shape almost instantly if the temperature is sufficiently high, but the molten mass formed is covered by a black crust for several seconds later. The crust then breaks up, and the brilliant surface of a liquid bath is seen. The muffle door should be closed immediately after the charging-in is completed, and kept closed until all the assays have thus "uncovered." The door is then opened a little way to let in a current of air by which the lead is oxidised, and the litharge, floating to the edge of the bath, is absorbed by the cupel, together with the oxides of other base metals, which are not taken up by bone-ash if they are in a state of purity. Uncombined metals are not readily absorbed by the cupel, and only traces of gold and silver are carried into it by the litharge. These quantities depend partly on the nature of the cupels, coarse-grained bone-ash absorbing more of the precious metals than fine-grained. By a rise in temperature the absorption of gold is increased. The presence of other base metals also has some influence on the amounts of precious metals absorbed, oxide of copper in particular carrying with it much gold into the cupel.

As the cupellation advances the lead bath is reduced in size by oxidation and absorption; reddish patches float slowly over its surface, appearing earlier in the richer assays; it becomes more convex and brighter; the red spots move more quickly and finally whirl round with great speed and then disappear; moving iridescent bands take their place for a moment and then disappear likewise, and the bead becomes suddenly much duller in appearance, thus indicating that the cupellation is at an end. The temperature must be raised towards the end of the operation to remove the last traces of lead, and the beads left for from three to ten minutes, after all apparent oxidation is at an end. The cupels may then be removed from the muffle, provided the ore is poor or has little silver in it. It must, however, be remembered that silver absorbs oxygen when molten and gives it off suddenly when solidifying, so that if the bead weighs more than 0·01 gramme (Rivot) little fountains of metal are thrown up and some part may be projected out of the cupel. This "sprouting," "spitting," or "vegetation" may take place in argentiferous-gold beads if the gold does not exceed one-third of the silver (Levol). At the Royal Mint it is found that a still larger percentage of gold does not prevent spitting unless a trace of copper is present. Where spitting is to be feared, therefore, either some copper is left in the bead (*vide infra*, p. 440), a course which is inadmissible if the silver is to be estimated, or else slow cooling is resorted to, the muffle being carefully closed and luted up and the fire withdrawn. The door is not opened again until the beads have solidified; under these circumstances, no sprouting occurs. Slow cooling may also be obtained by covering the cupel containing the silver bead with a red-hot empty cupel.

When cooled the beads often "flash"—*i.e.*, brighten suddenly at the moment of solidification. This is due to the fact that the latent heat of fusion being released raises the temperature of the bead enormously, the metal having been in a state of surfusion at a temperature many degrees below its melting point. The flashing of small beads can rarely be observed.

The proper temperature for cupellation of gold ores is higher than for that of silver, as loss by volatilisation is less to be feared. The muffle should be at a bright orange-red heat, the cupel red, and the melted lead much more luminous than the cupel; the fumes should rise slowly to the crown of the muffle. In Western America in both silver and gold assays the heat is kept low enough to enable crystals of litharge to form in a ring round the cupel, but the results thus obtained are too high, the lead not being completely eliminated. The whole of the litharge should be completely absorbed. The formation of scales, due to low temperature, is accompanied by a sluggish heavy movement of the fumes, which fall in the muffle. On the other hand, the heat is too great when the cupels are whitish, the fused metal is seen with difficulty, and the scarcely visible fumes rise rapidly in the muffle (Mitchell). If the assays are long in uncovering they may sometimes be started by dropping on them a little charcoal powder wrapped in tissue paper, or still better by placing a piece of charcoal near them. If one freezes before completion it is restarted in the same way, or fresh lead is added or the temperature is raised, but the results are not good.

The admission of the current of air to the muffle is carefully regulated. Too much air may cause the bath of lead to spirt and so occasion loss; too little air delays the operation and causes increased loss by volatilisation and absorption by the cupel.

The bead thus obtained should be well rounded and bright, loosely adherent to the cupel and slightly crystalline although malleable. If it contains lead it is more globular and brittle and its surface is very brilliant, while it does not adhere at all to the bone-ash. If copper is present the bead adheres firmly to the cupel, and in extreme cases its surface is blackened. Rhodium and iridium occasion black patches at the bottom of the bead; platinum makes the surface of the bead crystalline and rugose. If the bead cracks on being flattened Van Riemsdijk recommends a second cupellation with some more lead and 10 per cent. of cupric chloride. In this way the metals with volatile chlorides are eliminated.

Influence of Base Metals on Cupellation—Iron.—Its oxide is not readily fusible with lead oxide: the button is long in melting, and a brown scoria is left on the cupel, which may entangle lead globules and so contain gold. Care must be taken not to disturb the cupel during the operation, as, if the molten lead is moved so as to touch the scoria, part is entangled. The cupel is stained red.

Zinc burns with a blue flame at first and volatilises, taking gold with it; it forms a pale yellow scoria, having the same effects as that consisting of oxide of iron. The button is slightly crystalline.

Tin forms a grey scoria.

Copper carries gold into the cupel and is usually not wholly removed from the bead.

Nickel and Cobalt are not so easily oxidised as copper: they may form a dark green scoria and always stain the cupel green. The button is crystalline.

Antimony does not interfere if less than 1 per cent. is present

(Rivot). If there is more, in volatilising it takes gold with it, and it forms antimoniate of lead, giving a pale yellow scoria and causing the cupel to crack.

Arsenic has a similar effect. The last two metals mentioned are removed by scorification if present in large quantities.

Manganese causes black stains and corrosion of the cupel, and forms a dark scoria.

Chromium gives a brick-red stain and scoria on the cupel, and *aluminium* a grey scoria; both these metals delay the course of cupellation.

Cadmium causes a black sooty ring to form inside the cupel near its margin, and gives a brown scoria.

Tellurium causes loss by volatilisation, and also, in common with some other metals, has the effect of causing subdivision of the cupelled bead.

Bismuth has less prejudicial effects on cupellation than the metals mentioned above, and may be substituted for lead. However, according to E. A. Smith,[*] the losses experienced in this case, especially by absorption by the cupel, are much greater than if lead is used.

The loss during cupellation by volatilisation was proved by Makins. It is never absent, but is insignificant with ores assaying below 10 ozs. per ton, especially if highly volatile metals are absent. The absorption by the cupel is more serious (see under Bullion assaying, p. 448). Rivot states that gold is oxidised to some extent at a red heat in the presence of antimonic oxide, litharge or cupric oxide, and that it is the oxidised part which is absorbed by the cupel. This contention is as yet hardly supported by sufficient proof.

Inquartation and Parting.—The bead of silver and gold obtained by cupellation is squeezed between pliers, or flattened by a hammer on a clean anvil, to loosen the bone-ash adhering to its lower surface, and is then cleaned by a brush of wires or stiff bristles. It is then weighed, the silver removed by solution in nitric acid, and the weight of the residual gold taken, when the difference between the two weighings represents the silver. If the bead contains more than one-fourth its weight of gold, enough silver is added to it to make an alloy of about this composition, otherwise some of the silver will remain undissolved, being protected from the action of the acid by the outer layers of gold. The amount of silver to be added is calculated from the (approximately) known composition of the bead, or guessed from its colour. A pale yellow bead always contains more than 60 per cent. of gold, but a perfectly white bead may not "part" completely. The addition of the silver is effected in the case of small beads by fusion on charcoal by the blowpipe, but large beads are, together with the silver, wrapped in as small

---

[*] *Journ. Chem. Soc.*, vol. lxv., p. 624 (1894).

a piece of lead foil as possible and cupelled. This is called "*inquartation*." The resulting bead is cleaned, flattened by the hammer and, if it is large, by passing through the rolls, rolled up into a cornet (*vide* Bullion assaying) if of convenient size, and dropped into nitric acid. If the gold in the original bead did not exceed one-quarter of the whole, inquartation is omitted.

The parting is effected in a test tube or a porcelain crucible if the bead is small, in a "parting flask" if large. The strength of acid used for the first boiling varies with the composition of the alloy. Nitric acid of specific gravity 1·25 is used for the inquarted alloy, and more dilute acids for poorer alloys. In all cases the acids may be previously warmed with advantage, as the gold does not in that case break up into such fine particles as if cold acid is used. The acid must be of sufficient strength to attack the bead instantly, turning it black and giving off nitrous fumes. The temperature is gradually raised to boiling point, and maintained at it for two or three minutes, or until all apparent action of the acid on the metal has ceased. The acid is then poured off, the residue washed once with boiling water by decantation, and if the bead is very large fresh acid added of sp. gr. 1·32 (one part of strong nitric acid to one-half part of water). The boiling is then continued for five to ten minutes longer, when almost all the silver will have been dissolved. A very small amount of silver, weighing from ·05 to ·1 per cent. of the gold, obstinately resists the action of the acid, and remains as a surcharge which may be neglected in almost all ores. The gold may be left as a single piece if it constitutes not less than one part in 10 to 20 of the bead: if present in less than this proportion it breaks up, and the finer particles may float and be lost in decantation. Particles floating on the surface may be sunk by touching with a glass rod or by a drop of water let fall on them. Continued heating of the gold in acid makes it agglomerate to some extent, so that it is easier to wash. The second acid is rarely necessary.

After decantation of the second acid and washing twice with water, the water is drained off and the porcelain crucibles dried at a gentle heat, and then gradually raised to redness. The gold which was previously black and soft, being in a fine state of division, now resumes its usual yellow colour, and hardens so that it can be removed to the pan of a balance and weighed.

If a parting flask or test tube is used for the boiling, the parted gold is transferred to an unglazed Wedgewood crucible. To effect this the flask is filled with water, and the crucible placed over its mouth. On inverting both together the gold falls into the crucible, and the flask is removed in such a way as not to disturb the precious metal. The water which has filled the crucible is then poured off, and the crucible heated as before.

The chief difficulty in parting by the last-named method is encountered in transferring the gold from the glass vessel to the

Wedgewood crucible; minute particles of gold may adhere to the glass and are then left behind. The loss of these is the cause of the low results occasionally observed when this method is used. If the glazed porcelain crucibles are used for both boiling and annealing, no transference from one vessel to another of the gold in its soft state is necessary, so that the source of error mentioned above is avoided. On the other hand, the difficulty of boiling the acid in a small porcelain crucible without sustaining loss by projection may prevent the assayer from continuing the boiling for a sufficient length of time, and some silver may thus be left undissolved; in consequence of this the results obtained are sometimes too high by as much as 2 or 3 per cent. of the weight of the gold. Such errors would, however, be inappreciable in the assay of ores containing less than 1 ounce of gold per ton, and are the less serious for the reason that all the other sources of error (unclean slags, absorption by cupel, &c.), tend to make the result too low.

By whatever method the parting is performed, very finely divided gold may remain suspended in the liquids, and may thus be lost in the course of decantation.

To prevent "bumping," one or two small pieces of capillary glass tube or of clay pipe or other porous body are put into the acid together with the alloy to be parted.

By the operation of parting, silver, palladium and some platinum are removed in solution, but the greater part of the platinum, and all the rhodium, iridium, &c., remain with the gold. If the presence of these metals is suspected they must be looked for and removed by special methods (*vide infra*, p. 455).

Examination of Assay Materials.—The fluxes are usually free from the precious metals, but the litharge, red-lead, and granulated lead contain a small amount of silver and less gold (see p. 30). These are estimated by fusion with charcoal (or in the case of granulated lead by scorification) and cupellation.

Examination of the Cupel.—When rich ores are assayed, appreciable quantities of gold are carried into the cupel, especially if much copper is present in the lead button. To assay the cupel, all clean bone-ash is detached and thrown away, and the remainder crushed so as to pass through an 80-mesh sieve, and the charge made up as follows:—

Cupel,	100 parts.
Fluorspar,	75 ,,
Sand,	75 ,,
Soda carbonate,	100 ,,
Borax,	50 ,,
Litharge,	50 ,,
Charcoal,	4 ,,

The fusion is performed in an earthen crucible, and the resulting button of lead cupelled. The slag of the original fusion may be cleaned by adding it to the charge, when less fluxes will be required.

ASSAY BY SCORIFICATION.

As stated above, this process is especially applicable to (a) complex ores, (b) very rich ores, (c) ores which are mainly valuable for their silver contents. The losses are small since the slag is highly basic, consisting chiefly of litharge, while the bath of lead is always poor, so that there is little loss of the precious metals by volatilisation. Moreover the operations are easy to conduct, and need not be varied much for different classes of ore. For these reasons the process is generally preferred to the crucible process in Germany and the United States; if the ore is poor several assays are made and the lead buttons scorified together. The chief disadvantage of the process lies in the small quantity of ore that can be treated, so that the presence of one or two metallic particles of gold may cause the result to be erroneous.

Scorification is conducted in a muffle at a much higher temperature than that required for cupellation. It must be high enough to melt litharge when contaminated by silica and oxides of copper, iron, manganese, &c. A temperature of 1,050° to 1,100° C. is usually enough. The charge is placed in a *scorifier*, a shallow circular fire-clay dish 2 to 3 inches in diameter, which is charged in by the tongs shown at B, Fig. 61. The charge consists of about 50 grains of ore (or $\frac{1}{10}$ A.T.), 500 to 1,000 grains of granulated lead and a few grains of borax glass. The ore is mixed with half the granulated lead, the mixture put in the scorifier and smoothed down, the rest of the lead spread over evenly and the borax put on the top. The amount of lead to be used varies with the nature of the ore. Ricketts gives the following table* as a guide :—

Character of Gangue.	For one part Ore take	
	Parts of Granulated Lead.	Parts of Borax Glass.
Quartzose,	8	0
Basic,	8	0·25 to 1·00
Galena,	5·6	0·15
Arsenical,	16	0·10 to 0·50
Antimonial,	16	0·10 to 1·00
Fahlerz,	12 to 16	0·10 to 0·15
Iron pyrites,	10 to 15	0·10 to 0·20
Blende,	10 to 15	0·10 to 0·20

The borax lessens the corrosion of the scorifier and renders the

* *Notes on Assaying*, p. 69.

slag more liquid, but its quantity is kept as low as possible to prevent the slag from completely covering the bath of metal too soon. After charging-in, the door of the muffle is closed until fusion takes place. As soon as the lead is melted the door is opened, and a current of air allowed to pass over the bath of metal. Some of the ore is now seen to be floating on the surface of the lead, and is rapidly oxidised, partly by the air and partly by the litharge which immediately begins to form. The sulphur, arsenic, antimony, &c., are thus soon eliminated, while copper, iron and other bases slag off with the borax, and the silica, and other acids form fusible compounds with the litharge. Effervescence and spirting may occur, especially if the scorifier has not been well dried by warming before it is used. The slag soon forms a ring completely incircling the bath of metal. As oxidation of the lead proceeds, the litharge flows to the sides and increases the quantity of slag until at length the ring closes completely over the metal, leaving a "flat uniform surface" (Percy). This usually happens after from thirty to forty minutes. The slag should be "cleaned" before withdrawal. This is done by placing 3 grains of charcoal powder wrapped in tissue paper on the surface of the slag, with a pair of cupel tongs, and closing the muffle door. A number of globules of lead are formed by the reduction of the litharge, and these, falling through the slag, extract and carry down with them any gold and silver which it may still contain, and concentrate them in the molten lead below. The fusion being quiet again, the charge is poured into an iron mould by the scorifying tongs, and the lead button cleaned from slag by a hammer. If the button is too large to cupel at once it is re-scorified in the same dish, fresh lead being added if it is not soft and malleable. Less loss of the precious metals is incurred by scorifying lead than by cupelling it, and consequently it is better to reduce any lead button weighing more than 200 to 300 grains by re-scorification before cupellation.

The slag from the scorifier should always be separately retreated in the case of very rich ores, but seldom contains much gold.

The losses incurred in this process are chiefly due to improper temperatures. If the muffle is too cold at first, gold is retained by the slag. This initial low temperature is often indicated by the occurrence of white patches of sulphate of lead on the surface of the slag after pouring. If the slag is pasty, borax is added, but the slag is then rich. It is better to begin again with less ore or more lead. When extremely rich sulphides are assayed, results are often obtained which do not agree well. Stetefeldt[*] recommends in all such cases that the sulphides be attacked by nitric acid and the residues dried and scorified.

[*] *Lixiviation of Silver Ores*, New York, p. 106.

The subsequent treatment of the lead button is as already described. Tellurium ores often yield erroneous results, usually ascribed to volatilisation of the gold. Ricketts asserts* that it is not volatile in presence of tellurium (but see page 7), and that no difficulty occurs in either crucible or scorification method if plenty of litharge or lead is used. For rich tellurides 60 to 100 parts of lead are required, a large scorifier being used. The tellurium must be driven off before cupellation, as otherwise the gold button will be subdivided into minute particles on the cupel.

Detection of Gold in Minerals.—The detection of gold in loose alluvial ground has been already described, p. 44. A similar method may be employed in the case of auriferous quartz after grinding it. If the concentrates obtained in either case contain sulphides, these are collected, roasted or treated by nitric acid and reground. The light particles of oxide of iron can now be separated from any gold that may be present by washing. "Colour" may often be obtained thus when none could be seen after the first concentration. The washing is made easier by removing from the concentrates the magnetic oxides and iron from the grinding tools by a magnet. All the finest particles of gold are lost in the process of washing, and consequently many auriferous ores cannot be made to "show colour."

The following method devised by Wm. Skey, analyst to the Geological Survey of New Zealand, is said by him to give good results. The sample of ore is carefully roasted, then digested with an equal volume of an alcoholic solution of iodine for a length of time varying from twenty minutes to twelve hours, the longer time being allowed if the ore is poor. A piece of Swedish filter paper is then saturated with the clear supernatant liquid and afterwards burned to an ash; if gold is present in the ore the ash is coloured purple, and the colouring matter can be quickly removed by bromine. This method is said to show the presence of as little as 2 dwts. gold per ton in certain ores, but is not uniformly successful. It depends for its success on the solubility of gold in a solution of iodine, and the alcohol must be very pure to prevent reprecipitation of the gold. Iodine, however, is a slow and ineffective solvent for gold,† and the use of bromine is preferable, since it fails less frequently and acts more speedily than iodine. A mixture of 5 to 10 parts of bromine with 100 of water may be used to 100 parts of ore. The ore must be fine enough to pass an 80-mesh sieve and should be reground after roasting. After leaving the mixture to stand for some hours with occasional stirring, the liquid is filtered and the excess of bromine evaporated from the clear solution, which may then be tested by stannous chloride. Occasionally gold ores are met with which give negative results with all these methods, but yield small quantities of precious metal by crucible assay.

* *Notes on Assaying*, p. 79. † *Vide* p. 27.

Estimation of Gold in Dilute Solution.—A. Carnot has shown * that the rose colour produced in presence of arseniate of iron is very sensitive, and can be used for colorimetric estimation of small quantities of gold. To attain this end a neutral solution of chloride of gold of known strength is prepared, and some drops of solution of arsenic acid added slowly to it. Then after a time two or three drops of dilute ferric chloride and some hydrochloric acid are added. If the liquid is not acidulated, a flocculent purple precipitate forms; if it is too acid the reaction fails, and only a faint blue colour is seen. The liquid is made up to 100 c.c. with distilled water, a pinch of zinc dust added, and the mixture shaken in a flask. A colour is produced, varying from rose to purple, according to the amount of gold present. The solution is clear can be filtered unchanged, and kept without alteration for some time. If more than one milligramme of gold is present (1 in 100,000), the colour becomes too intense for small differences to be noticeable; if less than one-tenth of this amount is present (1 in 1,000,000), the colour becomes too faint. Between these proportions the amount of gold present in a liquid can be determined by comparison with a series of prepared coloured solutions.

For more dilute solutions, the test described on p. 26, depending on the use of stannous chloride, may be used. A number of precipitates are prepared from solutions of gold containing known amounts, and compared with that given by the solution to be estimated. Suitable volumes of liquid from which the test precipitates are obtained are as follows:—

Strength of Solution. Parts of Water present to each part of Gold.	Volume of Solution used to obtain a Precipitate.
1 to 5 millions,	100 c.c.
5 ,, 20 ,,	500 ,,
20 ,, 50 ,,	1,000 ,,
50 ,, 100 ,,	3,000 ,,

If more gold is present than one part per million, the colour of the precipitate is too intense for accurate measurement.

SPECIAL METHODS OF ASSAY.

1. Mixed Wet and Dry Method.—Complex minerals may be assayed by the method of Cumenge & Fuchs † as follows :— The ore is fused with grey antimonial-copper ore so as to form a

* *Bull. de la Soc. Chimie*, 1883.
† Frémy's *Encyclopædie Chimique*, T. iii., 16e cahier, p. 131.

matte or regulus. All the gold is found concentrated in the matte, which consists of sulphide of antimony, and forms about one-tenth of the weight of the original ore. The matte is fused with litharge and then dissolved in aqua regia, tartaric acid being added, and the solution diluted by boiling water. The whole of the silver separates out on cooling as AgCl, which is dried, fused with lead, and cupelled. The gold is precipitated by sulphurous acid or a little sodic hyposulphite, and is cupelled and weighed. The method is said by its devisers to be accurate, but grey antimonial-copper is difficult to obtain free from gold and silver.

2. **Amalgamation Assay.**—This is useful only in determining the amount of gold present in the ore capable of being extracted by mercury. The sample of ore is pulverised fine enough to pass through a 60 or 80-mesh sieve, made into a paste with a little water, mercury added, and the whole ground in an iron mortar with a pestle for from two to four hours, small additional amounts of water or mercury being added from time to time according to the appearance of the triturated mass. The consistency of the mass must be such that the globules of mercury do not sink in it but are broken up into very small particles. A little sodium amalgam dissolved in the mercury prevents it from flouring. The grinding is continued until the particles of ore are all impalpably fine. A machine for the purpose called the "arrastra" mortar has a large pear-shaped muller loosely fitting the inside of the mortar, and capable of being revolved in it by means of a handle. Complete amalgamation is performed in this machine much more rapidly than in an ordinary mortar. When the operator judges that amalgamation is complete, enough water is added to reduce the mass to a thin pulp and stirring is continued for a few minutes to collect the mercury at the bottom. The contents of the mortar are then "washed down" in a pan, the mercury collected and distilled, and the residue, consisting of gold, silver and base metals, cupelled and parted.

Mr. J. M. Merrick gives a similar method for the assay of pyrites.* He roasts from 8 ozs. to 1 lb. or more of the finely pulverised pyrites, mixed with marble-dust to prevent caking, and then amalgamates it in a stoneware pot or wooden bucket, collects the mercury and distils it. It is stated that gold may be detected in this way in samples showing none when assayed by fusion. Many auriferous pyritic ores, however, yield but little gold to mercury after roasting, so that the method is uncertain as well as laborious. Such ores, if very poor, are better treated either by chlorination or by Whitehead's method, which will be described subsequently (see p. 430).

3. **Assay by Chlorination.**—Plattner's method for the assay of roasted pyrites consists in placing the mineral moistened with

* Mitchell's *Manual of Assaying*, 1881, p. 694.

water in a glass cylinder, 200-250 millimetres deep and 20-30 millimetres in diameter, and introducing a current of chlorine gas at the bottom. When the odour of chlorine is noticed above the ore, a cover is put on, the stream of chlorine stopped and the whole left for twenty-four hours, after which the reaction is complete if chlorine is still in excess. Boiling water is now run through the ore until all soluble salts have been washed out, and the gold contained in the solution is precipitated by ferrous sulphate, collected, cupelled and parted. According to Balling this method fails if much silver is present, as the chloride of silver formed encrusts the gold and protects it from the action of the chlorine.* He recommends the addition of common salt to dissolve the chloride of silver, but found that the telluride ores of Nagyag yielded only 85 per cent. of their silver and 92 per cent. of their gold when successively treated with chlorine and sodium chloride.

A better method is to place the completely roasted sample in an ordinary soda-water bottle with enough water to make the whole of the consistency of thin mud. The ore and water should together occupy about two-thirds of the bottle. Bleaching powder and a thin glass bulb filled with dilute sulphuric acid are then added and the bottle securely closed. As cork is attacked by chlorine, glass or vulcanite stoppers are better, and the screw-stoppered bottles are most convenient; if corks are used they must be wired down. The bottle is then shaken so as to break the sulphuric acid bulb and mix its contents with the bleaching powder, when chlorine is evolved. The bottle is now left for several hours in a warm place, being shaken occasionally by hand to mix its contents. At the end of a period of eight to twelve hours the bottle is opened, and if excess of chlorine is still present the liquid is separated from the ore and the latter washed thoroughly by filtration or decantation. The liquid and washings, whether clear or muddy, are warmed to expel free chlorine and an excess of ferrous sulphate is then added to them. The precipitate is collected, scorified with lead and cupelled. In all cases it is better to keep the first liquid separate from the washings, which should be concentrated by evaporation, since, if this is not done, the precipitate of gold may be too fine to settle and will pass through filter-paper. A better method of precipitation is to boil the liquid with iron filings for a few minutes, decant through filter-paper, wash the filings and dissolve them in dilute sulphuric acid when a residue of gold is obtained which is easy to filter. Bromine may be used instead of the materials generating chlorine. The quantities of chemicals required will be such as are sufficient to generate a volume of chlorine equal to twice the capacity of the bottle used, or to make a solution of 2 per cent. of bromine in water. Only finely-divided gold is extracted by this method.

* Balling, *L'Art de l'Essayeur*, p. 433.

4. Whitehead's Method of Assay.—A useful method of determining small quantities of gold and silver in base metals, such as crude copper, or in mattes or base ores, is described by Mr. Whitehead,* as follows :—Weigh out from 1 to 4 A.T. of the auriferous material, place it in a beaker of 500 c.c. capacity, and add gradually enough nitric acid to dissolve it completely ; heat until red fumes cease to come off, dilute with water, and add 50 grammes of lead acetate, stir, and when dissolved add 1 c.c. dilute sulphuric acid and allow the lead sulphate to settle. Filter into a 1,000 c.c. flask and fill to the mark with distilled water. The filter contains the gold which has been collected and carried down by the sulphate of lead. The filter-paper and precipitate are dried, the paper burned, and the ash and lead sulphate scorified with test-lead. The button is cupelled and the gold with any trace of silver it may contain is weighed and then parted.

The solution is divided into two parts, and precipitated by sodium bromide with constant stirring. The precipitates, which consist of a mixture of the bromides of silver and lead, settle quickly, and filter and wash well, only cold water being used. When dry, the precipitate can easily be brushed from the paper, and so the trouble of burning the latter is avoided. The bromides are now mixed with three times their weight of carbonate of soda and a little flour or charcoal, placed in a small crucible, covered with borax, and melted down. The lead thus obtained should be free from copper, and is easy to cupel. Duplicate assays usually agree within two-tenths of an ounce of silver per ton. The use of the lead acetate is to cause the precipitates of gold and silver to settle quickly, and to enable them to be filtered effectively. Sodium bromide is used instead of the chloride, on account of the greater insolubility of the silver salt.

5. Assay of Pyrites.—*Schwartz's Method.*†—100 grammes of pyrites is fused with 46·6 grammes of iron turnings under a cover of salt. The protosulphide of iron formed is crushed and dissolved in dilute sulphuric acid; the gold remains undissolved. The liquid is filtered, the residue roasted, fused with borax and granulated lead, cupelled, and parted. It has been suggested that the iron turnings are really unnecessary, and only serve to increase the amount of regulus. Simple fusion with suitable fluxes gives a rich regulus which contains all the gold.

Stapff's Method.‡—The pyrites is fused with sulphur and an alkaline sulphate, and the alkaline aurosulphite thus formed is dissolved in water. The gold is precipitated from the filtered liquid by acidifying with sulphuric acid, and is cupelled and parted.

---

* *Chem. News*, 1892, vol. lxvi., p. 19.
† Dingler's *Polyt. Journal*, vol. ccxviii.
‡ Frémy's *Encyclopædie Chimique*, vol. iii., 16e cahier, p 202.

6. Assay of Purple of Cassius.—One part of purple of Cassius is fused with three parts of carbonate of soda, cooled and dissolved in water. The gold remains undissolved, and is collected on a filter and cupelled after incineration.*

7. Assay of a Mint Sweep.—The assay of sweepings of the floors of jewellers' workshops, refineries and mints, and that of pulverised crucibles, stirrers, &c., which have been used in melting gold bullion, are included in this section. The larger pieces of metal are removed from the sweepings by sieving and washing. The sweep then still contains a large number of metallic particles which will not pass through a fine sieve; a considerable percentage of sawdust, charcoal and graphite is also present. The sample must be selected with great care and with due regard to the fact that the lower parts of a heap are richer than the top, owing to subsidence of metallic particles through the mass. The amount of moisture present is first estimated, and the sample dried and passed through an 80-mesh sieve, after which it is roasted in a muffle and scorified with eight to twelve times its weight of lead and a few grains of borax. The metallics are scorified separately, or passed at once to the cupel. The remaining operations do not differ from those required in the case of a rich ore. Double parting is necessary, as both silver and gold are usually present. From $\frac{1}{4}$ to $\frac{1}{2}$ A.T. of the unroasted material is a convenient weight to take for assay, but the metallics must be separated from a much larger quantity and estimated by themselves.

CHAPTER XX.

THE ASSAY OF GOLD BULLION.

Assay of Gold Bullion.—The assay of gold bullion, as described in this chapter, has for its sole object the estimation of the percentage of gold present in the alloy, all other constituents being disregarded. In the first instance, the simple case of the assay of gold alloys containing appreciable quantities of only copper and silver will be dealt with. Refined gold ingots and the alloys used for coinage, and for almost all jewellery, come under this head. The effect of large quantities of other impurities and the precautions thereby rendered necessary will be discussed later.

The method universally employed is that of cupellation and subsequent parting. The gold bullion is cupelled with silver

* *Ann. de Pharmacie*, vol. xxxix.

and lead, by which the greater part of the base metals present is removed as oxides dissolved in litharge, and an alloy of gold and silver left on the cupel. This is "parted" by nitric acid, which dissolves the silver and leaves the gold unattacked.

In the following pages the practice at the Royal Mint is described, but the same description would apply, with very slight alterations, to the methods used at other mints and assay offices.

The "parting assay" was first mentioned in a decree of King Philippe of Valois, published in the year 1343.* The methods of procedure in the 17th century have been briefly described by Savot† and by J. Reynolds,‡ and more fully in the *Compleat Chymist*. In 1666 Pepys saw the parting assay being practised at the Mint in the Tower of London, and from his description it is clear that the method then employed bears a surprisingly strong resemblance to that of the present day. In 1829 a Royal Commission was appointed in France to examine into all questions relating to the methods of assaying gold and silver. The results of the labours of that Commission are to be found in great part in the Appendix to the *Report of the Select Committee on the Royal Mint*, 1837. The Committee arrived at the conclusion that the method adopted for assaying gold often overstated the amount of precious metal by 1 part per 1,000.§ The Mint Conference held in Vienna in 1857,‖ resulted in the almost universal adoption of a more uniform method of manipulation.

The degree of accuracy now attained in most assay offices reduces the probable error in the report of an assay to 0·1 per 1,000, but, to prevent the error from rising above this amount, all weighings must be correct to 0·05 per 1,000, which is not usually the case in ordinary bullion assays (see p. 451).

The system may be conveniently regarded as comprising six distinct operations, viz. :—

1. Selection of the sample.
2. Preparation of the assay piece for cupellation.
3. Cupellation.
4. Preparation of the assay piece for parting.
5. Parting and annealing the cornets.
6. The final weighing and reporting.

1. **Selection of the Sample.**—At the Royal Mint the contents of the pots are poured into moulds, each pot yielding about nine bars, and a piece is cut from the middle of the end of each of the first and last bars formed. Experience proves that these samples are usually representative of the whole mass—viz., 1,200

---

* *First Report of the Royal Mint*, 1870, p. 103.
† *Discours sur les Medalles Antiques*. Paris, 1627, p. 72.
‡ *A New Touchstone for Gold and Silver Wares*, London, 1679, p. 362.
§ *Report on the Royal Mint*, 1837, Appendix B, p. 123.
‖ *Kunst-und Gewerbeblatt Baiern*, 1857, p. 151.

ozs. In the case of the ingots, weighing 400 ozs., which are received from the Bank of England for coinage, a single sample is cut from the middle of one of the lower edges of the ingot. When the sample is not representative of the whole ingot (a somewhat rare event, judging from results), other base metals besides copper are probably present.* A more certain way of finding the composition of an ingot is to melt it under charcoal, stir well and dip out samples from the top and bottom of the mass, with an iron ladle. The metal thus obtained is granulated by pouring slowly in a thin stream into a porcelain-lined kettle filled with warm water, which must be constantly stirred. The dip-sample may also be taken by a little charcoal crucible fastened to an iron rod, a lid of charcoal being put on directly it emerges from the molten metal. Dip-samples should always be taken when very impure and base ingots are to be assayed. In place of taking a cutting from the edge of an ingot a drilling machine is used in the Paris Mint and in many other offices; by this instrument portions are taken for assay from any part of the ingot near its surface.

2. **Preparation of the Assay Piece for Cupellation.**—If the sample to be assayed is a single piece cut from a bar it is "flatted" on a clean anvil by a hammer with a rounded face, weighing about 11 lbs. A convenient thickness for the sample is about $\frac{1}{32}$ inch. The piece is then wrapped in paper which is marked distinctively, and an assistant then performs the operation of "bringing to weight," or obtaining a piece approximately $\frac{1}{2}$ gramme in weight. This he does by cutting with shears and filing.

The use of the shears can only be learned by practice, but the following remarks may be of use to a beginner. The metal to be cut is held firmly between the fore-finger and thumb of the left hand. Care must be taken, and considerable force exercised by the fingers if necessary, to keep the plane of the piece of metal to be cut perpendicular to the cutting faces of the shears, otherwise damage is done to the latter. Only clean portions of metal must be used.

The weights used in gold assaying are the $\frac{1}{2}$ gramme, which is stamped "1000," and decimal subsidiary weights stamped 900, 800, &c., and 90, 80, 70 and so on down to 0·5. These stamped numbers denote the number of $\frac{1}{2}$ milligrammes ("millièmes") contained in the weight. Ordinary weights in the gramme system may accordingly be used, each milligramme corresponding to two millièmes in the assay system. The report finally made gives the number of parts (in millièmes and tenths) of pure gold in 1,000 parts of the alloy.

Since the assay is reported to $\frac{1}{10000}$ part, it is evident that the balance used must clearly indicate a difference in weight of 0·1 per 1,000 or ·05 milligramme. It is convenient to have

* See under Liquation of Gold Alloys, p. 17.

the balance adjusted so that one subdivision on the ivory scale traversed by the pointer corresponds to this quantity. In some offices, however, "preparing" balances less delicate than this are used. The weight of alloy must not differ from the 1,000 by more than four subdivisions of the scale. The weighed portion is folded in a small square of paper and tucked inside the paper containing the rest, until it can be checked by the assayer. Weighing becomes a very rapid operation with practice, a skilful workman being able to prepare forty to fifty pieces in an hour.

The assayer now checks the weighed piece on a more delicate balance. Those used at the Royal Mint have beams only 6 inches long, weighing 156 grains, and requiring only ten seconds for a complete double oscillation. Suspension of the pans was formerly effected by means of two hard steel points instead of a knife-edge; in the year 1892 light agate knife-edges were introduced. Point suspension gave greater lightness, but the wear of the points was rapid, and slight blunting soon impaired the accuracy of the balance, necessitating repairs. These balances are released by a lever working in a vertical direction. The excess or deficiency in weight is recorded on the assay paper in tenths of a millième, the weight of the piece being adjusted by the assayer if it shows a deviation greater than four of these tenths. The piece is wrapped up in lead foil, which is very pure, and weighs 4 grammes or eight times as much as the assay piece. The lead is twisted into a form resembling that of a conical sugar bag, and the gold together with the silver necessary for parting is dropped in and wrapped up. The lead packets are put in order in the numbered compartments of the wooden tray shown at c, Fig. 67, p. 444, their position being noted on the assay paper. The corners of the packets are squeezed down so as to fit the cupels by pliers specially designed for the purpose, and the assays are then ready to be charged into the furnace.

The silver added to the assays is in the form of square pieces cut off by means of gauged shears from a strip rolled to uniform thickness; the squares weigh about $1\frac{1}{4}$ grammes each or two and a-half times the weight of the alloy. If silver is present in the alloy it must be allowed for. In the case of fine gold (used as *proofs*) or of any sample of gold of high standard, an alloy of silver and copper, 966·6 fine, is used instead of fine silver. The use of the copper is to make these assay pieces of about the same composition as the standard alloy (916·6 fine) worked with them. Fine silver is added to all alloys containing more than 50 millièmes of copper. All the silver added for purposes of parting must of course be free from gold.

It was formerly considered necessary for the metals to be present in the proportion of 1 of gold to 3 of silver, but, as early as the year 1627, Savot relates that the proportion of 1 to 2 was used, and strong acid employed in the boiling, "quand on veut faire quelque essay curieux et exact."* Both Chaudet and Kandelhardt recommend the proportion of 1 to 2½, on the ground that less silver is then retained by the cornet than in any other case. If much more than three parts of silver are present the gold breaks up in the acid. Pettenkofer† found that the proportion of 1 to 1·75 could be employed if the assays were boiled in concentrated nitric acid for some time.

The amount of lead used varies with the proportion of base metals present. The following table shows the proportions recommended by D'Arcet,‡ Cumenge & Fuchs,§ and Kandelhardt ∥ respectively:—

Gold in 1,000 parts: the Alloying Metal being Copper.	Amount of Lead employed for one part of Alloy.		
	D'Arcet.	Cumenge & Fuchs.	Kandelhardt.
1,000	1	1	8
900	10	14	16
800	16	20	20
700	22	24	24
600	24	28	24
500	26	32-34	28
400	34	32-34	28
300	34	32-34	32
200	34	32-34	32
100	34	30	32
50	34	28	32
0	...	11	32

Kandelhardt's table is modified to make it uniform with the others. It must be remembered that, although the above table gives the quantities of lead which will remove the greater part of the copper present in the alloy during cupellation, the last 2 or 3 millièmes are obstinately retained by the gold, and cannot be entirely eliminated even by a second cupellation with fresh lead. Instead of attempting to remove all the copper present in an alloy of low standard by one operation, using large quantities of lead as above, it saves time and gives more uniform results if a smaller amount of lead is used in two successive cupellations. The following table shows the proportions of lead employed by Mr. F. W. Bayly at the Royal Mint for gold-copper alloys. The object in view is not to remove all the copper by cupellation,

---

* *Discours sur les Medalles Antiques*, Paris, 1627, p. 73.
† *Bergwerksfreund*, 1849, vol. xii., p. 6.
‡ Pelouze et Frémy, *Traité de chimie générale*.
§ *Encyclopædie Chimique*, vol. iii, L'or, p. 154.
∥ *Gold-Probirverfahren*, Berlin, p. 3.

but to obtain a well-formed, clean and bright button suitable for parting: in each case silver is added equal to two and a-half times the weight of gold present.

Amount of Gold per 1,000 parts.	Amount of Lead used for one part of Alloy.	Proportion of Lead to Copper.
916·6	8	96 : 1
866	9·15	64 : 1
770	14·75 *	64 : 1
666	16 *	48 : 1
546	17·5 *	38 : 1
333	18·0 *	27 : 1

* In these cases two-thirds of the lead is charged in at first, and after the cupellation is finished the remainder is added in the form of a bullet. The results are good, and the absorption of gold by the cupel not excessive.

3. **Cupellation**—*Assay Furnace.**—That in use at the Royal

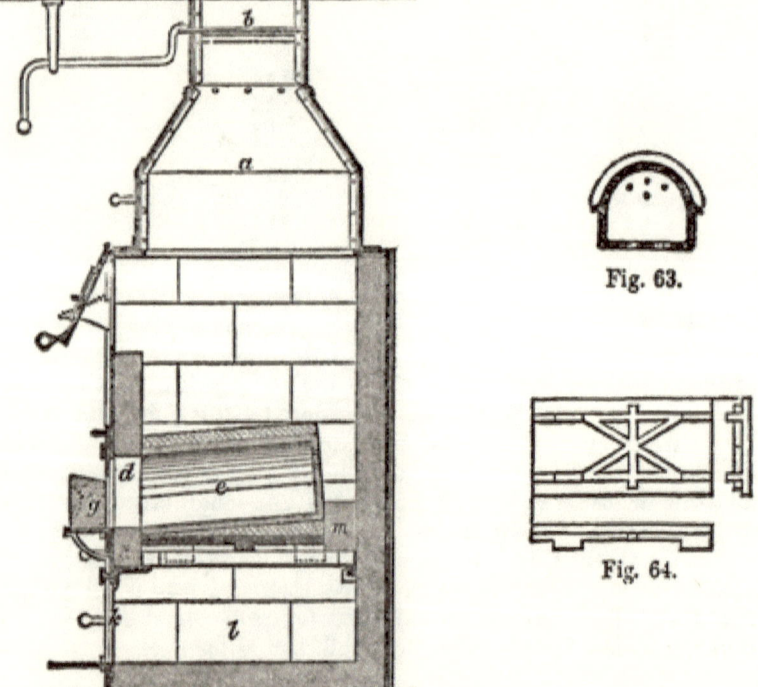

Fig. 63.

Fig. 64.

Fig. 62.   Scale, 1 in. = 1 ft. 5 ins.

* The description of this furnace is, in the main, the same as that given by Professor Roberts-Austen in Percy's *Metallurgy of Silver and Gold*, pp. 255-7.

Mint is shown in elevation in Fig. 62. It consists of an outer casing of wrought-iron plates about ⅛ inch thick, united by angle iron. This casing is connected with a chimney 60 feet high by means of a wrought-iron hood (a) and flue which is provided with a damper (b). The lining consists of Stourbridge fire-bricks, and there are five openings in the front of the furnace. Fuel is introduced through the uppermost one of these under the flap, shown partly open, the position of which can be varied by means of two saw-edged inclined planes. The opening (d) communicates directly with the muffle (e) and may be closed by sliding iron doors, or by the firebrick front (g) in conjunction with a sliding plate, not shown in the figure, covering the upper part of the opening (d). There are also two small openings, closed by sliding doors, at the side of (d), for introducing and withdrawing the fire-bars; these are seldom used to regulate the draught, for which latter purpose the doors (k) of the ash pit (l) are usually employed. The muffle is formed of fire-clay, and is pierced with four holes at the upper part of the back, by means of which a draught is established; these slope from within downwards, in order to prevent particles of fuel from finding their way into the muffle. The muffle is shown in front elevation in Fig. 63.

There are eleven fire-bars, but only the three outer ones on each side are covered with fuel. On the other five bars rest the cast-iron girder-plate (Fig. 64), which is flat on the upper surface, but is strengthened with ribs on the under-surface in order to prevent buckling. Above the girder-plate the muffle is fixed in an inclined position, so that all the cupels may be readily visible from the front. The muffle rests on a bed placed on the girder-plate consisting of pieces of fire-brick plastered over with fire-clay. It is convenient to cover the top of the muffle with a thick luting, composed of fire-clay and a little graphite in order to check radiation, and to protect the muffle from the coke let fall on it from above, and a fire-brick (m) is placed behind to prevent excessive heating of the back of the muffle. Charcoal, anthracite and coke might all be used in such a furnace, which would be built narrower if either of the first two were employed, but on account of the expense they are now rarely employed, and in the Royal Mint coke is always used. It is broken to about the size of hens' eggs and carefully screened.

Assay furnaces, heated by gas, have long been in use at the Berlin and Utrecht mints, at the Sheffield assay office and at many other places. Their use is becoming more general, especially in America.

Furnaces to burn soft coal are used in Western America where good coke is very expensive. They differ little from coke furnaces in construction, but have less space between the muffle and the side walls. The flame of the coal is chiefly instrumental in heating the muffle, a comparatively thin bed of fuel being employed.

The cupels in use at the Royal Mint are in sets of four, the outer margin being square, as shown in Fig. 65. This form was

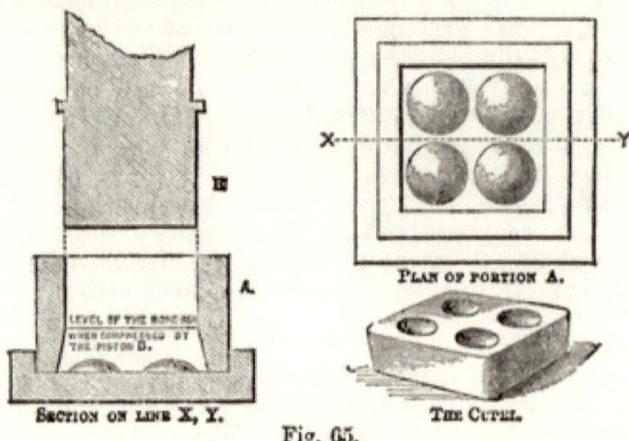

Fig. 65.

suggested by Mr. Joseph Groves of the Royal Mint, in 1872,* and is found to facilitate charging-in and withdrawal. It will be observed that the cupel mould A B is in three pieces, which are separated to take out the finished cupel. The furnace tools in use are shown in Fig. 66, where *a* represents the cupel tongs

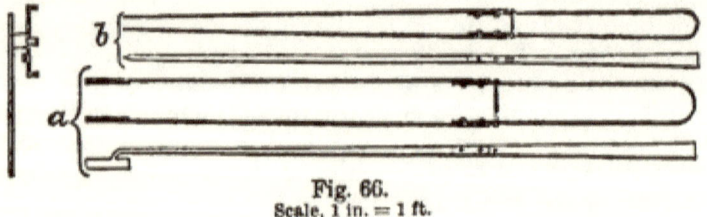

Fig. 66.
Scale, 1 in. = 1 ft.

and *b* the tongs used for charging-in the lead packets on to the cupel. The drawing at the left hand of the figure represents a section of the "peel," described on p. 446.

The furnace is lighted by faggots and charcoal, coke being added at intervals after the first few minutes. The full draught is used until the furnace has nearly attained the proper temperature. The floor of the muffle is lightly covered with bone-ash and the cupels introduced at least twenty minutes before charging-in is begun, some pieces of charcoal being laid near the front to assist in the attainment of a uniform temperature throughout the muffle. The furnace is filled with coke, well raked down but not packed too tightly, reaching to a height of about an inch

* *Third Annual Report of the Royal Mint*, 1872, p. 30.

above the top of the muffle. The coke is at a full red heat throughout before the cupellation is begun and no fresh fuel is added during the operation. The muffle is usually ready for work in about an hour from the lighting of the furnace. The charcoal is then removed from the muffle and all dust and ashes blown out by a pair of hand bellows.

The packets of lead containing the silver and gold are now transferred to the cupels, arranged in rows corresponding to those on the tray, the muffle being at a bright orange-red heat. This charging-in is performed by the tongs, b, in Fig. 66, over the top of the plumbago front (g, Fig. 62), the object of which is to render the temperature of the muffle as equable as possible. The cupels at the back are filled first, and by the time six assay pieces have been introduced, the first one should be melted and "uncovered" by the removal of the black crust which forms at first. At the Royal Mint the full fire of 72 assays takes about five minutes to charge-in. The slider is now put in place over the upper part of the opening (d), and the air supplied to the muffle increased by closing the damper. While the muffle door is open the indraught is diminished by opening this damper so as to prevent the cooling action of the current of air from proceeding too rapidly. The air supplied to the muffle and furnace may be entirely regulated by this damper. The furnace operation should be performed rapidly and in such a way that all the cupellations may be completed at as nearly as possible the same time.

Distinct stages may be noted in the action which now takes place on the cupel. Almost immediately the surface of the molten metal becomes covered with greasy-looking drops of litharge, which are rapidly absorbed by the porous cupel and replaced by others. They pass over the surface at first slowly, but as the operation continues move with greater rapidity. In from eight to fifteen minutes the metal suddenly becomes uniformly dull and glowing except for iridescent bands, produced by extremely thin films of fluid litharge, which are seen to pass over it. On the disappearance of these bands a bright liquid globule of a greenish tint is left, but the cupels are not withdrawn from the furnace until the expiration of another fifteen to twenty minutes so that the last traces of lead may be oxidised and absorbed. The completion of cupellation takes place first in the front rows and proceeds regularly backwards.

The cupels are withdrawn from the furnace while the assay pieces are still fluid, and "flashing" ensues in a few seconds. "Flashing" is most marked in the purer buttons, in which but little copper or lead remains. Slight effervescence may occur in these cases, but the buttons are never sufficiently freed from base metals for "sprouting" to take place.

Some assayers do not remove the assays from the furnace

until "brightening" or "flashing" has taken place and the buttons have become solid. The muffle door must be closed and the temperature reduced to effect this. Removal of liquid buttons saves much time and is always performed at the Royal Mint, but has two dangers. A cupel may be upset or "spitting" may take place. Under the conditions given above, however, a trace of copper is always retained by the cupelled button and all fear of spitting thus removed.

The "flashing" of gold assays was shown by Van Riemsdijk * to be due to solidification after superfusion. The temperature of the fluid metal falls until a certain point is reached when it solidifies and the sudden disengagement of the latent heat of fusion reheats the cooling globule to its true melting point—viz., 950°—and a peculiarly intense light is emitted which rapidly fades as the temperature again falls. A sudden jar at any moment causes the flashing to occur instantly. If the alloy contains a minute quantity of iridium, rhodium, ruthenium, osmium or osmium-iridium (the metals of the platinum group which are non-malleable, refractory to heat and resist the action of acids), the tendency of the cupelled metal to preserve its liquid state below the melting point, and therefore to flash during the final solidification is entirely prevented. These researches point to a simple method for detecting the presence of metals of the platinum group (except platinum and palladium) in ingots of commercial gold; for, if assays made on them solidify directly they are withdrawn from the muffle without flashing, it will be safe to conclude that the metal contains some of the above-named impurities.†

The buttons (which are of the form represented at a, Fig. 68, p. 444) are removed from the cupels, after cooling, by a pair of sharp-nosed pliers, cleaned by means of a stiff brush or by immersion in warm dilute hydrochloric acid, and are placed in the compartments corresponding to their cupels in the tray (d, Fig. 67). If the bone-ash is not completely removed from their lower surface it is of little moment since bone-ash is readily dissolved by nitric acid on parting. The surface of the cupels must be carefully examined for minute beads of metal due to spirting of the lead bath which sometimes happens if there is too strong a draught. If any such beads are found in a cupel the fact is noted and the assay repeated.

If traces of lead remain in the button it is more globular, separates more easily from the bone-ash of the cupel and has a brilliant steely surface. The effect of the presence of other metals is discussed on pp. 453-457.

Temperature.—The exact temperature suitable for cupellation can only be ascertained by practice, and the varying light of

* *Chemical News*, vol. xli., 1879, p. 126.
† Prof. Roberts-Austen in *Tenth Annual Mint Report*, 1879, p. 43.

the day may occasion error in judging the degree of heat. The remarks on temperature in the cupellation of buttons from ores (p. 419) apply here. Care should be taken to ensure that the heat is so high before "charging-in" that the chilling which necessarily takes place during this operation shall not cool the muffle below the requisite temperature. It is of more consequence that the muffle should be uniformly hot throughout than that any absolute degree should be attained, as the checks used eliminate uniform errors due to high temperature.

The measurement of the temperature of the assay muffle has not been frequently attempted. In a paper read before the Royal Society (*Phil. Trans.*, 1828, 79-96), Mr. James Prinsep, Assay Master of the Mint at Benares, gives an account of experiments on the subject by observing differences in the behaviour of a number of silver-gold and gold-platinum alloys when heated. He "made trials in different parts of the same (muffle) furnace. The disparity of heat," he remarks, "is greater than might be supposed, and where, as in assaying the precious metals, so much depends on the temperature at which the operation is performed, it would be useful to know every difference in this respect obtaining in different countries, and its effect upon the quality or standard of bullion." His results were as follows:—

		Maximum alloy melted.
Muffle of an assay furnace: front	. . .	pure silver.
,, ,, : middle (average)		{ silver, 70. { gold, 30.
,, ,, : behind (average)		{ silver, 50. { gold, 50.

The temperatures at which these alloys melt are

Silver,	. . .	943° (Violle).
Silver, 70 } Gold, 30 }	. . .	960°.
Silver, 50 } Gold, 50 }	. . .	990°.

No further experiments were made until the year 1892, when the temperature of the muffle in the furnace at the Royal Mint, described on p. 436, was measured by the author by means of the Le Chatelier pyrometer.* The arrangements were as follows:—The wires, sheathed in clay pipe-stems, were fixed in a porcelain tube, the junction being protected by plates of mica. This tube was introduced into the furnace by sliding through another of wider bore fixed in the brick door of the muffle. The inner end of the narrow tube was plugged by clay and coated over with a fusible mixture of sand, pipeclay and felspar. The galvanometer was shielded from radiation from the furnace, and remained at about a constant temperature, but the zero was frequently determined in order to detect any changes. In

* *Journ. Chem. Soc.*, vol. lxiii., 1893, p. 707.

calibrating the wire, the effects of heating various lengths of it were noted, and allowance made for the alterations in the deflections of the needle due to this cause. The junction was kept stationary at each point in the muffle until its temperature had become constant, as indicated by the cessation of movement of the spot of light on the scale.

The muffle is 15 inches long and 6½ inches wide, and its full charge consists of seventy-two assays arranged in twelve rows, the distance between the front and back rows being 10½ inches. The position of the junction was about 1 inch above the cupels in all cases. The error in observation in taking the readings was probably not more than 0·1 mm. on the scale, which was equivalent to 0·32°.

A.

Place.	Time.	Temperature in degrees above m. p. of gold-silver buttons left on the cupel, viz., 952°.
Middle of 1st (front) row of cupels,	10.19 a.m.	114·0°
,, ,, ,, ,,	10.20 ,,	113·0
,, 3rd row of cupels,	10.21 ,,	123·0
,, 4th ,,	10.22 ,,	124·0
,, 7th ,,	10.23 ,,	144·0
,, 3rd ,,	10.25 ,,	120·0
,, 11th ,,	10.28 ,,	140·0
Right side, 11th ,,	10.29 ,,	142·5
Left ,, 11th ,,	10.30 ,,	140·0
Middle of 11th ,,	10.31 ,,	139·0
,, 3rd ,,	10.32 ,,	112·0
,, 7th ,,	10.36 ,,	128·0

The assay pieces were charged in between 10.5 and 10.11; cupellation was complete at 10.23, and the charge drawn at 10.47 a.m.

B.

Place.	Time.	Temperature above 952°.
Middle of 3rd row,	11.3 a.m.	112·5°
,, 7th ,,	11.6 ,,	131·5
,, 11th ,,	11.8 ,,	143·0
,, 7th ,,	11.9 ,,	123·0
,, 3rd ,,	11.12 ,,	112·5

The assay pieces were charged in between 10.50 a.m. and 10.55; cupellation was complete at 11.8, and the charge was drawn at 11.30. The circuit was accidentally broken at 11.15.

C.

Place.					Time.	Temperature above 952°.
Middle of 3rd row,		.	.	.	11.56 a.m.	70°
,, 7th ,,		.	.	.	11.59 ,,	91
,, 11th ,,		.	.	.	12.0 noon	102
,, 7th ,,		.	.	.	12.2 p.m.	97
,, 3rd ,,		.	.	.	12.3 ,,	82
,, 11th ,,		.	.	.	12.8 ,,	105
,, 7th ,,		.	.	.	12.10 ,,	98
,, 3rd ,,		.	.	.	12.15 ,,	88
,, 3rd ,,		.	.	.	12.17 ,,	91

The assay pieces were charged in between 11.45 and 11.51; cupellation was complete at 12.8, and the charge drawn at 12.25. The muffle appeared to the eye to be decidedly cooler than usual.

These three experiments were conducted on different days. They show that the temperature gradually rises in passing from the front to the back of the muffle, the rate of change falling from 5° per row of cupels at the front to about 2° at the back, whilst along the sides the temperature is 1° or 2° higher than in the middle line. The mean difference in temperature between the 3rd and 7th rows is 15°·7, and between the 7th and 11th rows 10°·5, whilst the mean temperature of the muffle is 1070°. These differences seem to exercise an effect on the absorption of gold by the cupel. The mean "surcharge" (see p. 447) of the check assays in all the charges, 939 in number, worked off at the Royal Mint between January 1, 1880, and March 11, 1892, is as follows :—

Check in middle of		
(a) 3rd row.	(b) 7th row.	(c) 11th row.
+0·1828	+0·1533	+0·1347

Consequently, if, as is probable, the experiments indicate the ordinary conditions of the muffle, a rise of about 5° would correspond to a diminution in the surcharge of 0·01 per 1,000.

4. **Preparation of the Assay Buttons for Parting.**—The operation of "flatting" which now takes place requires some skill, as on it to a great extent depends the accurate and regular formation of the "cornets" finally obtained. A hammer weighing about 7 lbs., with a convex face, should be used. The anvil is kept quite bright and clean and used for this purpose only. A heavy blow is first delivered on the centre of the button, the diameter of which is thereby increased to nearly that of a threepenny piece. Two lighter blows are then given on opposite sides of the disc so as to elongate it, giving it the form shown at b, Fig. 68.

After being annealed in the iron tray, f, Fig. 67, which is placed in the muffle by the tool, g, and left until it is red hot, the flattened buttons are passed in succession through a pair of jeweller's rolls. The rolls are adjusted so that one passage

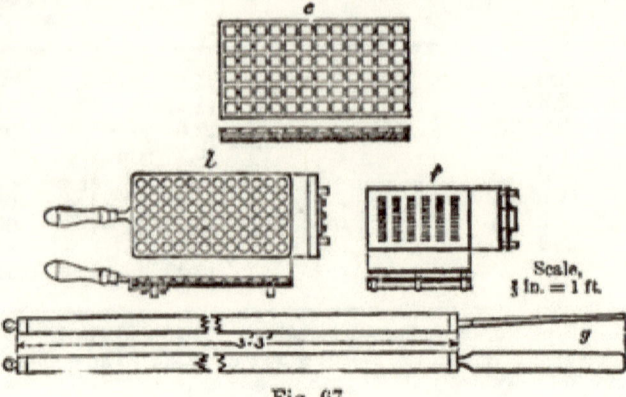

Fig. 67.

through them reduces the buttons to the required thickness—viz., about that of an ordinary visiting card. The "fillets" (c, Fig. 68)

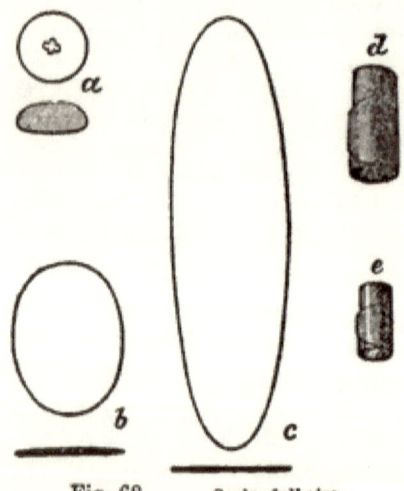

Fig. 68. Scale, full-size.

thus obtained should all be of uniform size and thickness, with "wire edges," as ragged edges expose them to loss during the boiling. After being rolled they are replaced in the tray, f, and annealed at a dull red heat. The object of the first annealing is to soften the buttons and facilitate their passage through the rolls, while that of the second is to enable the fillets to be rolled into "cornets" or spirals, d, between the finger and thumb. Care is taken in the latter operation to make that which was formerly the lower side of the button form the outer face of the coil, for a reason given below (p. 457). This face is easily recognised, as it is less brilliant than the other and is marked with undetached flattened particles of bone-ash. In the Mints of India, Scandinavia and some other countries, the fillets are rolled tightly round a glass rod and then slipped off the end. This operation occupies a longer time than the coiling between finger and thumb, and leaves the cornet less open to rapid attack by the acid than if the spiral form is used.

5. Parting.—This was formerly effected by boiling with nitric acid in glass "parting flasks." Platinum boiling trays

save time, and are now used whenever possible. The silver is dissolved by the acid which must be free from chlorine in any form, sulphuric and sulphurous acids, or sulphide from which sulphuric acid may be formed. A small quantity of silver is kept in solution, in the stock of acid, so that chlorine, if present, would instantly be detected.

When the flasks are used, 2 ozs. of nitric acid of specific gravity 1·2 are put into each flask and raised to boiling point. A cornet is then introduced and boiling continued for 15 or 20 minutes (*i.e.*, for about 10 minutes after nitrous fumes cease to be given off). Hot distilled water is then added, the solution of nitrate of silver decanted off, and the flask washed by filling with hot water, and decanting. Two ounces of hot nitric acid of specific gravity 1·3 is now poured in and the boiling continued for fifteen or twenty minutes, a parched pea or piece of charcoal being added to prevent bumping; after which decantation and washing is twice performed. Another boiling with acid of specific gravity 1·3 is recommended by Chaudet with the object of dissolving out the last traces of silver and leaving the gold quite pure. This practice has been adopted by many assayers, but is useless and causes loss of gold. If any small particles of gold have become detached from the cornet, time must be allowed for them to settle before each decantation. After the last decantation the flask is filled with hot water, the top covered by a small porous crucible, and the whole is carefully inverted; the pure gold, which is of a dark brown colour and exceedingly fragile, falls through the liquid and rests in the crucible, the water which enters with it being afterwards poured off. The crucible is dried and then annealed at a red heat over gas or in the muffle, when the gold shrinks greatly, though still preserving its shape, hardens and regains its ordinary pale yellow colour. It can now be weighed.

When a platinum boiler is used the cornets are put on platinum pins, as at the Sydney Mint, or more usually into platinum cups, one of which is shown in Fig. 69. These cups are supported in a platinum tray (which holds 144 cups at the Royal Mint) and the whole lowered by a platinum hook into a platinum vessel containing about 80 ozs. of hot nitric acid. Great attention must be paid to the temperature of the acid. At the Royal Mint acid of specific gravity 1·26 (in which, however, a small quantity of silver is already dissolved) is used, and the temperature at the moment of introduction of the tray is 90° C. If the acid is colder than this the cornets tend to break up, some pieces being usually detached in the boiling of 144 cornets, even if the temperature is only 1° or 2° lower than the correct point. If the acid is much hotter than 90°

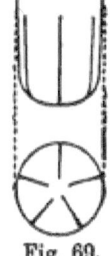

Fig. 69.

(93° or more) the action is so energetic that the vessel may boil over. If the temperature of the acid is as low as 85°, the cornets are not immediately attacked; they turn a coppery hue and then bubbles of gas form slowly on them. In such cases (where the action does not begin within 1 or 2 seconds after entrance into the acid) the only chance of saving the cornets is to withdraw them instantly and raise the temperature further. Boiling is kept up vigorously for twenty to thirty minutes, and the tray is then withdrawn, drained, washed by dipping vertically in and out of a vessel of hot distilled water, drained again, and placed in a second platinum boiler filled with boiling nitric acid of specific gravity 1·32 (also containing a little nitrate of silver). In this the cornets remain for a period of thirty to forty minutes, when they are drained and washed as before, and are then ready for annealing.

The platinum tray may be dried on a hot plate, or may be introduced into the muffle at once while still wet. The iron peel, shown in Fig. 66, is used to support the tray in the muffle. Rotating peels are sometimes used so that the temperature of annealing can be equalised by rotating the tray without withdrawing it from the furnace. A thin platinum plate may be placed between the tray and the peel to prevent scales of oxide of iron from entering the cups. The floor of the muffle should be scraped clean of bone-ash to prevent projection of the latter into the cups owing to the fall of drops of water from the wet tray when it is first introduced, or a clean tile may be put in the muffle to support the tray. The cornets are not injured by the slightly explosive evaporation of small quantities of water contained in their porous substance. Annealing should be conducted at as high a temperature as possible, consistent with the safety of the cornets. They fuse at 1,045° C. (having the same melting point as pure gold), at which temperature the muffle appears orange-red. If annealed at a low temperature, the cornets are rough in texture, dull and fragile, being crushed easily between the finger and thumb. In this condition they adhere to the platinum, and, in detaching them, fragments are often left sticking fast inside the cups. If annealed properly the cornets are smooth, lustrous and hard, showing signs of incipient fusion under a magnifying glass, and only yielding to considerable force exercised by the finger and thumb. Under these conditions they can always be detached from the platinum entire. By being annealed the cornets, which after boiling are very soft and fragile, and dark-brown in colour, shrink and harden, and regain the ordinary yellow colour of gold (e, Fig. 68).

Relative advantages of Parting in Flasks and in Platinum Boilers.—The use of platinum trays and boilers effects a great saving of time in decanting and washing, as one operation takes the place of as many as 144. If the standard of an alloy is unknown, so that it is not certain that the cornet will remain entire in the acid, it must, of course, be boiled separately, as, if one cornet in the tray breaks up, fragments may adhere to a number of others. The manipulation of the platinum tray is easier than that of parting flasks, and, in addition, the treatment of the cornets is more uniform, so that the correction afforded by the use of checks is more trustworthy. On the other hand, if platinum or palladium is present in an individual cornet, it imparts a straw-yellow or orange colour to the acid; but where a number of cornets are boiled together, it is obviously impossible to say from which the colour is derived, so that less information on the subject is obtained. A saving of acid is effected by the platinum vessels, 144 assays being boiled in 80 ozs. of acid at the Royal Mint, while 288 ozs. would be required if parting flasks were used.

6. **Weighing the Cornets.**—The final weighing is performed on a balance capable of indicating differences of 0·05 per 1,000, or one-fortieth of a milligramme, with a weight of $\frac{1}{2}$ gramme in each pan. The "checks" or "proofs" (*vide infra*) are weighed first, and their mean excess or deficiency in weight applied as a correction to all the cornets worked with them. The "weighing-in" correction (p. 434) is also allowed for, and the report is at once indicated by the marks on the weights without further calculation.

Surcharge.—The gold cornet does not actually contain the whole of the gold present in the original alloy and nothing else. Gold is lost by (*a*) volatilisation; (*b*) absorption by the cupel; (*c*) solution in the acid. On the other hand, the cornet always retains (1) some silver; (2) occluded gases. The algebraical sum of these losses and gains is called the "surcharge," since the cornet usually weighs more than the gold originally present in the assay piece; if the reverse is the case, the work is regarded as less accurate by some assayers. The various losses and gains are discussed in detail below.

Losses of Gold in Bullion Assaying.—The losses of gold can only be incurred in three ways, namely:—

1. Absorption by the cupel.
2. Volatilisation in the muffle.
3. Solution in the acid.

The two latter clauses of loss were discussed by the late Mr. G. H. Makins in a paper "On certain Sources of Loss of Precious Metals in some Operations of Assaying," which was read before the Chemical Society and published in their Journal (1860, 13, 77). He found gold and silver in the proportion of about 1 to 9 in the dust taken from a flue used only in gold and silver cupellation, but did not attempt to ascertain the percentage loss of gold by volatilisation. He also showed that large amounts of gold were dissolved by nitric acid in the course of assaying, and attributed the dissolution of gold to the presence of nitrous acid, but supposed that it would not be dissolved in the weaker acid, where nitrous acid was formed in larger quantities, owing to the protective action exercised by undissolved silver, which formed the positive element in the gold-silver couple. The fact that gold is actually dissolved in nitric acid, presumably as oxide, and remains dissolved even after dilution with water, was proved by Mr. A. H. Allen (*Chem. News*, vol. xxv., p. 85).

According to Bruno Kerl,* an alloy which requires from four to eight times its weight of lead for cupellation, sustains a loss which may be as high as 1 millième of gold; if sixteen to thirty-two parts of lead are required the loss reaches 2 millièmes.

According to Rössler the losses are heavier, varying from 1 to 3 millièmes of gold when four to five parts of lead are required for cupellation. As the result of a number of experiments conducted by the author at the Royal Mint,† it was found that the losses there in the assay of standard gold (916·6 fine) vary from 0·4 to 0·8 per 1,000, of which about 82 per cent. is absorbed by the cupel, 8 per cent. is dissolved in the acid, and the remaining 10 per cent., which is unaccounted for, is probably volatilised. These ratios, however, vary considerably, as a hot fire increases the loss by absorption and volatilisation, while by prolonged boiling in acid, and especially by annealing the fillets at a high temperature, the amount of gold dissolved in the acids is increased. An increase in the percentage of copper in the assay piece is accompanied by an increase in the loss under all three heads. This increase of loss has long been known. The following table of the results of experiments on synthetic alloys of gold and copper is given by Pelouze and Frémy‡ :—

Actual Standard of the Alloy.	Difference Surcharge.
900	+ 0·25
800	+ 0·50
700	0·00
600	0·00
500	− 0·50
400	− 0·50
300	− 0·50
200	− 0·50
100	− 0·50

The high surcharge assigned in this table to the 800-alloy is difficult to understand. The usual experience is that an increase of copper in an alloy causes greater loss of gold. The following table gives the relative surcharges obtained in the parting assay of gold of different standards; it is compiled from a number of results obtained at the Royal Mint by the author :—

* *Metallurgische Probirkunst.*, 1880.
† *Journ. Chem. Soc.*, lxiii, 1893, p. 710.
‡ *Traité de Chimie.*, 3me. Edition, vol. iii., p. 1230.

Standard of Alloy.	Surcharge on the Assay Piece.	Loss of Gold per 1,000 of Pure Gold.
916·6	+ 0·250	0·63
900·0	+ 0·225	0·65
800·0	− 0·075	1·00
666·6	− 0·2	1·20
546·0	− 0·7	2·20
333·3	− 2·8	9·30

The third column is obtained from the surcharge and the fineness in gold of the cornets obtained—viz., 999·1. The amount of lead used was eight times the weight of the alloy in the first three cases, and the quantities given in Mr. Bayly's table on p. 436 for the last three. This table, as already explained, only gives relative surcharges and losses of gold; the absolute amounts vary with the treatment.

The influence of the temperature of the furnace on the surcharge is pointed out by Prof. Roberts-Austen.* The following table is compiled from the results of his experiments:—

Temperature of Furnace.	Composition of Alloy.	Surcharge.
(a) Slightly lower than usual,	Gold 916·7 Copper 83·3	+ 0·05
(b) Ordinary temperature,		− 0·10
(c) Slightly higher than usual,		− 0·37

The loss of gold was found to be 0·645 per 1,000 in series (a), and 0·723 part in series (b). The more exact results of the author are given on p. 443.

Rössler † has shown that the loss of gold in cupellation increases with the amount of lead used and decreases as the amount of silver is increased.

Silver Retained in the Cornet.—It has been found at the Royal Mint that after boiling in the first acid the average amount of silver retained is about 2·5 parts per 1,000. The amount left undissolved by the stronger acid varies with the length of time of its action. Under normal conditions with a surcharge of 0·2 to 0·3, the silver left in the cornets is from 0·8 to 0·9, and this residue is not easily removed. If the acid continues to boil until its constant strength is reached, gold is dissolved to a considerable extent (0·5 per 1,000), while the proportion of silver present suffers little diminution, rarely falling below 0·6. If much gold is dissolved the results are not so uniform. When checks are used perfectly satisfactory results are obtained with a surcharge of + 1·0 per 1,000 or even more.‡

* Percy's *Metallurgy of Silver and Gold*, p. 275.
† *Dingl. Polytech. Journ.*, vol. ccvi., p. 185.
‡ T. K. Rose, Accuracy of Bullion Assay, *Journ. Chem. Soc.* (1893), p. 704.

The amount of silver retained by the cornets varies with the proportion of silver to gold in the fillet. Thus in an experiment conducted by the author* fillets in which the relative proportions were 2·47 : 1 and 2·4 : 1 yielded cornets containing respectively 0·6 and 0·7 per 1,000 of silver. This is a common source of error, as the variation of 0·1 would appear in the result, while such small differences in the proportion of parting silver added to the gold are usually disregarded. Even if the initial weight of silver is constant, the weight in the fillets may vary from unequal losses in the furnace. This loss was found to be about 1 per cent. when standard gold is being assayed, 1·6 per cent. in the case of gold 800 fine, and 3 per cent. if it is 666 fine.

Occluded Gases.—Graham proved† that cornets retained twice their volume of gases (mainly carbonic monoxide) in occlusion after annealing. This amounts to two parts by weight in 10,000, and is reckoned as silver in the preceding paragraph. According to Varrentrapp, the gas retained varies with the temperature at which annealing takes place.

Checks or Proofs.—Since the losses and gains detailed above are dependent on so many conditions, it is always necessary to subject check-pieces of known composition to the same operations as the alloys under examination. The use of checks in the Royal Mint was prescribed by law as early as the 14th century.‡ Standard trial plates (916·6 fine) were made and used for this purpose. Since, however, it is impossible to guarantee that a mass of alloyed metal shall have absolutely the same composition throughout, it is better to use pieces of pure gold, a corresponding amount of pure copper being added in order to make the assays absolutely comparable. The correction to be applied to a gold assay is given by the following formula:—

Let 1,000 represent the weight of the alloy originally taken.
  p the weight of the piece of gold finally obtained.
  x the actual amount of gold in the alloy expressed in thousandths.
  a the weight of pure gold taken as a check, approximately equal to x.
  b the "surcharge," i.e., the loss or gain in weight experienced by a during the process of assay expressed in thousandths.

Then $x \sim p$ or the corresponding loss or gain experienced by x is equal to $\frac{b + x}{a}$; so that $p = x \pm \frac{bn}{a}$ (i) or $x = \frac{pa}{a \pm b}$ (ii).

Example— Let $p = 916·7$ thousandths.
  $a = 1000·0$,,
  $b = 0·3$,, gain in weight.

Then b is a *positive* change, and therefore has the + sign.

Hence—
$$x = \frac{916·7 \times 1000}{1000 + 0·3}$$
$$= 916·424$$

---

* T. K. Rose, Accuracy of Bullion Assay, *Journ. Chem. Soc.* (1893), p. 706.
† *Phil. Trans. Roy. Soc.*, 1866, p. 433.
‡ *Mint Report for 1873*, p. 38.

This result would be reported as 916·4, and, therefore, the following rule is approximately correct; if there is a gain in weight by the checks, the amount is deducted from the weight of each of the other cornets; if a loss, it is added. A piece of pure gold weighing 1000 may thus be taken, without appreciable error, as a check on assays of all alloys over 900 fine, provided that the "surcharge" does not exceed ± 0·5. This error may be reduced to zero by taking as a check-piece an amount of gold equal in weight to that present in the alloy under examination, an unnecessarily laborious process when assay pieces of various degrees of fineness are present.

In the above calculation it was assumed that absolutely pure gold was used for proofs. It is doubtful whether it has ever been obtained. The method of preparation of proof gold in use at the Royal Mint is that given at p. 10. Since the assay of an alloy only gives the relative fineness of proof gold and the alloy, it follows that, if the proof gold is not quite pure, the amount found in the alloy will be in excess of the truth. If a sample of proof gold is less pure than the finest yet obtained, an allowance is made. Thus, if it is 999·9 fine, a deduction of 0·1 is made from all results of assays checked by it. This deduction is readily proved to be a very close approximation to the correct one.

Lastly, the "weighing-in" correction is applied. If the original weight taken was say 1000·4 (recorded as + 4), it is sufficient to deduct 0·4 from the final weighing. In this correction 0·4 of alloy is reckoned as fine gold, but the error is inappreciable, and will remain so if the difference from 1000 is kept less than 0·5.

Limits of Accuracy in Gold Bullion Assay.—Attention may here be drawn to the errors introduced by the lack of delicacy of the very finest assay balances in ordinary use. It has elsewhere been shown by the author[*] that by weighing in the way indicated above, errors not greater than 0·15 per 1,000 may be introduced. It is, therefore, clear that this amount represents the limit of accuracy when such balances are used. By weighing correctly to 0·01 per 1,000, however, and performing all other operations with scrupulous care, then in the determination of gold in high-standard alloys of gold and copper, or of gold, silver and copper, whether pure or contaminated by small quantities of lead, bismuth, zinc, antimony, nickel and some other elements, the error does not exceed ± 0·02 per 1,000, if the mean of three results is taken.

Parting by Sulphuric Acid.—The use of sulphuric acid of 66° B. instead of nitric acid for parting is recommended by some assayers on the ground that the losses of gold by dissolution in nitric acid are variable, while sulphuric acid does not dissolve gold. The inconveniences suffered by the use of sulphuric acid are that (1) lead and platinum are left undissolved by it; (2) violent bumping of the liquid occurs during ebullition; and (3)

[*] *Journ. Chem. Soc.*, vol. lxiii., pp. 700-714.

sulphate of silver is not very soluble in water, and the washing is consequently done with dilute sulphuric acid. However, less silver is left undissolved in the cornets than in the parting by nitric acid if the proportion of gold to silver is between 1 : 2 and 1 : 3.

Preliminary Assay.—If the composition of an alloy is quite unknown, a preliminary assay is necessary in order to determine the right quantities of silver, copper and lead to be added. This determination may be made by the touchstone, by considerations of the colour and hardness of the alloy, or by cupelling 2 grains of it with 6 grains of silver and 30 grains of lead, and parting the button in a flask. Simple cupellation with lead gives satisfactory results if silver is absent or insignificant in quantity; according to Frémy this method, in which parting is dispensed with, is accurate to 3 millièmes if carefully performed with proofs.

Assay of Gold by Means of Cadmium.—Balling has shown[*] that cadmium may be substituted for silver in the operation of parting. The $\frac{1}{2}$ gramme of gold alloy is placed in a porcelain crucible in which a little fragment of potassic cyanide has been previously fused in order to protect the metal from the air. Cadmium is then added in the proportion of $2\frac{1}{2}$ to 1 of gold. If silver is present in addition, the combined weight of cadmium and silver must be $2\frac{1}{2}$ times that of the gold. The whole is fused and then cooled and plunged into hot water to clean the button, which is then parted in nitric acid (specific gravity 1·3), dried and weighed. The silver, if any, can be estimated by precipitation as chloride from the acid solution. By this method the losses of gold and silver incidental to cupellation are entirely avoided. A similar method, employing zinc in place of cadmium, had previously been recommended by Güptner.[†]

Alloys of Gold, Silver and Copper.—These may be assayed by the method just given, the copper being estimated as difference; or the gold may be estimated as usual, and other assay pieces cupelled with enough lead to remove all the copper. The buttons thus obtained contain silver and gold only, and the proportion of silver is found by difference. The method of double cupellation, by which the button of silver and gold is weighed and then subjected to inquartation and parting is less accurate.

The cupellation designed to remove the copper is made with less lead than the quantities given on p. 435. If little gold is present, half the amount of lead there given is used, with increasing proportions as the amount of gold present increases. The temperature of cupellation must also be lower than for gold, approximating more to that used for silver. Proofs of similar composition must be used and the operations require much practice before the necessary skill is acquired. If less than

---

[*] *Oestr. Zeitsch. für Berg. und Hüttnwesn.*, 1879, p. 597.
[†] *Zeitsch. für Anal. Chem.*, 1879, p. 104.

30 per cent. of gold is present originally, the alloy may be parted at once without cupellation and the silver estimated by weighing as chloride.

Effects of the Presence of Other Metals on Gold Bullion Assay.—The effects on cupellation are the same as those given under ore assay, p. 420. In general, if a scoria is formed owing to the presence of large quantities of antimony, arsenic, cobalt, nickel, iron, tin, zinc, or aluminium, there is a loss of gold. The alloy should in that case be scorified with lead as a preliminary to cupellation. If mercury is present gold is volatilised (*vide* p. 455). The presence of tellurium is indicated by the formation of numbers of minute beads of precious metal dispersed over the surface of the cupel. Tellurium compounds are best analysed in the wet way, see p. 457.

If members of the platinum group are present they remain undissolved by the parting acid, and hinder the solution of the silver, and the assay is consequently rendered unreliable. The treatment of the alloys is discussed below, p. 455. The effects of the presence of small quantities of various metals on the surcharge in the ordinary parting assay is shown below. The table is the result of experiments made by the author.* The presence of 5 per cent. bismuth does not affect the surcharge. All assay pieces contained 1,000 parts of gold, 2,500 parts of silver, and 91 parts of copper, other metals being added in the proportions indicated.

Metal added.	None.	Antimony, 50 parts.	Zinc, 70 parts.	Tellurium, 33 parts.	Iron, 50 parts.	Nickel, 50 parts.
Surcharge,	+ 0·42 0·44 0·38	+ 0·28 0·28 ..	+ 0·35 0·38 ...	+ 0·02 0·12 ...	÷ 0·22 0·32 ...	+ 0·22 0·25 ...

These differences indicate the necessity of employing special checks containing these metals if such be present in the alloys.

It is often impracticable to apply the ordinary parting assay to the examination of low-standard alloys of gold with other metals. These are then tested by various other methods, of which a summary is given below, the alloys being grouped in four series for convenience:—

  A. Alloys requiring scorification.
  B. Amalgams.
  C. Alloys containing members of the platinum group.
  D. Tellurium compounds.

A. Scorification of Alloys.—Alloys of *arsenic* or *antimony* are reduced to a fine powder and scorified with thirty parts of

* *Journ. Chem. Soc.*, vol. lxiii. (1893), p. 706.

lead and a half part of borax. If the slag becomes pasty towards the end of the operation more borax is added, a little at a time. If the lead button obtained is hard, a second scorification is necessary, with the addition of more lead. There is considerable loss of gold from volatilisation, and therefore wet methods of analysis are preferable.

Iron or Manganese Alloys.—The operation is tedious and difficult with these alloys, as they are difficult to fuse, having higher melting points than pure gold,* and the oxides of iron do not form easily fusible compounds with the litharge. An extremely high temperature and much borax is required; ten parts of lead and one of borax usually suffice.

Cobalt and Nickel.—Twenty parts of lead are used, but no borax at first, so that the oxidation of the nickel may not be hindered. A very high temperature and the subsequent addition of two parts of borax are necessary. Several successive scorifications are required as nickel and cobalt are difficult to oxidise.

Zinc.—Oxide of zinc does not form a fusible mixture with litharge, and the slag is only rendered pasty by borax, unless it is added in large quantities. Gold is lost by volatilisation, but the loss is minimised by slagging off the zinc as rapidly as possible. Use fifteen to twenty parts of lead and two to three parts of borax added little by little, until the slag is fluid.†

Tin.—Twenty parts of lead are required; oxide of tin is rapidly formed, but the slag is not easily fusible. Large amounts of borax are necessary, or still better, borax mixed with potash which forms a fusible stannate with SnO_2.

Aluminium.—Alloys containing this metal cannot be assayed by scorification and cupellation. As soon as fusion takes place in the muffle, aluminium floats to the top of the bath, being of low density, and is rapidly oxidised, producing alumina which forms an exceedingly infusible scoria not easily removed by litharge. The production of the latter, moreover, is checked by the scum.

Mr. Edward Matthey observes‡ that the removal of aluminium by digestion in hydrochloric acid, and collection of the residual gold, does not yield satisfactory results. The process he recommends is as follows:—Accurately weighed portions of 50 grains each of the alloys are fused with litharge, under a flux of potassium carbonate and borax with a small proportion of powdered charcoal, and the resulting slag re-fused with a further small quantity of litharge and powdered charcoal. The lead buttons containing all the gold (the aluminium having combined with the fluxes employed) are cupelled, and the resulting gold cupelled with silver and parted with nitric acid in the usual

---

* Frémy, *Ency. Chim.*, T. iii., L'or, p. 147.
† *Op. cit.*, p. 148.
‡ *Phil. Trans. Roy. Soc.*, 1892, p. 643.

manner. The assays must be worked with checks or standards of fine gold and pure aluminium.

In the majority of the preceding cases it is better to analyse the alloys by wet methods, see p. 457.

B. **Amalgams.**—The alloy is placed in a weighed porcelain crucible and gradually heated so as to drive off the mercury. After the greater part of the mercury has been driven off the temperature is raised to a full red heat which is maintained for half an hour. About 0·1 per cent. of mercury still remains in the gold after this treatment and can only be completely removed by cupellation and parting. Checks must be used, as the loss of gold in the operation may amount to 1 per 1,000.

C. **Platinum Group.**—Cupellation must be performed at a higher temperature than usual, and the iridescent bands are seen to remain longer, although they are less numerous.

(1) *Platinum-Gold Alloys.*—The button obtained by cupellation is dull and crystalline. If the alloy contains as much as 7 or 8 per cent. of platinum, the cupellation proceeds slowly, brightening is only obtained at a very high temperature and the button appears flattened, and has a rough crystalline surface and a grey colour. If more than 10 per cent. is present, brightening does not occur at all, and the other features just mentioned are more strikingly exhibited. On parting, the platinum is dissolved with the silver if it does not form more than 10 per cent. of the original alloy, but the assay-piece must be boiled in acid for a long time, and the parting is less complete than in the simple case of a gold-silver button, from 0·05 to 0·4 per cent. of platinum being left in the cornets. This is completely removed by a second parting. If a larger proportion of platinum is present pure gold is added, or else the operations of inquartation and parting are repeated two or three times, the loss of gold showing a corresponding increase. Unless proofs are made up of similar composition the results are not satisfactory if more than 1 or 2 parts of platinum are present per 1,000 of alloy.

According to results recently obtained in the laboratory at the Royal Mint by Mr. A. Stansfield, it is impossible to free buttons containing platinum from lead at the ordinary temperatures attainable in a muffle. It is necessary to employ the oxyhydrogen blowpipe for the purpose, with the result that the losses are irregular. In the cupellation of pure platinum, buttons weighing from 0·002 to 0·01 gramme retain about 10 per cent. of their weight of lead, and those weighing from 0·04 to 2 grammes retain about 33 per cent.

If parting is effected by boiling in concentrated sulphuric acid instead of in nitric acid, almost all the platinum remains undissolved with the gold.

Chaudet * recommends the following proportions of silver for this purpose:—

* *L'art de l'essayeur*, Paris, 1835.

Ratio of Platinum to Gold.	Ratio of Silver to Platinum and Gold together.
Less than 1 to 10.	2 to 1
1 ,, 1.	1·5 ,, 1
More than 10 ,, 1.	1·25 ,, 1

The reason for the reduction, shown in the table, in the amount of silver when the platinum is increased is that, according to Chaudet and Haindl, considerable quantities of platinum are dissolved, and the error caused thus is corrected by leaving some silver undissolved in the cornet. The losses of platinum, however, are very irregular, and the method is not a good one. The reason for the variation is probably the lack of homogeneity in the buttons and prills, as gold-platinum alloys show a considerable amount of liquation.

The platinum is not dissolved by nitric acid if zinc or cadmium is substituted for silver.

Mr. Edward Matthey recommends the following method for alloys of platinum 900, and gold 100 parts :—50 grains each are taken of the alloys and treated with an excess of nitrohydrochloric acid which gradually dissolves the whole. The resulting solutions of platinum-gold chloride are then evaporated nearly to dryness to drive off the free acid and diluted with distilled water to about 20 c.c., a degree of strength ascertained by experiment to be the best for the precipitation of the gold. The metallic gold is thrown down by means of crystals of oxalic acid, and is carefully washed, dried and weighed. It is necessary to use checks.

(2) *Palladium-Gold Alloys.*—The palladium is dissolved in parting if the weight of silver is at least three times that of the gold, yielding an orange coloured solution. Matthey recommends double parting. Separation may also be effected by fusion with six to eight parts of potassic bisulphate, and dissolving out the dark brown palladious sulphate by boiling water.* A second fusion is usually necessary to render the gold residue quite pure.

As in the case of platinum, palladium may be separated from gold by dissolving the alloy in aqua regia, evaporating to dryness, taking up with water and precipitating the gold by oxalic acid from a very dilute solution: the palladium remains in solution.

(3) *Rhodium- and Iridium-Gold Alloys.*—Iridium, if present, always sinks to the bottom of the cupelled button as it is very dense (specific gravity = 21·38), and is not usually fused at the temperature of the muffle, but occurs in the state of fine black crystalline particles. Hence, when the button is rolled into a

* H. Rose.

cornet with the lower face outwards (p. 444), iridium occurs as black sooty spots or streaks which are seen by a lens to fill up depressions in the surface of the gold.

Both rhodium and iridium are almost insoluble in aqua regia. If gold alloys containing both of them are parted in the ordinary way with nitric acid, only a small quantity of rhodium goes into solution with the silver. The residue consisting of the gold, the iridium and most of the rhodium may then be attacked by dilute aqua regia when the gold is dissolved together with only traces of the other metals. These may be separated by evaporating the solution to dryness and heating to dull redness, when the reduced metals being no longer alloyed may be completely separated by dissolving the gold in aqua regia.

According to d'Hennin,* iridium may be separated from gold by fusing it with fluxes, the charge being made up as follows :—

Iridic gold, ...	12·5 grains.
Sodic arseniate, ...	3 ,,
Black flux, ...	18 ,,
Ordinary flux (consisting of borax, cream of tartar, charcoal and litharge), . .	20 ,,

The iridium forms a speiss with the iron and the arsenic, and the lead button formed at the bottom of the fused mass contains all the gold.

D. Tellurium Compounds.—These must be treated by wet methods. An aqua regia solution containing both gold and tellurium is evaporated with a large excess of hydrochloric acid until no more chlorine is given off, when both gold and tellurium are readily precipitated by a current of sulphur dioxide gas. On attacking the precipitate with nitric acid the tellurium is dissolved in the state of tellurous acid, and the gold residue may be dried and weighed, and its purity ascertained by inquartation and parting. Other methods for separating gold and tellurium will present themselves on studying the properties of the latter.

Wet Methods of Assay of Gold Alloys, Compounds, &c.—Assays or complete analyses of gold bullion, natural minerals, &c., can be made by the ordinary chemical methods given by Fresenius, Crookes and others. From 1 to 5 grammes of bullion are usually enough, but 10 grammes are necessary if the alloy is nearly pure gold. In general the residue left after prolonged action of nitric or sulphuric acid is not sufficiently pure to weigh as gold, and complete solution in aqua regia is usually necessary. From the solution the gold may be precipitated by (*a*) ferrous sulphate, (*b*) sulphurous acid, (*c*) oxalic acid, (*d*) sulphuretted hydrogen, (*e*) ammonic sulphide, followed by the addition of hydrochloric acid. The following remarks may be

* *Dingl. Polyt. Journ.*, vol. cxxxvii., p. 443.

of value in aiding the chemist in his choice of a precipitant in any particular case.

Nitric acid must always be expelled from the solution by warming with successive additions of hydrochloric acid. The acid solution must not be heated too strongly or loss of gold chloride by volatilisation occurs. Some other chlorides escape more freely. Ferrous sulphate and sulphurous acid act well in strongly acid (HCl) solutions; oxalic acid, sulphuretted hydrogen and ammonic sulphide best in presence of small quantities of HCl.* The solution should be dilute (say 1 part of gold in 300 of water), so that other metals may not be carried down by the gold. Sulphate of iron gives a very finely divided precipitate which is difficult to wash by decantation without loss; precipitation is slow in cold solutions. Oxalic acid causes plates and scales to form which are readily washed and are very pure; it acts best in boiling liquids, but a temperature of 80° for forty-eight hours suffices; in the cold or in the presence of much hydrochloric acid or alkaline chlorides the action is very slow and partial; a large excess of the precipitant must be present. Oxalic acid is used for solutions containing metals of the platinum group, which are not precipitated by it. Alkaline oxalates act better than the free acid.

Sulphurous acid is an excellent precipitant for solutions from which all other metals but gold have been removed. It acts rapidly and completely in the cold, but unfortunately precipitates many other metals. Sulphuretted hydrogen is used in the absence of all other metals whose sulphides are insoluble in hydrochloric acid.

In all cases careful consideration must be given to the nature of the base metals present, and the precipitant, which will not render any of them insoluble, must be selected.

Other Methods of Bullion Assay.—Among methods which have been proposed at various times, and which may still be of service occasionally in particular cases, may be mentioned:—(1) The trial by the touchstone (a method formerly more extensively used by jewellers than at present); the assay by means of considerations as to (2) the colour, and (3) the density of alloys; (4) spectroscopic assay; (5) assay by electrolysis, and (6) by the induction balance. A brief description of each of these methods is appended.

1. Trial by the Touchstone.—This is the oldest method of assay. It consists in rubbing the gold bullion to be tested on a hard smooth stone, and comparing the appearance and colour of the streak with those made by carefully prepared needles of known composition. The effect of the action of nitric acid and dilute aqua regia on these streaks is also noted. Touchstones usually consist of Lydianstone or of silicified wood, and black or

* H. Rose.

dark green stones are best. Only alloys of gold and copper or of gold and silver can be thus tested. The trial is more sensitive for alloys below 750 fine than for higher standards. The amount of gold in alloys between 700 and 800 fine can be determined correct to 5 parts per 1,000.

2. **Colour and Hardness of Alloys.**—These properties form a guide to the composition of copper-gold alloys, an increase of copper corresponding to a heightening of the colour and an increase of the hardness as tested by shears or a knife. On heating the alloy to redness in air, the degree of blackening of the surface is a further indication of the percentage composition, if compared with plates of known fineness.

3. **Density of Gold-copper Alloys.**—The determination of the fineness of these alloys by taking their densities was investigated by Professor Roberts-Austen at the Royal Mint in 1876.* He showed that the densities found by experiment were nearly equal to those obtained by calculation on the assumption that the union of the two metals was accompanied neither by contraction nor expansion. The alloys examined ranged from 860 to 1000 fine, and were made into discs which were all compressed to the same extent. The conclusion arrived at was that the fineness of large masses of gold can be deduced from their densities correct to $\frac{1}{10000}$ part. In the case of individual coins the results are only approximate.

4. **Assay by means of the Spectroscope.**—This method of determining the composition of gold-copper alloys was investigated by Lockyer & Roberts-Austen.† The spectrum of pure gold was shown to be altered by successive additions of copper, and near the English standard (916·6 fine) a difference of two or three-tenths of a millième in composition could be readily detected, but the amount of metal volatilised is so small that it cannot be made to represent with certainty the average composition of the mass, which is never perfectly homogeneous.

The detection of traces of gold in alloys or ores by means of the spectroscope, though sometimes attempted, is not remarkable for its delicacy or certainty. The method of procedure ‡ is to dissolve the auriferous material in aqua regia, evaporate off the nitric acid, and pass induction sparks through the surface film of liquid, when the spectrum shows some narrow bands and some nebulous bands. The latter only are seen if a drop of the solution is placed in a Bunsen flame. The method may sometimes be useful when complex minerals are being examined. Mr. Capel has shown that $\frac{1}{1000}$ of a milligramme of gold will show a spectrum if the spark be passed through a weak solution

---

* *Seventh Annual Mint Report*, 1876, p. 41.
† *Phil. Trans. Roy. Soc.*, 164, part ii., p. 495, and *Mint Report for 1874*, p. 38.
‡ Frémy, *Ency. Chim.*, T. iii., L'or, p. 134.

of the pure metal. But when operating on a slip of alloy formed of

Silver,	708
Copper,	254
Gold,	38
	1000

the spectra of copper and silver alone were visible. In an alloy of gold and copper containing from 200 to 250 parts in the thousand of the precious metal, the gold spectrum is barely visible. On the other hand, in an alloy of gold and copper containing traces only (·01 per cent.) of the latter, the copper spectrum was distinctly shown. Alloys of gold are rarely so perfectly homogeneous that the particles of metal volatilised and giving the spectrum represent the whole mass.

5. **Assay by Electrolysis.**—Solutions of gold are readily decomposed by low tension currents, a difference of potential of one volt sufficing in the case of chloride of gold. Consequently, gold can be separated thus in solutions of chlorides from copper, lead, iron, &c. The process differs little from the electrolytic assay of copper.* The special precautions to be observed are (a) to replace the positive platinum pole by one of carbon; (b) to cover the platinum cone (negative pole) with a thin deposit of silver or copper, so as to render the gold easily detachable; (c) to avoid an excess of HCl or HNO_3, while having a few drops of free H_2SO_4 present. The precipitation is effected at 50° with one Bunsen cell; the gold is washed, detached from the cone by nitric acid, and weighed with the usual precautions.

6. **Assay by the Induction Balance.**—Full descriptions of the instrument employed may be found in the published papers of Professors Hughes and Roberts-Austen,† but the principle on which it depends may be briefly stated as follows :—The balance of two rapidly-intermittent induced currents of equal strength, flowing in two coils connected by a wire, is disturbed by the presence of metal within one of the coils, and the fact of this disturbance is made evident by a telephone. The balance can be restored and the telephone silenced by introducing into the opposing coil an identical mass of metal, and it is found that widely different effects are produced by equal volumes of different metals and alloys. In testing discs of the gold copper alloys, the balance is maintained by a tapered and graduated rod of zinc, which is moved in and out of one coil; this method, however, is not adapted for examining alloys which differ but slightly in composition, as is shown by the following table :—

---

* Crookes's *Select Methods.*
† *Proc. Roy. Soc.*, vol. xxix. (1879), p. 56; *Phil. Mag.* [5], vol. viii. (1879), p. 50; and *Proc. Phys. Soc.*, vol. iii. (1879), p. 81.

Standard by Assay.	Readings on Zinc Scale.	Standard by Assay.	Readings on Zinc Scale.
1000·0	200	913·7	66
990·9	138	911·8	66
950·8	78	901·3	65
925·0	70	878·0	64
920·3	69	859·2	62
918·1	67	709·2	60
916·3	66		

By a different arrangement* the instrument may be made more sensitive, but its indications are not trustworthy, as the presence of traces of impurity in the alloy and even its physical condition have great influence on the readings obtained, which depend on electrical resistance. The instrument is not used in practice although it is capable of distinguishing between most counterfeit and genuine coins.

* *Tenth Report on the Royal Mint*, 1880, p. 47.

CHAPTER XXI.

ECONOMIC CONSIDERATIONS.

Management of Gold Mills.—The cost of extracting gold from its ores in the mill is chiefly of interest to the producer when it is combined with the cost of mining. Nevertheless, it is convenient that the two items should be ascertained separately. For this purpose, each truck load of ore delivered from the mine to the mill, and the amount actually treated in the latter, should be weighed carefully. Materials, implements, &c., should not be served out to the mill from the stores without an exactly-worded order. The cost of transport of the ore from mine to mill is often ascertained separately, but if this is not done it is better to include it in the cost of mining. The cost of superintendence must usually be distributed between the various operations, and this course must also be sometimes resorted to in the case of power, lighting, and other items. When the total expenses of milling are ascertained for any period, say for one month, it is necessary to allow for depreciation of the plant each time. It is also advisable not to neglect the question of interest on the capital sunk in providing the mill, and of a sinking fund, as the ore may come to an end before the machinery is discarded for other reasons.

Similarly, exactness in ascertaining the percentage of gold extracted is in the highest degree desirable. Each load of ore on its way to the mill should be automatically sampled, and frequent assays made on mixtures of these samples, the value of the tailings being determined with equal care. From the results thus obtained, not only is a watch kept on the relative success of the treatment from day to day, but an additional check is afforded on the amount of bullion produced in the mill. A well-appointed assay office in connection with the works is obviously necessary in order to carry out these tests, and laboratory extraction trials should also be made at frequent intervals in order to determine how far the efficiency of the mill is being maintained.

Details concerning the cost of treating gold ores by the various processes have already been given separately in the chapters respectively devoted to them.

Cost of Production of Gold.—In view of the fact that the value of money is measured in almost all gold-producing countries by the metal itself, it would be of special interest to estimate the average cost per ounce of its production. This was done by Prof. Roberts-Austen in the case of silver in the year 1887,* and all subsequent computations have served to show that a high degree of accuracy was attained by him. In the case of gold, further difficulties are encountered in the endeavour to frame a trustworthy estimate, owing to the differences in the mode of treating silver and gold ores. Thus the greater part of the silver produced annually is derived from the output of large mills or smelting establishments, where the total cost of treatment is well known to the managers, even though they withhold it from the public. On the other hand, a large amount of gold is even now extracted in small mills or by individuals, particularly in the case of placer deposits, and the exact cost is frequently a matter of doubt to the proprietors themselves. Moreover, both silver and gold mining companies are usually reticent as to their costs, although the advantages of this course of action to the proprietors, as distinguished from the managers, are not easy to understand. By certain large companies, systematic accounts are published, and from these the following results are extracted:—

At the Alaska Treadwell Company's mill about 20,000 tons of ore, containing 3 dwts. of gold per ton, are treated monthly by crushing and amalgamation, followed by concentration and chlorination of the sulphides. The cost (including both mining and milling) per ounce of gold extracted was £2 1s. 1d. in 1890-91, and £2 4s. 8d. in 1891-92. (The value of 1 ounce of pure gold is £4 4s. 11½d.) At the El Callao mine, Venezuela, the cost per ounce was formerly as low as 25s., but rose to £3 10s. 9d. in 1891, and £3 1s. 8d. in 1892. The cost to the Montana Company was £3 10s. 6d. per ounce in 1891, and as much as £4 8s. in 1892. In some of the large hydraulicking companies the cost was similar to that of the Alaska Treadwell mine. Thus, at the La Grange Company's workings in 1874-76, the cost was £2 4s. 3d. per fine ounce of gold, and at the New Bloomfield Company's Claim the cost was £3 0s. 10d. in 1875, and £2 1s. in 1876 and in 1877. The above companies, however, were thriving, and the average cost per ounce in America, even at the mines and mills which pay their way, is probably not less than £3. The average cost of production in California is estimated by A. G. Charleton † at about £2 2s. 6d. per ounce. In South Africa, on the Witwatersrand, the cost of production has been greatly reduced during the last few years. In 1892 the cost of production per fine ounce of gold at those mills which were working more or less continuously seems to have been about £3 10s., and in 1895, for the mines situated on the outcrop of the Main Reef, Hatch and Chalmers estimate it at

---

* *Nineteenth Annual Report of the Royal Mint*, 1888, p. 56.
† *Trans. Fed. Inst. Mng. Eng.*, 1893, p. 217.

about £2 14s. At Barberton, the cost of gold obtained at the Sheba mine is said to be £1 4s. per ounce. In Mysore, where fuel is costly and the climate bad, the cost is said to be about £1 18s. 10d. per ounce of gold extracted.* At the placer mine at Saint-Elie in French Guiana, which produces about 1,600 ounces of gold per month, the cost of production was £2 1s. 6d. in 1887, and at the Siberian placers the cost is £1 17s. 2d. at Berezovsk, £1 17s. 8d. at Nijni-Tassnil, and £2 15s. 8d. at Tchernaia-Retchka.

Judging from the published results generally, it would appear that the average cost of production of gold is somewhat less than £3 per ounce. It must be remembered, however, that the interest on the initial cost of discovering and developing the mine, and of the machinery used in both mining and milling is, in many cases, not included in the so-called cost of production. It has, moreover, been frequently pointed out, particularly by Mr. A. Del Mar, that if the working costs of unsuccessful mines, and the money spent in fruitless prospecting, and in the erection of mills unsuitable to the ore to be treated, were to be included in the cost of the production of the gold, it would doubtless appear that a heavy loss is annually experienced by the gold-mining industry. It has been well observed that, no matter what the market prices of gold, silver, copper, &c., may be, mines and ores will be worked which leave no profit, in the never-failing expectation that the market will rise again, or that the ore will become more abundant or richer.

The "market price" of gold at any one time must, of course, be derived from the mean price in gold of all, or at any rate a large number of, other commodities, when compared with this mean price at some other time.

Annual Production of the Gold Mines of the World.—The production of gold in ancient times cannot be closely estimated, but, judged from a modern standpoint, it was probably very small. In the middle ages, however, between the fall of Rome and the discovery of America, the production was far smaller than before, and Jacob observes that in this period "the precious metals were sought not by exploring the bowels of the earth, but by the more summary process of conquest, tribute, and plunder." Even after the exploitation of the New World began, the output of gold was for many years much too small to satisfy the cupidity of the conquerors. The development of the mining industry was prevented by the ruin and destruction of the natives, and by the almost incessant irregular warfare waged against the Spaniards in America in the 16th century, first by the Dutch and later by the English. Fifty years elapsed after Columbus discovered America, before the annual production of gold reached £1,000,000, and even at the end of

* *Ibid.*, p. 218.

the 17th century Soetbeer estimates that it was only £1,500,000. The discovery and working of the rich Brazilian placers during the next half century raised the annual product to over £3,500,000 in the period 1740-1760 (Soetbeer), but as these deposits became exhausted, the output again fell off, and in the period 1810-1820 had again sunk to about £1,500,000 per annum. The gradual development of the Siberian placers was the main cause of the subsequent steady increase in production up to an average of £7,500,000 per annum in the period 1841-1850, and this was followed by a sudden rise consequent on the discoveries in California and Australia. The maximum output from the rich placers of these countries was reached in 1853, when the world's production of gold is estimated by Sir Hector Hay to have been £38,000,000. After falling to £21,000,000 in 1867, the output remained nearly stationary until about the year 1888, when, from various causes mentioned below, the production again began to increase, and in 1895 reached £41,000,000, the greatest amount on record.

The doubling of the world's production which has taken place in the course of the last eight years is due (1) to the discovery and development of new districts, (2) to the progress in the art of metallurgy, and (3) to the increased attention given to gold mining consequent on the fall in the price of silver. This fall administered a severe check to the silver mining industry and set free a considerable amount of skilled labour and of capital which were largely diverted to gold mining. The most important new districts which have been developed are the Rand, Cripple Creek in Colorado, and Coolgardie in Western Australia. The output of gold at the Rand was valued at £81,000 in 1887, and £7,838,000 in 1895; Cripple Creek had not been discovered in 1887, and produced about £1,420,000 in 1895, while Western Australia produced £20,000 in gold in 1887, and £880,000 in 1895. The total increase in annual output on these three fields was therefore about £10,000,000 between 1887 and 1895.

The increase of production due to improved metallurgical methods is difficult to estimate. Most of it must certainly be put down to the account of the cyanide process by which gold valued at about £3,000,000 was produced in 1895, though the greater part of this was produced on the Rand, and is included in the amount already given. The increase of about £1,500,000 which took place in the Russian output between 1888 and 1895 is partly due to improvements in the methods of washing the gravels, and partly to the extension of the Siberian railway and the greater attention now being paid to vein mining. The third cause of the increase in output of gold has been more felt in the United States and Mexico than elsewhere. Of the increase in the output of the United States from little more than £6,000,000 in 1887 to over

£9,000,000 in 1895, about half is attributable to the output from the new field at Cripple Creek as already noted, and the remainder is mainly due to the changes in the conditions of silver mining and their ultimate effects. The increase of about £1,000,000 in Mexico is also to be ascribed to the last-named cause. The output of Australasia (excluding Western Australia) was £2,000,000 more in 1895 than in 1887, and although local causes were largely responsible for this increase, as, for example, the encouragement given by the government of New South Wales in 1894 and 1895 to "fossicking," nevertheless improved methods of extraction were also of great effect.

Speaking generally, it is clear from the above considerations that of the increase in the annual output of gold of some £20,000,000 which took place between 1887 and 1895, about half is due to the discovery of new fields, and the other half to the development of districts already known. The credit for having caused the latter part of the increase may be about equally divided between improvements in metallurgical processes and the greater attention which has been given to gold mining during the last few years.

As regards the future production, it is obviously impossible to predict what will be the effect of the opening up of gold fields as yet undiscovered. Even without these, however, it seems almost certain that the production will continue to rise rapidly for some years to come. The most conservative estimates place the output of the Rand in five years time at from £12,000,000 to £14,000,000 per annum, and when the labour problem has been fully solved, even this huge output might conceivably be doubled soon afterwards. The output in Western Australia may similarly be expected to increase largely when, if ever, the difficulties due to lack of water have been overcome. Siberia, developed by French engineers with French capital, may also contribute much more than at present to the world's production, and the improvements in methods of extraction which have been so marked of late years will doubtless help to swell the total. It is, therefore, no exaggeration to say that present indications point to an annual production of gold of not less than £55,000,000 to £60,000,000 in the year 1900.

The following table gives details of the production of gold in various countries in the years 1894 and 1895:—

ECONOMIC CONSIDERATIONS. 467

NAME OF COUNTRY.	1894.		1895.		Percentage of Total Production.	
	Ounces.	Value.	Ounces.	Value.	1894.	1895.
Australasia,	2,243,716	£8,450,000	2,355,562	£8,900,000	22·77	21·74
United States,	1,923,619	8,250,020	2,170,000	9,220,000	22·18	22·50
Africa,	2,365,853	8,100,000	2,605,882	8,875,000	21·83	21·68
Russia,	1,337,400	5,680,800	1,647,000	7,000,000	15·31	17·08
China,	290,930	1,235,800	290,000	1,230,000	3·33	3·00
Mexico,	217,700	923,000	259,000	1,100,000	2·48	2·68
British India,	209,247	773,900	250,000	925,000	2·08	2·25
Colombia,	139,950	594,400	141,200	600,000	1·60	1·46
British Guiana,	138,500	506,350	141,500	520,000	1·37	1·27
Brazil,	107,350	456,690	110,000	460,000	1·24	1·12
Germany,	106,580	452,700	106,000	450,000	1·22	1·10
Austria Hungary,	81,500	346,200	82,400	350,000	·94	·85
French Guiana,	64,300	273,100	65,000	275,000	·74	·67
Canada,	52,980	225,060	55,000	230,000	·61	·56
Venezuela,	38,990	165,640	38,000	160,000	·45	·39
Dutch Guiana,	28,030	119,080	28,300	120,000	·32	·29
Japan,	23,700	100,650	23,700	100,000	·27	·24
Central American States,	22,700	96,680	23,500	100,000	·26	·24
Korea,	22,600	96,030	19,000	80,000	·26	·19
Chili,	22,440	95,430	24,000	100,000	·26	·24
France,	8,970	38,080	8,950	37,000	·10	·09
Uruguay,	6,850	29,040	6,850	30,000	·08	·07
Italy,	5,700	24,000	5,700	24,000	·06	·06
Argentine,	4,600	19,480	4,600	19,500	·05	·05
Peru,	3,600	15,250	3,600	15,250	·04	·04
Great Britain,	4,235	14,811	4,500	15,000	·04	·04
Equador,	3,310	14,020	3,300	14,000	·04	·036
Bolivia,	3,250	13,759	3,250	13,750	·036	·03
Sweden,	3,020	12,809	3,000	12,090	·03	·03
Turkey,	380	1,600	380	1,600	·004	·004
	9,482,060	£37,103,731	10,479,174	£40,977,190	100·000	100·000

The product of each of the Colonies of Australasia was as follows:—

	In 1894.	In 1895.
Queensland,	679,511 ozs.	623,000 ozs.
Victoria,	716,955 ,,	740,036 ,,
New South Wales,	324,787 ,,	360,165 ,,
New Zealand,	221,615 ,,	293,491 ,,
Western Australia,	207,131 ,,	231,513 ,,
Tasmania,	57,873 ,,	59,964 ,,
South Australia,	35,844 ,,	47,343 ,,
Total,	2,243,716 ,,	2,355,562 ,,

The product of the individual States in the U.S.A. for the year 1894 was as follows:—

California,	636,468 fine ozs.
Colorado,	461,969 ,,
Montana,	176,637 ,,
South Dakota,	159,594 ,,
Idaho,	100,682 ,,
Arizona,	96,313 ,,
Oregon,	68,792 ,,
Nevada,	55,042 ,,
Alaska,	53,868 ,,
Utah,	41,991 ,,
New Mexico,	27,465 ,,
Eleven other States,	24,793 ,,
Total,	1,923,619 ,,

The output in Africa was divided as follows:—

	In 1894.	In 1895.
Witwatersrand,	2,024,163 ozs.	2,278,110 ozs.
De Kaap,	92,577 ,,	63,046 ,,
Klerksdorp and Potchefstroom,	77,714 ,,	90,841 ,,
Zoutpansberg,	10,629 ,,	10,379 ,,
Lydenberg, Malmani, &c.,	60,770 ,,	63,506 ,,
Gold fields not in the Transvaal (estimated),	100,000 ,,	100,000 ,,
Total,	2,365,853 ,,	2,605,882 ,,

In 1895, the product of the Witwatersrand field was 2,278,110 ozs., valued at £7,837,779. To obtain this amount, 3,456,575 tons of ore were crushed, the average number of stamps at work being about 2,600; the yield on the plates was 8·86 dwts. per ton, from the concentrates the yield was 0·61 dwt., and from the tailings by the use of cyanide 3·70 dwts. per ton of the original ore, making a total of 13·17 dwts. of gold per ton.

In the table given above, the figures for 1894 are for the most part taken from the table given by the Director of the United States Mint, and published in his *Annual Report on the Production of Gold and Silver for 1894*. The output of Australasia in both years is that given in the government reports of the separate colonies. The African product in each year consists of the returns for the Transvaal made by the Witwatersrand Chamber of Mines, together with a rough estimate of the gold produced in the rest of the Continent. The estimates of the product of most of the other countries in 1895 are those compiled by the *Engineering and Mining Journal*, and consist in some cases of official returns for a part of the year, together with an estimate of the remainder based on these returns. Thus, for example, the Russian product consists of official returns for $10\frac{1}{2}$ months in the year, and estimates of the remainder. The output in Mexico consists of returns for the first 6 months, and an estimate of the remainder, and the estimate for British Guiana includes 11 months' returns. In other cases, such as most of the South American States and China, the estimates are merely based on the output of the previous year. The last two columns giving the percentages have been added by the author.

Without attempting to make an exact estimate of the amount of gold won by each metallurgical method, it is obvious from the fact that almost all the Russian product, about one-fourth of that from Australasia, and probably over one-fourth of that from the United States, is derived from placer washings, that these deposits yield little less than 30 per cent. of the world's product. The yield of gold from quartz crushing is now nearly double as much as that from placers, although, up to a few years ago, it had always been less. Gold obtained by smelting (chiefly in the United States) may be about 3 per cent. of the total produce of the world, while that by the wet methods (viz., treatment by chlorine and by cyanide of potassium) is probably over 10 per cent., the reagent last named being instrumental in obtaining about two-thirds of this amount. It is probable that over nine-tenths of the gold is won by methods best suited to the ores dealt with in each case, taking into account the conditions prevailing in the district. Alterations in these conditions, such as the improvement of transport or changes in the cost of labour or materials, may render some other method preferable to the one in use. The great problem has long been to find a cheap method of treating large quantities of low-grade pyritic ores which do not readily yield their gold to the action of mercury. Such ores can usually be treated by chlorination or by smelting, but the expense of either method is in many cases too great to afford a profit. It seems quite possible that the cyanide process will come into general use for these ores.

Consumption of Gold.—Many attempts have been made at various times to estimate the amount of gold used annually for purposes other than that of additional coinage. In 1831, Jacob[*] estimated the annual amount of gold "converted into

[*] *History of the Precious Metals*, vol. ii., p. 322.

utensils and ornaments" at about £4,500,000, while the production was under £2,000,000, without counting, however, imports from India and China, which were supposed to be considerable. In 1881, Dr. Soetbeer estimated the annual industrial consumption of gold in the world, after making deductions for old material employed, as amounting to 84,000 kilogrammes or £11,500,000, and in 1885 he put it at 90,000 kilogrammes or £12,300,000, against a production of £21,000,000. In 1891, the same authority gave the industrial consumption of gold added to the amount hoarded and that exported to the East as 120,000 kilogrammes or £16,400,000, against a production of £24,000,000. In 1894, Ottomar Haupt estimated the industrial consumption at only 280,000,000 francs or £11,200,000, but the exports to the East are not included in this amount. In the same year the Director of the United States Mint* estimated the industrial consumption at $50,177,300 or about £10,300,000. Doubtless the industrial consumption of gold is increasing in amount with the increase of population and wealth, but nevertheless it is fair to assume that no such rapid growth has taken place of late years in consumption as has been already pointed out to have taken place in production.

Taking the annual consumption of gold, including that exported to the East, as being, on the basis of the estimates given above, from £10,000,000 to £15,000,000, it would appear that the amount of gold added to the world's coinage, or retained as reserves in the form of bullion, was from £5,000,000 to £10,000,000 per annum previous to the year 1890, and has been gradually increasing ever since, so that in 1895 it was probably over £25,000,000. Now, in 1894, it was estimated that the total available stock of gold in the world was $3,965,900,000 or about £800,000,000, so that the annual addition to this stock is now probably at the rate of between 3 and 4 per cent. per annum, and may be expected to increase gradually for some time to come. With regard to the effect which this increase may have on prices and on commercial prosperity, it is to be observed that in the years succeeding 1848, when the available stock of gold is estimated to have been less than £300,000,000, an annual addition (not including the industrial consumption) was made to the available stock of about £20,000,000, or 6·7 per cent. of the total amount. This addition is usually supposed to have stimulated trade and inaugurated a long period of commercial prosperity. Jevons estimated that the purchasing power of gold was depreciated by from 9 to 15 per cent., and that it was only kept from falling lower by the absorption of vast quantities of gold for use in currency by several countries which had previously mainly employed silver or paper. Similar results in the near future are now looked for by some economists.

* *Production of the Precious Metals*, 1893, p 53.

BIBLIOGRAPHY.

It has been found impracticable to enumerate the articles and paragraphs relating to the metallurgy of gold, which have from time to time appeared in the various periodicals, as very little search results in the accumulation of thousands of such references. In general, therefore, only the names of some of the publications which contain important or interesting matter on the subject are given; but a few exact references on special points have been added, and many others occur in the footnotes to the text.

PERIODICAL LITERATURE.

Anales de la Mineria Mexicana, ó sea: Revista de Minas. Mexico. From 1861.
Anales de Minas. Madrid. From 1841.
Annalen der Berg-und Hüttenkunde. Salzburg, 1802-5.
Annales des Mines. Paris. From 1816.
Annuaire du Journal des Mines de Russie. St. Petersburg. First published in 1840.
Annuaire des Mines et de la Métallurgie Françaises. Paris. First published in 1876.
Annual Reports of the Ballarat School of Mines. From 1882.
Annual Reports of the Californian State Mineralogist. Sacramento. From 1881.
Annual Reports of the Deputy-Master of the Royal Mint. London. From 1870.
Annual Reports of the Director of the United States Mint. Washington.
Annual Reports on Gold Mining. Victoria, British Columbia. From 1875.
Berg-und Hüttenmännisches Jahrbuch. Vienna. From 1866.
Berg-und Hüttenmännische Zeitung. Freiberg. From 1842.
Biennial Reports of the Nevada State Mineralogist. From 1871.
Boletin oficial de minas Madrid, 1844-5.
Bulletin de l'Association amicale des anciens élèves de l'École des Mines. Paris. From 1869.
Dingler's Polytechnisches Journal. From 1815.
Engineering and Mining Journal. New York. From 1866.
Jahrbuch für Berg-und Hüttenwesen. Freiberg. From 1827.
Journal des Mines de Freiberg. Koehler. Freiberg, 1788-1793.
Journal des Mines de Russie. 1832 to 1835.
La Mineria. Mexico, 1843.
Mining and Scientific Press. San Francisco. From 1864.
Mining and Smelting Magazine. London, 1862-65.
Mining Journal. London. From 1836.
Mining Review. Denver, Colorado, 1873-76.
Mining World. London. From 1871.
Nouveau Journal des Mines de Freiberg. Koehler & Hoffmann. Freiberg, 1795-1804.

Oesterreich Zeitschrift für Berg-und Hüttenwesen. Otto Freihern. Vienna. From 1853.
Precious Metals of the United States, Annual Reports on. Washington. From 1880.
Reports of the Mining Commissioner of New Zealand. Wellington, New Zealand. From 1871.
Revista Minera y Metalurgica. Madrid. From 1850.
School of Mines Quarterly. Columbia, U.S. From 1879.
Scientific American. New York. From 1846.
Silliman's American Journal of Science and the Arts. New Haven and New York. From 1816.
Transactions of the American Institute of Mining Engineers. Philadelphia. From 1871.

GENERAL METALLURGY OF GOLD.

Geber, the Works of. Translated by R. Russell. London, 1686.
Biringuccio. De la Pirotechnia. Venice, 1540. French translation, Rouen, 1627.
Agricola (Georgius). De re metallica. Bale, 1556.
Michaelis (Johannis). De Oro. Leipzig, 1630.
Barba (Alphonzo). Arte de los metales. Madrid, 1639. French translations, Paris, 1751, and La Haye, 1782.
Schluter. Principles of Metallurgy and Assaying. Brunswick, 1738.
Vargas (Perez de). Traité singulier de metallurgie. Translated from the Spanish. Paris, 1743.
Lewis (Wm.) Commercium Philosophico-Technicum. London, 1763.
Valerius. Grundriss der metallurgie. Ulm, 1768.
Cramer. Principes de metallurgie et docimasie. Blankenburg, 1774.
Jars. Voyages metallurgiques. Paris, 1774-81.
Karsten. System der metallurgie. Breslau, 1818.
Kiessling. Die metallurgie. Dresden, 1841.
Ansted (D. T.) The Gold-Seekers' Manual. London, 1849.
Landrin (H.) Traité de l'or. Paris, 1850 and 1863.
Rammelsberg. Lehrbuch der chemischen metallurgie. 1850.
Phillips (J. A.) Gold Mining and Assaying. London, 1852.
Phillips (J. A.) Encyclopædia metropolitana. Article on Metallurgy. London, 1854.
Rivot (L.) Principes generaux du traitement des minerais metalliques. Paris, 1859.
Crookes (Wm.) and Röhrig (E.) Treatise on Practical Metallurgy, translated from the German. London, 1860.
Kerl (B.) Die Rammelsberger Hüttenprozesse. Clausthal, 1861.
Kustel (G.) Processes of Gold and Silver Extraction. San Francisco, 1863.
Kerl (B.) Handbuch der metallurgischen Hüttenkunde. Clausthal, 1865.
Phillips (J. A.) The Mining and Metallurgy of Gold and Silver. London, 1867.
Overman (F.) A Treatise on Metallurgy. New York, 1868.
Blake (W. P.) Report on the Precious Metals. Washington, 1869.
Raymond (R. W.) Mineral Resources West of the Rocky Mountains. 7 vols. Washington, 1869-74.
Raymond (R. W.) Mines, Mills, and Furnaces of the Pacific States. New York, 1871.
Schiern (F.) Sur l'origine de la tradition des fourmis qui ramassent l'or. Copenhagen, 1873.
Greenwood (W. H.) Manual of Metallurgy, vol. ii. London, 1875.
Cox (S. H. F.) Treatment of Gold Ores. *Mining Journal*, vol. xlvii., p. 1388, 1877.

Attwood (G.). The Batea, the Milling of Auriferous Veinstones, and the Mineralisation of Gold. Articles in *Alta California.* San Francisco, Sept., 1878.
Encyclopædia Britannica. 9th Edition. Article on "Gold." London, 1879.
Kerl (B.)   Grundriss der Metallhüttenkunde.   Leipzig, 1880.
Simonin (L.)   L'Or et l'Argent   Paris, 1880.
Industrial Progress in Gold Mining.   Philadelphia, 1880.
Percy (John).   Metallurgy of Silver and Gold, vol. i.   London, 1880.
Ryan (J.)   Gold Mining in India.   London, 1881.
Lock (A. G.)   Gold: Its Occurrence and Extraction.   With a Bibliography.   London, 1882.
Egleston (T.)   The Progress of the Metallurgy of Gold and Silver in the United States.   New York, 1882.
Balch (W. R.)   Mines, Miners, and Mining Interests of the United States in 1882.   Philadelphia, 1882.
Restrepo.   Estudio sobre las minas de oro y Plata de Colombia.   Bogota, 1884.
Gore (G.)   Art of Electro-Metallurgy.   New York, 1884.
Zoppeti.   L'electrolisi in metallurgica.   Milan, 1885.
Phillips (J. A.) and Bauerman.   The Elements of Metallurgy.   London, 1887.
Egleston (Thos.)   Metallurgy of Silver, Gold, and Mercury in the United States.   2 vols.   London, 1887-90.
Balling (C.)   Grundriss der electrometallurgie.   Stuttgard, 1888.
Frémy.   L'Or dans le laboratoire.   *Encyclopædie Chimique,* vol. iii.   Cahier 16c.   Paris, 1888.
Watt.   Electro-Metallurgy Practically Considered.   London, 1889.
Lock (G. W.)   Practical Gold Mining.   London, 1889.
Eissler (M.)   The Metallurgy of Gold.   London, 1891.
Frémy.   L'Or dans les centres de travail et de l'industrie.   *Encyclo. Chim.,* vol. v.   Cumenge and Fuchs.   Paris, 1891.
Raymond (R. W.)   Gold and Silver: Report of the 11th Census of the U.S.   New York, 1892.
Hatch and Chalmers.   Gold Mines of the Rand.   London, 1895.
De la Coux.   L'or.   Gites auriferes—Extraction de l'or.   Paris, 1895.

CHAPTERS I. AND II.—PROPERTIES OF GOLD, ITS ALLOYS AND COMPOUNDS.

Budelius (R.)   De monetis.   Coloniæ Agrip, 1591.
Savot.   Discours sur les Medalles Antiques.   Paris, 1627.
Potier (M.)   Philosophica Chemica.   Francfort, 1648.
Borrichius.   Hermetes Ægyptiorum et Chemicorum Sapientia.   Copenhagen, 1674.
Gobet.   Les anciens mineralogistes du royaume de France.   Paris, 1679.
Gellert (C. E.)   Metallurgic Chemistry.   Translation, London, 1796.
Hatchett (J.)   Wear of Coins.   *Phil. Trans. Roy. Soc.*, 1803, p. 43.
Hatchett.   Experiences et observations sur l'or.   Paris, 1804.
D'Arcet.   L'art de dorer le bronze.   Paris, 1818.
Jacob (Wm.)   A History of the Precious Metals.   2 vols.   London, 1831.
Schmieder.   Geschichte der alchemie.   Halle, 1832.
Boué (P.)   Traité d'orfèvrerie, bijouterie, et jouaillerie.   Paris, 1832.
Levol (C.)   Liquation.   *Ann. de Chimie et de Phys.*, vol. xxxix., p. 163.   1853.
Ansell (G. F.)   A Treatise on Coining.   London, 1862.
Rossignol (J. P.)   Les metaux dans l'antiquité.   Paris, 1863.

Watt's Dictionary of Chemistry and Supplements. Articles on "Gold." London, 1864, 1872, 1875, and 1881.
Wilm (E.) *Dictionnaire de chimie, Wurtz.* Article on "L'Or." Paris, 1868.
Ronchaud (L. de). Dictionnaire des antiquités grecques et romaines. Article, "aurum."
Mommsen. Histoire de la monnaie romaine. Paris, 1868.
Jevons (S.) Wear of Coin. *Journ. of Statistical Society.* London, 1868.
Report on European Mints. London, 1870.
Skey (Wm.) Gold and Platina in Solutions of Alkaline Sulphates. *Trans. N.Z. Inst.*, vol. iv., p. 313. Wellington, 1872.
Skey (Wm.) On the Mode of Producing Alloys by Wet Processes. *Trans. N. Z. Inst.*, vol. v., 1873.
Skey (Wm.) On the Oxidation of Gold in the Presence of Water. *Trans. N. Z. Inst.*, vol. viii., p. 339, 1876.
Peligot. Alloys used for Coinage. *Comptes Rendus*, vol. lxxvi. (1873), p. 1441.
Booth (J. C.) and Wm. E. Dubois. Condemnation of Ingot Melts. Washington, 1875.
Lepsius. Les metaux chez les Egyptiens. Paris, 1877.
Wright (C. R. A.) Metals. London, 1878.
Lenormant. La Monnaie dans l'antiquité. 3 vols. Paris, 1878.
Gee (G. E.) The Goldsmith's Handbook. London, 1879.
Pollen (J. H.) Ancient and Modern Gold and Silversmith's Work. London, 1879.
Noback (Fr.) Münz-, Maass- und Gewichtsbuch. Leipzig, 1879.
Cripps. Old English Plate. London, 1878, 1891.
Douan. Inoxydation, dorure, et platinage des métaux. Paris, 1880.
Krüss (G.) Untersuchungen über das Atomgewicht des Gold. Munich, 1880.
Brandis. Das Münz-mass und Gewichtwesen in Vorder-Asien bis auf Alexander den Grossen. Berlin.
Del Mar (Alex.) History of the Precious Metals. London, 1880.
Wagner (A.) Gold, Silber, und Edelslema. Vienna, Leipzig, and Pesth. 1881.
Wheatley and Delamotte. Art Work in Gold and Silver, London, 1881.
Bloxam (C. L.) Metals: their Properties and Treatment. London, 1882.
Martin (W.) Wear of Coins. *Journ. Inst. Bankers.* London, June, 1882.
Achiardi (T.) Metalli. Milan, 1883.
Kenyon (R. L.) The Gold Coins of England. London, 1884.
Roberts-Austen (W. C.) Alloys used for Coinage. *Cantor Lectures, Soc. of Arts.* London, 1884.
Blake (W. P.) Crystalline Forms of Gold. *Precious Metals of the United States.* Washington, 1884.
Berthelot (M.) Origines de l'alchemie. Paris, 1885.
Streeter (E. W.) Gold: the Standards of all Countries. London, 1885.
Kopp (H.) Die Alchemie in älterer und neuerer Zeit. Heidelberg, 1886.
Rochas (A. de). L'or Alchemique. Article in *La Nature.* Paris, 1886.
Schaefer (H. W.) Die Alchemie, &c. Flensborg, 1887.
Roberts-Austen (W. C.) Crystallisation in Gold. *Phil. Trans. Roy. Soc.*, vol. clxxix. (1888), p. 339.
Peligot. Liquation. *Bull. de la Soc. d'encouragement*, vol. iv. (1889), p. 171.
Riche (A.) Monnaie, Medailles et Bijoux. Paris, 1889.
Brannt (W. T.) Metallic Alloys. London, 1889.
Guettier (A.) Practical Guide for the Manufacture of Metallic Alloys, 1865. Translated from the French by A. A. Fesquet. New York, 1890.
Hiorns (A. H.) Mixed Metals. London, 1891.
Wagner (A.) Gold, Silber und Edelsteine. Handbuch für Gold-, Silber-, Bronze-Arbeiter und Juweliere. Leipzig, 1895.

CHAPTER III.—MODE OF OCCURRENCE AND DISTRIBUTION OF GOLD. PRODUCTION OF GOLD.

Holzchul. Remarques sur l'or des mines de Saxe. Penig, 1805.
Atkinson (S.) Discoverie and Historie of the Gold Mines in Scotland. Edinburgh, 1825.
Miers (J.) Travels in Chile and La Plata. London, 1828.
Humboldt. Fluctuations in the Supplies of Gold. London, 1839.
Dupont (S. C.) De la Production des Metaux precieux au Mexique: consideree dans ses rapports avec la metallurgie, &c. Paris, 1843.
Papers relating to the Discovery of Gold in Australia. 2 vols. London, 1852-57.
Reports on the Gold Returns of Victoria of 1859-62. Melbourne.
Ansted (D. T.) Gold in Wales. *Min Journ.*, vol. xxviii., p. 241, 1858.
Clarke (W. B.) Researches in the Southern Gold Fields of New South Wales. Sydney, 1860.
Readwin (T. A.) The Gold Discoveries in Merionethshire; and a Mode for its Economic Extraction. Manchester, 1860.
Davison (S.) Geognosy of Gold Deposits in Australia. London, 1861.
Rosales (H.) Essay on the Origin and Distribution of Gold in Quartz Veins. Melbourne, 1861.
Dubois (Wm. E.) The Dissemination of Gold. *Trans. Am. Phil. Soc.* Philadelphia, June, 1861.
Clarke (W. B.) Auriferous and Non-auriferous Quartz Reefs of Australia. *Geol. Mag.*, vol. iii., p. 561, 1866.
Brown (J. R.) Mineral Resources of the United States. Washington, 1867.
Smyth (R. Brough). Gold Fields of Victoria. Melbourne, 1867.
Lovell (J.) Gold Fields of Nova Scotia. Montreal, 1868.
Forbes (D.) Gold from Clogau. *Geol. Mag.*, vol. v., p. 224, 1868.
Bankart (H.) Gold Fields of Uruguay. *Journ. Roy. Geog. Soc.*, vol. xxxix., p. 339. London, 1869.
Mackay (J.) Report on the Thames Gold Fields. Wellington, N.Z., 1869.
Cotta (B. von). Treatise on Ore Deposits. Translated. New York, 1870.
Bateman (A. W.) South African Gold Fields. *The Times*, Sept. 28, 1874.
Domeyko (J.) Ensayo sobre los Depositos Metaliferos de Chile. Santiago, 1876.
Bain (A. G.) Gold Regions of S.E. Africa. London and Cape Colony, 1877.
Ott (A.) Nature and Distribution of Gold in Metallic Sulphides. *John Franklin Institute Journal.* 3rd series, vol. lvii., pp. 129-132.
Phillips (J. A.) Ore Deposits. London, 1878.
Daintree (R.) Modes of Occurrence of Gold in Australia. *Journ. Geol. Soc.*, vol. xxxiv., p. 431, 1878.
Jenney (W. P.) Mineral Resources of the Black Hills of Dakota. Washington, 1880.
Ball (Prof. B.) Diamonds, Coal, and Gold in India: their Occurrence and Distribution. London, 1881.
Blake (W. P.) Geology and Mineralogy of California. Sacramento, 1881.
Jervis (G.) Dell'Oro in Natura. Rome, 1881.
Burton (R. F.) Gold on the Gold Coast. *Journ. Soc. of Arts*, vol. xxx., p. 785, 1882.
Ratte (A. F.) Descriptive Catalogue (with notes) of the General Collection of Minerals in the Australian Museum. Sydney, 1885.
Emmons (S. F.) and Becker (G. F.) Precious Metals: being vol. xiii. of U.S. Census Reports of 1880. Washington, 1885.

Handbook of New Zealand Mines. Wellington, 1887.
Liversedge (J.) Minerals of New South Wales. London, 1888.
Mathers (E. P.) Gold Fields of South Africa revisited. London, 1889.
Anderson (J. W.) The Prospector's Handbook; a guide for the Prospector and Traveller in search of metal-bearing and other valuable minerals. London, 1889.

CHAPTERS IV. AND V.—PLACER MINING.

Moneeram. Native Account of Washing for Gold in Assam. *Journ. Asiat. Soc. Bengal*, vol. vii., p. 621, 1838.
Abbott (Capt. J.) Account of the process employed for obtaining gold from the sand of the river Beyass: with a short account of the gold mines of Siberia. *Journ. Roy. Asiat. Soc. Bengal*, vol. xvi., pp. 266-272, 1847.
Delesse. Gisement et exploitation de l'or en Australie. Paris, 1853.
Blake (W. P.) Hydraulic Mining in Georgia. *Am. Journ. Sci. and Art.* 2nd series, vol. xxvi., p. 278, 1858.
Report of the Royal Commission appointed to inquire into the best methods of removing sludge from the gold fields. Melbourne, 1859.
Sauvage (Edw.) On Hydraulic Gold Mining in California. *Proc. Inst. C.E.*, vol. xlv., p. 321, 1859.
Radde (Gustav). Reisen im Süden von Ost-Sibirien in den Jahren, 1855-59. St. Petersburg, 1863.
Debombourg (G.) Gallia aurifera. Études sur les alluvious auriféres de la France. Lyons, 1868.
Wilkinson (C.) Formation of Gold Nuggets in Drift. *Trans. Roy. Soc. Victoria*, vol. viii., p. 11, 1872.
Newbery (J. Cosmo). Formation of Nuggets in Auriferous Deposits. *Trans. Roy. Soc. of Victoria*, vol. ix., pp. 52-60, 1873.
Skey (Wm.) Formation of Gold Nuggets in Drift. *Trans. N.Z. Inst.* vol. v., p. 377, 1873.
Christy (S. B.) Ocean Placers of San Francisco. *Proc. Cal. Acad. Sci.*, August, 1878.
Egleston (T.) Hydraulic Mining in California. London, 1878.
Goodyear (W. A.) Auriferous Gravels of California. *Proc. Cal. Acad. Sci.* San Francisco, 1879.
Whitney (J. D.) Auriferous Gravels of the Sierra Nevada of California. Cambridge, U.S., 1880.
Egleston (T.) Formation of Nuggets and Placer Deposits. *Trans. Am. Inst. Mng. Eng.*, vol. ix., p. 633, 1881.
Hammond (J. H.) Auriferous Gravels of California and the Methods of Drift Mining. *Prod. of Gold and Silver in U.S. for 1881.* Washington, 1882.
Newbery (J. Cosmo). Genesis and Distribution of Gold. *The School of Mines Quarterly*, vol. iii., 1882.
Randall (P. M.) The Miner's Inch. *Prod. of Gold and Silver in U.S. for 1884.* Washington, 1885.
Bowie, Jr. (A. J.) Practical Treatise on Hydraulic Mining in California. New York, 1885.
Bowie, Jr. (A. J.) Mining Debris in Californian Rivers. San Francisco, 1887.
Gould (E. S.) Practical Hydraulic Formulæ for the Distribution of Water through long Pipes. New York, 1891.
Kirkpatrick (T. S. G.) Hydraulic Gold Miner's Manual. New York, 1891.
Wagenen (T. F. von). Manual of Hydraulic Gold Mining. New York, 1891.

CHAPTERS VI., VII., AND VIII.—QUARTZ CRUSHING AND AMALGAMATION.

De Born. Amalgamation des minerais d'or et d'argent. Vienna, 1786. English translation, 1791.
Sonneschmied. L'Amalgame espagnol. Leipzig, 1811.
Sonneschmied. Traité sur l'amalgamation. Ronneburg, 1811.
Rivot. Nouveau procédé de traitement des minerais d'or et d'argent (a manuscript in the archives of the École des Mines de Paris). Paris, 1818.
Ortmann. Kurze Geschichte der Amalgamation in Sachsen. Freiberg, n.d.
Lawson (G.) Improvements in Amalgamation. *Trans. Nova Scotia Inst.* 1866.
Hague (J. D.) Gold Mining in Colorado. *Report on the Fortieth Parallel*, vol. iii. Washington, 1870.
Keith (N. S.) Amalgamated Copper Plates. *Eng. and Mng. Journ.*, vol. xi., p. 270, 1871.
Blake (W. P.) Mining Machinery. New Haven, 1871.
Fonseca. Memoire sur l'amalgamation Chilienne. Paris, 1872.
Bergmann (E. von). Die Anfänge des Geldes in Ægypten. Vienna, 1872.
Thompson (H. A.) Extraction of Gold. *Trans. Roy. Soc. Vict.*, vol. viii., pp. 15-26, 1872.
Skey (Wm.) Electromotive Power of Certain Metals in Cyanide of Potassium with reference to Gold Milling. *Trans. N.Z. Inst.*, vol. viii., p. 334, 1876.
Attwood (G.) Chile Vein Gold Works, South America. *Proc. Inst. C.E.*, vol lvi., p. 244, 1879.
Cumenge and Fuchs. Effect of Antimony and Arsenic on Amalgamation. *Comptes Rendus*, March 17, 1879.
Egleston (T.) Californian Stamp Mills. London, 1880.
Randall (P. M.) Quartz Operators' Handbook. New York, 1880.
Habermann (I.) The Heberle Mill. *Oesterr. Ztschr. für Berg. und Hüttenwesen*, 1880.
Egleston (T.) Treatment of Gold Quartz in California. London, 1881.
Egleston (T.) Losses in Amalgamation. *Trans. Am. Inst. Mng. Eng.*, vol. ix., p. 633, 1881.
Egleston (T.) Causes of Rustiness in Gold. *Trans. Am. Inst. Mng. Eng.*, vol. ix., p. 646. New York, 1881.
Yale (Charles G.) Mining Machinery. *Prod. of Gold and Silver in the U.S.*, 1881.
Richards (J.) Quartz Crushing Machinery. *Prod. of Gold and Silver in U.S.* Washington, 1881.
Attwood (G.) Milling of Gold Quartz. *Prod. of Gold and Silver in U.S.*, 1881.
Reed (S. A.) Ore Sampling. *School of Mines Quarterly*, vol. iii., p. 253., 1881.
M'Dermott and Duffield. Gold Amalgamation and Concentration. London and New York, 1890.
Lock (G. Warnford). Gold Amalgamation. *Proc. Inst. Mng. and Met.*, Session ii., 3rd meeting. London, Dec., 1892.
Curtis (A. Harper). Gold Quartz Reduction. *Proc. Inst. C.E.*, vol. cviii. (1892), part ii.
Charleton (A. G.) Coarse and Fine Crushing. *Trans. Fed. Inst. Mng. Eng.*, 1892-3. Four Papers.
Louis (H.) Handbook of Gold-Milling. London, 1894.
Rickard (T. A.) Variations in Gold Milling. New York, 1895.

CHAPTER IX.—CONCENTRATION OF GOLD ORES.

Rittinger (P. von). Lehrbuch der Aufbereitungskunde. Berlin, 1867, 1870, and 1873.

Smyth (Sir W. W.) Dressing, or the Mechanical Preparation of Gold Ores. Lectures on Gold. *Mng. Journ.*, 1873.

Full descriptions of the different kinds of round buddles are given in the following papers:—

>   **Teague (W., Jun.)** "On Dressing Tin Ores." *Proc. Mining Inst. Cornwall*, vol. i., No. 3 (Truro, 1877).
>   **Ferguson (H. T.)** "On the Mechanical Appliances used for Dressing Tin and Copper Ores in Cornwall." *Proc. Inst. Mech. Eng.*, 1873.

Schmidt (A. W.) Der Schlamfänger auch Kornfänger genannt. Dillenburg, 1877.

Reytt (C. von). Comparison of the Hand Buddle and the Rotary and Percussion Tables. *Berg-u. Hüttenmänn. Jahrbuch*, vol. xxii., 1877.

Habermann (J.) Comparison of the Salzburg Table, the Rittinger Table, and the Hand Buddle. *Oesterr. Zeitschr. für Berg-und Hüttenwesen*, 1879, No. 8.

Cazin (F. M. F.) Dynamical Metallurgy or Mechanical Ore Concentration. *Mining Record*, 1881-2.

Richards (R. H.) A new Hydraulic Separator to prepare Ores for Jigging and Table Work. *Trans. Am. Inst. Mng. Eng.*, 1883.

Reytt (C. von). Salzburg Percussion Table. *Berg-und Hüttenmanisches Jahrbuch*, xxxiii., p. 3, 1883.

Callon (J.) Lectures on Mining. Translated by Le Neve Foster and W. Galloway, vol. iii. London, 1886.

The Settling of Solid Particles in Liquids. Bulletin No. 36. *United States Geological Survey*. Washington, 1886.

Ore Dressing in California. *Sixth Report of the Cal. State Min.*, 1886. This is a full account of the machines actually at work.

Reytt (C. von). The Linkenbach Table and Hand-Buddle compared. *Oesterr. Ztsch.*, Oct. 6, 1888.

Lock (C. G. W.) Mining and Ore Dressing Machinery. London, 1891.

Kunhardt (W. B.) The Practice of Ore Dressing in Europe. New York, 1891.

Scheidel (A.) New Process of Dressing the Gold Ores of the Thames Valley. *New Zealand Mng. Comm. Rept.*, 1891, p. 28.

Meier (J. W.) Concentration at Pribram, Bohemia. *Eng. and Mng. Journ.*, vol. liv. (1892), p. 665, vol. lv. (1893), pp. 5, 28, and 52.

Commans. Concentration and Sizing of Crushed Ore. *Proc. Inst. Civil Eng.*, 1894.

Rosales (H.) Report on the Loss of Gold in the reduction of Auriferous Veinstone in Victoria. Melbourne, 1895.

CHAPTER XI.—ROASTING OF GOLD ORES.

Plattner (C. F.) Die metallurgische Röstprozesse. Freiburg, 1856.

Dixon (Wm.) Methods of extracting Gold, Silver, and other Metals from Pyrites. *Chemical News*, vol. xxxviii., pp. 281, 293, and 301; and vol. xxxix., p. 7. 1879.

Küstel (G.) Roasting of Gold and Silver Ores. San Francisco, 1880.

Cumenge (E.) Note relative a l'emploi de la vapeur d'eau dans certaines operations metallurgiques. *Annales des mines*, vol. i., p. 1852. Paris, 1882.

Stetefeldt (C. A.) On Salt Roasting. *Trans. Am. Inst. Mng. Eng.*, vol. xiii. New Haven, 1885.

Christy (S. B.) Losses of Gold in Roasting. *Trans. Am. Inst. Mng. Eng.*, 1888.

Producer Gas for Roasting. *Tenth Cal. State Min. Report*, p. 897. San Francisco, 1890.

Adams (W. H.) Pyrites: Practical Methods for Extraction of Gold, &c. New York, 1892.

Stetefeldt (C. A.) Taylor's Gas Producer for Roasting. *Eng. and Mng. Journ.*, vol. lvi. (1893), p. 124.

CHAPTERS XII., XIII., AND XIV.—CHLORINATION.

The earliest researches on the chlorination of gold were made by—
 (1) Duflos (Dr.) *Die schles. gesell. Uebersicht.* Breslau, 1848.
 (2) Richter (Theo.) *Journ. für Prak. Chem.*, vol. li. (1849), p. 151.
 (3) Lange (Herr.) *Karsten's Archiv*, vol. xxiv., pp. 396-429. 1852.
 (4) Plattner (C. F.) Probirkunst, p. 570. 1853.
 (5) Percy (J.) *Phil. Mag.*, vol. xxxvi. (1853), pp. 1-8.

Whelpley and Storer. Method of separating Metals from Sulphurets. Boston, 1866.

Küstel (G.) Concentration and Chlorination. San Francisco, 1868.

Bleasdale (J. J.) On Chlorine as a Solvent for Gold. *Trans. Roy. Soc. Victoria*, vol. vi., pp. 47-52. 1870.

Pyrites: Report of the Board appointed to report on the methods of treating Pyrites and Pyritous Veinstuff, as practised on the Gold Fields. Melbourne, 1874.

Attwood (G. M.), and Aaron (C. H.) *Prod. of Gold and Silver in the United States.* Washington, 1881.

France (Ch. de). Extraction par voie humide du cuivre, de l'argent, et de l'or. Brussels, 1882.

Egleston (T.) Leaching Gold and Silver Ores in the West. New York, 1883.

Egleston (T.) Leaching Gold Ores containing Silver. London, 1886.

Stetefeldt (C. A.) Lixiviation of Silver Ores. New York, 1888. This work gives many details applicable to all wet processes.

O'Driscoll (F.) Notes on the Treatment of Gold Ores. London, 1889.

The Pollok Process of Chlorination. *Eng. and Mng. Journ.*, vol. xlix. (1890), Feb. 15.

Burfeind (J. H.) Chlorination of Gold Ores. *Eng. and Mng. Journ.*, vol. li. (1891), p. 446.

Precipitation of Gold from Chloride Solution. Letters and Articles by John E. Rothwell, W. Langguth, C. H. Aaron, L. D. Godshall, J. T. Blomfield, and T. K. Rose in *Eng. and Mng. Journ.*, vol. li. (1891), pp. 74, 112, 165, 204, 229, 282, 347, 373, 465; vol. lii. (1891), p. 211; and *Mining Journal*, March, 18, 1893.

Vautin (C.) Decomposition of Auric Chloride. *Proc. Inst. Mining and Met.*, Session ii., 5th meeting. London, Feb. 15, 1892.

Langguth (W.) Chlorination of Gold. *Trans. Am. Inst. Mng. Eng.* June, 1892.

Barrel Chlorination. *Eng. and Mng. Journ.*, vol. lv. (1893), pp. 244 and 269.

Rothwell (J. E.) Recent improvements in Chlorination. *Mineral Industry for 1892*, p. 236. New York, 1893.

Godshall (L. D.) Modern Chlorination. *Eng. and Mng. Journ.*, vol. lvii. (1894), Jan. 6 and 13.

CHAPTERS XV. AND XVI.—CYANIDE PROCESS.

Scheidel (A.) The Cyanide Process. Sacramento, 1894.
Janin, Jr. (L.) Article in *Mineral Industry* for 1892, pp. 239-270.
Reunert (T.) Diamonds and Gold in South Africa. London, 1894.
Butters (C.) and Smart (E.) Plant for the extraction of Gold by the Cyanide Process. *Proc. Inst. Civil Eng.*, vol. cxx. 1895.
Eissler (M.) The Cyanide Process. London, 1895.

CHAPTER XVII.—PYRITIC SMELTING.

Austin (W. L.) Matting Dry Auriferous Sulphides. *Trans. Am. Inst. Mng. Eng.*, 1889.
The Austin System of Pyritic Smelting. Denver, Colo., 1892.
Pyritic Smelting. Articles and letters on the subject have appeared in the *Engineering and Mining Journal* under the following dates:—Vol. l. (1890), Aug. 2 and Nov. 15; vol. li., p. 229 (Feb. 21, 1891); vol. lii., pp. 174, 471, 721 (Aug. 8, Oct. 24, and Dec. 26, 1891); vol. lv., pp. 28, 99, 244, 292, 339, and 364 (Jan. 14, Feb. 4, Mar. 18, April 1, April 15, April 22, 1893).
Lang (H.) Matte Smelting. New York, 1896.

CHAPTER XVIII.—REFINING AND PARTING OF GOLD BULLION.

Goddard (Jonathan). Experiments on Refining Gold with Antimony. *Phil. Trans. Roy. Soc.*, 1676.
Leibius (A.) Separation of Gold from Silver Chloride in Miller's Process. *Trans. Roy. Soc. of New South Wales*, 1872, p. 67.
Egleston (T.) Parting Gold and Silver in California. New York, 1877.
Egleston (T.) Parting Gold and Silver by means of Iron at Lautenthal. New York, 1885.
Egleston (T.) The Separation of Silver and Gold from Copper at Oker. Washington, 1885.
Hugon. Etude sur le raffinage electrolytique du cuivre noir. Paris, 1885.
Egleston (T.) Treatment of Gold and Silver at the United States Mint. London, 1886.
Skidmore (W.) Parting Gold and Silver in the United States. *Ninth Report of the Cal. State Min.*, 1889, pp. 67-90. This is a complete account of the methods in use in the United States.

CHAPTERS XIX. AND XX.—ASSAYING.

Carranza (A.) El Ainstamieto i Proporcion de las Monedas de Oro, Plata i Cobre, i la reduccion distos Metales a su Debida estimaccion. Madrid, 1629.
Badrock (Wm.) A new Touchstone for Gold and Silver Wares. London, 1651. 2nd edition, 1679.
Le Febure (N. R.) Compleat body of chymistry. 1670.
Petty (Sir John). Laws of Nature in Assaying Metals. Translated in part from the German of L. Erckern. London, 1683.
Cramer (J. A.) Elementa artis docimasticæ. Lugduni Batavorum (Leyden), 1744.
Symonds (W.) Essay on the Weighing of Gold, &c. London, 1756.
Cramer, M.D. (J. A.) Elements of the Art of Assaying Metals. London, 1764.

Pouchet. Le nouveau titre des matières d'or et d'argent. Rouen, 1798.
Becquerel. Gold and Electricity. *Ann. de Chimie et de Physique*, vol. xxiv. (1823); also Article by Oersted, d° vol. xxxix. (1828), p. 274.
Chaudet. L'Art de l'essayeur. Paris, 1835.
Bodemann (Th.) Anleitung zur Berg-und Hüttenmannischen Probirkunst. Clausthal, 1845.
Berthier. Traité des essais par la voie séche. Paris, 1847.
Pettenkofer. *Bergwerksfreund*, vol. xii. (1849). Article on Gold Bullion Assaying.
Watherston (J. H.) The Gold Valuer. London, 1852.
Plattner (C. F.) Probirkunst. Freiberg, 1853.
Bodemann and Kerl. Treatise on Assaying. Translated by W. A. Goodyear. New York, 1868.
Domeyko (J.) Tratado de Ensayes, tanto por la via seca comopor la via humeda. Chile, 1873.
Foord (G.) Mechanical Assay of Quartz. *Trans. Roy. Soc., Victoria*, vol. x., pp. 139-147. 1874.
Broch (Dr. O.) Assay of Gold by means of its Density. *Norwegian Nyt. Mag. für Naturvsk*. Christiania, 1876.
Ricketts (P.) Notes on Assaying. New York, 1876.
Kerl (B.) Metallurgische Probirkunst. 1880.
Attwood (G.) Practical Blowpipe Assaying. London, 1880.
Chapman (E. J.) Assay Notes. Practical instructions for the determination by furnace assay of Gold and Silver in rocks and ores. Toronto, 1881.
Mitchell (W.) Manual of Practical Assaying. Edited by Wm. Crookes. London, 1881.
Balling (C.) L'Art de l'essayeur. Paris, 1881.
Rössler (H.) Article on Gold Bullion Assaying. *Dingler's Polyt. Journ.*, vol. ccvi. (1884).
Black (J. G.) Chemistry of the Gold Fields. Dunedin, N.Z., 1885.
Aaron (C. H.) Manual of Assaying. San Francisco, 1885.
Hiorns (A. H.) Practical Metallurgy and Assaying. London, 1888.
Frémy. L'or dans le Laboratoire. Cumenge and Fuchs. *Ency. Chim.*, vol. iii., c. 16c. Paris, 1888.
Ross (W. A.) Blowpipe Analysis. London, 1889.
Brown and Griffiths. Manual of Assaying of Gold, Silver, &c. London, 1890.
Beringer (J. J. & C.) Manual of Assaying. London, 1890.
Lieber (O. M.) Assayer's Guide. New York.
Plattner. Blowpipe Analysis. Enlarged by Richter (Th.) Translated by H. B. Cornwall. New York, 1890.
Riche (A.) L'Art de l'essayeur. Paris, 1892.
Furman. Practical Assaying. New York, 1894.

INDEX.

Aaron, C. H., 127, 274, 278, 412.
Africa, Gold in, 40, 207, 468.
Agitation in cyanide process, 312.
Agricola, 1, 89, 93, 371.
Alaska Treadwell Mine, 239, 286, 463.
Albertus Magnus, 371.
Alchemy, 1.
Allen, A. H., 448.
Allotropic forms of gold, 11.
Alloys, Assay of, 453.
  ,,   Crystalline, 13, 14.
  ,,   Gold, 12.
  ,,   Gold and copper, 16.
  ,,   Gold and silver, 15.
  ,,   Scorification of, 453.
Aluminium Alloys, Assay of, 454.
Amalgam, Cleaning of, 131.
  ,,   Composition of, 133.
  ,,   Retorting of, 133.
Amalgamated plates, *see* Plates, Amalgamated.
Amalgamation, 122.
  ,,   assay, 428.
  ,,   Causes of prevention of, 144.
  ,,   Designolle process of, 129.
  ,,   Effect of chemicals on, 124.
  ,,   ,, hammering on, 140.
  ,,   ,, temperature on, 116.
  ,,   Methods of causing, 141.
  ,,   pans, 158.
Amalgams, 14.
  ,,   Assay of, 455.
Ammonium carbonate used in treating gold ores, 307.
Ancient rivers of California, 63, 67.
Annealing cornets, 446.
  ,,   crucibles, 360.
  ,,   fillets, 444.
Annual production of gold, 464, 468.

Antimony alloys, Assay of, 453.
  ,,   in roasting furnace, 231, 232.
  ,,   used for parting, 370.
Apron plates, 119.
Arborescent gold, 8, 34.
Arnold, J. O., 19.
Arrastra, 90.
  ,,   for prospecting, 93.
Arsenic alloys, Assay of, 453.
  ,,   in roasting furnace, 231.
Artificial crystals of gold, 9.
Asbestos filtering cloth, 292, 295.
Assay by amalgamation, 428.
  ,, ,, blowpipe, 407.
  ,, ,, bromine, 429.
  ,, ,, cadmium, 452.
  ,, ,, chlorination, 428.
  ,, ,, colour and hardness, 459.
  ,, ,, crucible method, 410.
  ,, ,, density, 459.
  ,, ,, electrolysis, 460.
  ,, ,, induction balance, 460.
  ,, ,, scorification, 424, 453.
  ,, ,, spectroscope, 459.
  ,, ,, touchstone, 458.
  ,, Fluxes used in, 413.
  ,, furnaces, 410, 436.
  ,, General charges in, 411.
  ,, materials, Examination of, 423.
  ,, of base ores, 417.
  ,, ,, bleaching powder, 270.
  ,, ,, complex materials, 427.
  ,, ,, cupel, 423.
  ,, ,, cyanide of potassium, 345.
  ,, ,, gold bullion, 431.
  ,, ,, ,, Accuracy of, 451.
  ,, ,, ,, Lead used in, 435.
  ,, ,, ,, Losses of gold in, 447.
  ,, ,, ,, Use of proofs in, 450.
  ,, ,, gold ores, 407.
  ,, ,, ,, Cleaning slag in, 417.

INDEX.

Assay of gold ores, Cupellation in, 417
,, ,, Mint sweep, 431.
,, ,, purple of Cassius, 431.
,, ,, pyrites, 428, 430.
,, ,, rich sulphides, 425.
,, ,, tellurides, 426.
,, Preliminary, 452.
,, Roasting in, 446.
,, Special methods of, 427.
,, Wet methods of, 457.
,, Whitehead's method of, 430.
Assay-ton, 411.
Attwood, G. M., 146.
Atwood's amalgamator, 190.
Aurates, 29.
Auric bromide, 27.
,, chloride, 20.
,, ,, Decomposition of, 23.
,, ,, Volatilisation of, 21.
,, oxide, 28.
Auricyanide of potassium, 28.
Auriferous quartz, Formation of, 32, 41.
Aurocyanide of potassium, 27.
Aurosilicates, 32.
Aurosulphites, 29.
Aurous chloride, 20.
,, cyanide, 27.
,, oxide, 28.
Austin, W. L., 353.
,, process, 354.
Australia, Battery practice in, 199.
,, Gold in, 40, 468.
Automatic feeding machines, 113.
Available chlorine, 269.
,, cyanide, 350.

Bagration, 333.
Balances for bullion assay, 433, 434, 447.
Ball, E. J., 415.
,, mills, 156, 158.
,, stamp, 151.
Balling, C., 429, 452.
Banket, 207.
Bar mining, Deep, 57.
Barba, 89.
Barrel for cleaning up, 131, 202.
,, process of chlorination, 262, 287.
,, ,, Advantages of, 264.
,, ,, Amount of chemicals used in, 270.
,, ,, Amount of water used in, 268.
,, ,, History of, 262.
Basalt overlying auriferous deposits, 63.

Base bullion, 356.
,, Assay of, 430.
,, ores, Assay of, 417.
Bassick mine, 37.
Batea, 44.
Battery, Stamp, *see* Stamp battery.
Bayly, F. W., 435.
Bazin's centrifugal amalgamator, 129.
Beach mining, 61.
Beckmann, 93, 371.
Becquerel, 3.
Berdan pan, 205
Berthelot, 333.
Berzelius, 2.
Bettel, W., 328, 346.
Bibliography, 471.
Biringuccio, 89, 371.
Black concentrates, 48.
Blackhawk, Stamp mills at, 195.
Blake crusher, 96, 222.
,, W. P., 9, 35.
Blake-Marsden crusher, 97, 208.
Blanching, 17.
Blanket strakes, 172, 201.
Bleaching powder, 269.
,, ,, Assay of, 270.
Blomfield, J. T., 278, 295.
Blowpipe assay, 407.
Blue gravels, 65.
,, lead theory, 65.
,, Spur Company, 77, 84.
Bone-ash for cupels, 418.
,, to cover bullion, 361.
,, to protect muffle, 418.
Booming, 53.
Booth, J., 18.
Borax for refining bullion, 361.
,, in crucible assay, 413.
,, in scorification, 424, 454.
Boss, M. P., 159.
,, system of pan-amalgamation, 162.
Boxes, Puddling, 47.
,, Sluicing, 49.
Boyle, 4.
British Isles, Gold in, 38.
Brittle gold, 18, 363, 364, 403.
Bromide of gold, 27.
Bromination of ores, 299.
Bromine, Action of, on gold, 255.
,, Assay by, 429.
Brough, B. H., 94, 326.
Brown and Griffiths, 412.
Brown, R. E., 67.
Brückner furnace, 233.
Büchner, 2.
Buckboard, 409.
Bucyrus steam shovel, 62.
Buddle, Cornish, 173.

INDEX.

Buddle, German, 171.
Buisson, 2.
Bullion, Gold, Assay of, *see* Assay of gold bullion.
,, Casting of, 364.
,, Composition of, 357, 381, 387, 400, 402, 403.
,, Definition of, 356.
,, Granulation of, 371.
,, Losses of, 366.
,, Melting of, 360.
,, Parting of, 369.
,, Refining of, 357, 361.
,, Skimming of, 362.
,, Toughening of, 363.
Bunker Hill mine, 240, 265.
Butters, C., 257, 259, 260, 284, 311, 318, 320, 339, 345.

Cadmium, Assay by, 452.
Calaverite, 37.
Caledonia Mill, 206.
California, Battery practice in, 190.
,, Chlorination in, 284.
,, Deep placers in, 63.
Cam-pulley, 108.
,, -shaft, 106.
Cams, 106.
Canada, Gold in, 40.
Capel, 459.
Carnelley, 3.
Carnot, A., 427.
Carolina, Chlorination in, 290.
Cassel process, 283.
Cassius, Purple of, 2, 26, 33, 34.
,, ,, Assay of, 431.
Casting ingots, 364.
Cement gravel, 54, 65.
,, ,, treated by arrastra, 92.
,, mill, 82.
,, pan, 83.
Cementation, 370.
Challenge feeder, 115, 208.
Chalmers, *see* Hatch.
Charcoal, Precipitating gold by, 277.
Charging scoop for bullion, 360.
Charleton, A. G., 463.
Chaudet, 435, 445, 455, 456.
Checks in bullion assay, 450.
Chemicals for stamp battery, 124.
Chemistry of cyanide process, 347.
,, of oxidising roasting, 229.
Chester, 9.
Chilian Mill, 90.
China, Gold in, 39.
Chlor-auric acid, 26.
Chloride of gold, 20.
,, ,, Decomposition of, 23.
,, ,, Melting point of, 23.

Chloride of gold, Volatility of, 21, 236.
,, of silver, *see* Silver, chloride of.
Chlorination at Alaska-Treadwell Mine, 286.
,, at Deloro, 287.
,, at Golden Reward Mill, 291.
,, at Mt. Morgan, 305.
,, at Rapid City, 299.
,, at Robinson Mine, 326.
,, Barrel process, 283.
,, Cassel's process, 283.
,, Greenwood process, 283.
,, in California, 284.
,, in Carolina, 290.
,, Julian process, 283.
,, Munktell process, 281.
,, Newbery-Vautin process, 280.
,, Pollok process, 280.
,, Vat process, 249, 284.
Chlorine, Action of, on gold, 254.
,, ,, on gold alloys, 255.
,, ,, on organic matter, 256.
,, ,, on oxides of metals, 256.
,, ,, on sulphates, 256.
,, ,, on sulphides, 253, 255.
,, Amount required for ores, 257.
,, Generation of, 252.
,, Parting by, 386.
,, Refining by, 403.
Christy, S. B., 235, 236, 238, 389.
Chrome-steel, 105.
Clarkson, T., 188.
Classification of ores, 169.
,, of parting processes, 369.
Claveus, Gasto, 4.
Clay crucibles, 360.
Clean-up barrel, 131, 202.
,, in cyanide process, 318.
,, in hydraulicking, 79.
,, in sluicing, 52.
,, of stamp battery, 130.
,, pan, 132.
Cleaning slags, 417, 425.
Clennel, 318, 320, 339.
Cobalt alloys, Assay of, 454.
Cohesion of gold, 3.
Coinage, Alloys used in, 16.
Collins, H. F., 354.
Colorado, Battery practice in, 193.
Colour of alloys, Assay by, 459.
,, of gold, 1.

Composition of bullion, 357, 381, 387.
,, ,, from chlorination process, 357.
,, from cyanide process, 320.
,, ,, placers, 358.
,, ,, stamp battery, 358.
,, of gold slimes, 319.
,, of native gold, 38.
Compounds of gold, 19.
,, ,, Natural, 37.
Concentrates, Amalgamation of, 163, 190, 437.
,, Black, 48.
,, Chlorination of, 284, 286.
,, Cyaniding of, 327.
,, Grey, 48, 59.
Concentration, 163, 166.
Concentrator, Centrifugal, 174.
,, Clarkson & Stanfield's, 189.
,, Duncan, 174.
,, Embrey, 183.
,, Frue vanner, 176.
,, Gilpin County, 175.
,, Golden Gate, 193.
,, Hendy, 174.
,, Lührig vanner, 184.
,, Raising Gate, 173.
,, Triumph, 184.
Conductivity of gold, 4.
Consumption of cyanide, 324, 340, 347.
,, of gold, 469.
Copper amalgamated plates, 117.
,, ,, discoloration of, 123.
,, ,, for cement gravel, 82.
,, ,, in sluicing, 52, 61.
,, Oxide of, in toughening bullion, 364.
,, sulphate, Crystallisation of, 379.
Cornets, 444.
,, Silver in, 449.
,, Weighing the, 447.
Corrosive sublimate, in toughening gold, 363.
Cost of barrel chlorination, 291, 298, 302.
,, cyanide process, 214, 325, 331.
,, Munktell process, 282, 283.
,, parting. 375, 376, 380, 381, 386, 401, 407.
,, Plattner process, 261, 287.
,, production of gold, 462.
,, stamp amalgamation process, 214.
,, working placers, 88.
,, ,, deep placers, 87.

Cost of working shallow placers, 85.
Coyoting, 43.
Cradle, 45.
Crawford Mill, 156.
Creuzbourg, 2.
Crinoline hose, 73.
Cripple creek, 37, 465.
Croesus Mine, 209.
Crookes, W., 137, 460.
Crosse, A. F., 348.
Crown Mine, 214.
Crucibles for assaying. 411.
,, ,, melting bullion, 360.
,, Size of, 365.
Crushing before chlorination, 218, 302.
,, ,, cyanide process, 308.
,, in stamp battery, 88.
Crystalline alloys of gold, 13, 14.
Crystallisation of copper sulphate, 379.
,, gold, 8, 34.
,, silver sulphate. 383, 384.
Cumenge, 32, 427, 435.
Cupellation in assay of bullion, 436.
,, ,, ores, 417.
,, Influence of base metals on, 420.
,, ,, temperature on, 440.
Cupels, 418, 438.
,, Assay of, 423.
Curtis, A. H., 152. 219, 223.
Cyanide of gold, 27.
,, of mercury, 349, 350.
,, of potassium, action on gold and other metals, 333.
,, ,, action on salts and minerals, 340.
,, ,, action on sulphides, 147, 341.
,, ,, Assay of, 345.
,, ,, Commercial, 316.
,, ,, Consumption of, 324, 347.
,, ,, Decomposition of, 324, 338, 340, 343.
,, ,, in stamp battery, 124, 325.
,, ,, Selective action of, 307, 341.
,, ,, Solubility of gold in, 336.
,, ,, Solubility of silver in, 338.
,, ,, Solubility of various metals in, 335.

INDEX. 487

Cyanide of zinc, 339.
,, process, 306.
,, ,, Chemistry of, 333.
,, ,, Direct treatment in, 325.
,, ,, Disposal of tailings in, 316.
,, ,, Double treatment in, 311, 328.
,, ,, MacArthur-Forrest, see MacArthur-Forrest process.
,, ,, Results of, 331.
,, ,, Siemens-Halske, 329.
,, ,, Sulman-Teed, 351.
Cyanogen bromide, 351.
Cyclops mill, 158.

Daintree, R., 68, 146.
Dakota, Battery practice in, 206.
,, Bromination in, 299.
,, Chlorination in, 291.
,, Pyritic smelting in, 354.
,, Roasting in, 244.
Dams, in river mining, 55.
D'Arcet, C., 375, 435.
Davis, W. M., 277.
De Lacy, 262.
De Morveau, Guyton, 2.
Dead-leaf gold, 15.
Debray, 20, 34.
Decomposition in zinc boxes, 339.
,, of gold chloride, 23.
,, of potassium cyanide, 324, 338, 340, 343.
Deep bar mining, 57.
,, placer deposits, 62.
,, ,, Method of working, 70.
Deetken, G. F., 190, 194, 216, 250, 255, 262.
Del Mar, A., 42.
Deloro, Mears process at, 287.
Dendritic gold, 8, 34.
Dennes, D., 222, 273, 293, 299, 301.
Density, Assay by, 459.
,, of gold, 3.
,, of alloys of gold, 16.
Designolle process, 129.
Desmarest, 2.
Detection of gold in alluvium, 44.
,, ,, ores, 426.
,, ,, solution, 26, 427.
Deville, 4.
D'Hennin, 457.
Dibromide of gold, 27.
Dies, 105.
Diffusion of gold, 14.
Dingler, 430.

Diodorus Siculus, 89.
Dip sample, 433.
Dissemination of gold, 35.
Distribution of gold, 34, 38.
,, ,, in gravels, 66.
Ditches in hydraulicking, 72.
Dixon, 348.
Dodge crusher, 97.
Double discharge in stamp battery, 112.
Double treatment in cyanide process, 311.
Dredging for gold, 57.
Drift mining, 81.
Drop of stamps, 110.
Drops in sluice, 51.
Dry crushing, 218.
,, diggings, 54.
Drying ore, 224.
Dubois, W. E., 35.
Ductility of gold, 2.
Duffield, 143, 183.
Duflos, 215, 262.

Egleston, T., 68, 140, 367.
Eissler, 284, 353.
El Callao Mine, 463.
Electricity in amalgamation, 128.
Electrolysis, Assay by, 460.
,, Parting by, 404.
Electrum, 15.
Elephant stamp, 151.
Elevator, Hydraulic, 58, 83.
Elkington, 333.
Elsner, 333.
Embrey concentrator, 183.
Emmons, 35.
Endlich, F. M., 307.
Equal falling particles, 167.
Esson, 24.
Eureka rubber, 190.
Europe, Gold in, 38.
Expansion of gold, 4.
Extra solution, Russell's, 30.

Faraday, 2, 333.
Farrar, S. H., 207.
Feather river, 56, 80.
Feeding in stamp battery, 113.
Feix, J., 367.
Feldtmann, 340.
Ferry, N. A., 260.
Figuier, 2.
Filter beds for leaching, 249, 251, 292, 295, 310.
,, press, 320, 323.
Fineness of bullion produced by Miller's process, 402.
Finkener, 346.

Flashing, 419, 439, 440.
Float gold, 142, 143.
Flouring of Mercury, 136.
Flumes for hydraulicking, 72.
Fluviatile theory of Californian placer deposits, 62.
Fluxes in assaying, 412, 413.
,, in pyritic smelting, 354.
,, in refining bullion, 362.
Fly catchers, 55.
Foliated tellurium, 37.
Formation of auriferous quartz, 32.
,, of nuggets, 68.
Forms of gold found in nature, 34.
Forrest, W., 307.
,, R. W., 307.
Foster, Le Neve, 116.
Foundations of stamp battery, 103.
Framework of stamp battery, 103.
Freezing point of gold, 3, 12.
Frémy, 427, 430, 435, 448, 454, 459.
Fresenius, 342, 346.
Fromm, 13.
Frue vanner, 176.
,, Riffled belt for, 183.
Fuchs, 427, 435.
Fulminating gold, 29.
Furnaces, Assay, 410, 436.
,, Brückner, 243.
,, Chlorination, 393.
,, Gas Assay, 437.
,, Hofmann, 244.
,, Mechanical, 238.
,, Melting, 359.
,, Muffle, 436.
,, O'Hara, 238.
,, Pearce Turret, 240.
,, Refining, 359.
,, Reverberatory, 226.
,, Revolving cylindrical, 243.
,, Rotating-bed, 240.
,, Shaft, 238.
,, Spence, 239.
,, White, 246.
,, White-Howell, 247.
Fusibility of gold, 3.
Future production of gold, 466.

Gas, Chlorine, method of production, 252.
,, ,, used in Plattner process, 252, 254.
,, Roasting by, 248.
Gates crusher, 98, 208.
Geber, 350.
Geographical distribution of gold, 38.
Geological ,, ,, 35.
Gernet, von, 330.

Gilpin County, Battery practice at, 195.
,, ,, concentrator, 175.
Godshall, L. D., 302.
Gold, Action of chlorine on, 255.
,, ,, of potassium cyanide on, 336.
,, bullion, see Bullion, Gold.
,, Detection of, see Detection of gold.
,, Losses of, see Loss of gold.
,, ores, Assay of, 407.
,, Properties of, 1.
,, residues, Treatment of, 373, 278, 383, 385.
,, slimes, 319.
,, ,, Treatment of, 320.
,, Solubility of, 9, 31, 255, 336.
Golden Gate concentrator, 193.
,, Reward Mill, 291.
Gooseneck in Mears process, 263.
Gore, G., 334, 335, 404.
Grade of plates, 125.
,, sluices, 50, 59, 78.
Granulation of bullion, 371, 375.
Graphic tellurium, 37.
Graphite crucibles, 360.
Gravel, Cemented, 54.
,, Method of washing, 44.
Gravels, Blue and red, 65.
Gravitation stamps, 102.
Green gold, 15.
Greenwood process, 283.
Grey concentrates, 48, 59.
Griffiths, H. D., 185.
Grizzly, 51.
Grommestetter, 93.
Ground sluicing, 53.
Groves, J., 438.
Grüner, 226.
Guides in stamp battery, 109.
Güptner, 452.
Gutters in ancient rivers, 66.
Gutzkow, F., 381, 383, 386.
,, process, 381.
,, ,, New, 383.

Haile Mine, 129, 290.
Haindl, 456.
Hall-marking, 15.
Halogen compounds of gold, 20.
Hammond, J. H., 191.
Hankey, J., 144.
Hanks, 362.
Harcourt, 24.
Hardness of alloys, 459.
,, of gold, 2.
Hare, R., 4.
Harris, 91.

INDEX.

Harscher, C., 93.
Hartz jigs, 187.
Hatch and Chalmers, 210, 212, 213, 311, 325, 328, 463.
Heat of formation of chlorides, 20
 ,, ,, of cyanides, 335.
Hendy's challenge feeder, 115.
Heycock and Neville, 12, 13.
Hickock, 244.
Hofmann furnace, 244.
Holleman, 11.
Homburg, 4.
Homestake Mill, 206.
Hood, J. J., 350.
 ,, process, 349.
Horn spoons, 45.
Howe, H. M., 232.
Howell-White furnace, 246.
Huggins, Dr., 3.
Hughes, Prof., 460.
Huntington Mill, 152.
Husband's Stamp, 148.
Hydrates of gold, 28.
Hydraulic elevator, 58, 83.
 ,, sizing boxes, 170.
Hydraulicking, 56, 70.
Hydrocyanic acid, 346.
Hydrogen amalgamator, 164.
Hydrolytic decomposition of potassium cyanide, 338.
Hyposulphites of gold, 30.

Inch, Miners', 73.
Independence Mill, 293, 295.
India, Gold in, 39.
Induction balance, 460.
Ingots, Casting, 364.
Inquartation, 372, 421.
Iodide of gold, 27.
Iridium alloys, Assay of, 456, 457.
Iridosmine, 69.
Iron alloys, Assay of, 454.
 ,, pipes in hydraulicking, 72.
 ,, riffle bars, 77.
 ,, Sulphate of, used to precipitate gold, 259, 274, 286, 287.
 ,, vessels for parting, 374, 381.

Jacob, 464, 469.
Janin, Jun., L., 18, 129, 146, 147, 307, 337, 342.
Japan, Gold in, 39.
Jevons, 470.
Jigs, Hartz, 187.
 ,, Pneumatic, 188.
Johnson's filter press, 320, 323.
Jordan's amalgamator, 165.
Julian, 283, 310.
 ,, process, 283.

Kandelhardt, 435.
Kasentseff, 14.
Kedzie, G. E., 340.
Keith, N. S., 348.
Kennel, California, Chlorination works at, 284.
Kerl, B., 216, 448.
Kerr, Prof., 35.
Knox pan, 132, 190.
Kohlrausch, 11.
Krafft, 9.
Krom, S. R., 188, 219.
 ,, rolls, 219.
Krüss, 2, 4, 24, 25, 28, 33.
Kumara tail sluice, 54.
Kunckel, 375.
Küstel, G., 226, 229, 235, 255.

Lang, H., 353, 355, 356.
Lange, 215.
Langguth, W., 275.
Langlaagte estate, 317.
Latent heat of gold, 3.
Latta, 146.
Laurie, 4, 13.
Lava above auriferous gravels, 63.
Law, R., 395.
Le Chatelier pyrometer, 441.
Leaching by decantation, 271.
 ,, ,, pressure, 273, 294, 300.
 ,, ,, vacuum, 272, 280.
 ,, Difficulties of, 272.
 ,, in chlorination barrel, 293.
 ,, ,, cyanide process, 309.
 ,, ,, ,, with agitation, 314.
 ,, in Plattner process, 258.
 ,, vats in chlorination process, 249.
 ,, ,, cyanide process, 309.
Lead, amount used in bullion assay, 435, 436.
 ,, ,, reduced in ore assay, 413.
 ,, lining for chlorination barrel, 268.
Leibius, A., 5, 389.
Lever or jack in stamp battery, 109.
Levol, 15, 31, 33, 419.
Lime used in cyanide process, 344.
Lindet, 20.
Liquation of gold alloys, 17.
Liversedge, 69.
Lockyer, 459.
Long Tom, 46.
Loss of gold by volatilisation, 367.
 ,, ,, in amalgamation, 138.

Loss of gold in assaying, 447.
,,   ,,   ,, refining, 366.
,,   ,,   ,, roasting, 235.
Loss of mercury, 203.
Lossen, 10.
Louis, H., 3, 89.
Lowe, 23.
Lührig vanner, 184, 203.

MacArthur, J. S., 307, 308, 316, 320, 336, 341, 342, 347.
MacArthur-Forrest process, 307-328.
　At the George and May, 324.
　At the New Primrose, 328.
　At the Sylvia mine, 326.
　Composition of bullion from, 320.
　Consumption of cyanide in, 329, 347.
　Cost of, 331.
　Direct treatment of ore in, 324.
　in S. Africa, 331.
　Leaching in, 311.
　Ores suitable for, 352.
　Plant required in, 322.
　Production of bullion in, 319.
　Results of, 331.
　Strength of solution for, 315, 337.
　Treatment of concentrates by, 326.
M'Cutcheon, J., 133, 401.
M'Dermott, 143, 169, 182, 183.
M'Dougal, G., 142.
Maclaurin, J. S., 333, 337.
Mactear, 145.
Magnetism of gold, 3.
Makins, G. H., 5, 447.
Maldonite, 38.
Malleability of gold, 2.
Mallet, 4.
Maltitz, Count von, 93.
Management of gold mills, 462.
Manganese alloys, Assay of, 454.
Maray, 89.
Materials for crushing surfaces, 100.
Mathison, 375.
Matthey, E., 7, 19, 454, 456.
Matthiessen, 13.
Mattison, Edw., 70.
Maynard, G., 405.
Mears, Dr. H., 262, 263.
　,, process, 263, 287.
Mechanical furnaces, 238.
Medina, 89.
Mein, 311.
Meinecke, 113.
Melting gold bullion, 360.
Mercur Gold Mine, 308, 312.
Mercury, 14, 125.
　,, Chloride of, used to toughen gold, 363.

Mercury cyanide, Use of, in cyanide process, 348, 350.
　,, Flouring of, in stamp batteries, 136.
　,, Loss of, in stamp battery, 137.
　,, Sickening of, 136.
　,, Use of, 14, 79, 89.
　,, ,, in sluicing, 52.
　,, wells in stamp batteries, 128, 142, 201, 206.
Merrick, J. M., 428.
Merrifield Mine, 237.
Metallics, 409.
Methylamine, 339.
Mexico, Gold in, 40.
Miers, 89.
Mill, Site for, 119.
Miller, F. B., 387.
　,, process, 386.
Minerals in placer deposits, 69.
　,, occurring with gold, 36.
Miner's inch, 73.
　,, pan, 44.
Mint, Method of assaying in, 432.
　,, sweep, Assay of, 431.
Miocene Ditch Company, 72.
Mitchell, 411, 420, 428.
Modderfontein Mine, 209.
Mode of occurrence of gold, 34.
Moebius, B., 404.
　,, process, 404.
Moldenhauer, 334.
Molloy's amalgamator, 164.
　,, method of precipitating gold cyanide, 321.
Monitor, 74.
Monochloride of gold, 20.
Mortar for crushing gold ores, 89.
　,, stamp battery, 103.
Morton, Prof., 147.
Mould for ingots, 364.
Mt. Morgan Mine, 38, 305.
　,, ,, Precipitation of gold at, 277.
　,, ,, Roasting at, 145, 272.
Muffle furnace, 436.
　,, Temperature of, 441.
Mühlenberg, 307.
Müller, 24, 30.
Munktell process, 281.
Muntz metal plates, 126.
Mylius, 13.

Nagyagite, 37.
Napier, 4, 5.
Native crystals of gold, 8.
　,, gold, 38.
Natural compounds of gold, 37, 38.

INDEX. 491

Natural forms of gold, 34.
Neville, *see* Heycock.
New Primrose Mine, Cyanide process at, 328.
New South Wales, Gold in, 41.
New Zealand, Battery practice in, 199.
" " Gold in, 40.
Newbery, J. Cosmo, 136, 186, 262, 280.
Newbery-Vautin process, 280.
Nickel alloys, Assay of, 454.
Nicklès, 10.
Nitric acid, Action of, on metals, 373.
" Parting by, 371.
" Solubility of gold in, 447.
North Bloomfield Mine, 71, 79.
Nuggets, 36, 69.
" Formation of, 68.

Occurrence of gold in nature, 34.
O'Driscoll, F., 262.
O'Hara furnace, 238.
Ore bins, 121.
Ores, Assay of, 407.
Origin of deep placer deposits, 62.
" gold in gravels, 67.
" " in ores, 41.
Osmiridium, Separation of, from gold, 368.
Oxides of gold, 28.
Oxidised pyrites, Action of potassium cyanide on, 343.
" " Amalgamation of, 146.
Oxidising roasting, 229.
Oxygen, action on dissolution of gold by potassium cyanide, 333, 337, 348.

Palladium alloys, Assay of, 456.
Pan-amalgamation, 159.
Pan, Boss, 162.
" Knox, 132.
" Patton, 161.
" Siberian, 60.
" Washing by the, 42.
Paracelsus, 1.
Parting, 369.
" assay, 431.
" " by sulphuric acid, 451.
" " double, 452.
" " History of, 432.
" by cementation, 370.
" " chlorine, 386.
" " combined process, 380.
" " electrolysis, 404.

Parting by Gutzkow process, 381.
" " new " " 383.
" " nitric acid, 371.
" " sulphide of antimony, 370.
" " sulphur, 370.
" " sulphuric acid, 375.
" flask, 422.
" in bullion assay, 444.
" " ore assay, 421.
" " platinum tray, 445.
" " porcelain crucible, 423.
" " test tube, 423.
" processes, Classification of, 369.
Patio process, 90.
Pearce, 14, 15, 17.
Peel, 446.
Peligot, 17.
Pelouze, 435, 448.
Pepys, 432.
Percussion tables, 175.
Percy, J., 215, 277, 370, 376, 381, 401, 412, 417, 425, 449.
" collection, 9, 14, 35, 277.
Pestarena Mine, 92.
Pettenkofer, 435.
Petzite, 37.
Philadelphia Mint, Liquation at, 18.
" " Parting at, 380.
Philippe of Valois, 432.
Phœnix Mine, Chlorination at, 290.
Pinos Altos, Treatment of bullion at, 406.
Placer deposits, Deep, 62.
" " Shallow, 42.
" gold, 36, 66, 68.
" mining, Cost of, 41.
Plates, Amalgamated, 117.
" " Copper, *see* Copper plates.
" " Corrugated, 128.
" " Grade of, 125.
" " in sluicing, 52, 61.
" " in Transvaal, 210.
" " Muntz metal, 126.
" " Preparation of, 117.
" " Shaking, 127.
" " Treatment of, 122, 133, 211.
Platinum alloys, Assay of, 455.
" Tray for parting, 445.
Plattner, C. F., 215, 235, 407, 428.
" process, 215.
" " Amount of water used in, 251.
" " at Reichenstein, 216.
" " at various mills, 284-287, 326.

Plattner process, Cost of, 261, 287.
  ,,   ,, Method of working, 249.
  ,,   ,, Reactions in, 255.
  ,,   ,, Vats used in, 249.
Pliny, 89, 370.
Pliocene gravels of California, 63.
Plymouth Consolidated Mill, Chlorination at, 285.
Pneumatic jig, 188.
Poggendorf, 334.
Pointed boxes, 170.
Pollok chlorination process, 280.
Poussee process, 362.
Power, F. R., 392.
Prat, 20.
Precipitation of gold chloride, 24, 273.
  ,, by charcoal, 277.
  ,, by iron sulphate, 259, 274, 286, 287.
  ,, by metallic sulphides, 278.
  ,, by metals, 279.
  ,, by organic substances, 275.
  ,, by sulphuretted hydrogen, 275, 289, 276.
  ,, by sulphurous acid, 11, 296.
  ,,   ,,   ,, Cost of, 303.
  ,,   ,,   ,, Use of molasses in, 260.
  ,, of gold alloys, 13.
  ,, of gold cyanide by sodium, 322.
  ,,   ,,   ,, by electricity, 330.
  ,,   ,,   ,, by zinc, 317, 340.
  ,,   ,,   ,, ,, Influence of temperature on, 340.
  ,,   ,, in leaching vats, 345.
  ,,   ,, from solution, 24, 68.
  ,, of silver by salt, 374.
  ,,   ,, by copper, 379.
Pressure leaching, 273, 294, 300.
Price, T., 402.
Primitive methods of crushing quartz, 88.
Prinsep, James, 441.
Production of gold, Causes of increase of, 465.
  ,,   ,, Cost of, 462.
  ,,   ,, in the world, 464.
  ,,   ,, by cyanide, 331.
  ,,   ,, in the future, 466.
  ,,   ,,   ,, past, 464.
  ,,   ,, in various countries, 467.
Proofs in bullion assay, 450.
Properties of gold, 1.

Proportion of gold and silver for parting, 372.
  ,,   ,, to gangue in ores, 37.
Prospecting trough, 45.
Protobromide of gold, 26.
Proust, 2.
Prussian blue in cyanide process, 347.
Puddling tub, 46.
Pure gold, Preparation of, 10.
Purple alloy, 13.
  ,, gold, 2.
  ,, of Cassius, 2, 26, 33.
  ,,   ,, Assay of, 431.
Pyrites, Assay of, 430.
  ,, Condition of gold in, 145.
Pyritic smelting, 353.
  ,, Austin process, 353, 354.
Pyrometer, Le Chatelier, 441.

Quartz crushing in stamp battery, 88.
Quicksilver, *see* Mercury.

Rae, J. H., 306.
Randall, P. M., 73, 117.
Rapid City Mill, 222, 299.
Raymond, 172.
Reactions in chlorination vat, 255.
Recovery of gold lost in refining, 366.
Reduction of silver chloride by zinc, 374, 391.
  by iron, 392.
Reduction of silver sulphate by charcoal, 385.
  by copper, 379.
  by iron, 383.
  by sulphate of iron, 383.
Refining of gold bullion, 357, 361.
  by borax, 361.
  by chlorine, 386, 403.
  by nitre, 361.
  by sulphur, 368.
Regnault, 3.
Reichenstein, Plattner process at, 216.
Reservoirs for hydraulicking operations, 72.
Retorting, 133.
  furnace, 134.
Reverberatory furnace, 226.
Reynolds, J., 432.
Rhodio-platinum couple, 441.
Rhodium alloys, Assay of, 456.
Richards, J., 224.
Rickard, T. A., 198, 200.
Ricketts, 417, 424, 426.
Riffle bars, 49, 59, 77.
  ,, blocks, 76.

Riffled belt in Frue vanner, 183.
," sluice, 173.
Rim-rock, 66.
Riotte, E. N., 273.
River mining, 55.
Rivot, 415, 419.
Roasting at Deloro, 284.
," at Kennel, California, 284.
," at Treadwell Mine, 286.
," by producer gas, 248.
," Cost of, 291, 298, 302.
," furnaces, *see* Furnaces, roasting.
," ores, 225.
," ," for assay, 416.
," oxidising, Chemistry of, 229.
," with salt, 233.
Roberts-Austen, 3, 10, 13, 14, 17, 226, 229, 403, 436, 440, 449, 459, 460.
Robinson Mine, 214, 312, 317, 326.
Rock-breakers, 96.
Rock-cut sluices, 78.
Rock pavements for sluices, 76.
Rocker for washing gold, 45.
Roller feeder, 116.
Rolls for crushing, 219.
," compared with stamps, 223.
Rosales, H., 41, 169.
Rose, G., 3, 38.
Rose, H., 346, 456, 458.
Rose, T. K., 4, 5, 18, 23, 26, 441, 449, 450.
Rössler, 448, 449.
Rotary drying furnace, 225.
Rothschild's refinery, 375.
Rothwell, J. E., 221, 247, 288, 291, 298.
Royal College of Science, 324, 415.
Royal Mint, Bullion assay at, 432.
," Cupels at, 438.
Ruby gold, 2.
Russell process, 31.
Russia, Gold in, 39.
," Placer mining in, 47, 58.
Rusty gold, 144.

St. John del Rey Mine, 172.
Sal ammoniac used to toughen bullion, 362, 363.
Salamander crucibles, 360.
Salt used in ore assay, 414, 416.
," ," roasting, 233.
Sampling cone, 408.
," of bullion, 432.
," ," ores, 408.
," tin, 408.
Sandberger, F. von, 41.
Savot, 372, 432, 435.

Scheidel, A., 326.
Schönfeld, 371.
Schranz crusher, 99.
Schwartz's assay of pyrites, 438.
Scorification, Assay of alloys by, 453.
," ," ores by, 424.
Screens, 110.
Selective action of cyanide, 307, 341.
Selwyn, A. C., 68.
Settling boxes, 168.
Shaft mining, 81.
Shaking copper plates, 127.
Shallow placer deposits, 42.
Sheba Mine, 211.
Shovel, Steam, for placer gravels, 61.
Siberia, Gold in, 39.
," Method of washing in, 47, 58.
Siberian pan, 60.
," sluice, 59.
," trommel, 60.
," trough, 47.
Sickening of mercury, 136.
Siemens-Halske process, 329.
Silicate of gold, 32.
Silver chloride, Reduction of, 374, 391.
," Compounds of, dissolved by potassium cyanide, 342.
," dissolved by nitric acid, 373.
," ," potassium cyanide, 338.
," ," sulphuric acid, 377, 381, 384.
," filter, 385.
," for gold bullion assay, 434.
," Gold separated from chloride of, 389.
," in cornets, 449.
," ," gold coins, 15.
Simmer & Jack, cyanide works, 312.
Simpson, J. W., 306.
Sizing boxes, 170.
Skey, Wm., 68, 144, 349, 426.
Skimming bullion, 362.
Slime presses, 295.
," separator, 311.
Slimed ore, Treatment of, by cyanide, 323.
Slimes, Gold, 319.
Sluice plates, 119.
Sluicing, 49, 76.
Smith, E. A., 36, 421.
Soda, Caustic, use in amalgamation, 125.
," ," cyanide process, 344.
Sodium amalgam, 137.

Sodium carbonate, used to separate gold from silver chloride, 389.
,, hyposulphite, Action on gold of, 31.
,, used to precipitate gold cyanide, 321.
Soetbeer, 465, 470.
Solubility of gold, 9, 31, 349, 447.
,, ,, in potassium cyanide, 333, 336.
,, of silver chloride, 11.
,, ,, ,, sulphate, 377.
,, ,, sulphates, 378.
,, ,, various metals in potassium cyanide, 335, 336, 338.
Solutions of gold, Test for, 25.
Solvents of gold, 9, 31, 447.
Sonstadt, 35.
South America, Gold in, 40.
,, Clunes United Mill, 201
Spanish Mine, 155.
Specific gravity of gold, 3.
,, heat ,, 3.
Spectroscope, Assay by, 459.
Spectrum of gold, 3.
Spence furnace, 239.
Splash box, 105.
Spring, 10.
Stamp battery, 82, 88, 93, 102.
,, ,, Chemicals used in, 124.
,, ,, Foundations of, 103.
,, ,, Framework of, 103.
,, ,, General arrangement of, 119.
,, ,, practice, 190.
,, Californian, 102, 108.
,, Elephant, 151.
,, German, 94.
,, Husband's 148.
,, Steam, 149.
,, Weight of parts of, 109, 210.
Stamps, Order of fall of, 113.
,, Weight of, in California, 191.
Standard gold of various countries, 16.
Stanfield, 188.
Stanford's self feeder, 114.
Stansfield, A., 455.
Stapff's method of assaying pyrites, 430.
Steam shovel for placer gravel, 61.
,, stamp, 149.
Stelzner, A., 41.
Stetefeldt, C. A., 164, 225, 235, 315, 320.
,, drying kiln, 225, 249.
,, furnace, 238.

Stock of gold, 470.
Stone-breakers, 96.
Strength of cyanide solution, Method of testing, 345.
Strength of cyanide solution required for ores, 347.
Sulman, H. L., 351.
,, -Teed process, 351.
Sulphate of copper, Crystallisation of, 379.
Sulphide of gold, 30, 31, 32, 33.
Sulphides, Action of potassium cyanide on, 147, 341.
,, Amalgamation of, 145.
,, Assay of 425.
,, Gold in, 145.
,, Oxidation of, 343.
,, Roasting of, 227.
,, Smelting of, 353.
,, used as fuel, 354.
Sulphites of gold, 29.
Sulphur as fuel, 354.
,, in parting, 370.
,, in refining, 368.
Sulphuretted hydrogen as precipitant for gold, 275, 296.
Sulphuric acid, Action on metals of 372.
,, ,, Assay by, 455.
,, ,, Parting by, 375.
Sulphurous acid used to destroy chloride, 275.
Surcharge, 447.
,, Effect of base metals on, 453.
,, Influence of temperature on, 443, 449.
,, Variations in, 448.
Sweating copper plates, 133.
Swedish chlorination process, 281.
Sweep, Mint, Assay of, 431.
Swinging amalgamated plates, 119.
Sydney Mint, Parting at, 387.
Sylvanite, 37.
Sylvia Mine, Cyanide process at, 326.

Tail race, 53.
,, sluices, 54.
Tailings, Examination of, 143.
,, from hydraulicking, 80.
Tappet, 106.
Teed, Dr., 351.
Tellurides of gold, 37.
,, Assay of, 426, 457.
Tenacity of gold, 2.
Test for gold in solution, 25.
Testing ores by chlorine, 429.
,, ,, cyanide, 323.
,, solutions of gold, 259.

Testing toughness of bullion, 363.
Thames Valley, N.Z., Battery practice in, 203.
Thies, A., 262, 265.
  ,, process, 265, 280.
  ,, ,, Cost of, 291.
Thiosulphates of gold, 30.
Thompson, L., 386.
Thomsen, 20, 335.
Thorpe, 4.
Tin alloys, Assay of, 454.
Tools used in assaying, 414, 418, 438.
Touchstone, 458.
Transvaal, Cyanide process in, 331.
  ,, Production of gold in, 40, 214, 468.
  ,, Stamp amalgamation in, 207.
Trapiche, 90.
Tribromide of gold, 27.
Trichloride ,, 20.
Triumph concentrator, 184.
Trommel, Siberian, 60.
Trough, Prospecting, 45.
  ,, Siberian, 47.
Tub, Puddling, 46.
Tulloch feeder, 116.
Tunnels in drift mining, 81, 82.
Turret furnace, 240.

Undercurrents, 51.
United States, Gold in, 40.
Ural Mountains, Gold mining in, 39.

Vacuum pump for dredging, 57.
Van Riemsdijk, 420, 440.
Vanner, Frue, 176.
  ,, Lührig, 184.
Van't Hoff, 23.
Vat process of chlorination, 249.
Vats for leaching processes, 249, 309.
Vautin, C., 262, 274.
Victoria, Gold Mining in, 41.
Violle, 3, 5.
Volatilisation of alloys of gold, 5, 7.
  ,, of gold, 4, 367, 447.
  ,, ,, Effect of base metals on, 5, 7.

Volatilisation of gold, Effect of temperature on, 5.
Volatility of gold chloride, 21, 236.
Voltaic order of metals in potassium cyanide, 334.

Wales, Gold mining in, 38.
Washing gravel by the pan, 44.
  ,, ,, in sluices, 50.
Water supply for hydraulicking, 71.
  ,, used in barrel chlorination, 268.
  ,, ,, cyanide process, 322.
  ,, ,, sluicing, 50.
  ,, ,, stamp battery, 116, 210.
  ,, ,, vat chlorination, 251.
Watson-Denny pan, 205.
Watts, 346.
Wear of screens, 112.
  ,, shoes and dies, 105.
Weight of parts of stamps, 109, 210.
Weights, Assay-ton, 411.
  ,, for bullion assaying, 433.
Weinberg, 305.
Wells, J. S. C., 340.
Welman dredge, 57.
Whisk brooms for copper plates, 124.
White furnace, 246.
White-Howell furnace, 247, 292.
Whitney, 63, 69.
Wiebe, 4.
Wilm, 12.
Wingdams, 55.
Witwatersrand, *see* Transvaal.
Worcester mill, 329.
World's production of gold, 464.

Zinc, Action of potassium cyanide on, 339.
  ,, alloys, Assay of, 454.
  ,, Assay by, 452.
  ,, Precipitation of gold cyanide by, 317, 340.
  ,, Reduction of silver chloride by, 374, 391.

www.ingramcontent.com/pod-product-compliance
Lightning Source LLC
Chambersburg PA
CBHW051158300426
44116CB00006B/359